Stem Cells in Domestic Animals: Applications in Health and Production

Editors

Eleonora Iacono
Barbara Merlo

MDPI • Basel • Beijing • Wuhan • Barcelona • Belgrade • Manchester • Tokyo • Cluj • Tianjin

Editors
Eleonora Iacono
University of Bologna
Italy

Barbara Merlo
University of Bologna
Italy

Editorial Office
MDPI
St. Alban-Anlage 66
4052 Basel, Switzerland

This is a reprint of articles from the Topical Collection published online in the open access journal *Animals* (ISSN 2076-2615) (available at: https://www.mdpi.com/journal/animals/topical_collections/stemcells_animals_health_production).

For citation purposes, cite each article independently as indicated on the article page online and as indicated below:

LastName, A.A.; LastName, B.B.; LastName, C.C. Article Title. *Journal Name* **Year**, *Volume Number*, Page Range.

ISBN 978-3-0365-5909-4 (Hbk)
ISBN 978-3-0365-5910-0 (PDF)

Contents

animals

MDPI

Editorial

Stem Cells in Domestic Animals: Applications in Health and Production

Eleonora Iacono [1,2,*] and Barbara Merlo [1,2]

1 Department of Veterinary Medical Sciences, University of Bologna, Via Tolara di Sopra 50, Ozzano Emilia, 40064 Bologna, Italy
2 Interdepartmental Centre for Industrial Research in Health Sciences and Technologies, University of Bologna, Via Tolara di Sopra 41/E, Ozzano Emilia, 40064 Bologna, Italy
* Correspondence: eleonora.iacono2@unibo.it; Tel.: +39-051-209-7567

Citation: Iacono, E.; Merlo, B. Stem Cells in Domestic Animals: Applications in Health and Production. *Animals* **2022**, *12*, 2753 . https://doi.org/10.3390/ani12202753

Received: 20 September 2022
Accepted: 10 October 2022
Published: 13 October 2022

In the last decade, researchers described Mesenchymal Stem/stromal cells (MSCs) as a possible population of cells for cell-based therapies in regenerative medicine, both for humans and animals.

The aim of this first article (and the aim of the next books in this collection) is to gather high-quality research and review articles that could broaden knowledge regarding the role of MSCs in domestic animals' health and production.

Nowadays, in veterinary medicine, the owners require their animals to be treated with sophisticated and new treatments with the aim to improve the patient's life quality but also, in the case of livestock animals, to improve the quality of products, aiming to preserve human health [1]. MSCs therapy could be then considered as an opportunity for researchers, veterinary practitioners, and animal owners for contributing to animal and human health and well-being.

Moreover, despite the fact that the mouse remains the proof-of-principle and allows to test a wide variety of therapeutic protocols, its homogeneous genetic background is not the same as that of humans, and the knockout model of pathology, experimentally induced, is not always a mirror of spontaneous pathology. In this context, domestic animals can be considered spontaneous models, both from a pathogenetic and therapeutic point of view, of hereditary and acquired pathologies. Moreover, especially regarding pets (i.e., dogs, cats), which share the same living environment as humans and are often subjected to the same stressful agents. For the reason listed above, domestic animals could be considered an important suitable model for human spontaneous diseases, as already stated in the guidelines emitted by the U.S. Food and Drug Administration (FDA) and by the European Medicines Agency (EMA). Regarding the role of domestic animals as in vivo models for human diseases, La Mantia et al. [2] in their systematic review reported the use of stem-cell treatment against acute or chronic ischemic cardiomyopathies in large animal models with regard to Left Ventricular Ejection Fraction (LVEF). The meta-analysis reported by the Authors showed that stem-cell therapy may improve heart function in large animal models and that the swine species is confirmed as a relevant animal model in the cardiovascular field. In this context, there is also the study of Garcia-Mendivil et al. [3] regarding the development of in vitro cellular models using ovine MSCs for prion neurodegenerative disorders affecting both humans and animals, particularly ruminants. Indeed, the response of ovine bone marrow-derived MSCs and their neuron-like derivatives to prion infection allowed us to find that BM-MSC-derived neuron-like cells could be a good candidate for developing in vitro studies.

As reported by Svoradova et al. [4], MSCs can be used as an avian culture model to better understand osteogenic, adipogenic, and myogenic pathways; moreover, chicken MSCs could also be used as a model for in vitro meat culture.

On the other hand, canine and equine species can be considered as both patients and clinical models. As reported by Prislin et al. [5] and Cequier et al. [6] in their reviews, canine and equine MSCs have been used for treating different pathologies, not only regarding the musculoskeletal system; but also involving ophthalmology, reproduction, gastroenterology, metabolic and neurologic disorders, and respiratory and integumentary systems.

To date, in canine and equine regenerative medicine, adult tissue, such as bone marrow (BM) and adipose tissue (AT) represent the most used sources of MSCs. Usually, these cells are cultured in a culture medium added with 10% of FBS (Foetal Bovine Serum). Despite that it is critically discussed for its ethical and healthy implications, FBS is still the gold standard for in vitro cultivation of MSCs. However, the trend in cell culture points to the use of xeno-free culture supplements, for which blood products, like platelet lysate (PL) from the same species, appear most promising. In PL the platelet-derived growth factors have already been released and cell membranes removed, thus it can be stored for a long time in the freezer. Moreover, positive and synergistic effects of PL might not only be achieved in cell culture but also in the subsequent therapeutic application when combined with MSCs. Hagen et al. [7], cultured canine and equine ATMSCs with 2.5% ad 10% of autologous PL. Cells cultured with 10% of FBS were used as control. It was found that PL did not support stem cell culture in dogs in the same beneficial way observed in the horse, revealing that using analogous canine and equine biologicals does not entail the same results. In fact, canine ATMSC cultured in medium supplemented with 2.5% and 10% of autologous PL changed their morphology, showed decreased metabolic activity, and increased apoptosis and necrosis; however, at passage five canine ATMSCs showed less genetic aberrations when cultured with 10% of PL than with FBS. It was concluded that, even if 10% of PL seems not lead to cell damage, considering the strong alteration observed in cell morphology and expansion, the use of PL cannot be recommended for canine ATMSC culture in its current form [7].

Due to invasive cell harvesting, donor site morbidity, cell amount, and characteristics related to donor age [8–12] connected to the use of BM and ATMSCs, in the last years researchers have directed their attention towards the study of new sources.

In canine regenerative medicine, an alternative to MSCs could be a stromal vascular fraction (SVF) non-cultured MSCs, separated from adipose tissue (AT). In recent clinical trials freshly isolated primary Stromal Vascular Fraction (SVF) cells have been used instead of cultured ATMSCs [13–15]. Hendawy et al. [16] demonstrated that in middle-aged and old dogs, the peri-ovarian harvesting site yielded higher SVF viability percentage, and viable cell number/gm fat than that of the other harvesting sites, such as subcutaneous abdominal fat and falciform ligament. In this study SVF cells from periovarian AT recorded revealed a higher expression of MSC markers (CD90, CD44, and CD29) compared to the other sites, with weak CD45 and CD34 expressions. Furthermore, the positive OCT-4 expression of SVF cells isolated from periovarian AT demonstrated their pluripotency, indicating them as a valid alternative to ATMSCs for cell therapy in canines. Similar data have been reported by Prislin et al. [5]. As reported by the Authors, canine SVF and ATMSCs treatments provide many benefits, in degenerative orthopedic pathologies, both in skin, bowel, and eye diseases [5].

Foetal fluids (amniotic fluid, umbilical cord blood), and foetal adnexa (Wharton's jelly, amniotic membrane) have been identified as ideal alternative sources of MSCs in different animal species, such as horse [17–19], cattle [20,21], goat [22,23], and others. The benefits of these cells compared to adult MSCs are due to their origin from extraembryonic tissues; in fact, because they are at the maternal—foetal interface, these cells present low immunogenicity and immunomodulatory properties, making them a good candidate for allo- and xenotransplantation [24]. Iacono et al. [25], in their review, observed that, like reported in human and other animal species, also in dog MSCs derived from foetal fluid and adnexa may have an attraction compared to other established SCs in different clinical approaches, although more in vitro studies on their metabolism and clinical applications are needed to fully understand their properties and to establish the future clinical use in the treatment of various diseases. In this contest, Humenik et al. [26] described the effective protocols for the isolation of MSCs from canine bone marrow, adipose tissue, and amnion membrane, showing differences in yield of isolation, morphology, phenotype, multilineage potential, and proliferation activity.

While Humenik et al. compared canine MSCs isolated from AT, BM, and amniotic membrane, Merlo et al. compared equine ATMSCs and WJ (Wharton's jelly) MSCs [26]. Due to the difficulties encountered by the practitioners in skin wound healing and the role of integrin in the reparative process, in this pilot study, the authors analyzed the effect of an $\alpha4\beta1$ integrin agonist on cell adhesion of equine AT and WJ-derived MSCs and investigated their adhesion ability to GM18 incorporated poly L-lactic acid (PLLA) scaffolds. The preliminary results reported in this paper represent a first step in the study of MSCs adhesion to PLLA scaffolds containing GM18, suggesting that WJ-MSCs might be more suitable than AT-MSCs. However, the results need to be confirmed by increasing the number of samples before drawing definite conclusions.

Additionally, the olfactory mucosa is a promising candidate for both humans and animals [27–29]. Mollichella et al. [30] evaluate the feasibility of collecting, purifying, and amplifying olphactory-ecto (OE) MSCs from the cat nasal cavity. The OEMSCs were isolated from biopsies and their stemness features as well as their mesodermal differentiation capabilities were characterized. This report shows for the first time that the isolation of OE-MSCs from cat olfactory mucosa is possible. These cells showed stemness features and multilineage differentiation capabilities, indicating they may be a promising tool for autologous grafts and feline regenerative medicine.

Beyond natural sources that are limited by stem cell availability, immune intolerance and lineage specification, bioengineered stem cells, such as induced pluripotent stem cells (IPSCs) have been developed [31]. In canine species, several reports have described the generation of IPSCs using retroviral or lentiviral transduction using Yamanaka's factors [32–34]. Regarding viral reprogramming, different studies have shown that it can induce genomic integration and increase cell tumorigenic potential [35,36], so viral reprogramming is not suitable for clinical applications. In the study of Kim et al. [37], the 13-year-old canine fibroblasts were reprogrammed using a non-integrating Venezuelan equine encephalitis (VEE) RNA virus replicon, which has four reprogramming factors (collectively referred to as T7-VEE-OKS-iG and comprised of hOct4, hKlf4, hSox2, and hGlis1) and co-transfected with the T7-VEE-OKS-iG RNA and B18R mRNA. The derived colonies of putative canine IPSCs showed a resemblance to naïve iPSCs in their morphology (dome-shaped). The expression of endogenous pluripotency markers such as Oct4, Nanog, and Rex1 transcripts was confirmed, suggesting that induced cells were in the late intermediate stage of reprogramming. The reported research is the first of this type in canine species and, despite the good results obtained, it is a preliminary study and requires repeating with quantitative methodologies.

For therapeutic use, MSCs need to be isolated and expanded in vitro to obtain a sufficient amount for clinical application. Sometimes second or third applications could be needed, but long-term cultivation before therapeutic use is not recommended, since the cells may lose their stemness features and bacterial contamination may occur. For these reasons, it is very useful to cryopreserve these cells in order to gain a ready and controlled source of abundant autologous stem cells that maintain unaltered characteristics of the freshly isolated cells by preserving their vitality and maintaining their pluripotent phenotype. Di Bella et al. [38], evaluated the effects of 7-year-long cryopreservation using 10% DMSO and different FBS concentrations (from 10 to 90%). The Phenotype morphology, cell viability, differentiation, and proliferative potential, the expression of pluripotency markers in both fresh and thawed cells were analyzed. This study demonstrated that canine adipose tissue MSCs cryopreserved with more than 50% FBS and thawed after 7 years showed similar proliferative ability and morphological and molecular characteristics as fresh cells.

Usually, fresh or frozen-thawed cells after in vitro expansion in the laboratory are sent back to attending clinicians. As reported above, preserving MSCs characteristics en route from the laboratory to the clinic is fundamental for the success of the therapy. Due to the importance of this topic for veterinary regenerative medicine, in the last 10 years, different storage solutions, temperatures, and periods have been tested [39–42]. For equine MSCs isolated from AT and Wharton's jelly (WJ), Iacono et al. [43] demonstrated that different types of MSCs react differently to the storage conditions frequently used for shipping them from the laboratory to the clinic. These conditions influence the viability and, depending on the cell type, they can also influence different MSCs characteristics. Particularly, equine WJMSCs need to be used quickly to maintain their viability. However, data recovered in vitro need to be compared with results obtained in vivo using cells shipped under tested conditions and with data obtained using frozen-thawed cells implanted directly.

Finally, among domestic animal species, camelids are an important source of both food and sport, as racing animals. In this case, they can present osteoarticular damages and the treatment with MSCs could be useful for accelerating the healing process. In this contest, Son et al. [44], for the first time, isolated, expanded, and studied cells isolated from BM and Synovial Fluid (SF) of Camelus dromedaries (camel). Due to the observed chondrogenic ability of SF-MSCs, they could be considered as a target cell source for future use in therapeutic cartilage regeneration in this species.

The contributors published in this first book collection, "Stem Cells in Domestic Animals: Applications in Health and Production," are excellent examples of recent advances made in the field of stem/stromal cell research in veterinary medicine. We would like to thank the Authors for their excellent contributions and acknowledge Sandra Spatariu and the *Animals* Editorial Office for their support.

The Collection is open for submission of original manuscripts and reviews authored by outstanding experts in any aspect of stromal cell biology.

Author Contributions: E.I.: writing—original draft preparation; B.M.: writing—review and editing. All authors have read and agreed to the published version of the manuscript.

Funding: This work received no external funding.

Conflicts of Interest: The authors declare no conflict of interest.

References

1. Merlo, B.; Gugole, P.M.; Iacono, E. An Update on Applications of Cattle Mesenchymal Stromal Cells. *Animals* **2022**, *12*, 1956. [CrossRef] [PubMed]
2. La Mantia, D.; Bernardini, C.; Zannoni, A.; Salaroli, R.; Wang, C.; Bencivenni, S.; Forni, M. Efficacy of Stem Cell Therapy in Large Animal Models of Ischemic Cardiomyopathies: A Systematic Review and Meta-Analysis. *Animals* **2022**, *12*, 749. [CrossRef]
3. García-Mendívil, L.; Mediano, D.R.; Hernaiz, A.; Sanz-Rubio, D.; Vázquez, F.J.; Marín, B.; López-Pérez, Ó.; Otero, A.; Badiola, J.J.; Zaragoza, P.; et al. Effect of Scrapie Prion Infection in Ovine Bone Marrow-Derived Mesenchymal Stem Cells and Ovine Mesenchymal Stem Cell-Derived Neurons. *Animals* **2021**, *11*, 1137. [CrossRef] [PubMed]
4. Svoradova, A.; Zmrhal, V.; Venusova, E.; Slama, P. Chicken Mesenchymal Stem Cells and Their Applications: A Mini Review. *Animals* **2021**, *11*, 1883. [CrossRef] [PubMed]
5. Prišlin, M.; Vlahović, D.; Kostešić, P.; Ljolje, I.; Brnić, D.; Turk, N.; Lojkić, I.; Kunić, V.; Karadjole, T.; Krešić, N. An Outstanding Role of Adipose Tissue in Canine Stem Cell Therapy. *Animals* **2022**, *12*, 1088. [CrossRef] [PubMed]
6. Cequier, A.; Sanz, C.; Rodellar, C.; Barrachina, L. The Usefulness of Mesenchymal Stem Cells beyond the Musculoskeletal System in Horses. *Animals* **2021**, *11*, 931. [CrossRef]
7. Hagen, A.; Holland, H.; Brandt, V.-P.; Doll, C.U.; Häußler, T.C.; Melzer, M.; Moellerberndt, J.; Lehmann, H.; Burk, J. Platelet Lysate for Mesenchymal Stromal Cell Culture in the Canine and Equine Species: Analogous but Not the Same. *Animals* **2022**, *12*, 189. [CrossRef]
8. D'Ippolito, G.; Schiller, P.C.; Ricordi, C.; Roos, B.A.; Howard, G.A. Age-Related Osteogenic Potential of Mesenchymal Stromal Stem Cells from Human Vertebral Bone Marrow. *J. Bone Miner. Res. Off. J. Am. Soc. Bone Miner. Res.* **1999**, *14*, 1115–1122. [CrossRef]
9. Mueller, S.M.; Glowacki, J. Age-Related Decline in the Osteogenic Potential of Human Bone Marrow Cells Cultured in Three-Dimensional Collagen Sponges. *J. Cell. Biochem.* **2001**, *82*, 583–590. [CrossRef]
10. Stenderup, K.; Justesen, J.; Clausen, C.; Kassem, M. Aging Is Associated with Decreased Maximal Life Span and Accelerated Senescence of Bone Marrow Stromal Cells. *Bone* **2003**, *33*, 919–926. [CrossRef]
11. Choudhery, M.S.; Badowski, M.; Muise, A.; Pierce, J.; Harris, D.T. Donor Age Negatively Impacts Adipose Tissue-Derived Mesenchymal Stem Cell Expansion and Differentiation. *J. Transl. Med.* **2014**, *12*, 8. [CrossRef] [PubMed]

12. Alves, H.; van Ginkel, J.; Groen, N.; Hulsman, M.; Mentink, A.; Reinders, M.; van Blitterswijk, C.; de Boer, J. A Mesenchymal Stromal Cell Gene Signature for Donor Age. *PLoS ONE* **2012**, *7*, e42908. [CrossRef]
13. Upchurch, D.A.; Renberg, W.C.; Roush, J.K.; Milliken, G.A.; Weiss, M.L. Effects of Administration of Adipose-Derived Stromal Vascular Fraction and Platelet-Rich Plasma to Dogs with Osteoarthritis of the Hip Joints. *Am. J. Vet. Res.* **2016**, *77*, 940–951. [CrossRef] [PubMed]
14. Quaade, M.L.; Jensen, C.H.; Andersen, D.C.; Sheikh, S.P. A 3-Month Age Difference Profoundly Alters the Primary Rat Stromal Vascular Fraction Phenotype. *Acta Histochem.* **2016**, *118*, 513–518. [CrossRef] [PubMed]
15. Kemilew, J.; Sobczyńska-Rak, A.; Żylińska, B.; Szponder, T.; Nowicka, B.; Urban, B. The Use of Allogenic Stromal Vascular Fraction (SVF) Cells in Degenerative Joint Disease of the Spine in Dogs. *Vivo Athens Greece* **2019**, *33*, 1109–1117. [CrossRef]
16. Hendawy, H.; Uemura, A.; Ma, D.; Namiki, R.; Samir, H.; Ahmed, M.F.; Elfadadny, A.; El-Husseiny, H.M.; Chieh-Jen, C.; Tanaka, R. Tissue Harvesting Site Effect on the Canine Adipose Stromal Vascular Fraction Quantity and Quality. *Animals* **2021**, *11*, 460. [CrossRef]
17. Iacono, E.; Brunori, L.; Pirrone, A.; Pagliaro, P.P.; Ricci, F.; Tazzari, P.L.; Merlo, B. Isolation, Characterization and Differentiation of Mesenchymal Stem Cells from Amniotic Fluid, Umbilical Cord Blood and Wharton's Jelly in the Horse. *Reprod. Camb. Engl.* **2012**, *143*, 455–468. [CrossRef]
18. Merlo, B.; Teti, G.; Lanci, A.; Burk, J.; Mazzotti, E.; Falconi, M.; Iacono, E. Comparison between Adult and Foetal Adnexa Derived Equine Post-Natal Mesenchymal Stem Cells. *BMC Vet. Res.* **2019**, *15*, 277. [CrossRef]
19. Iacono, E.; Pascucci, L.; Rossi, B.; Bazzucchi, C.; Lanci, A.; Ceccoli, M.; Merlo, B. Ultrastructural Characteristics and Immune Profile of Equine MSCs from Fetal Adnexa. *Reprod. Camb. Engl.* **2017**, *154*, 509–519. [CrossRef]
20. Corradetti, B.; Correani, A.; Romaldini, A.; Marini, M.G.; Bizzaro, D.; Perrini, C.; Cremonesi, F.; Lange-Consiglio, A. Amniotic Membrane-Derived Mesenchymal Cells and Their Conditioned Media: Potential Candidates for Uterine Regenerative Therapy in the Horse. *PLoS ONE* **2014**, *9*, e111324. [CrossRef]
21. Rossi, B.; Merlo, B.; Colleoni, S.; Iacono, E.; Tazzari, P.L.; Ricci, F.; Lazzari, G.; Galli, C. Isolation and in Vitro Characterization of Bovine Amniotic Fluid Derived Stem Cells at Different Trimesters of Pregnancy. *Stem Cell Rev. Rep.* **2014**, *10*, 712–724. [CrossRef] [PubMed]
22. Kumar, K.; Agarwal, P.; Das, K.; Mili, B.; Madhusoodan, A.P.; Kumar, A.; Bag, S. Isolation and Characterization of Mesenchymal Stem Cells from Caprine Umbilical Cord Tissue Matrix. *Tissue Cell* **2016**, *48*, 653–658. [CrossRef] [PubMed]
23. Pratheesh, M.D.; Dubey, P.K.; Gade, N.E.; Nath, A.; Sivanarayanan, T.B.; Madhu, D.N.; Somal, A.; Baiju, I.; Sreekumar, T.R.; Gleeja, V.L.; et al. Comparative Study on Characterization and Wound Healing Potential of Goat (Capra Hircus) Mesenchymal Stem Cells Derived from Fetal Origin Amniotic Fluid and Adult Bone Marrow. *Res. Vet. Sci.* **2017**, *112*, 81–88. [CrossRef]
24. Le Blanc, K.; Tammik, C.; Rosendahl, K.; Zetterberg, E.; Ringdén, O. HLA Expression and Immunologic Properties of Differentiated and Undifferentiated Mesenchymal Stem Cells. *Exp. Hematol.* **2003**, *31*, 890–896. [CrossRef]
25. Iacono, E.; Marcoccia, R.; Merlo, B. Current Status on Canine Foetal Fluid and Adnexa Derived Mesenchymal Stem Cells. *Animals* **2021**, *11*, 2254. [CrossRef]
26. Humenik, F.; Maloveska, M.; Hudakova, N.; Petrouskova, P.; Hornakova, L.; Domaniza, M.; Mudronova, D.; Bodnarova, S.; Cizkova, D. A Comparative Study of Canine Mesenchymal Stem Cells Isolated from Different Sources. *Animals* **2022**, *12*, 1502. [CrossRef] [PubMed]
27. Delorme, B.; Nivet, E.; Gaillard, J.; Häupl, T.; Ringe, J.; Devèze, A.; Magnan, J.; Sohier, J.; Khrestchatisky, M.; Roman, F.S.; et al. The Human Nose Harbors a Niche of Olfactory Ectomesenchymal Stem Cells Displaying Neurogenic and Osteogenic Properties. *Stem Cells Dev.* **2010**, *19*, 853–866. [CrossRef]
28. Girard, S.D.; Devéze, A.; Nivet, E.; Gepner, B.; Roman, F.S.; Féron, F. Isolating Nasal Olfactory Stem Cells from Rodents or Humans. *J. Vis. Exp. JoVE* **2011**, e2762. [CrossRef]
29. Veron, A.D.; Bienboire-Frosini, C.; Feron, F.; Codecasa, E.; Deveze, A.; Royer, D.; Watelet, P.; Asproni, P.; Sadelli, K.; Chabaud, C.; et al. Isolation and Characterization of Olfactory Ecto-Mesenchymal Stem Cells from Eight Mammalian Genera. *BMC Vet. Res.* **2018**, *14*, 17. [CrossRef]
30. Mollichella, M.-L.; Mechin, V.; Royer, D.; Pageat, P.; Asproni, P. Isolation and Characterization of Cat Olfactory Ecto-Mesenchymal Stem Cells. *Animals* **2022**, *12*, 1284. [CrossRef]
31. Nelson, T.J.; Martinez-Fernandez, A.; Yamada, S.; Ikeda, Y.; Perez-Terzic, C.; Terzic, A. Induced Pluripotent Stem Cells: Advances to Applications. *Stem Cells Cloning Adv. Appl.* **2010**, *3*, 29–37. [CrossRef]
32. Nishimura, T.; Hatoya, S.; Kanegi, R.; Wijesekera, D.P.H.; Sanno, K.; Tanaka, E.; Sugiura, K.; Hiromitsu Tamada, N.K.; Imai, H.; Inaba, T. Feeder-Independent Canine Induced Pluripotent Stem Cells Maintained under Serum-Free Conditions. *Mol. Reprod. Dev.* **2017**, *84*, 329–339. [CrossRef] [PubMed]
33. Shimada, H.; Nakada, A.; Hashimoto, Y.; Shigeno, K.; Shionoya, Y.; Nakamura, T. Generation of Canine Induced Pluripotent Stem Cells by Retroviral Transduction and Chemical Inhibitors. *Mol. Reprod. Dev.* **2010**, *77*, 2. [CrossRef] [PubMed]
34. Betts, D.H.; Tobias, I.C. Canine Pluripotent Stem Cells: Are They Ready for Clinical Applications? *Front. Vet. Sci.* **2015**, *2*, 41. [CrossRef] [PubMed]
35. Debowski, K.; Warthemann, R.; Lentes, J.; Salinas-Riester, G.; Dressel, R.; Langenstroth, D.; Gromoll, J.; Sasaki, E.; Behr, R. Non-Viral Generation of Marmoset Monkey IPS Cells by a Six-Factor-in-One-Vector Approach. *PLoS ONE* **2015**, *10*, e0118424. [CrossRef]

36. Schlaeger, T.M.; Daheron, L.; Brickler, T.R.; Entwisle, S.; Chan, K.; Cianci, A.; DeVine, A.; Ettenger, A.; Fitzgerald, K.; Godfrey, M.; et al. A Comparison of Non-Integrating Reprogramming Methods. *Nat. Biotechnol.* **2015**, *33*, 58–63. [CrossRef]
37. Kim, M.; Hwang, S.-U.; Yoon, J.D.; Jeong, Y.W.; Kim, E.; Hyun, S.-H. Optimized Approaches for the Induction of Putative Canine Induced Pluripotent Stem Cells from Old Fibroblasts Using Synthetic RNAs. *Animals* **2020**, *10*, 1848. [CrossRef]
38. Di Bella, S.; Cannella, V.; Mira, F.; Di Marco, P.; Lastra, A.; Gucciardi, F.; Purpari, G.; Guercio, A. The Effect of a 7 Year-Long Cryopreservation on Stemness Features of Canine Adipose-Derived Mesenchymal Stem Cells (CAD-MSC). *Animals* **2021**, *11*, 1755. [CrossRef]
39. Garvican, E.R.; Cree, S.; Bull, L.; Smith, R.K.; Dudhia, J. Viability of Equine Mesenchymal Stem Cells during Transport and Implantation. *Stem Cell Res. Ther.* **2014**, *5*, 1. [CrossRef]
40. Bronzini, I.; Patruno, M.; Iacopetti, I.; Martinello, T. Influence of Temperature, Time and Different Media on Mesenchymal Stromal Cells Shipped for Clinical Application. *Vet. J. Lond. Engl. 1997* **2012**, *194*, 121–123. [CrossRef]
41. Mercati, F.; Pascucci, L.; Curina, G.; Scocco, P.; Tardella, F.M.; Dall'aglio, C.; Marini, C.; Ceccarelli, P. Evaluation of Storage Conditions on Equine Adipose Tissue-Derived Multipotent Mesenchymal Stromal Cells. *Vet. J. Lond. Engl. 1997* **2014**, *200*, 339–342. [CrossRef] [PubMed]
42. Espina, M.; Jülke, H.; Brehm, W.; Ribitsch, I.; Winter, K.; Delling, U. Evaluation of Transport Conditions for Autologous Bone Marrow-Derived Mesenchymal Stromal Cells for Therapeutic Application in Horses. *PeerJ* **2016**, *4*, e1773. [CrossRef] [PubMed]
43. Iacono, E.; Lanci, A.; Gugole, P.; Merlo, B. Shipping Temperature, Time and Media Effects on Equine Wharton's Jelly and Adipose Tissue Derived Mesenchymal Stromal Cells Characteristics. *Animals* **2022**, *12*, 1967. [CrossRef] [PubMed]
44. Son, Y.-B.; Jeong, Y.I.; Jeong, Y.W.; Hossein, M.S.; Olsson, P.O.; Tinson, A.; Singh, K.K.; Lee, S.-Y.; Hwang, W.S. Cell Source-Dependent In Vitro Chondrogenic Differentiation Potential of Mesenchymal Stem Cell Established from Bone Marrow and Synovial Fluid of Camelus Dromedarius. *Animals* **2021**, *11*, 1918. [CrossRef] [PubMed]

animals

MDPI

Article

Optimized Approaches for the Induction of Putative Canine Induced Pluripotent Stem Cells from Old Fibroblasts Using Synthetic RNAs

Mirae Kim [1,2], Seon-Ung Hwang [1,2], Junchul David Yoon [1,2], Yeon Woo Jeong [3], Eunhye Kim [1,*] and Sang-Hwan Hyun [1,2,*]

1 Veterinary Medical Center and College of Veterinary Medicine, Laboratory of Veterinary Embryology and Biotechnology (VETEMBIO), Chungbuk National University, Cheongju 28644, Korea; kmr9309@naver.com (M.K.); ghkdsun@hanmail.net (S.-U.H.); jdyoon86@hanmail.net (J.D.Y.)

2 Institute of Stem Cell & Regenerative Medicine (ISCRM), Chungbuk National University, Cheongju 28644, Korea

3 Abu Dhabi Biotech Research Foundation, 64 Kyungin-ro, Guro-gu, Seoul 08359, Korea; doctorj1@adbrf.org

* Correspondence: ehkim@cbu.ac.kr (E.K.); shhyun@cbu.ac.kr (S.-H.H.); Tel.: +82-43-249-1746 (E.K.); +82-43-261-3393 (S.-H.H.)

Received: 7 September 2020; Accepted: 7 October 2020; Published: 11 October 2020

Simple Summary: A non-integrating and self-replicating *Venezuelan equine encephalitis* RNA replicon system can potentially make a great contribution to the generation of clinically applicable canine induced pluripotent stem cells. Our study shows a new method to utilize the synthetic RNA-based approach for canine somatic cell reprogramming regarding transfection and reprogramming efficiency.

Abstract: Canine induced pluripotent stem cells (ciPSCs) can provide great potential for regenerative veterinary medicine. Several reports have described the generation of canine somatic cell-derived iPSCs; however, none have described the canine somatic cell reprogramming using a non-integrating and self-replicating RNA transfection method. The purpose of this study was to investigate the optimal strategy using this approach and characterize the transition stage of ciPSCs. In this study, fibroblasts obtained from a 13-year-old dog were reprogrammed using a non-integrating *Venezuelan equine encephalitis* (VEE) RNA virus replicon, which has four reprogramming factors (collectively referred to as *T7-VEE-OKS-iG* and comprised of *hOct4*, *hKlf4*, *hSox2*, and *hGlis1*) and co-transfected with the *T7-VEE-OKS-iG* RNA and *B18R* mRNA for 4 h. One day after the final transfection, the cells were selected with puromycin (0.5 μg/mL) until day 10. After about 25 days, putative ciPSC colonies were identified showing TRA-1-60 expression and alkaline phosphatase activity. To determine the optimal culture conditions, the basic fibroblast growth factor in the culture medium was replaced with a modified medium supplemented with murine leukemia inhibitory factor (mLIF) and two kinase inhibitors (2i), PD0325901(MEK1/2 inhibitor) and CHIR99021 (GSK3β inhibitor). The derived colonies showed resemblance to naïve iPSCs in their morphology (dome-shaped) and are dependent on mLIF and 2i condition to maintain an undifferentiated phenotype. The expression of endogenous pluripotency markers such as *Oct4*, *Nanog*, and *Rex1* transcripts were confirmed, suggesting that induced ciPSCs were in the late intermediate stage of reprogramming. In conclusion, the non-integrating and self-replicating VEE RNA replicon system can potentially make a great contribution to the generation of clinically applicable ciPSCs, and the findings of this study suggest a new method to utilize the VEE RNA approach for canine somatic cell reprogramming.

Keywords: canine; reprogramming; iPSCs; VEE RNA; integration-free

1. Introduction

The derivation of induced pluripotent stem cells (iPSCs) using four reprogramming factors known as Yamanaka's factors (*Oct4*, *Sox2*, *Klf4*, and *c-Myc*) opens a new avenue for patient-specific regenerative medicine therapies [1–3]. As public understanding of animal welfare and health issues increases, this technology can be also applied in veterinary medicine. Dogs, the most representative companion animals, are very useful model for the development of new animal therapeutics such as gene and stem cell-based therapies [4–6]. Canine iPSCs (ciPSCs) show great potential not only for veterinary regenerative medicine but also for translational medicine as a disease model [7,8]. While many studies have used rodent models for human disease, they are poor representations of the human system [9], but dogs are a more applicable model for a number of reasons. Most practically, the lifespan of dogs is longer than that of rodents, which allows long-term studies [10]. Dogs also have a similar size and similar genomic, anatomical, and physiological characteristics to those of humans [11]. It is important to note that the average age of companion dogs has increased in recent years, and that dogs share some disease-related genes with humans. In particular, common canine diseases share similarities with human diseases, such as cancers, autoimmune diseases, and diabetes [12–15].

To date, several reports have described the generation of ciPSCs using retroviral or lentiviral transduction using Yamanaka's factors [7,8,16]. Although viral reprogramming is very useful and conventionally performed, numerous studies have shown that it can induce genomic integration and increase the tumorigenic potential [17,18]. Due to these potential problems associated with integration, viral reprogramming is not suitable for clinical applications. For instance, one reprogramming factor *c-Myc* can increase the reprogramming efficiency, but it is a proto-oncogene that can induce tumor formation [19–21]. Consequently, it has long been debated whether *c-Myc* can be safely used as a reprogramming factor. Instead, the glis-family zinc finger 1 (*Glis1*) gene, which encodes a Krüppel-like protein, was found to have the ability to directly reprogram somatic cells into iPSCs more fully than the *c-Myc* gene [22]. Unlike *c-Myc*, the *Glis1* gene poses no increased risk for tumor formation and efficiently suppresses the proliferation of colonies that have not been fully reprogrammed. Hence, using the *Glis1* gene for reprogramming may be a good alternative instead of the *c-Myc* gene.

Recently, many kinds of integration-free methods for generating iPSCs without *c-Myc* have been developed [18,23]. For example, the *Sendai* virus, episome (Epi), and synthetic mRNA-transfection-based methods can avoid potential integration problems. Especially, there have recently been two reports of canine somatic cell reprogramming into iPSCs using the replication-defective and persistent *Sendai* virus (SeVdp) vector [24,25]. However, it is not easy to use SeVdp vectors for the regulation of reprogramming gene expressions, and they are too expensive to use frequently. In previous studies, a polycistronic and synthetic self-replicating RNA system was developed to generate human iPSCs using the RNA replicon from of *Venezuelan equine encephalitis* (VEE) virus [26,27]. The VEE replicon is a single-stranded positive sense RNA that contains a 5′ cap and poly (A) tail, similar to cellular mRNAs. The VEE replicon has no potential for problems associated with genomic DNA integration, because it does not use a DNA intermediate [28–30]. It is also easier and more productive to synthesize RNA than to produce *Sendai* virus particles with higher biosafety in the laboratory.

Although several reports have described the generation of ciPSCs [7,16,24,25,31,32], little is known about the characteristics of the early stages in the process of canine cell reprogramming from somatic cells to pluripotent stem cells. Unlike mouse and human cells, a few reports have characterized the final pluripotent stage of ciPSCs [10,31]. In mice, a genetic study was conducted to analyze progression from the transition stage to naïve embryonic stem cells [33], which reaffirmed the intermediate phase of pluripotency and helped to understand the molecular dynamics of the transition state. Likewise, several studies have analyzed the generation of authentic human pluripotent stem cells from pre-pluripotent stem cells [34,35]. Data from these studies have shown that it is necessary to characterize transition-stage cell lines before applying iPSCs in clinical research. Furthermore, there remain several major limitations for the use of the generated ciPSCs as a reliable cell source. First, canine somatic cell reprogramming protocols were not well established and consistent with each other, resulting in an insufficient reproducibility [16]. Second, it has often been reported that they are difficult to maintain in long-term culture [8,36,37]. Particularly, there has been a continuous problem in the long-term culture of ciPSCs derived from the fibroblasts of a 13-year-old dog because the senescence easily occurs at passage 7 or more [38]. Third, the number of established ciPSC colonies is too low to be suitable for applications for the development of directed iPSC differentiation for broader clinical use (heart, neuron, muscle, etc.) [24,37].

To overcome these limitations, we have tried to reprogram canine adult fibroblasts (CAFs) into a pluripotent state by using an RNA transfection-based method, which is more reproducible and efficient for reprogramming [39]. Here, we first tried to canine somatic cell reprogramming using a synthetic and self-replicating RNA-based approach. We transfected in CAFs using an RNA strategy involving a non-infectious, self-replicating, and integration-free VEE virus that expresses four reprogramming open reading frames (ORFs) (*OKS-iG; hOct4, hKlf4, hSox2*, and *hGlis1*). CAFs were also co-transfected with *B18R* mRNA, which acts as an inhibitor of type-1 interferons to reduce the strong immune responses induced by VEE RNA [40,41]. Furthermore, we investigated the optimal conditions for the transfection and selection steps, and characterized the transition stage of putative ciPSCs.

2. Materials and Methods

2.1. Ethics Statement

This study was carried out in strict accordance with the recommendations in the Guide for the Care and Use of Laboratory Animals of the National Veterinary and Quarantine Service. The protocol was approved by the Committee on the Ethics of Animal Experiments of the Chungbuk National University (permit number: CBNUA-1415-20-02). All sacrifice was performed under isoflurane anesthesia, and all efforts were made to minimize suffering.

2.2. Chemicals

All the reagents were purchased from Sigma–Aldrich Chemical Company (St. Louis, MO, USA), unless stated otherwise.

2.3. Cell Culture

Canine skin tissue was isolated from a 13-year-old female Jindo dog. Primary CAFs were obtained from the Abu Dhabi Biotech Research Foundation (Seoul, Korea). The CAFs were cultured at 37 °C in Dulbecco's Modified Eagle's Medium (DMEM, high glucose) containing 10% fetal bovine serum (FBS), 1 × MEM non-essential amino acids, 1 × GlutaMAX, 0.1 mM β-mercaptoethanol, and 1 × antibiotic–antimycotic (all from Gibco, Carlsbad, CA, USA) in air containing 5% CO_2.

2.4. Plasmid Construction and RNA Synthesis

To generate the *T7-VEE-mCherry* construct, *T7-VEE-OKS-iG* (#58974; Addgene, Cambridge, MA, USA) was linearized by NdeI/NotI digestion and modified by removing the *OKS-iG* sequence [42]. This vector was renamed as *T7-VEE*. *pSicoR-Ef1a-mCh-Puro* (#31845; Addgene) was digested with the same restriction enzymes, and then the mCherry-Puro sequence was introduced to the *T7-VEE* vector. The *pTNT-B18R* vector (#58978; Addgene) was amplified by PCR, and the amplicon was purified by gel extraction. These linearized DNA templates were used for in vitro transcription. The synthesis of *T7-VEE-mCherry* and *pTNT-B18R* RNA was performed with the RiboMAX Large Scale RNA Production System-T7 Kit (Promega, Madison, WI, USA). The ScriptCap m7G Capping System (Epicentre, Madison, WI, USA) was used for 5′ capping, which confers mRNA stability and efficient translation. After the 5′ mRNA- capping reaction, a poly (A) tail was added using poly (A) polymerase (Epicentre). These individual RNAs were purified by ammonium acetate and isopropanol precipitation, resuspended in the RNase-free water, and stored at −80 °C. A PCR was performed to confirm that both mRNAs were properly synthesized.

2.5. Determination of Optimal Puromycin Concentration

To determine the optimal puromycin concentration for selecting *VEE-OKS-iG* transfected cells, a puromycin resistance test was first performed on CAFs. CAFs were plated at a density of 5×10^4 cells/mL in a 6-well dish and cultured in DMEM containing 0.25, 0.5, 0.8, 1, 1.5, and 2 μg/mL of puromycin (Puromycin dihydrochloride, P8833) for 7 days. The cell death of CAFs was observed by culturing with various concentration of puromycin for up to 7 days.

2.6. Measurement of Transfection Efficiencies by Flow Cytometry

Initially, CAFs were seeded in a 6-well plate at a density of 1×10^4 cells/mL 24 h before transfection (Day 0). On day 1, the cells were transfected with *T7-VEE-mCherry* mRNA using Lipofectamine MessengerMAX (Thermo Fisher Scientific, Waltham, MA, USA) or the RiboJuice™ transfection reagent (EMD Millipore, Burlington, MA, USA). Both transfection reagents were used at 4 μL each and the RNA template was used at 1 μg each. The expression of *T7-VEE-mCherry* was measured by flow cytometry on day 2. CAFs were then washed once with DPBS (LB 001-02, Welgene, Gyeongsangbuk-do, Korea) and incubated with 0.25% trypsin–EDTA (Gibco) for 2 min in a 37 °C incubator. After dissociation and trypsin inactivation, these cells were resuspended with DPBS containing 1% FBS. The transfection efficiency was measured by a flow cytometry analysis using the SH800S cell sorter (Sony Imaging Products & Solutions, Tokyo, Japan) and analyzed with the cell sorter software version 1.8.3 (Sony Imaging Products and Solutions).

2.7. Preparation of Feeder Cells

Mouse embryonic fibroblasts (MEFs) were isolated from embryonic day 13.5 fetuses of the Institute of Cancer Research (ICR) strain. To inactivate the MEFs, they were treated with 10 μg/mL of mitomycin C (Roche, Basel, Switzerland) at passage 3 for 2 to 2.5 h before being used as feeder layers. The inactivated MEFs were plated at a density of 2.5×10^5 cells/mL in 6-well dishes coated with 0.1% gelatin (EMD Millipore).

2.8. Preparation of Mouse Embryonic Fibroblast Conditioned Medium (MEF-CM)

To prepare the MEF-CM, inactivated MEFs were seeded at a concentration of 4×10^5 cells/mL in 6-well dishes coated with 0.1% gelatin (EMD Millipore). The inactivated MEFs were cultured at 37 °C in DMEM containing 10% FBS, 1 × GlutaMAX, and 1 × antibiotic-antimycotic (all from Gibco) in air containing 5% CO_2. The MEF-CM was collected on the next day and filtrated through a 0.2 μm syringe filter (Corning Incorporated, Corning, NY, USA).

2.9. Generation and Culture of the Putative ciPSCs

Reprogramming was performed using the Simplicon™ Reprogramming Kit (EMD Millipore). On day 0, the CAFs were seeded at a density of 5×10^4 cells/mL in a 6-well plate coated with 0.1% gelatin (EMD Millipore). To minimize strong immune responses, CAFs were pre-treated with 200 ng/mL of human recombinant B18R protein (EMD Millipore) for 20 min before transfection. Transfection was performed on day 1 (1 × Tfx), on days 1 and 2 (2 × Tfx), or on days 1 to 4 (4 × Tfx). The cells were transfected with both *VEE-OKS-iG* RNA (EMD Millipore) and *B18R* RNA (EMD Millipore) using the RiboJuice™ mRNA Transfection Kit (EMD Millipore). After 4 h, the transfection medium was changed to stage 1 medium, consisting of Advanced-DMEM (Gibco) with 10% FBS (Gibco), 1 × GlutaMAX (Gibco), and 200 ng/mL of human recombinant B18R protein (EMD Millipore). The transfected cells were grown in stage 1 medium and selected with puromycin (0.5 μg/mL) until day 10. At day 10 post-transfection, the transfected cells were passaged onto inactivated MEFs using Accutase (Biowest,

Riverside, MO, USA) in a 6-well plate (1×10^5 cells/mL in each well). Stage 1 medium was used through day 10 and then it was replaced with stage 2 medium, comprised of MEF CM containing 10 ng/mL of basic fibroblast growth factor (bFGF; Bio-Bud, Gyunggido, Korea), 1 × human iPS reprogramming boost supplement II (TGF-β RI kinase inhibitor IV, sodium butyrate, and PS48) from the Simplicon™ Reprogramming Kit (EMD Millipore), and 200 ng/mL of B18R protein (EMD Millipore). The stage 2 medium was changed every other day. B18R protein (EMD Millipore) was added until small iPSC colonies appeared. Putative ciPSC colonies were mechanically isolated on days 22 to 30 and replated on inactivated MEFs in stage 3 medium, which consisted of DMEM/F12 medium, 20% KnockOut Serum Replacement medium (KSR), 1 × GlutaMAX, 1 × MEM NEAA, 0.1 mM of β-mercaptoethanol, and 1 × antibiotic–antimycotic (all from Gibco). Putative ciPSC colonies were cultured in stage 3 medium containing 10 ng/mL of bFGF (Bio-Bud) or 1000 units/mL of murine leukemia inhibitory factor (mLIF; ESGRO, Millipore, Billerica, MA, USA), plus 0.5 μM of MEK1/2 inhibitor (PD0325901; Stemgent, Cambridge, MA, USA) and 3 μM of GSK3β inhibitor (CHIR99021; Stemgent).

2.10. Primary iPSC Colony Staining with Alkaline Phosphatase (AP) and TRA-1-60

ciPSCs were washed with PBS three times after removing the iPSC culture medium from each 6-well plate, after which they were fixed in PBS containing 4% paraformaldehyde for 5 min at room temperature. Fixed cells were washed three times in PBS and stained with 5-bromo-4-chloro-3-indolyl phosphate/nitroblue tetrazolium (BCIP/NBT; Roche, Basel, Switzerland) solution for 30 min at room temperature in the dark. AP-positive cells were stained a dark purple color and visualized under a light microscope. For the TRA-1-60 staining, ciPSC colonies were washed with PBS three times, and colonies were stained using an Alexa Fluor 488-conjugated TRA-1-60 antibody (1:50) (Life Technologies, Carlsbad, CA, USA). After 1 h, the staining solution was removed and the cells were gently washed three times with FluoroBrite™ DMEM (Life Technologies). Images were acquired using a fluorescent microscope.

2.11. Gene Expression Analysis of Putative ciPSCs by Reverse Transcription PCR (RT-PCR)

The expression levels of *RPL13A* (a reference gene), *VEE-hOct4, VEE-hKlf4, VEE-hSox2, VEE-hGlis1, Rex1, Oct4*, and *Nanog* mRNA in ciPSCs were analyzed by RT-PCR. All the cDNA samples of ciPSC-like colonies were stored at −80 °C until PCR analysis. Total RNA was extracted from putative ciPSCs using the Trizol reagent (Invitrogen, Carlsbad, CA, USA). Complementary DNA (cDNA) was synthesized using Moloney murine leukemia virus (MMLV) reverse transcriptase (Invitrogen) and random primers (Invitrogen). All the procedures were performed in accordance with the manufacturer's instructions. RT-PCR analysis was conducted using cDNA from putative ciPSCs. cDNA was amplified in a 20 μL PCR mixture consisting of 10 pmol of forward and reverse primers, 2 units of Taq polymerase, 2 μL of 10x PCR buffer, 5 pmol of dNTP mixtures (all from iNtRON Biotechnology, SungNam, Korea), and template DNA. The oligonucleotide primer sequences are presented in Table S1. PCR amplification was performed for 30 cycles of denaturation at 95 °C for 30 s, annealing at 57 °C for 30 s, and extension at 72 °C for 90 s. The reaction products were analyzed on a 1.25% agarose gel pre-stained with RedSafe™ Nucleic Acid Staining Solution (iNtRON Biotechnology).

2.12. Statistical Analysis

The statistical analyses were performed using SPSS 17.0 (SPSS, Inc., Chicago, IL, USA). The percentage data were compared using a one-way ANOVA followed by Tukey's multiple range tests. All the results are expressed as the mean ± standard error of mean (SEM). $p < 0.05$ were considered to be statistically significant.

3. Results

3.1. Determination of Optimal Puromycin Concentration

Since the synthetic VEE RNA used in the experiment has a puromycin resistance gene, puromycin selection was performed to isolate the transfected cell line. Prior to transfecting synthetic RNAs, it is important to determine the optimal puromycin concentration for selecting transfected cells. For this, CAFs were plated at a density of 5×10^4 cells/mL in a 6-well plate, after which they were cultured with various puromycin concentrations on day 1. After the addition of puromycin, varying levels of cell death were observed for 7 days. Surviving cells were not observed when exposed to puromycin concentrations higher than 0.5 μg/mL from day 4 (Figure 1A). The surviving cell number of the 0.8 μg/mL puromycin-treated group is significantly lower than that of the 0.5 μg/mL puromycin-treated group (Figure 1B). Therefore, it was determined that the treatment with 0.5 μg/mL of puromycin for 7 days enabled the most efficient selection of transfected CAFs.

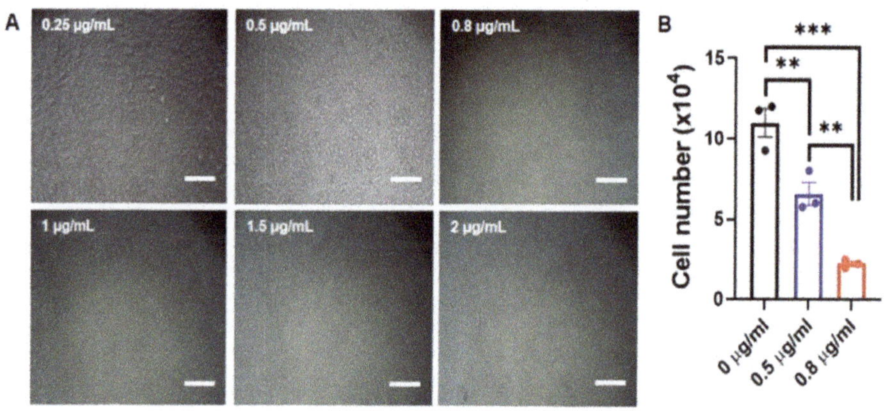

Figure 1. Determination of the optimal puromycin concentration for selecting transfected cells. (**A**) Representative images of various puromycin concentration-treated fibroblasts from a 13-year-old dog. Scale bars = 200 μm. (**B**) The number of cells counted on 4 days after the puromycin treatment. The value represents mean ± the standard error of the mean (SEM). Data were analyzed by one-way analysis of variance (ANOVA). Asterisks indicate statistical significance (** $p < 0.01$, *** $p < 0.005$).

3.2. Synthesis of RNA by In Vitro Transcription (IVT) and Measurement of Transfection Efficiencies

To perform the mRNA-based reprogramming, the first step for the reprograming of CAFs is the production of synthetic RNA molecules encoding the reprogramming factors. The synthetic modified mRNAs containing each gene were produced as shown in Figure 2A. Synthetic *T7-VEE-mCherry* mRNA and *B18R* mRNA were synthesized by in vitro transcription (IVT). A non-infectious and self-replicating VEE RNA replicon was modified by removing the *OKS-iG* sequence. The VEE plasmid was linearized by NdeI/NotI digestion, and the fluorescent mCherry gene was introduced 3′ of the replicon ORF. The *B18R* plasmid was amplified by PCR, and the amplicon was purified by gel extraction. These linearized plasmids served as templates for RNA synthesis. RNA transcription was performed for 4 h at 37 °C with the RiboMAX Large Scale RNA Production System-T7 Kit. A 5′ capping enzyme and poly (A) polymerase were used to promote efficient translation. After the purification of the PCR product and the IVT, the cDNAs correctly synthesized from these RNAs were analyzed using agarose gel electrophoresis to determine the specific length and purity. The detected bands showed the expected lengths of the plasmid DNA (control) and PCR products (cDNA) from the RNA templates (Figure 2B). The sizes of the vectors used in the experiment are as follows: *T7-VEE-OKS-iG* is 16.86 kb, *T7-VEE-mCherry* is 11.477 kb, and *B18R* is 3.907 kb.

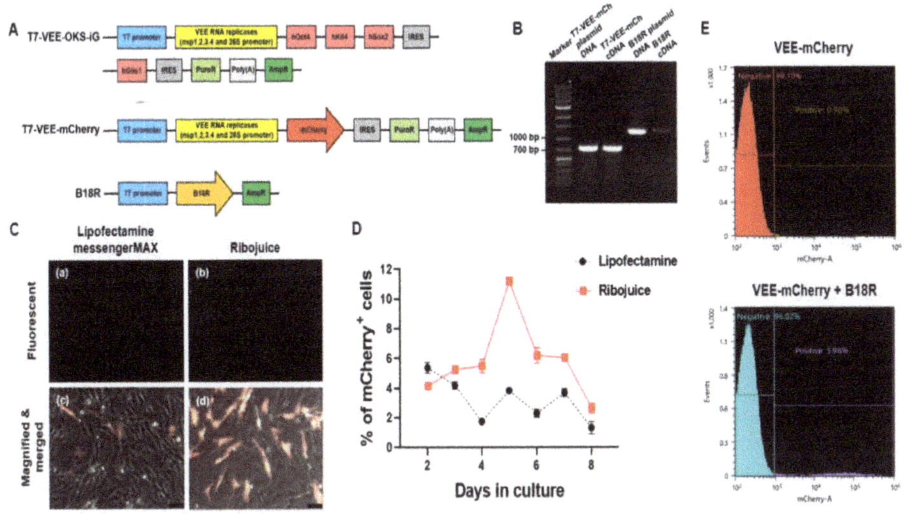

Figure 2. Construction of synthetic mRNAs and the measurement of transfection efficiencies. (**A**) Maps of *T7-VEE-OKS-iG*, *T7-VEE-mCherry*, and *B18R* vectors. *Venezuelan equine encephalitis* (VEE) RNA replicases are represented as bright yellow boxes. (**B**) Confirmation of plasmid DNA (control), and polymerase chain reaction (PCR) products (complementary DNA; cDNA) from RNA templates by reverse transcription (RT)-PCR analysis. (**C**) Representative images of *T7-VEE-mCherry* RNA-transfected canine fibroblasts comparing Lipofectamine messengerMAX and Ribojuice tranfection reagents. Scale bars = 250 μm for (a, b) and 75 μm for (c, d). (**D**) Percentage of *mCherry*$^+$ transfected cells using the Lipofectamine MessengerMAX or RiboJuice transfection reagent. The value represents mean ± SEM. (**E**) Representative flow cytometry histogram showing the mCherry fluorescence of the *mCherry*$^+$ transfected cell population. Transfected with VEE RNA alone (upper); co-transfected with the T7-VEE-mCherry RNA replicon and B18R (bottom). All the experiments were replicated at least three times.

To determine which transfection reagent was more efficient, CAFs were transfected with *T7-VEE-mCherry* RNA on day 1, using the Lipofectamine MessengerMAX or the RiboJuice™ transfection reagent (Figure 2C). The results showed that the RiboJuice™ transfection reagent showed a higher percentage of mCherry$^+$ cells than that of the Lipofectamine MessengerMAX-treated group (Figure 2D), which suggests that the use of the RiboJuice™ transfection reagent is a more optimized approach for the generation of ciPSCs. Additionally, CAFs were transfected with *T7-VEE-mCherry* RNA alone or co-transfected with *B18R* RNA using the RiboJuice™ transfection reagent. Because of the strong immune response induced by VEE RNA, few mCherry-positive cells were observed after transfection with *T7-VEE-mCherry* RNA alone. However, those cells co-transfected with the *T7-VEE-mCherry* RNA replicon and *B18R* RNA have shown more mCherry expression (3.98%) than the cells transfected with VEE RNA alone (0.9%) (Figure 2E). These results suggested that *B18R* is required for the more efficient transfection of VEE RNA replicons.

3.3. Generation of Putative ciPSCs Using Fibroblasts from an Aged Dog

The synthetic mRNA-based reprogramming required the daily transfection of the cells to maintain a constant level of reprogramming factor expression. In this study, transfection was performed with the *T7-VEE-OKS-iG* replicon and *B18R* RNA on day 1 (1 × Tfx), days 1 and 2 (2 × Tfx), or days 1 to 4 (4 × Tfx) (Figure 3A). One day after the final transfection, the cells were selected with puromycin (0.5 μg/mL) until day 10. Then, 200 ng/mL of *B18R* protein was added to minimize the immune response until iPSC colonies appeared. To find the optimal reprogramming conditions, we have attempted to reprogram using a variety of culture media such as MEF-CM, Advanced-DMEM, or DMEM/F12. Among these conditions, the most effective culture condition was to use MEF-CM until the first colony appeared (for about 19 days) after VEE-OKS-iG mRNA transfection (data not shown). Twelve putative ciPSC colonies were derived from the CAFs and isolated on days 22 to 30. These initial colonies were of two distinct types, which are a spherical type and a radial type (Figure 3B). Two colonies showed a spherical-type morphology and 10 colonies retained a radial-type morphology (Figure 3C). Interestingly, when they were subcultured, all the spherical-type colonies (clone #4 and #8) either differentiated or died, whereas all the radial-type colonies proliferated except for clones #2 and #3 (these two cell lines died during maintenance).

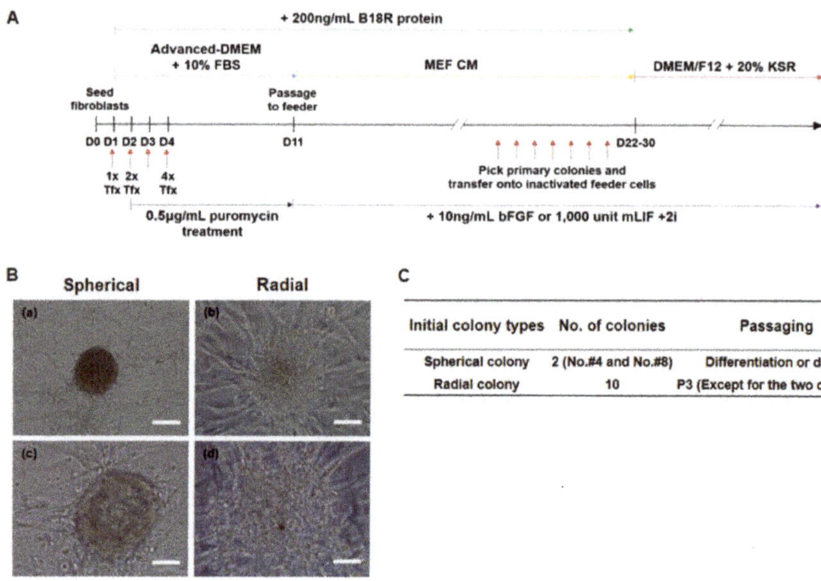

Initial colony types	No. of colonies	Passaging
Spherical colony	2 (No.#4 and No.#8)	Differentiation or death
Radial colony	10	P3 (Except for the two cell lines)

Figure 3. Two distinct types of initial canine induced pluripotent stem cell (ciPSC) colonies. (**A**) Timeline of synthetic mRNA-mediated reprogramming using a method of co-transfecting *T7-VEE-OKS-iG* and *B18R* mRNAs. (**B**) Representative images of the initial ciPSC colonies after introducing four reprogramming factors by VEE RNA on day 25. (a) and (c) Typical images of a spherical colony. (b) and (d) Typical images of a radial colony. Scale bars = 200 μm for (a, b); 50 μm for (c, d). (**C**) Information of the initial ciPSCs classified based on the colony morphology.

3.4. Characterization of Pluripotency Markers in Primary ciPSC Colonies

To evaluate the quality of the putative primary ciPSC colonies, the expression of the pluripotent stem cell marker TRA-1-60 was assessed by live-cell imaging. Both the spherical- and radial-type colonies expressed the TRA-1-60 pluripotency marker, as shown in (Figure 4A). AP activity was also observed in a spherical colony (Figure 4B).

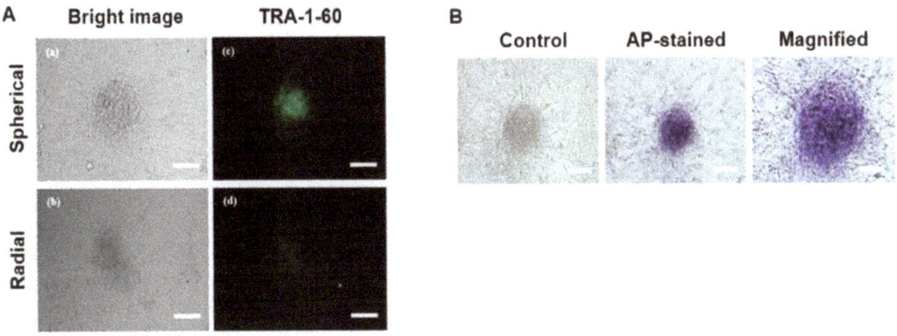

Figure 4. Immunocytochemistry of pluripotency markers in primary canine induced pluripotent stem cell (ciPSC) colonies. (**A**) Expression of the pluripotency marker TRA-1-60 in ciPSCs, as assessed by immunocytochemical analysis. (a) and (c) A spherical colony (No. #4 colony). (b) and (d) A radial colony (No. #2 colony). Scale bars = 50 µm. (**B**) Image of a spherical domed ciPSC colony (No. #4 colony) after introducing four factors (*VEE-hOct4*, *VEE-hKlf4*, *VEE-hSox2*, and *VEE-hGlis1*) by mRNA transfection on day 27 (left); scale bars = 200 µm. Representative image of an AP-positive colony (middle); scale bars = 200 µm. Magnified image of an AP-positive colony (right); scale bars = 50 µm.

The radial colonies were subcultured and maintained well on the feeder layers. When the radial colonies were subcultured, almost of them have changed their morphology from a flat shape to a round shape (Figure 5A), and several colonies spontaneously differentiated in the presence of only 10 ng/mL of bFGF.

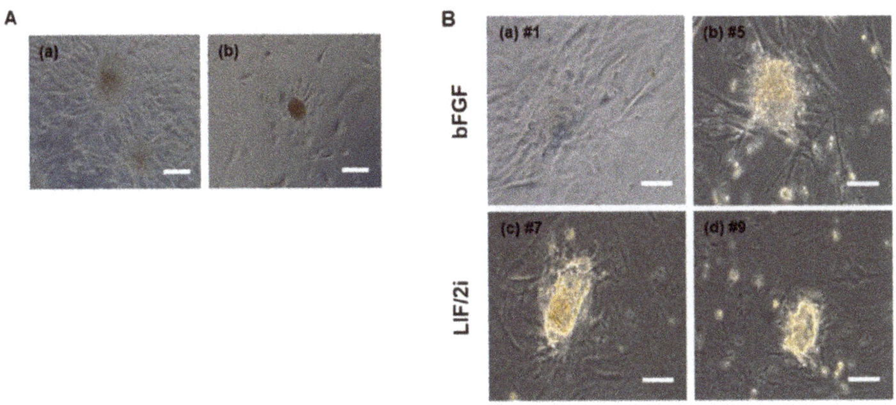

Figure 5. Morphological changes of canine induced pluripotent stem cell (ciPSC). (**A**) Representative images of radial colonies. (a) Radial colonies of putative ciPSCs were appeared at day 26 after VEE RNA transfection. (b) At passage 1, a flat and rounded colony appeared. Scale bars = 100 µm. (**B**) Morphological changes of ciPSCs culture in the presence of basic fibroblast growth factor (bFGF) or leukemia inhibitory factor (LIF)/ PD0325901 and CHIR99021 (two inhibitor; 2i). (a, b) The attached clumps appeared under bFGF-only conditions. (c, d) The culture conditions were changed to LIF/2i. Scale bars = 200 µm.

To determine the optimal culture conditions, the 10 ng/mL of bFGF culture condition was changed to 1000 units of mLIF + 0.5 µM MEK1/2 inhibitor (PD0325901) + 3 µM of GSK3β inhibitor (CHIR99021) (LIF/2i). Both culture conditions were compared to investigate which condition was more appropriate. Six days later, the flat or round-shaped colonies had changed in appearance to dome-shaped colonies and gradually proliferated under the LIF/2i condition (Figure 5B). However, these results should be further examined to determine whether such ciPSCs are LIF-dependent. These data indicated that the primary ciPSC lines generated using RNA-based reprograming were maintained well under the LIF/2i condition.

3.5. Gene Expression Analysis of Putative ciPSCs by RT-PCR

The total eight radial colonies were cultured to passage 3, so we investigated the gene expression levels by RT-PCR. The putative ciPSCs in the transition or intermediate stage did not grow well after passage 3, so it was difficult to obtain a sufficient cDNA concentration to perform a gene expression analysis. Of the total eight cell lines maintained until passage 3, we analyzed the level of gene expression by PCR using the ciPSC #5 cell line and the ciPSC #7 cell line, with a relatively sufficient cDNA concentration. To characterize putative ciPSCs, the expression levels of *RPL13A* (a reference gene), *VEE-hOct4*, *VEE-hKlf4*, *VEE-hSox2*, *VEE-hGlis1*, *Oct4*, *Nanog*, and *Rex1* mRNA were measured by RT-PCR (Figure 6 and Table S1). Among the exogenous *VEE-hOct4*, *VEE-hKlf4*, *VEE-hSox2*, and *VEE-hGlis1* transcripts, only *VEE-hOct4* was expressed in the putative ciPSC #5 cell line. The expression of the endogenous *Klf4*, *Sox2*, and *Glis1* genes was not detected (data not shown), but the endogenous *Oct4* and *Nanog* genes were expressed in both cell lines. The putative ciPSC #7 cell line also showed the expression of *Rex1*, which is a well-known ground-state (naïve) pluripotent marker gene [43]. The results of this study suggest that the ciPSC lines generated using RNA-based reprograming were in the late intermediate stage of reprogramming.

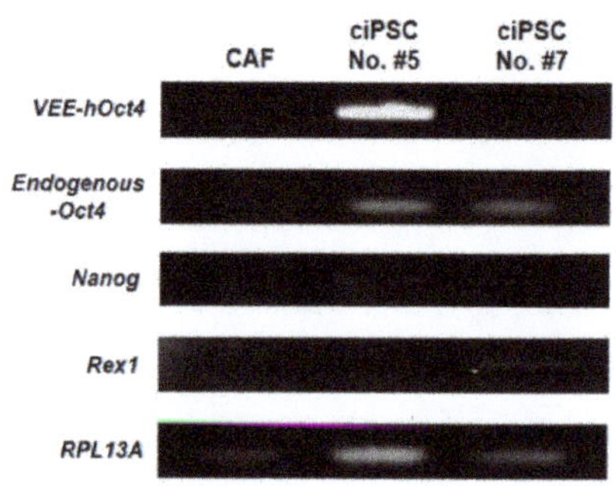

Figure 6. Gene expression analysis of the putative canine induced pluripotent stem cell (ciPSC) lines. The expression of *RPL13A*, *VEE-hOct4*, endogenous *Oct4*, *Nanog*, and *Rex1* mRNAs was analyzed in the putative ciPSC colonies on day 17 by RT-PCR. The *RPL13A* gene served as a reference gene; *VEE-hOct4* served as an exogenous gene; *Oct4*, *Nanog*, and *Rex1* served as pluripotency marker genes.

4. Discussion

The generation of iPSCs using an integration-free VEE RNA replicon system has great potential for clinical research [26,27,44] due to concerns over the integration of DNA vectors into the genome. To generate integration-free iPSCs from the fibroblasts of a 13-year-old dog, we used an efficient VEE RNA-based iPSC generation strategy. Our findings suggest a new method to utilize the VEE RNA approach for canine somatic cell reprogramming regarding the transfection and reprogramming efficiency (Figure 7).

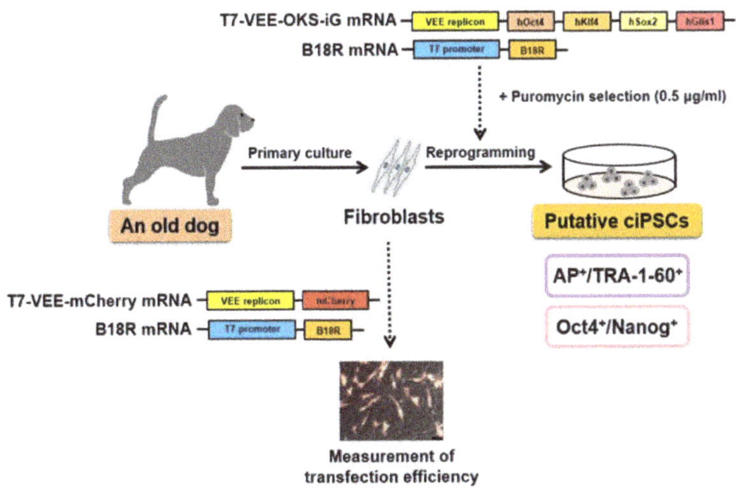

Figure 7. Schematic diagram showing the generation of putative induced pluripotent stem cells (iPSCs) from canine adult fibroblasts (CAFs) using synthetic RNAs. In this study, the fibroblasts obtained from an old dog were reprogrammed using a non-integrating VEE RNA virus replicon which has four reprogramming factors (collectively referred to as *T7-VEE-OKS-iG*) and co-transfected with the *B18R* mRNA. To evaluate the transfection efficiency, CAFs were co-transfected with *T7-VEE-mCherry* mRNA and *B18R* mRNA on day 1, using the RiboJuice™ transfection reagent. One day after the final transfection, the cells were selected with puromycin (0.5 µg/mL) until day 10. After about 25 days, the putative canine iPSC colonies were identified showing TRA-1-60 expression and alkaline phosphatase (AP) activity. Finally, the expressions of endogenous pluripotency markers such as the *Oct4* and *Nanog* transcripts were confirmed, suggesting that the generated canine iPSCs were in the late intermediate stage of reprogramming.

The VEE RNA replicon used in the experiments contains the puromycin resistance gene. Thus, the transfected cell lines have been isolated by puromycin selection. The typical puromycin working concentration range is 0.5 to 10 µg/mL for mammalian cells [45], and the optimal concentration should be determined empirically before selecting transfected cells. Hong et. al. have reported that stably transfected canine fetal fibroblasts were selected with 2 µg/mL of puromycin for 2 weeks [46]. Techangamsuwan et al., on the other hand, have transfected canine schwann cells and olfactory ensheathing cells, and the stable transfectants were selected with 0.4 µg/mL of puromycin [47]. Therefore, the puromycin sensitivity varies according to the target cell type, so an appropriate concentration should be determined prior to antibiotic selection. In this study, the optimal puromycin concentration for the selection of CAFs transfected with VEE-OKS-iG synthetic mRNA was found to be 0.5 µg/mL. In addition, a previous study showed that the activation of innate immune responses enhanced the efficiency of iPSC generation by repeated mRNA transfection [48]. It has also been shown that VEE RNA induces especially strong interferon (IFN)-α/β innate immune responses [26]. To reduce the innate immune response to VEE RNA, the B18R protein was used, which binds to and neutralizes type 1 IFNs [40]. Interestingly, the cells co-transfected with *T7-VEE-mCherry* RNA and *B18R* RNA expressed mCherry at higher levels than the cells transfected with *T7-VEE-mCherry* alone. Therefore, the B18R protein mitigated VEE RNA-induced immune responses and increased the transfection efficiency.

According to previous studies, the pluripotent states were classified into two states, referred to as the "naïve" and "primed" states [33,49,50]. The naïve PSCs typically retain small, compact, dome-shaped colonies, whereas the primed PSCs display a large and flattened monolayer morphology. In this study, 12 colonies were derived from CAFs following only one transfection experiment using the

VEE RNA system. On the basis of their morphology, these initial colonies were classified into two types, spherical and radial. Neither are a typical morphology of ciPSC colonies, composed of tightly packed dome-shaped cells surrounded by a distinct border similar to canine ESCs. Several researchers have identified those colonies, showing unclear margins and radial colony morphologies as a partially reprogrammed cell [51–53]. Even though both the spherical and radial colonies stained positive for the pluripotency marker TRA-1-60 by immunocytochemistry, they might be partially reprogrammed [10,54,55]. In addition, David et. al. have reported that AP is detected during the early intermediate stage of reprogramming [56]. Because our results have also shown AP activity in initial colonies through AP staining, these cell lines can be considered to have passed the early intermediate stage of reprogramming. Especially, all the spherical colonies (No. #4 and No. #8) differentiated or died when they were subcultured, whereas all the radial colonies proliferated after subculture except for clones #2 and #3, suggesting that the radial-type colonies were maintained better than the spherical-type colonies.

Naïve and primed PSCs differ substantially in their major pluripotency-related signaling pathways: the LIF-dependent naïve state and the bFGF-dependent primed state, both of which are commonly reported in mouse and human cells [57]. Even though both signaling pathways seemed to be mutually exclusive, several reports have shown that most ciPSCs require both bFGF and LIF supplementation for proliferation in the undifferentiated state [32,36,37,54], whereas only one report has shown that ciPSCs were dependent on LIF alone [10]. The LIF-dependent ciPSCs seemed to be established in the naïve state. Whitworth et al. [10] also reported that ciPSCs could be cultured well in a cocktail medium supplemented with a cocktail of small molecules containing not only bFGF and LIF, but also GSK3β inhibitor, MEK1/2 inhibitor, TGF-β antagonist, and valproic acid. Therefore, we also compare the numerous different culture conditions for RNA-based reprogramed ciPSCs established in this study. Although additional study should be performed to determine whether these ciPSCs are LIF-dependent or not, the evaluation of different culture media for the ciPSC lines revealed that those based on the LIF/2i condition are more suitable for both the isolation and expansion of ciPSCs compared to the bFGF culture condition.

Recently, RNA-based iPSC approaches using four individual reprogramming-factor mRNAs generated by IVT were noted [41,58,59]. However, due to the rapid degradation of reprogramming factor mRNAs, this RNA-based reprogramming method required daily transfections during the reprogramming period or that the mRNA transfection be continued for a few days to increase the transfection efficiency [60]. Therefore, the transfection efficiency was determined as a function of the transfection frequency in this study. As the frequency of transfection increased, primary colonies appeared to be reprogrammed more efficiently because they were observed at a much higher percentage, whereas the cells transfected only once were speculated to have reached the transition state (data not shown). Since there is a self-replicative VEE replicon in the vector used in the experiment, only one transfection was attempted without continuous transfection. Using this system, the reprogramming factor mRNAs replication and transcription may continue to occur in transfected cells once [61]. Even though mRNA delivery to CAFs has been conducted several times, the initial colonies were speculated to have been partially reprogrammed and were potentially in the transition state. The main cause of entry into the transition state by these primary colonies is as follows. Despite the potential of efficient mRNA transfection for reprogramming, cellular senescence served as a fundamental barrier against fully reprogramming fibroblasts from aged dogs to iPSCs [62,63]. Moreover, a more efficient transgene-delivery system was required for successful reprogramming. Previous data has shown that electroporation was more suitable for producing stable cell lines than lipofection [64].

Recent reports have shown the cellular changes that occur during the reprogramming process of MEFs [33,65,66]. The reprogramming process can be classified into two phases—namely, an early "stochastic" and a following "deterministic" phase [65]. If the reprogrammable cells enter an intermediate state (i.e., a transition state), then the cells go through a stochastic activation of pluripotency genes [67], the transient activation of developmental regulators [68], and the activation of glycolysis [69,70]. One thing to be sure is that an intermediate phase of morphological changes from somatic cells to iPSCs during reprogramming must occur, during which reprogrammable cells undergo transcriptional and epigenetic genetic changes [61,71]. The stably expressed reference genes differ with each cell line, and care should be taken when selecting an appropriate reference gene for RT-PCR. Although glyceraldehyde-3-phosphate dehydrogenase (*GAPDH*) is the most commonly used reference gene, many reports have demonstrated that this gene is not stably expressed in different tissues [72,73]. Instead, it was reported that *RPL13A* is more stably expressed as a reference gene in whole canine skin than *GAPDH* [72]. Using *RPL13A* as a reference gene, the expression levels of exogenous (*VEE-hOct4, VEE-hKlf4, VEE-hSox2*, and *VEE-hGlis1*) and endogenous (*Oct4, Nanog*, and *Rex1*) genes were determined in the putative ciPSC colonies and CAFs. In this study, endogenous *Oct4, Klf4, Sox2*, and *Glis1* genes were analyzed in the putative ciPSCs by RT-PCR, but those endogenous genes were not fully activated in these cell lines except for the *Oct4* gene. Generally, many researchers have reported that the expression of the endogenous expression of *Oct4* might be detected from the early transitional stage of reprogramming [65,71,74,75], whereas the endogenous expression of *Sox2* could be detected only after transition stage, particularly in the stabilization stage of reprogramming [56,71,76]. However, little is known about the transitional state of canine reprogramming process. The expressions of the *Oct4* and TRA-1-60 with AP activity without the endogenous expression of the *Sox2* in the ciPSCs generated by the VEE-RNA system suggest that they are in a late intermediate or transition state, and not in the stabilization stage of reprogramming. To our knowledge, we are the first group demonstrating a significant improvement in ciPSCs in a transitional state.

In previous studies, the expression of endogenous factors such as *Oct4, Klf4, Sox2*, and *c-Myc* was analyzed in ciPSCs [7,31], but these primer sequences are not suitable for exogenous and endogenous gene expression analysis because they overlap with those of humans. Unlike mouse and human cells, the development of canine-specific markers is insufficient and some gene sequences are very similar between dogs and humans, making it difficult to perform gene-expression analysis because only a few certain canine gene sequences are found in the GenBank® (database; http://www.ncbi.nlm.nih.gov). In addition, only the exogenous *VEE-hOct4* gene was expressed in the putative ciPSC #5 cell line, suggesting that this cell line was more partially reprogrammed than the No. #7 cell line because the overexpressed exogenous Oct4 genes remain. Although the endogenous *Oct4* (strong expression) and *Nanog* (weak expression) genes were expressed in both cell lines, *Rex1*, known to be important for the naïve pluripotent state [43,77], was clearly observed only in the No. #7 cell line. Although all of the colonies generated in this study showed a strong expression of *Oct4*, almost of the colonies (eight cell lines including #5) showed the weak expression of *Rex1*, except for No. #7 ciPSCs. The weak expression level of *Rex1* with the strong expression of "*Oct4*" in these colonies suggests that those colonies are not in the ground state of pluripotency [49], unlike the #7 ciPSCs, which show the most "naïve" pluripotency. These results indicate that the almost of the ciPSC colonies generated using RNA-based reprograming were in the transition stage of reprogramming. The lack of silencing of exogenous transgenes has been reported as a well-known feature of incomplete reprogramming [78]. This indicates that VEE RNA was not fully transfected into CAFs because the size of VEE replicon is very large (11 kb), and a strong immune response was induced by the modified mRNA during reprogramming. Taken together, these findings suggest that the putative ciPS cell lines may be in a late intermediate stage of reprogramming, and in particular the No. #7 cell line is closer to the ground state of pluripotency than the No. #5 line. A limitation of our findings is that the ciPSCs generated in this study subcultured well up to five serial passages while keeping their incomplete pluripotency. Several researchers have previously described that a self-replicating VEE RNA vector is another transgene delivery method for viral-free iPSCs [26,44,79]. However, further studies are required to set up the complete reprogramming and optimize the culture conditions [80] for more long-term culture. It is also necessary to further study the mechanism from the transition state to the fully pluripotent state. This is a preliminary study and requires repeating with quantitative methodologies.

5. Conclusions

This work describes a new method for transfecting dog somatic cells. To our knowledge, we are the first group to report synthetic RNA-based reprogramming for ciPSCs from the fibroblasts of an old dog. We performed reprogramming with the somatic cells of a Korean native dog, an old Jindo dog, and it was possible to determine the optimal concentration of puromycin treatment suitable for the selection of transfected individual cells and to present the results of an RNA transfection efficiency measurement. It is speculated that the reason why the established putative ciPSC lines remained only in the early passage was that they were partially reprogrammed or at an intermediate reprogramming stage, but further experimentation is needed to prove this. Although further studies are needed to characterize the conversion of canine somatic cells to ciPSCs using synthetic RNAs, these results may be useful for developing canine-specific markers to characterize iPSCs in the transition or deterministic phase. Furthermore, the reprogramming method using non-integrating, self-replicative VEE RNA will help us to generate clinically applicable ciPSCs.

Supplementary Materials: The following are available online at http://www.mdpi.com/2076-2615/10/10/1848/s1: Table S1: Primers list for RT PCR.

Author Contributions: Conceptualization, M.K., S.-U.H., and S.-H.H.; Methodology, M.K., S.-U.H., and J.D.Y.; Investigation, M.K.; Validation, M.K., E.K., and S.-H.H.; Formal analysis, M.K., S.-U.H., and E.K.; Resources, Y.W.J.; Writing—original draft preparation, M.K., E.K., and S.-H.H.; Writing—review and editing, M.K., E.K., and S.-H.H.; Funding acquisition, S.-H.H. All authors have read and agreed to the published version of the manuscript.

Funding: This work was supported, in part, by a grant from the "National Research Foundation of Korea Grant funded by the Korean Government (2020R1A2C2008276)", "Korea Institute of Planning and Evaluation for Technology in Food, Agriculture, Forestry and Fisheries (IPET)" through Agri-Bio industry Technology Development Program (grant number: 318016-5), Agri-food R&D Performance Follow-up Support Program (grant number: 819029-2) and Agriculture, Food and Rural Affairs Convergence Technologies Program for Educating Creative Global Leader (grant number: 320005-4), funded by Ministry of Agriculture, Food and Rural Affairs (MAFRA)", "The Global Research and Development Center (GRDC) Program through the National Research Foundation of Korea (NRF) funded by the Ministry of Education, Science and Technology (2017K1A4A3014959)" and "Business for Cooperative R&D between Industry, Academy, and Research Institute funded Korea Small and Medium Business Administration in 2017 (Grants No. 2017020681010101)", Republic of Korea.

Acknowledgments: We would like to thank the officials of the International Society for Stem Cell Research (ISSCR) conference who accepted a poster presentation in June 2018 that included some findings on "Morphological changes of putative canine induced pluripotent stem cells in early stage".

Conflicts of Interest: The authors declare no conflict of interest.

References

1. Takahashi, K.; Tanabe, K.; Ohnuki, M.; Narita, M.; Ichisaka, T.; Tomoda, K.; Yamanaka, S. Induction of pluripotent stem cells from adult human fibroblasts by defined factors. *Cell* **2007**, *131*, 861–872. [CrossRef] [PubMed]
2. Park, I.-H.; Zhao, R.; West, J.A.; Yabuuchi, A.; Huo, H.; Ince, T.A.; Lerou, P.H.; Lensch, M.W.; Daley, G.Q. Reprogramming of human somatic cells to pluripotency with defined factors. *Nature* **2008**, *451*, 141. [CrossRef] [PubMed]
3. Takahashi, K.; Yamanaka, S. Induction of pluripotent stem cells from mouse embryonic and adult fibroblast cultures by defined factors. *Cell* **2006**, *126*, 663–676. [CrossRef] [PubMed]
4. Zucconi, E.; Vieira, N.M.; Bueno, D.F.; Secco, M.; Jazedje, T.; Ambrosio, C.E.; Passos-Bueno, M.R.; Miglino, M.A.; Zatz, M. Mesenchymal stem cells derived from canine umbilical cord vein—a novel source for cell therapy studies. *Stem Cells Dev.* **2010**, *19*, 395–402. [CrossRef]
5. Shearin, A.L.; Ostrander, E.A. Leading the way: Canine models of genomics and disease. *Dis. Models Mech.* **2010**, *3*, 27–34. [CrossRef]
6. Tsai, K.L.; Clark, L.A.; Murphy, K.E. Understanding hereditary diseases using the dog and human as companion model systems. *Mamm. Genome* **2007**, *18*, 444–451. [CrossRef]
7. Nishimura, T.; Hatoya, S.; Kanegi, R.; Wijesekera, D.P.H.; Sanno, K.; Tanaka, E.; Sugiura, K.; Tamada, H.; Kawate, N.; Imai, H. Feeder-independent canine induced pluripotent stem cells maintained under serum-free conditions. *Mol. Reprod. Dev.* **2017**, *84*, 329–339. [CrossRef]

8. Shimada, H.; Nakada, A.; Hashimoto, Y.; Shigeno, K.; Shionoya, Y.; Nakamura, T. Generation of canine induced pluripotent stem cells by retroviral transduction and chemical inhibitors. *Mol. Reprod. Dev.* **2010**, *77*, 2. [CrossRef]
9. Hainsworth, A.H.; Allan, S.M.; Boltze, J.; Cunningham, C.; Farris, C.; Head, E.; Ihara, M.; Isaacs, J.D.; Kalaria, R.N.; Oberstein, S.A.L. Translational models for vascular cognitive impairment: A review including larger species. *BMC Med.* **2017**, *15*, 16. [CrossRef]
10. Whitworth, D.J.; Ovchinnikov, D.A.; Wolvetang, E.J. Generation and characterization of LIF-dependent canine induced pluripotent stem cells from adult dermal fibroblasts. *Stem Cells Dev.* **2012**, *21*, 2288–2297. [CrossRef]
11. Tecirlioglu, R.T.; Trounson, A.O. Embryonic stem cells in companion animals (horses, dogs and cats): Present status and future prospects. *Reprod. Fertil. Dev.* **2007**, *19*, 740–747. [CrossRef] [PubMed]
12. Hytönen, M.K.; Lohi, H. Canine models of human rare disorders. *Rare Dis.* **2016**, *4*, e1006037. [CrossRef] [PubMed]
13. Wood, S.H.; Ollier, W.E.; Nuttall, T.; McEwan, N.A.; Carter, S.D. Despite identifying some shared gene associations with human atopic dermatitis the use of multiple dog breeds from various locations limits detection of gene associations in canine atopic dermatitis. *Vet. Immunol. Immunopathol.* **2010**, *138*, 193–197. [CrossRef] [PubMed]
14. Paoloni, M.; Khanna, C. Translation of new cancer treatments from pet dogs to humans. *Nat. Rev. Cancer* **2008**, *8*, 147–156. [CrossRef] [PubMed]
15. Robinson, R.A.; Pugh, R.N. Dogs, zoonoses and immunosuppression. *J. R. Soc. Promot. Health* **2002**, *122*, 95–98. [CrossRef] [PubMed]
16. Betts, D.H.; Tobias, I.C. Canine Pluripotent Stem Cells: Are They Ready for Clinical Applications? *Front. Vet. Sci.* **2015**, *2*, 41. [CrossRef] [PubMed]
17. Debowski, K.; Warthemann, R.; Lentes, J.; Salinas-Riester, G.; Dressel, R.; Langenstroth, D.; Gromoll, J.; Sasaki, E.; Behr, R. Non-viral generation of marmoset monkey iPS cells by a six-factor-in-one-vector approach. *PLoS ONE* **2015**, *10*, e0118424. [CrossRef]
18. Schlaeger, T.M.; Daheron, L.; Brickler, T.R.; Entwisle, S.; Chan, K.; Cianci, A.; DeVine, A.; Ettenger, A.; Fitzgerald, K.; Godfrey, M. A comparison of non-integrating reprogramming methods. *Nat. Biotechnol.* **2015**, *33*, 58–63. [CrossRef]
19. Hoffman, B.; Liebermann, D.A. The proto-oncogene c-myc and apoptosis. *Oncogene* **1998**, *17*. [CrossRef]
20. Hoffman, B.; Amanullah, A.; Shafarenko, M.; Liebermann, D.A. The proto-oncogene c-myc in hematopoietic development and leukemogenesis. *Oncogene* **2002**, *21*, 3414–3421. [CrossRef]
21. Vafa, O.; Wade, M.; Kern, S.; Beeche, M.; Pandita, T.K.; Hampton, G.M.; Wahl, G.M. c-Myc can induce DNA damage, increase reactive oxygen species, and mitigate p53 function: A mechanism for oncogene-induced genetic instability. *Mol. Cell* **2002**, *9*, 1031–1044. [CrossRef]
22. Maekawa, M.; Yamaguchi, K.; Nakamura, T.; Shibukawa, R.; Kodanaka, I.; Ichisaka, T.; Kawamura, Y.; Mochizuki, H.; Goshima, N.; Yamanaka, S. Direct reprogramming of somatic cells is promoted by maternal transcription factor Glis1. *Nature* **2011**, *474*, 225–229. [CrossRef] [PubMed]
23. Nakagawa, M.; Koyanagi, M.; Tanabe, K.; Takahashi, K.; Ichisaka, T.; Aoi, T.; Okita, K.; Mochiduki, Y.; Takizawa, N.; Yamanaka, S. Generation of induced pluripotent stem cells without Myc from mouse and human fibroblasts. *Nat. Biotechnol.* **2008**, *26*, 101. [CrossRef]
24. Chow, L.; Johnson, V.; Regan, D.; Wheat, W.; Webb, S.; Koch, P.; Dow, S. Safety and immune regulatory properties of canine induced pluripotent stem cell-derived mesenchymal stem cells. *Stem Cell Res.* **2017**, *25*, 221–232. [CrossRef] [PubMed]
25. Tsukamoto, M.; Nishimura, T.; Yodoe, K.; Kanegi, R.; Tsujimoto, Y.; Alam, M.E.; Kuramochi, M.; Kuwamura, M.; Ohtaka, M.; Nishimura, K. Generation of footprint-free canine induced pluripotent stem cells using auto-erasable Sendai virus vector. *Stem Cells Dev.* **2018**, *27*, 1577–1586. [CrossRef]
26. Yoshioka, N.; Gros, E.; Li, H.-R.; Kumar, S.; Deacon, D.C.; Maron, C.; Muotri, A.R.; Chi, N.C.; Fu, X.-D.; Benjamin, D.Y. Efficient generation of human iPSCs by a synthetic self-replicative RNA. *Cell Stem Cell* **2013**, *13*, 246–254. [CrossRef]
27. Yoshioka, N.; Dowdy, S.F. Enhanced generation of iPSCs from older adult human cells by a synthetic five-factor self-replicative RNA. *PLoS ONE* **2017**, *12*, e0182018. [CrossRef]

28. Kinney, R.M.; Johnson, B.J.; Welch, J.B.; Tsuchiya, K.R.; Trent, D.W. The full-length nucleotide sequences of the virulent Trinidad donkey strain of Venezuelan equine encephalitis virus and its attenuated vaccine derivative, strain TC-83. *Virology* **1989**, *170*, 19–30. [CrossRef]
29. Greenway, T.E.; Eldridge, J.H.; Ludwig, G.; Staas, J.K.; Smith, J.F.; Gilley, R.M.; Michalek, S.M. Enhancement of protective immune responses to Venezuelan equine encephalitis (VEE) virus with microencapsulated vaccine. *Vaccine* **1995**, *13*, 1411–1420. [CrossRef]
30. Paessler, S.; Ni, H.; Petrakova, O.; Fayzulin, R.Z.; Yun, N.; Anishchenko, M.; Weaver, S.C.; Frolov, I. Replication and clearance of Venezuelan equine encephalitis virus from the brains of animals vaccinated with chimeric SIN/VEE viruses. *J. Virol.* **2006**, *80*, 2784–2796. [CrossRef]
31. Gonçalves, N.; Bressan, F.; Roballo, K.; Meirelles, F.; Xavier, P.; Fukumasu, H.; Williams, C.; Breen, M.; Koh, S.; Sper, R. Generation of LIF-independent induced pluripotent stem cells from canine fetal fibroblasts. *Theriogenology* **2017**, *92*, 75–82. [CrossRef] [PubMed]
32. Luo, J.; Cibelli, J.B. Conserved Role of bFGF and a Divergent Role of LIF for Pluripotency Maintenance and Survival in Canine Pluripotent Stem Cells. *Stem Cells Dev.* **2016**, *25*, 1670–1680. [CrossRef] [PubMed]
33. Kalkan, T.; Olova, N.; Roode, M.; Mulas, C.; Lee, H.J.; Nett, I.; Marks, H.; Walker, R.; Stunnenberg, H.G.; Lilley, K.S. Tracking the embryonic stem cell transition from ground state pluripotency. *Development* **2017**, *144*, 1221–1234. [CrossRef] [PubMed]
34. Moon, J.-H.; Heo, J.S.; Kwon, S.; Kim, J.; Hwang, J.; Kang, P.J.; Kim, A.; Kim, H.O.; Whang, K.Y.; Yoon, B.S. Two-step generation of induced pluripotent stem cells from mouse fibroblasts using Id3 and Oct4. *J. Mol. Cell Biol.* **2011**, *4*, 59–62. [CrossRef] [PubMed]
35. Chan, E.M.; Ratanasirintrawoot, S.; Park, I.-H.; Manos, P.D.; Loh, Y.-H.; Huo, H.; Miller, J.D.; Hartung, O.; Rho, J.; Ince, T.A. Live cell imaging distinguishes bona fide human iPS cells from partially reprogrammed cells. *Nat. Biotechnol.* **2009**, *27*, 1033–1037. [CrossRef]
36. Lee, A.S.; Xu, D.; Plews, J.R.; Nguyen, P.K.; Nag, D.; Lyons, J.K.; Han, L.; Hu, S.; Lan, F.; Liu, J. Preclinical derivation and imaging of autologously transplanted canine induced pluripotent stem cells. *J. Biol. Chem.* **2011**, *286*, 32697–32704. [CrossRef] [PubMed]
37. Luo, J.; Suhr, S.T.; Chang, E.A.; Wang, K.; Ross, P.J.; Nelson, L.L.; Venta, P.J.; Knott, J.G.; Cibelli, J.B. Generation of leukemia inhibitory factor and basic fibroblast growth factor-dependent induced pluripotent stem cells from canine adult somatic cells. *Stem Cells Dev.* **2011**, *20*, 1669–1678. [CrossRef]
38. You, S.; Moon, J.-H.; Kim, T.-K.; Kim, S.-C.; Kim, J.-W.; Yoon, D.-H.; Kwak, S.; Hong, K.-C.; Choi, Y.-J.; Kim, H. Cellular characteristics of primary and immortal canine embryonic fibroblast cells. *Exp. Mol. Med.* **2004**, *36*, 325. [CrossRef]
39. Steinle, H.; Weber, M.; Behring, A.; Mau-Holzmann, U.; Schlensak, C.; Wendel, H.P.; Avci-Adali, M. Generation of iPSCs by nonintegrative RNA-based reprogramming techniques: Benefits of self-replicating RNA versus synthetic mRNA. *Stem Cells Int.* **2019**. [CrossRef]
40. Alcamí, A.; Symons, J.A.; Smith, G.L. The vaccinia virus soluble alpha/beta interferon (IFN) receptor binds to the cell surface and protects cells from the antiviral effects of IFN. *J. Virol.* **2000**, *74*, 11230–11239. [CrossRef]
41. Warren, L.; Manos, P.D.; Ahfeldt, T.; Loh, Y.-H.; Li, H.; Lau, F.; Ebina, W.; Mandal, P.K.; Smith, Z.D.; Meissner, A. Highly efficient reprogramming to pluripotency and directed differentiation of human cells with synthetic modified mRNA. *Cell Stem Cell* **2010**, *7*, 618–630. [CrossRef] [PubMed]
42. Mirae Kim, S.-U.H.; Hyun, S.-H. An integration-free method for the generation of putative induced pluripotent stem cells derived from aged dogs. In Proceedings of the ISSCR 2018 Annual Meeting, Melbourne, Australia, 20 June 2018; p. 282.
43. Kalkan, T.; Smith, A. Mapping the route from naive pluripotency to lineage specification. *Philos. Trans. R. Soc. B Biol. Sci.* **2014**, *369*, 20130540. [CrossRef] [PubMed]
44. Sosa, E.; Kim, R.; Rojas, E.J.; Hosohama, L.; Hennebold, J.D.; Orwig, K.E.; Clark, A.T. An integration-free, virus-free rhesus macaque induced pluripotent stem cell line (riPSC89) from embryonic fibroblasts. *Stem Cell Res.* **2016**, *17*, 444–447. [CrossRef] [PubMed]
45. Watanabe, S.; Iwamoto, M.; Suzuki, S.-i.; Fuchimoto, D.; Honma, D.; Nagai, T.; Hashimoto, M.; Yazaki, S.; Sato, M.; Onishi, A. A novel method for the production of transgenic cloned pigs: Electroporation-mediated gene transfer to non-cultured cells and subsequent selection with puromycin. *Biol. Reprod.* **2005**, *72*, 309–315. [CrossRef] [PubMed]

46. Hong, S.G.; Oh, H.J.; Park, J.E.; Kim, M.J.; Kim, G.A.; Koo, O.J.; Jang, G.; Lee, B.C. Production of transgenic canine embryos using interspecies somatic cell nuclear transfer. *Zygote* **2012**, *20*, 67–72. [CrossRef]
47. Techangamsuwan, S.; Kreutzer, R.; Kreutzer, M.; Imbschweiler, I.; Rohn, K.; Wewetzer, K.; Baumgärtner, W. Transfection of adult canine Schwann cells and olfactory ensheathing cells at early and late passage with human TERT differentially affects growth factor responsiveness and in vitro growth. *J. Neurosci. Methods* **2009**, *176*, 112–120. [CrossRef] [PubMed]
48. Lee, J.; Sayed, N.; Hunter, A.; Au, K.F.; Wong, W.H.; Mocarski, E.S.; Pera, R.R.; Yakubov, E.; Cooke, J.P. Activation of innate immunity is required for efficient nuclear reprogramming. *Cell* **2012**, *151*, 547–558. [CrossRef]
49. Weinberger, L.; Ayyash, M.; Novershtern, N.; Hanna, J.H. Dynamic stem cell states: Naive to primed pluripotency in rodents and humans. *Nat. Rev. Mol. Cell Biol.* **2016**, *17*, 155–169. [CrossRef]
50. Silva, J.; Nichols, J.; Theunissen, T.W.; Guo, G.; van Oosten, A.L.; Barrandon, O.; Wray, J.; Yamanaka, S.; Chambers, I.; Smith, A. Nanog is the gateway to the pluripotent ground state. *Cell* **2009**, *138*, 722–737. [CrossRef]
51. Wen, Y.; Wani, P.; Zhou, L.; Baer, T.; Phadnis, S.M.; Reijo Pera, R.A.; Chen, B. Reprogramming of fibroblasts from older women with pelvic floor disorders alters cellular behavior associated with donor age. *Stem Cells Transl. Med.* **2013**, *2*, 118–128. [CrossRef]
52. Neely, M.D.; Tidball, A.M.; Aboud, A.A.; Ess, K.C.; Bowman, A.B. Induced Pluripotent Stem Cells (iPSCs): An emerging model system for the study of human neurotoxicology. In *Cell Culture Techniques*; Springer: New York, NY, USA, 2011; pp. 27–61.
53. Rosselló, R.A.; Chen, C.-C.; Dai, R.; Howard, J.T.; Hochgeschwender, U.; Jarvis, E.D. Mammalian genes induce partially reprogrammed pluripotent stem cells in non-mammalian vertebrate and invertebrate species. *eLife* **2013**, *2*, e00036. [CrossRef] [PubMed]
54. Koh, S.; Thomas, R.; Tsai, S.; Bischoff, S.; Lim, J.-H.; Breen, M.; Olby, N.J.; Piedrahita, J.A. Growth requirements and chromosomal instability of induced pluripotent stem cells generated from adult canine fibroblasts. *Stem Cells Dev.* **2012**, *22*, 951–963. [CrossRef] [PubMed]
55. Nishimura, T.; Hatoya, S.; Kanegi, R.; Sugiura, K.; Wijewardana, V.; Kuwamura, M.; Tanaka, M.; Yamate, J.; Izawa, T.; Takahashi, M. Generation of functional platelets from canine induced pluripotent stem cells. *Stem Cells Dev.* **2013**, *22*, 2026–2035. [CrossRef] [PubMed]
56. David, L.; Polo, J.M. Phases of reprogramming. *Stem Cell Res.* **2014**, *12*, 754–761. [CrossRef]
57. Hassani, S.-N.; Totonchi, M.; Gourabi, H.; Schöler, H.R.; Baharvand, H. Signaling roadmap modulating naive and primed pluripotency. *Stem Cells Dev.* **2014**, *23*, 193–208. [CrossRef]
58. Angel, M.; Yanik, M.F. Innate immune suppression enables frequent transfection with RNA encoding reprogramming proteins. *PLoS ONE* **2010**, *5*, e11756. [CrossRef]
59. Warren, L.; Ni, Y.; Wang, J.; Guo, X. Feeder-free derivation of human induced pluripotent stem cells with messenger RNA. *Sci. Rep.* **2012**, *2*, 657. [CrossRef]
60. Poleganov, M.A.; Eminli, S.; Beissert, T.; Herz, S.; Moon, J.-I.; Goldmann, J.; Beyer, A.; Heck, R.; Burkhart, I.; Barea Roldan, D. Efficient reprogramming of human fibroblasts and blood-derived endothelial progenitor cells using nonmodified RNA for reprogramming and immune evasion. *Hum. Gene Ther.* **2015**, *26*, 751–766. [CrossRef]
61. Pushko, P.; Bray, M.; Ludwig, G.V.; Parker, M.; Schmaljohn, A.; Sanchez, A.; Jahrling, P.B.; Smith, J.F. Recombinant RNA replicons derived from attenuated Venezuelan equine encephalitis virus protect guinea pigs and mice from Ebola hemorrhagic fever virus. *Vaccine* **2000**, *19*, 142–153. [CrossRef]
62. Lapasset, L.; Milhavet, O.; Prieur, A.; Besnard, E.; Babled, A.; Aït-Hamou, N.; Leschik, J.; Pellestor, F.; Ramirez, J.-M.; De Vos, J. Rejuvenating senescent and centenarian human cells by reprogramming through the pluripotent state. *Genes Dev.* **2011**, *25*, 2248–2253. [CrossRef]
63. Heng, B.C.; Heinimann, K.; Miny, P.; Iezzi, G.; Glatz, K.; Scherberich, A.; Zulewski, H.; Fussenegger, M. mRNA transfection-based, feeder-free, induced pluripotent stem cells derived from adipose tissue of a 50-year-old patient. *Metab. Eng.* **2013**, *18*, 9–24. [CrossRef] [PubMed]
64. Tabar, M.S.; Hesaraki, M.; Esfandiari, F.; Samani, F.S.; Vakilian, H.; Baharvand, H. Evaluating electroporation and lipofectamine approaches for transient and stable transgene expressions in human fibroblasts and embryonic stem cells. *Cell J. (Yakhteh)* **2015**, *17*, 438.

65. Buganim, Y.; Faddah, D.A.; Jaenisch, R. Mechanisms and models of somatic cell reprogramming. *Nat. Rev. Genet.* **2013**, *14*, 427–439. [CrossRef] [PubMed]
66. Hackett, J.A.; Dietmann, S.; Murakami, K.; Down, T.A.; Leitch, H.G.; Surani, M.A. Synergistic mechanisms of DNA demethylation during transition to ground-state pluripotency. *Stem Cell Rep.* **2013**, *1*, 518–531. [CrossRef] [PubMed]
67. Buganim, Y.; Faddah, D.A.; Cheng, A.W.; Itskovich, E.; Markoulaki, S.; Ganz, K.; Klemm, S.L.; van Oudenaarden, A.; Jaenisch, R. Single-cell expression analyses during cellular reprogramming reveal an early stochastic and a late hierarchic phase. *Cell* **2012**, *150*, 1209–1222. [CrossRef] [PubMed]
68. Polo, J.M.; Anderssen, E.; Walsh, R.M.; Schwarz, B.A.; Nefzger, C.M.; Lim, S.M.; Borkent, M.; Apostolou, E.; Alaei, S.; Cloutier, J. A molecular roadmap of reprogramming somatic cells into iPS cells. *Cell* **2012**, *151*, 1617–1632. [CrossRef] [PubMed]
69. Folmes, C.D.; Nelson, T.J.; Martinez-Fernandez, A.; Arrell, D.K.; Lindor, J.Z.; Dzeja, P.P.; Ikeda, Y.; Perez-Terzic, C.; Terzic, A. Somatic oxidative bioenergetics transitions into pluripotency-dependent glycolysis to facilitate nuclear reprogramming. *Cell Metab.* **2011**, *14*, 264–271. [CrossRef]
70. Panopoulos, A.D.; Yanes, O.; Ruiz, S.; Kida, Y.S.; Diep, D.; Tautenhahn, R.; Herrerías, A.; Batchelder, E.M.; Plongthongkum, N.; Lutz, M. The metabolome of induced pluripotent stem cells reveals metabolic changes occurring in somatic cell reprogramming. *Cell Res.* **2012**, *22*, 168–177. [CrossRef]
71. Van Den Hurk, M.; Kenis, G.; Bardy, C.; Van Den Hove, D.L.; Gage, F.H.; Steinbusch, H.W.; Rutten, B.P. Transcriptional and epigenetic mechanisms of cellular reprogramming to induced pluripotency. *Epigenomics* **2016**, *8*, 1131–1149. [CrossRef]
72. Wood, S.H.; Clements, D.N.; McEwan, N.A.; Nuttall, T.; Carter, S.D. Reference genes for canine skin when using quantitative real-time PCR. *Vet. Immunol. Immunopathol.* **2008**, *126*, 392–395. [CrossRef]
73. Vandesompele, J.; De Preter, K.; Pattyn, F.; Poppe, B.; Van Roy, N.; De Paepe, A.; Speleman, F. Accurate normalization of real-time quantitative RT-PCR data by geometric averaging of multiple internal control genes. *Genome Biol.* **2002**, 3, research0034.1. [CrossRef] [PubMed]
74. Peñalosa-Ruiz, G.; Mulder, K.W.; Veenstra, G.J.C. The corepressor NCOR1 and OCT4 facilitate early reprogramming by suppressing fibroblast gene expression. *PeerJ* **2020**, *8*, e8952. [CrossRef] [PubMed]
75. Radzisheuskaya, A.; Chia, G.L.B.; Dos Santos, R.L.; Theunissen, T.W.; Castro, L.F.C.; Nichols, J.; Silva, J.C. A defined Oct4 level governs cell state transitions of pluripotency entry and differentiation into all embryonic lineages. *Nat. Cell Biol.* **2013**, *15*, 579–590. [CrossRef] [PubMed]
76. Golipour, A.; David, L.; Liu, Y.; Jayakumaran, G.; Hirsch, C.L.; Trcka, D.; Wrana, J.L. A late transition in somatic cell reprogramming requires regulators distinct from the pluripotency network. *Cell Stem Cell* **2012**, *11*, 769–782. [CrossRef] [PubMed]
77. Nichols, J.; Smith, A. Naive and primed pluripotent states. *Cell Stem Cell* **2009**, *4*, 487–492. [CrossRef] [PubMed]
78. Soto, D.A.; Ross, P.J. Pluripotent stem cells and livestock genetic engineering. *Transgenic Res.* **2016**, *25*, 289–306. [CrossRef] [PubMed]
79. Rony, I.; Baten, A.; Bloomfield, J.A.; Islam, M.; Billah, M.; Islam, K. Inducing pluripotency in vitro: Recent advances and highlights in induced pluripotent stem cells generation and pluripotency reprogramming. *Cell Prolif.* **2015**, *48*, 140–156. [CrossRef]
80. Brouwer, M.; Zhou, H.; Kasri, N.N. Choices for induction of pluripotency: Recent developments in human induced pluripotent stem cell reprogramming strategies. *Stem Cell Rev. Rep.* **2016**, *12*, 54–72. [CrossRef]

animals

MDPI

Communication

Tissue Harvesting Site Effect on the Canine Adipose Stromal Vascular Fraction Quantity and Quality

Hanan Hendawy [1,2], Akiko Uemura [1], Danfu Ma [1], Ryosuke Namiki [1], Haney Samir [3,4], Mahmoud F. Ahmed [2], Ahmed Elfadadny [5], Hussein M. El-Husseiny [1,6], Cheng Chieh-Jen [1] and Ryou Tanaka [1,*]

1 Laboratory of Veterinary Surgery, Tokyo University of Agriculture and Technology, Tokyo 183-8509, Japan; hanan_attia@vet.suez.edu.eg (H.H.); anco@vet.ne.jp (A.U.); dandanma1000@gmail.com (D.M.); nxwcm554@ybb.ne.jp (R.N.); s195162s@st.go.tuat.ac.jp (H.M.E.-H.); john_199328@yahoo.com.tw (C.C.-J.)

2 Department of Veterinary Surgery, Faculty of Veterinary Medicine, Suez Canal University, Ismailia 41522, Egypt; mahmoud_ali@vet.suez.edu.eg

3 Department of Theriogenology, Faculty of Veterinary Medicine, Cairo University, Giza 12211, Egypt; haneyvet360@yahoo.com

4 Laboratory of Veterinary Physiology, Department of Veterinary Medicine, Tokyo University of Agriculture and Technology, Tokyo 183-8509, Japan

5 Department of Animal Medicine, Faculty of Veterinary Medicine, Damanhur University, Damanhur, El-Beheira 22511, Egypt; ahmed.elfadadny@vetmed.dmu.edu.eg

6 Department of Surgery, Anesthesiology, and Radiology, Faculty of Veterinary Medicine, Benha University, Elqaliobiya 13736, Egypt

* Correspondence: ryo@vet.ne.jp; Tel.: +81-042-367-5904

Citation: Hendawy, H.; Uemura, A.; Ma, D.; Namiki, R.; Samir, H.; Ahmed, M.F.; Elfadadny, A.; El-Husseiny, H.M.; Chieh-Jen, C.; Tanaka, R. Tissue Harvesting Site Effect on the Canine Adipose Stromal Vascular Fraction Quantity and Quality. *Animals* **2021**, *11*, 460. https://doi.org/10.3390/ani11020460

Academic Editors: Salvatore Desantis and Eleonora Iacono

Received: 22 December 2020
Accepted: 5 February 2021
Published: 9 February 2021

Simple Summary: Adipose stromal vascular fraction (SVF) cells are freshly isolated non-cultured mesenchymal stem cells, which have been recently applied in the treatment of several musculoskeletal inflammatory conditions in dogs. However, the best adipose tissue (AT) sampling site is still challenging. This study first addressed the ideal AT harvesting site in canines ranging between middle and old age, the most susceptible age to chronic musculoskeletal problems. Our results showed that the peri-ovarian region is the best AT harvesting site, which yields high amounts of SVF cells with enough adipose-derived stem cells. These data may help the further set-up of cell-based regenerative therapies at the preclinical and experimental level in canines.

Abstract: Mesenchymal stem cells (MSCs) constitute a great promise for regenerative therapy, but these cells are difficultly recovered in large amounts. A potent alternative is the stromal vascular fraction (SVF), non-cultured MSCs, separated from adipose tissue (AT). We aim to evaluate AT harvesting site effect on the SVF cells' quantity and quality in dogs. Subcutaneous abdominal fat, falciform ligament and peri-ovarian fat were sampled. After SVF isolation, the trypan blue exclusion test and a hemocytometer were used to assess the cell viability and cellular yield. SVF cells were labeled for four surface antigenic markers, clusters of differentiation CD90, CD44, CD29, and CD45, and then examined by flow cytometry. Semi-quantitative RT-PCR was used to evaluate the gene expression of the former markers in addition to OCT-4 and CD34. SVF cells in the peri-ovarian AT recorded the highest viability% ($99.63 \pm 0.2\%$), as well as a significantly higher cellular yield ($36.87 \pm 19.6 \times 10^6$ viable cells/gm fat, $p < 0.001$) and a higher expression of adipose-derived mesenchymal stem cells AD-MSCs surface markers than that of other sites. SVF cells from the peri-ovarian site revealed a higher expression of MSC markers (CD90, CD44, and CD29) and OCT-4 compared to the other sites, with weak CD45 and CD34 expressions. The positive OCT-4 expression demonstrated the pluripotency of SVF cells isolated from different sites. To conclude, the harvesting site is a strong determinant of SVF cells' quantity and quality, and the peri-ovarian site could be the best AT sampling site in dogs.

Keywords: canine; stromal vascular fraction; harvest site; adipose-derived MSCs; flow cytometry

1. Introduction

The use of mesenchymal stem cells (MSCs) in regenerative medicine holds a great promise for repairing damaged tissues in both acute conditions, such as injuries of ligaments, tendons, cartilage or bone, and chronic conditions, such as osteoarthritis [1]. These cells have been separated from many tissues, including bone marrow, adipose tissue, blood, cartilage, and muscle [2]. Although studies initially focused on bone marrow-derived MSCs, the high concentration of MSCs in adipose tissues (100–1000 times that in bone marrow) triggered adipose-derived MSCs' use in regenerative stem cell therapy [3–5]. Adipose tissue (AT) is a practical and reasonable source for both freshly isolated stromal vascular fraction and cultured adipose-derived stem cells (AD-MSCs) [6]. Stromal vascular fraction (SVF) cells can be directly separated from minced AT tissue by incubation with collagenase enzyme followed by centrifugation [7,8]. SVF cells are a heterogeneous cell population from adipose stromal cells, progenitor cells, hematopoietic stem cells, and endothelial cells. This cell population produces a homogeneous cell population of plastic adherent AD-MSCs when expanded on culture [9]. Both cell types, primary SVF and cultured AD-MSCs, represent an important therapeutic target [10].

Recent clinical trials have proposed using freshly isolated primary SVF cells instead of cultured AD-MSCs [11–13] for many reasons. Cultivation of the SVF cells leads to changes in the AD-MSCs' phenotype and reduces the differentiation potential of these cells. Thus, the primary SVF cells have a larger regenerative capacity than cultured AD-MSCs [14,15]. Additionally, the popularity of freshly isolated SVF cells over cultured AD-MSCs in the veterinary practice may be due to the high-expenses, time-consuming isolation and the in vitro serial expansion of AD-MSCs, which may lead to contamination, loss of differentiation ability, and the neoplastic transformation of cells [6]. It is beneficial to isolate autologous SVF cells and re-use them as a therapy in a single surgical procedure. Therefore, SVF cells may have the chief prospective for future stem cell therapy [12].

Dogs are a good preclinical animal model to study several degenerative and traumatic diseases in humans [16,17]. Recently, adipose SVF has been used for the treatment of several inflammatory and immune-mediated conditions in canines, such as osteoarthritis, tendinopathy, hip dysplasia and multiple sclerosis [10,11,18,19]. Despite the wide use of canine SVF cells in veterinary clinics, only two studies investigated the cellular component of freshly isolated canine SVF cells. Astor et al. [20] discussed AT collection from three different sites (subcutaneous caudal to the scapula, falciform, and inguinal region). However, Astor et al. [20] did not assess the immune phenotyping of the freshly isolated SVF cells. Sullivan et al. [21] conducted another study, in which dogs were less than 2 years of age, and AT samples were collected from two sites only (subcutaneous AT caudal to the scapula and falciform ligament). However, Astor et al. [20] and Sullivan et al. [21] did not investigate the peri-ovarian site as a promising source for AT harvest. Despite this expansion, SVF cells have not been characterized in dogs compared to those reports in humans or bone marrow cells [21]. Our study aims to investigate the best AT harvesting site in middle-aged to older dogs regarding the quantity and quality of the SVF cells. Thus, we isolated AT from abdominal subcutaneous fat, falciform ligament, and peri-ovarian sites. Then, we compared the isolated SVF cells in cell viability, cellular yield, and AD-MSC surface markers' expression using the trypan blue exclusion test, a hemocytometer, and flow cytometry. Flow cytometry results were validated by semi-quantitative RT-PCR.

2. Materials and Methods

2.1. Ethical Statement

All experiments were compliant with ethical standards and safety guidelines of regenerative medicine and cell therapy in dogs and cats designed by the Japanese Society for Veterinary Regenerative Medicine. We had extracted adipose tissue samples from dogs only after their owners provided written informed consent.

2.2. Animals

This study involved ten healthy female dogs that were admitted to the Animal Medical Center at Tokyo University of Agriculture and Technology, Japan for spaying. As the most susceptible age to chronic musculoskeletal problems, middle to old age dogs (mean 8.9 ± 1.1 years) were used for AT extraction, with the mean body weight of 9.55 ± 4.86 kg. These chosen dogs belonged to different breeds: Beagle (n = 2), Chihuahua (n = 2), Miniature Dachshund (n = 1), Pug (n = 1), Golden Retriever (n = 1), Jack Russell Terrier (n = 1), Pomeranian (n = 1), and Mix (n = 1). All dogs were examined for complete blood count, urine analysis, and serum biochemistry. Before extracting AT samples, dog anesthesia was initiated by I/V injection of Propofol (6 mg/kg, Propofol 1%, Nichi-Iko, Toyama, Japan) and maintained by 2% isoflurane (Isoflu; Dainippon Sumitomo Pharma, Chuo-ku, Osaka, Japan) intubation [22,23]. Then, the surgical field was aseptically prepared and draped.

2.3. AT Harvesting

AT samples were collected from three different sites: abdominal subcutaneous, falciform ligament, and the peri-ovarian region. Subcutaneous AT was extracted from surgical wound edges, while the falciform ligament fats were immediately harvested after a midline celiotomy incision. The peri-ovarian region AT was collected from the uterine broad ligament-enclosing fats by monopolar electrocautery according to the standard surgical technique. All dogs were further monitored for any surgical complications.

2.4. Isolation of SVF Cells

We isolated SVF cells from AT samples under complete aseptic conditions using procedures described by Zuk et al. [5], with little modifications. Briefly, AT samples were collected in 50 mL conical tubes (Falcon®, Corning Inc., Tewksbury, MA, USA) and weighed. Extensive washing was performed with phosphate-buffered saline (PBS). Then, tissues were placed in sterile dishes and minced into small pieces (1–3 mm) with a sterile scalpel. A 0.2% collagenase (Gibco; Thermo Fisher Scientific, Inc., Waltham, MA, USA)/Hanks' Balanced Salt Solution (HBSS; Wako Pure Chemical Industries, Ltd., Chuo-ku, Osaka, Japan) mixture (mL) was added to the minced AT (cm^3) at the ratio of 1:1 and incubated at 37 °C with shaking (120 rpm, 30 min). The enzymatic activity was neutralized by cold HBSS. Following centrifugation (800× g, 10 min), the cell pellet containing SVF cells was collected, rewashed with HBSS, and successively filtered using 100 µm and 40 µm nylon meshes to remove any cellular debris. Freshly isolated SVF cells were used in assessing the cell viability and quantity and in flow cytometry. However, the remaining cells were suspended in a freezing medium, containing 10% Dulbecco's Modified Eagle's medium (DMEM/F-12, Gibco; Thermo Fisher Scientific, Inc.), 10% dimethyl sulfoxide (DMSO, Wako Pure Chemical Industries, Ltd.), and 80% fetal bovine serum (FBS, Gibco; Thermo Fisher Scientific, Inc.) and cryopreserved at −80 °C using a Bicell cryopreservation device (Nihon-freezer, Tokyo, Japan) until use for RT-PCR.

2.5. Assessment of SVF Cells Viability and Quantity

Cell viability and quantity were determined using a hemocytometer combined with the routine trypan blue exclusion test. SVF cell suspension (10 µL) was diluted 1:1 with 0.4% trypan blue solution (Gibco; Thermo Fisher Scientific, Inc., Waltham, MA, USA) and loaded into the hemocytometer chamber. We used the average of two full squares to calculate the percentage of viable cells. SVF cellular yield or concentration was calculated by dividing the total number of viable cells per gm fat (dry fat digested by collagenase). Data were represented as the number of viable cells × 10^6/gm of dry fat ± standard deviation [1].

2.6. Immunophenotyping of the Potential AD-MSCs in SVF Samples

We used flow cytometry to assess the potential of AD-MSC subpopulations from the freshly isolated SVF samples from different AT extraction sites. SVF cells were labeled with a panel of monoclonal antibodies against mesenchymal (CD90, CD44, and CD29) and hematopoietic (CD45) stem cell markers according to Krešić et al. [24] and Yaneselli et al. [25]. Briefly, SVF cells were suspended in PBS and incubated at 4 °C for 30 min with Phycoerythrin (PE)-conjugated antibodies against CD90, CD44, and CD29, and fluorescein isothiocyanate (FITC)-conjugated antibody against CD45, listed in Table 1. Flow cytometry analyses were performed using the CytoFLEX Flow Cytometer (Beckman Coulter, Brea, CA, USA) equipped with a blue laser (488 nm). The percentage of each marker was separately detected in SVF samples. The resulting data were further analyzed using CytExpert Software v1.2.

Table 1. The list of antibodies used for flow cytometry.

Cell Surface Marker	Antibody Clone	Species Reactivity	Clonality	Antibody Quantity	Cat. No	Source
CD90 PE	5E10	Dog	Monoclonal	10 μL/10^6 cells	ARG54208	Arigo Biolaboratories
CD44 PE	IM7	Dog	Monoclonal	10 μL/10^6 cells	GTX80086	GeneTex
CD29 PE	MEM-101A	Dog	Monoclonal	10 μL/10^6 cells	1P219T025	EXBIO antibodies
CD45 FITC	YKIX716.13	Dog	Monoclonal	10 μL/10^6 cells	GTX43583	GeneTex

CD: Cluster of Differentiation, PE: Phycoerythrin, FITC: fluorescein isothiocyanate.

2.7. Semi-Quantitative RT-PCR

The phenotypic expression of different MSC markers was assessed by performing Reverse-Transcriptase-PCR (RT-PCR) as previously described [26]. Table 2 shows the details of analyzed genes and the sequence of specific primers. All primers were manufactured by the FASMAC Company, Midorigaoka, Kanagawa, Japan. Total RNA was extracted from SVF cells (1.5 × 10^6) isolated from different harvesting sites using RNeasy Isolation Kit (Qiagen AG, Garstligweg, Hombrechtikon, Switzerland) with DNase I treatment following the manufacturer's protocol. The RNA quantity was measured by a NanoDrop ND-1000 spectrophotometer (NanoDropTechnologies, Wilmington, NC, USA). Agarose gel electrophoresis was used to assess the RNA integrity. First-strand complementary DNA cDNAwas synthesized from 1 μg of total RNA with a PrimeScript RT Master Mix (Takara, Kusatsu, Shiga, Japan) according to the manufacturer's protocol. EmeraldAmp MAX PCR Master Mix (Takara, Kusatsu, Shiga, Japan) was used for RT-PCR. The final RT-PCR mixtures contained 25 μL EmeraldAmp MAX PCR Master Mix, 2 μL template cDNA, 0.2 μM of each specific forward and reverse primer, and ddH_2O up to 50 μL. Cycling protocols were as follows: 35 cycles of denaturation at 94 °C for 30 s, annealing at 55 °C for 30 s, and extension at 72 °C for 1 min. RT-PCR products were examined on ethidium bromide-stained agarose gel (1.5%) in TAE (Tris Acetate-EDTA) buffer. Amplicons were visualized using a UV Transilluminator, and images were captured by a Canon digital camera. All analyses were repeated with two replicates for each AD harvesting site sample. For semi-quantitative analysis of MSC markers expression, we used the ImageJ image processing software to evaluate the optical density of each positive band normalized to that of the endogenous housekeeping gene (β-actin).

Table 2. Primers used in RT-PCR.

Target Gene	Accession Number	Primers	Amplicon Size	Reference
Oct-4	XM_538830.1	Fw: AAGCCTGCAGAAAGACCTGRv: GTTCGCTTTCTCTTTCGGGC	286 bp	Ivanovska et al. [26]
CD90	NM_001287129.1	Fw: AAGCCAGGATTGGGGATGTGRv: TGTGGCAGAGAAAGCTCC TG	285 bp	Ivanovska et al. [26]
CD44	NM_001197022.1	Fw: CCCATTACCAAAGACCACGARv: TTCTCGAGGTTCCGTGTCTC	408 bp	Ivanovska et al. [26]
CD29	XM_005616949.1	Fw: AGGATGTTGACGACTGCTGGRv: ACCTTTGCATTCAGTGTTGTGC	356 bp	Ivanovska et al. [26]
CD45	XM_005622282.1	Fw: TGTTTCCAGTTCTGTTTCCCCARv: TCAGGTACAAAGCCTTCCCA	432 bp	Ivanovska et al. [26]
CD34	NM_001003341.1	Fw: GAGATCACCCTAACGCCTGGRv: GGCTCCTTCTCACACAGGAC	383 bp	Ivanovska et al. [26]
β-actin	XM_544346	Fw: GAGACCTGACCGACTACCTRv: GCT GCCTCCAGACAACAC	553 bp	Qiu et al. [27]

2.8. Statistical Analysis

All data were expressed as mean ± standard deviation values (SD). Statistical analysis was carried out using GraphPad Prism software version 6 (GraphPad Software, Inc., La Jolla, CA, USA). Data were analyzed by one-way analysis of variance (ANOVA) followed by Tukey's post hoc test to evaluate differences between groups. Statistically significant differences were considered at a p value less than 0.05.

3. Results

3.1. Effect of Harvesting Site on AT yield

Table 3 shows samples' weights for each adipose tissue submission. The mean weight of collected AT samples was 2.7 ± 1.6, 9.47 ± 2.9, and 8.13 ± 4.3 gm for the subcutaneous abdominal, falciform ligament, and peri-ovarian AT, respectively. Compared to the subcutaneous abdominal and falciform ligament sites, the recovered AT weight was highly variable at the peri-ovarian site. Moreover, AT harvesting from both falciform ligament and peri-ovarian sites was much easier than that of subcutaneous abdominal AT harvesting, which required an extensive dissection. No clinical complications were recorded in dogs either during or after AT harvesting from all sites.

3.2. Effect of AT Harvesting Site on SVF Cell Viability and Quantity

Cell viability was examined after the extraction of SVF cells from different sites. Among all, SVF cells isolated from peri-ovarian AT showed the highest viability percentage, but without significant difference. The mean SVF cell viability% was 94.94 ± 2.9, 94.58 ± 4.1, and 99.63 ± 0.2% for the subcutaneous abdominal, falciform ligament, and peri-ovarian AT, respectively (Figure 1a). We did not find any significant difference in the number of viable cells/gm fat between the falciform ligament and subcutaneous abdominal sites. However, peri-ovarian AT showed the highest number of viable cells per gram fat ($36.87 \pm 19.6 \times 10^6$) at a significant level ($p < 0.001$) compared to the subcutaneous abdominal site ($4.18 \pm 8.25 \times 10^6$) and the falciform ligament site ($5.71 \pm 3.09 \times 10^6$), as shown in Figure 1b.

Table 3. Data of animals used in adipose tissue harvesting.

Breed	Body Weight (Kg)	Age (Years)	Sex	Adipose Tissue Harvest (gm)	Viability%	Viable Cell/Gram × 10^6
Beagle	9	8	Female	1 S 7.3 F 9.5 P	99.8 90 99.8	0.24 4.2 46.2
Beagle	10	8	Female	1.3 S 14.7 F 2 P	92 90.9 99.7	2.4 3 24
Chihuahua	3.7	9	Female	1 S 8 F 2.5 P	96.6 98.7 99.8	27.5 5.4 28
Chihuahua	6.14	8	Female	5.5 S 6.5 F 2.6 P	95.5 90.9 99.6	0.43 8.95 29
Miniature Dachshund	7.7	10	Female	4.7 S 10 F 7 P	97.1 97.9 99.7	2 3.3 44.6
Pug	6	8	Female	1 S 8.3 F 6 P	95.2 95 99.6	2.4 3.6 31
Golden Retriever	20.6	8	Female	3 S 10 F 12 P	97.2 98.4 99.3	1.6 8 87.5
Jack Russell Terrier	9.72	10	Female	2.6 S 8.2 F 2 P	89.7 98.8 99.7	4.2 1.6 19.6
Pomeranian	8	11	Female	3.4 S 14.7 F 14 P	93.4 97 99.8	0.78 8.1 32.8
Mix	14.7	9	Female	4.2 S 7 F 5.8 P	92.9 88.2 99.2	0.29 11 26

S: subcutaneous abdominal; F: falciform ligament; P: peri-ovarian region.

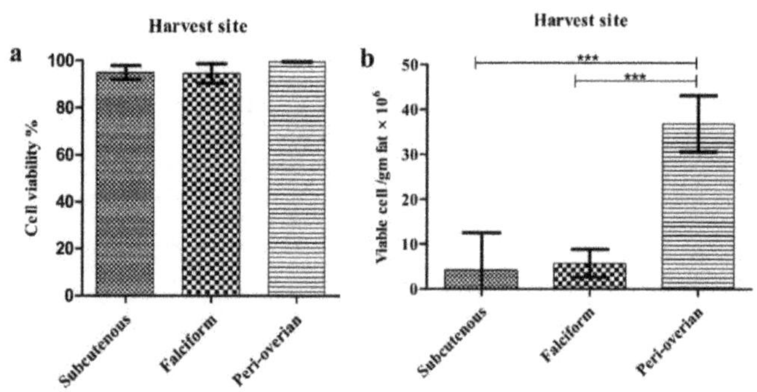

Figure 1. The effect of harvesting site on stromal vascular fraction (SVF) cell viability and quantity. (**a**) Viability percentage of the SVF cells from different harvesting sites. SVF cells isolated from per-ovarian adipose tissue (AT) showed the highest viability percentage; (**b**) Number of viable cells per gram fat from different harvesting sites. SVF cells isolated from the per-ovarian AT showed the highest number of viable cells per gram fat at *** $p < 0.001$. Data are expressed as the mean ± standard deviation ($n = 10$). Statistical significance was tested using one-way ANOVA followed by Tukey's post hoc test for multi-group comparisons. *** $p < 0.001$.

3.3. The Potential AD-MSCs in the Freshly Isolated SVF Cells from Different Harvesting Sites

We assessed the harvesting site effect on the phenotype of the potential AD-MSCs within the SVF cells by calculating the mean percentage of cells positive to MSCs markers and negative to the hematopoietic marker. The percentage of SVF cells with $CD90^{+}$ve expression for the peri-ovarian site (49.56 ± 4.9%) was significantly higher ($p < 0.01$) than that of the abdominal and falciform ligament (34.32 ± 2.55%, 17.65 ± 5.52%, respectively) (Figure 2a). Similarly, the highest percentage of CD44+ve cells was recorded in the peri-ovarian SVF cells (45.25 ± 3.55%) (Figure 2b). CD29+ve cells showed no significant differences among the three sites and were 32.34 ± 0.94% for the subcutaneous abdominal site, 32.68 ± 0.8% for the falciform ligament, and 38.00 ± 62.7% for the peri-ovarian site (Figure 2c). To confirm cell identity, the CD45 surface marker was used to identify hematopoietic cell contamination (Figure 2d). The percentages of CD45-ve SVF cells were significantly higher ($p < 0.001$) in the cells obtained from the subcutaneous abdominal site (89.77 ± 1.62%) and the peri-ovarian site (88.58 ± 2.25%) than those recovered from the falciform ligament site (70.35 ± 6.33%).

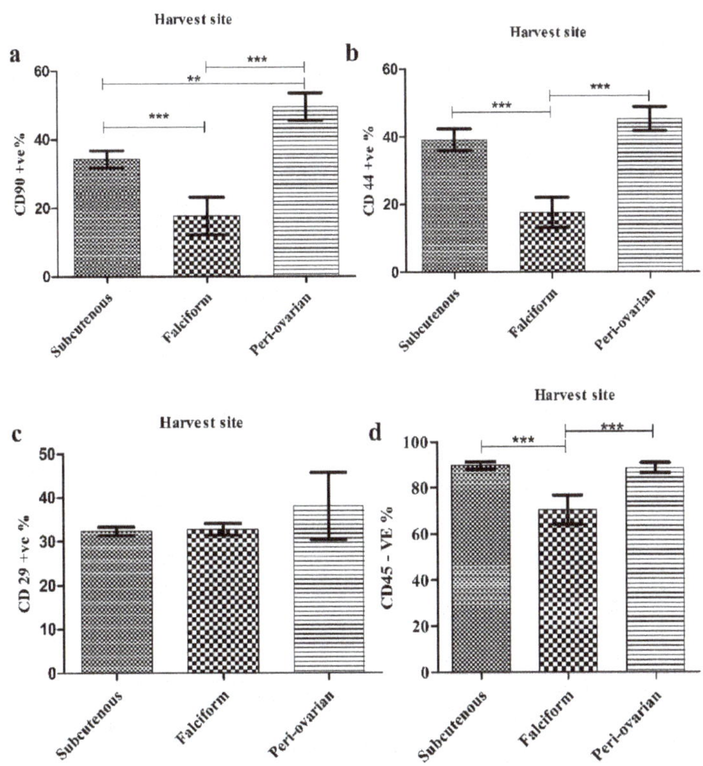

Figure 2. The effect of harvesting site on the mean percentage of adipose-derived mesenchymal stem cells surface marker expression in SVF cells. (**a**) The mean percentage of CD90 positive cells from different harvesting sites; (**b**) The mean percentage of CD44 positive cells from different harvesting sites; (**c**) The mean percentage of CD29 positive cells from different harvesting sites; (**d**) The mean percentage of CD45 negative cells from different harvesting sites. Data are represented as the mean percentage ± standard deviation ($n = 10$). ** $p < 0.01$: a significant difference; *** $p < 0.001$: a highly significant difference.

Figures 3–5 show representative histograms for using the flow cytometer to detect the potential AD-MSCs within the SVF cells isolated from the subcutaneous abdominal, falciform ligament, and peri-ovarian AT, respectively.

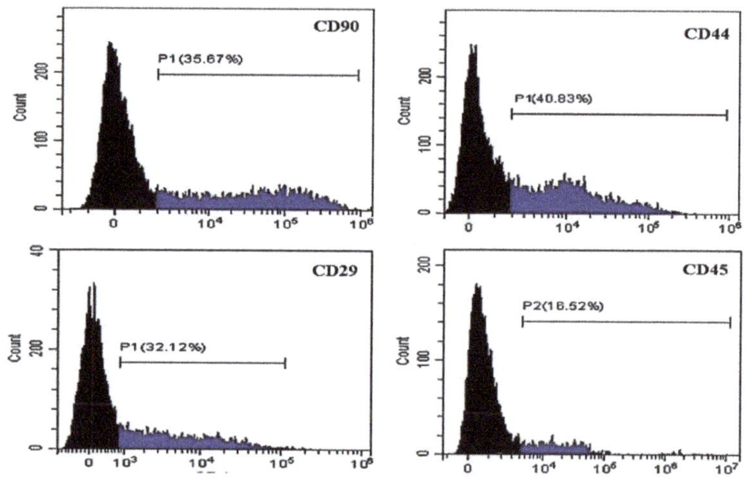

Figure 3. Results of flow cytometry assay of canine SVF cells isolated from the subcutaneous abdominal AT stained with PE-conjugated anti-CD90, anti-CD44, anti-CD29, and FITC-conjugated anti-CD45. Black areas indicate unstained cells. The percentage represents gated positive areas.

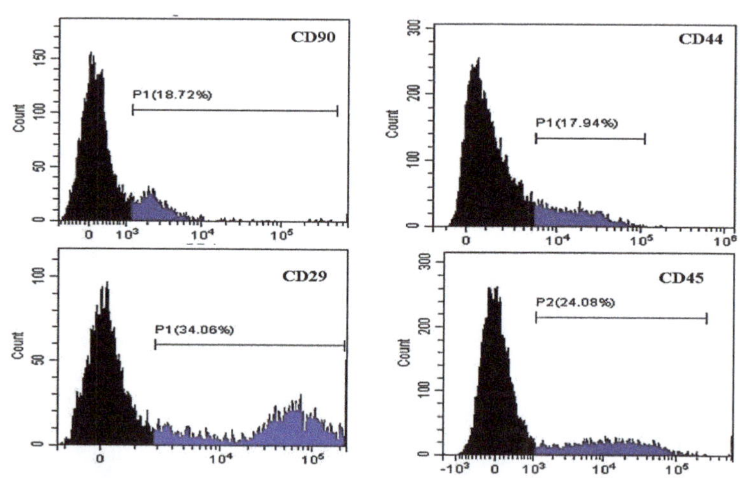

Figure 4. Results of flow cytometry assay of canine SVF cells isolated from the falciform ligament AT stained with PE-conjugated anti-CD90, anti-CD44, anti-CD29, and FITC-conjugated anti-CD45. Black areas indicate negative areas. The percentage represents gated positive areas.

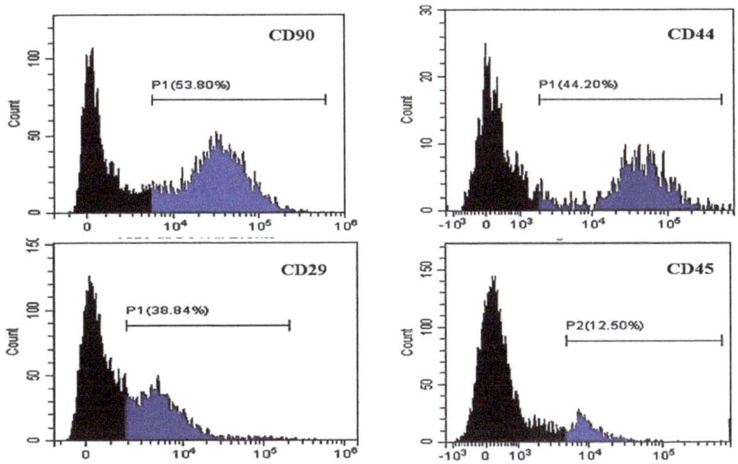

Figure 5. Results of flow cytometry assay of canine SVF cells isolated from the peri-ovarian AT stained with PE-conjugated anti-CD90, anti-CD44, anti-CD29, and FITC-conjugated anti-CD45. Black areas indicate unstained cells. The percentage represents gated positive areas.

3.4. Semi-Quantitative RT-PCR Results

We used RT-PCR to evaluate the expression of AD-MSC surface marker genes (CD90, CD44, and CD29), hematopoietic markers (CD45 and CD34), and the pluripotent transcription factor OCT-4 in SVF cells from different harvesting sites. Our results showed that AD-MSC genes were expressed in SVF cells from all sites, but peri-ovarian SVF cells showed the highest expression level. SVF samples from all sites showed weak expressions for CD45 and CD34. The OCT-4 gene was mainly expressed in peri-ovarian and subcutaneous abdominal fats but weakly expressed in falciform ligament fats (Figures 6 and 7).

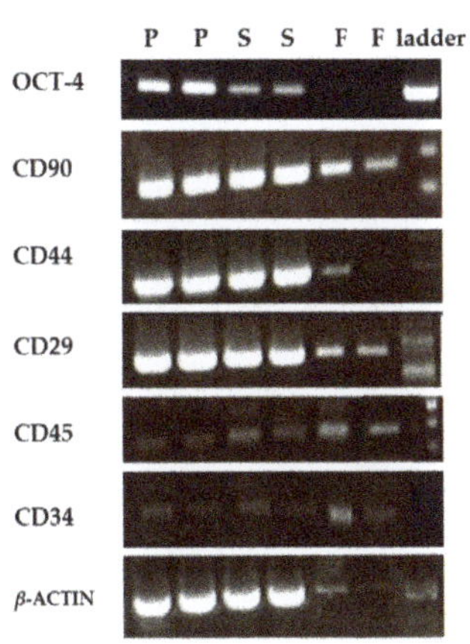

Figure 6. RT-PCR analysis of OCT-4, CD90, CD44, CD29, CD45, CD34, and β-ACTIN genes in SVF cells isolated from the peri-ovarian (P), subcutaneous abdominal (S), and falciform ligament (F) fats.

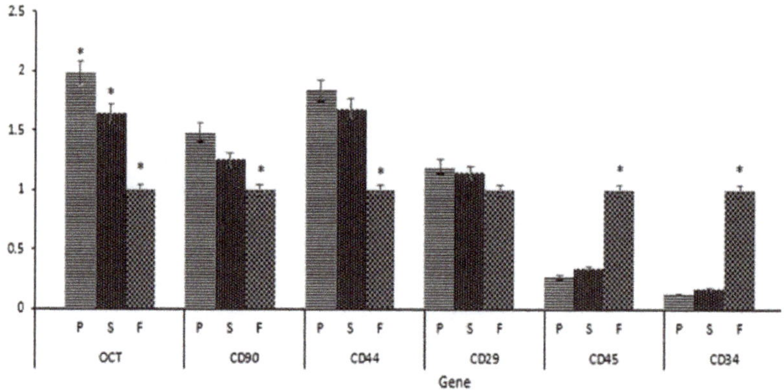

Figure 7. Semi-quantitative data from the gels using ImageJ software. P: peri-ovarian region; S: subcutaneous abdominal; F: falciform ligament; * $p < 0.05$.

4. Discussion

Adipose tissue is an appealing cell source for regenerative and engineering medicine due to easy harvesting and the abundance of stem cell populations. Autologous adipose SVF injection has gained popularity in the orthopedic field because it is a favorable, minimally invasive, and non-surgical alternative for the handling of musculoskeletal disorders [28]. Despite the increasing number of pre-clinical and clinical studies on the potential role of SVF cells to treat osteoarthritis and/or cartilage lesions, clear findings are missing due to the insufficient standardization of SVF cell isolation and characterization [29]. This study addressed the best AT sampling sites for SVF cell isolation in canines as an excellent model for humans. SVF cells were isolated from different AT sampling sites (subcutaneous abdominal, falciform ligament, and peri-ovarian fats). Then, we compared the isolated SVF cells regarding the cell viability and cellular yield using trypan blue staining and a hemocytometer, as well as AD-MSC surface markers' expression by flow cytometry.

Our study results revealed that the peri-ovarian site is an excellent and suitable source for AT harvest, with a mean weight of 8.13 ± 4.3 gm, because ovariohysterectomy, as a routine surgery, is a relatively easy method to obtain high amounts of AT during surgery without significant risk for the donor's tissues. This fact should be considered when the donor is emaciated or has chronic or nutritional diseases. However, subcutaneous abdominal AT harvests required more dissection and relied on higher body condition scores of the donor animals [20,30], which may explain the lower mean weight of subcutaneous abdominal AT harvests (2.7 ± 1.6 gm) in our study. Although a high AT weight (9.47 ± 2.9 gm) was harvested from the falciform ligament site in our study, the falciform AT collection may cause some complications, such as postoperative pain related to intra-abdominal adhesion, celiotomy, seroma, abdominal incision dehiscence, or incision site infection [31]. The mean weight of falciform AT samples in the current study was lower than that reported by Astor et al. [20] (91.42 ± 48.55 gm). This difference may be attributed to the large sample of client-owned dogs used in the study of Astor et al. [20]. The larger scale of falciform fat samples evaluated in the study of Sullivan et al. [21] may represent complete removal of the falciform ligament from euthanized dogs, while the small scale of the falciform sample collected from live dogs in the present study was aimed to conserve materials.

Site-specific properties of the adipose tissue plus paracrine interactions between adipose harvests and contiguous tissues have been considered in the previous studies with the perinodal AT around lymph nodes [32], perivascular AT [33], and pericardial AT [34]. The differences in cell isolation from various anatomical locations recorded in both the current study and the previous literature may be attributed to the different degrees of vascularization of AT and the harvest sites. Different canine breeds included in this study showed homogenous results related to the studied parameters. We did not include the body condition score of donors in this study due to its non-significant effect on the viable cells% per gram fat as previously proposed by Astor et al. [20]. Additionally, AD-MSC yield was not correlated with body mass index in humans [35].

Precise determination of cell viability and concentration in the freshly isolated adipose SVF is critical to accomplish the clinical research outcomes [1]. SVF cells isolated in this study from different sites showed a very good viability exceeding 90%. However, the highest SVF cell viability percentage was recorded from the peri-ovarian AT (99.63 ± 0.2%). These results disagreed with the study of DePompeo et al. [31], who reported a lower cell viability percent. The lower percentage of cell viability recorded by DePompeo et al. [31] may be related to sample storage for 20 h before tissue digestion, resulting in a 10–20% decrease in the viability.

Here, we examined the impact of AT harvesting sites on the number of viable cells per gram fat, which is essential for the presence of sufficient cells for treatment procedures. Among the examined sites, the peri-ovarian harvest showed the highest concentration of viable cells/gram of digested fat at *** $p < 0.001$. Moreover, a non-significant difference in viable cell number/gm fat was noticed between the subcutaneous abdominal and falciform ligament AT, which was consistent with the results of Guercio et al. [15]. By contrast, Astor et al. [20] reported that viable cells/gm of the SVF isolated from falciform AT were lower than that of the subcutaneous fat in spayed/neutered dogs. This disagreement could be explained by using non-spayed female donors in the present study, and Astor et al. [20] speculated that hormones may influence the viable cells per gram of tissue at the falciform location.

Adipose SVF has been used to treat multiple inflammatory and immune-mediated disorders in canines [10,11,18,19]. However, the presence of sufficient AD-MSCs, with a differentiation capacity, in SVF isolates is crucial for a successful treatment. Thus, the characterization of SVF cells is an important aspect of quality control for use in regenerative therapies. Flow cytometry can identify the different cell types within the adipose SVF [36]. Moreover, immunophenotyping is frequently achieved by flow cytometry to identify individual cells that simultaneously express the key MSC markers and lack the expression of hematopoietic markers. These cell surface and intracellular markers belong to the cluster of differentiation (CD) group [37]. To identify AD-MSCs, we labeled SVF samples with known stem cell markers. Although we could not define SVF cells with multiple MSC marker criteria due to the lack of suitable facilities, our results showed that the isolated SVF from different sites expressed AD-MSC surface markers (CD90, CD44, and CD29) and lacked the CD45 hematopoietic stem cell marker. This expression profile agreed with previous studies [24,25], showing that most AD-MSCs are CD90+, CD44+, CD29+ and CD45−. Positive markers consist of members of the integrin family, such as CD90 surface marker, which present on a high proportion of MSCs (71.4% ± 15.8%). The activation of CD90 stimulates T cell activation in addition to the regulation of various biological mechanisms, such as cell–cell and cell–matrix cellular interactions in axon regeneration, adhesion, apoptosis, migration, fibrosis, and cancer [38]. CD29, the Very Late Activation antigen, participates in the mechanism of cell adhesion [39]. Another MSC positive marker is the hyaluronate receptor CD44, which is a non-integrin cell surface marker essential for the adhesion of different leukocytes to endothelia and T-lymphocyte activation [40].

We studied the effect of the AT harvesting site in canines on the potential yield of AD-MSCs in the isolated SVF. The mean percentage of CD90+ viable cells in this study varied between samples from different harvesting sites. Cells isolated from the peri-ovarian site showed the highest CD90+ expression (49.56 ± 4.9%), with a significant difference. Similarly, the highest proportion of CD44 + cells was recorded in the peri-ovarian SVF cells. These data were significantly different from the falciform ligament site but not from the subcutaneous abdominal site. Although CD29+ve cells showed no significant differences among the three sites, the peri-ovarian SVF cells had the highest

CD29+ mean percentage (38.00 ± 7.62%). CD45 negative cells from the subcutaneous abdominal and peri-ovarian sites were nearly equal and significantly higher than those from the falciform ligament site, indicating a lower contamination with hematopoietic cells in the first two sites. Although the falciform CD45− cells' mean percentage in our study approached that of Sulvian et al. [21], our CD90+ and CD44+ cells were lower than those of Sulvian et al. [21]. This difference is possibly due to the variation in SVF cell isolation protocols and in the donors' ages. Quaade et al. [12] observed that younger rats had more MSC cells in SVF than aged ones, suggesting that the animal age affected cell type relative distribution in the SVF cell population. Here, we did not cultivate SVF to address the differentiation potential because we mainly focused on studying the freshly isolated uncultivated SVF as a point-of-care therapy. Culturing SVF cells for even one passage would alter their cellular composition and differentiation potential [41]. Additionally, our cryopreserved SVF samples were not suitable for assessing the differential potential according to Duan and Lopez [42], who reported that cryopreservation alters AD-MSCs ultrastructure and immunophenotype. Further studies are needed to evaluate the therapeutic effects of SVF cells isolated from different sites on musculoskeletal problems. However, our flow cytometry data showed that cells isolated from the peri-ovarian site had an AD-MSC concentration higher than that from the falciform ligament and subcutaneous abdominal sites.

We used RT-PCR to evaluate the gene expression of the same markers analyzed by flow cytometry as well as Oct-4 and CD34. The RT-PCR results agreed with results obtained by flow cytometry. SVF cells from the peri-ovarian site maintained stable expression of MSC markers (CD90, CD44, and CD29) and pluripotent transcription factor OCT-4 in a higher level compared to the other sites. The positive expression of Oct-4 demonstrated the pluripotency of AD-MSC as well as it has already been described in canines [26]. The core pluripotent transcription factors, such as Oct-4, Sox-2, and Nanog, regulate the self-renewal ability and differentiation abilities of AD-MSCs [43]. SVF cells from all sites revealed weak expression of hematopoietic markers (CD45 and CD34). CD34 is a physiological niche-specific marker of immature/early progenitor cells, which is lost in the in vitro condition. CD34 marks different progenitor cell types, such as different MSCs and vascular endothelial progenitor cells [29]. Together, our flow cytometry and RT-PCR results suggest that the peri-ovarian site AT harvest site may have a higher potential for use in regenerative therapy.

5. Conclusions

The peri-ovarian AT harvesting site is a favorable and suitable source for AT harvest in middle-aged and old dogs, without substantial risk to the donor tissues. This harvesting site could be valuable in emaciated donors. The peri-ovarian harvesting site yielded a higher SVF viability percentage, viable cell number/gm fat and AD-MSC marker expression than that of the other harvesting sites, indicating a high potential for application in regenerative therapy. This study may lay the foundation for further studies in the set-up of cell-based regenerative therapies at the preclinical and experimental level in the canine model.

Author Contributions: Conceptualization, H.H., A.U. and R.T.; methodology, H.H. and A.U.; data curation, H.H., D.M. and R.N.; software, H.H., H.S. and C.C.-J.; formal analysis, A.E.; investigation H.H. and A.E., resources, R.T.; writing—original draft, H.H.; writing—review and editing, H.S., M.F.A., A.E., H.M.E.-H., and R.T.; supervision, R.T. All authors have read and agreed to the published version of the manuscript.

Funding: This research received no external funding.

Institutional Review Board Statement: The study was conducted according to the guidelines of the Declaration of Helsinki, and approved by the Animal Ethics Committee of Institute of Rescue Animal Medicine (Permit Number 19-012, Tokyo University of Agriculture and Technology, Japan).

Informed Consent Statement: Informed consent was obtained from all subjects involved in the study.

Data Availability Statement: Data are contained within the article.

Acknowledgments: We are grateful to Masahiro Kaneda from Tokyo University of Agriculture and Technology for his help in RT-PCR.

Conflicts of Interest: The authors declare no conflict of interest regarding the publication of this paper.

References

1. Chan, L.L.-Y.; Cohen, D.A.; Kuksin, D.; Paradis, B.D.; Qiu, J. Automated enumeration and viability measurement of canine stromal vascular fraction cells using fluorescence-based image cytometry method. *J. Fluoresc.* **2014**, *24*, 983–989. [CrossRef]
2. Mohal, J.S.; Tailor, H.D.; Khan, W.S. Sources of adult mesenchymal stem cells and their applicability for musculoskeletal applications. *Curr. Stem Cell Res. Ther.* **2012**, *7*, 103–109. [CrossRef]
3. Nakao, N.; Nakayama, T.; Yahata, T.; Muguruma, Y.; Saito, S.; Miyata, Y.; Yamamoto, K.; Naoe, T. Adipose tissue-derived mesenchymal stem cells facilitate hematopoiesis in vitro and in vivo: Advantages over bone marrow-derived mesenchymal stem cells. *Am. J. Pathol.* **2010**, *177*, 547–554. [CrossRef] [PubMed]
4. Choudhery, M.S.; Badowski, M.; Muise, A.; Harris, D.T. Comparison of human mesenchymal stem cells derived from adipose and cord tissue. *Cytotherapy* **2013**, *15*, 330–343. [CrossRef] [PubMed]
5. Zuk, P.A.; Zhu, M.; Mizuno, H.; Huang, J.; Futrell, J.W.; Katz, A.J.; Benhaim, P.; Lorenz, H.P.; Hedrick, M.H. Multilineage cells from human adipose tissue: Implications for cell-based therapies. *Tissue Eng.* **2001**, *7*, 211–228. [CrossRef] [PubMed]
6. Marx, C.; Silveira, M.D.; Nardi, N.B. Adipose-derived stem cells in veterinary medicine: Characterization and therapeutic applications. *Stem Cells Dev.* **2015**, *24*, 803–813. [CrossRef] [PubMed]
7. Markarian, C.F.; Frey, G.Z.; Silveira, M.D.; Milani, A.R.; Ely, P.B.; Horn, A.P.; Nardi, N.B.; Camassola, M. Isolation of adipose-derived stem cells: A comparison among different methods. *Biotechnol. Lett.* **2014**, *36*, 693–702. [CrossRef]
8. Yoshimura, K.; Shigeura, T.; Matsumoto, D.; Sato, T.; Takaki, Y.; Aiba-Kojima, E.; Sato, K.; Inoue, K.; Nagase, T.; Koshima, I. Characterization of freshly isolated and cultured cells derived from the fatty and fluid portions of liposuction aspirates. *J. Cell. Physiol.* **2006**, *208*, 64–76. [CrossRef]
9. De Girolamo, L.; Lucarelli, E.; Alessandri, G.; Antonietta Avanzini, M.; Ester Bernardo, M.; Biagi, E.; Teresa Brini, A.; D'Amico, G.; Fagioli, F.; Ferrero, I. Mesenchymal stem/stromal cells: A new' cells as drugs paradigm. Efficacy and critical aspects in cell therapy. *Curr. Pharm. Des.* **2013**, *19*, 2459–2473. [CrossRef]
10. Marx, C.; Silveira, M.D.; Selbach, I.; Da Silva, A.S.; Braga, L.M.G.D.M.; Camassola, M.; Nardi, N.B. Acupoint injection of autologous stromal vascular fraction and allogeneic adipose-derived stem cells to treat hip dysplasia in dogs. *Stem Cells Int.* **2014**, *2014*, 391274. [CrossRef]
11. Upchurch, D.A.; Renberg, W.C.; Roush, J.K.; Milliken, G.A.; Weiss, M.L. Effects of administration of adipose-derived stromal vascular fraction and platelet-rich plasma to dogs with osteoarthritis of the hip joints. *Am. J. Vet. Res.* **2016**, *77*, 940–951. [CrossRef] [PubMed]
12. Quaade, M.L.; Jensen, C.H.; Andersen, D.C.; Sheikh, S.P. A 3-month age difference profoundly alters the primary rat stromal vascular fraction phenotype. *Acta Histochem.* **2016**, *118*, 513–518. [CrossRef] [PubMed]
13. Kemilew, J.; Sobczyńska-Rak, A.; Żylińska, B.; Szponder, T.; Nowicka, B.; Urban, B. The use of allogenic stromal vascular fraction (SVF) cells in degenerative joint disease of the spine in dogs. *In Vivo* **2019**, 33, 1109–1117. [CrossRef]
14. Lee, K.S.; Kang, H.W.; Lee, H.T.; Kim, H.-J.; Kim, C.-L.; Song, J.-Y.; Lee, K.W.; Cha, S.-H. Sequential sub-passage decreases the differentiation potential of canine adipose-derived mesenchymal stem cells. *Res. Vet. Sci.* **2014**, *96*, 267–275. [CrossRef]
15. Guercio, A.; Di Bella, S.; Casella, S.; Di Marco, P.; Russo, C.; Piccione, G. Canine mesenchymal stem cells (MSCs): Characterization in relation to donor age and adipose tissue-harvesting site. *Cell Biol. Int.* **2013**, *37*, 789–798. [CrossRef]
16. Bergknut, N.; Rutges, J.P.; Kranenburg, H.J.C.; Smolders, L.A.; Hagman, R.; Smidt, H.J.; Lagerstedt, A.S.; Penning, L.C.; Voorhout, G.; Hazewinkel, H.A.; et al. The dog as an animal model for intervertebral disc degeneration? *Spine* **2012**, *37*, 351–358. [CrossRef] [PubMed]
17. McMahill, B.G.; Borjesson, D.L.; Sieber-Blum, M.; Nolta, J.A.; Sturges, B.K. Stem cells in canine spinal cord injury-promise for regenerative therapy in a large animal model of human disease. *Stem Cell Rev. Rep.* **2015**, *11*, 180–193. [CrossRef]
18. Albano, D.; Messina, C.; Usuelli, F.G.; De Girolamo, L.; Grassi, M.; Maccario, C.; Bignotti, B.; Tagliafico, A.; Sconfienza, L.M. Magnetic resonance and ultrasound in Achilles tendinopathy: Predictive role and response assessment to platelet-rich plasma and adipose-derived stromal vascular fraction injection. *Eur. J. Radiol.* **2017**, *95*, 130–135. [CrossRef]
19. Abdallah, A.N.; Shamaa, A.A.; El-Tookhy, O.S. Evaluation of treatment of experimentally induced canine model of multiple sclerosis using laser activated non-expanded adipose derived stem cells. *Res. Vet. Sci.* **2019**, *125*, 71–81. [CrossRef]
20. Astor, D.E.; Hoelzler, M.G.; Harman, R.; Bastian, R.P. Patient factors influencing the concentration of stromal vascular fraction (SVF) for adipose-derived stromal cell (ASC) therapy in dogs. *Can. J. Vet. Res.* **2013**, *77*, 177–182.
21. Sullivan, M.O.; Gordon-Evans, W.J.; Fredericks, L.P.; Kiefer, K.; Conzemius, M.G.; Griffon, D.J. Comparison of mesenchymal stem cell surface markers from bone marrow aspirates and adipose stromal vascular fraction sites. *Front. Vet. Sci.* **2016**, *2*, 82. [CrossRef]
22. Watkins, S.; Hall, L.; Clarke, K. Propofol as an intravenous anaesthetic agent in dogs. *Vet. Rec.* **1987**, *120*, 326–329. [CrossRef]
23. Tomihari, M.; Nishihara, A.; Shimada, T.; Yanagawa, M.; Miyoshi, M.; Miyahara, K.; Oishi, A. A comparison of the immunological effects of propofol and isoflurane for maintenance of anesthesia in healthy dogs. *J. Vet. Med. Sci.* **2015**, *77*, 1227–1233. [CrossRef]

24. Krešić, N.; Šimić, I.; Lojkić, I.; Bedeković, T. Canine adipose derived mesenchymal stem cells transcriptome composition alterations: A step towards standardizing therapeutic. *Stem Cells Int.* **2017**, *2017*, 4176292. [CrossRef] [PubMed]
25. Yaneselli, K.M.; Kuhl, C.P.; Terraciano, P.B.; De Oliveira, F.S.; Pizzato, S.B.; Pazza, K.; Magrisso, A.B.; Torman, V.; Rial, A.; Moreno, M. Comparison of the characteristics of canine adipose tissue-derived mesenchymal stem cells extracted from different sites and at different passage numbers. *J. Vet. Sci.* **2018**, *19*, 13–20. [CrossRef]
26. Ivanovska, A.; Grolli, S.; Borghetti, P.; Ravanetti, F.; Conti, V.; De Angelis, E.; Macchi, F.; Ramoni, R.; Martelli, P.; Gazza, F. Immunophenotypical characterization of canine mesenchymal stem cells from perivisceral and subcutaneous adipose tissue by a species-specific panel of antibodies. *Res. Vet. Sci.* **2017**, *114*, 51–58. [CrossRef] [PubMed]
27. Qiu, C.; Lin, D.; Wang, H.; Qiao, C.; Wang, J.; Zhang, T. Quantification of VEGF-C expression in canine mammary tumours. *Aust. Vet. J.* **2008**, *86*, 279–282. [CrossRef] [PubMed]
28. Senesi, L.; De Francesco, F.; Farinelli, L.; Manzotti, S.; Gagliardi, G.; Papalia, G.F.; Riccio, M.; Gigante, A. Mechanical and enzymatic procedures to isolate the stromal vascular fraction from adipose tissue: Preliminary results. *Front. Cell Dev. Biol.* **2019**, *7*, 88. [CrossRef]
29. Bora, P.; Majumdar, A.S. Adipose tissue-derived stromal vascular fraction in regenerative medicine: A brief review on biology and translation. *Stem Cell Res. Ther.* **2017**, *8*, 145. [CrossRef]
30. Zhang, N.; Dietrich, M.A.; Lopez, M.J. Canine intra-articular multipotent stromal cells (MSC) from adipose tissue have the highest in vitro expansion rates, multipotentiality, and MSC immunophenotypes. *Vet. Surg.* **2013**, *42*, 137–146. [CrossRef]
31. DePompeo, C.M.; Giassetti, M.I.; Elnaggar, M.M.; Oatley, J.M.; Davis, W.C.; Fransson, B.A. Isolation of canine adipose-derived mesenchymal stem cells from falciform tissue obtained via laparoscopic morcellation: A pilot study. *Vet. Surg.* **2020**, *49*, O28–O37. [CrossRef]
32. Pond, C.M.; Mattacks, C.A. Interactions between adipose tissue around lymph nodes and lymphoid cells in vitro. *J. Lipid Res.* **1995**, *36*, 2219–2231. [CrossRef]
33. Löhn, M.; Dubrovska, G.; Lauterbach, B.; Luft, F.C.; Gollasch, M.; Sharma, A.M. Periadventitial fat releases a vascular relaxing factor. *FASEB J.* **2002**, *16*, 1057–1063. [CrossRef] [PubMed]
34. Fox, C.S.; Gona, P.; Hoffmann, U.; Porter, S.A.; Salton, C.J.; Massaro, J.M.; Levy, D.; Larson, M.G.; D'Agostino, R.B.; O'Donnell, C.J. Pericardial fat, intra-thoracic fat, and measures of left ventricular structure and function: The Framingham heart study. *Circulation* **2009**, *119*, 1586. [CrossRef]
35. Jurgens, W.J.; Oedayrajsingh-Varma, M.J.; Helder, M.N.; Zandieh Doulabi, B.; Schouten, T.E.; Kuik, D.J.; Ritt, M.J.; Van Milligen, F.J. Effect of tissue-harvesting site on yield of stem cells derived from adipose tissue: Implications for cell-based therapies. *Cell Tissue Res.* **2008**, *332*, 415–426. [CrossRef] [PubMed]
36. Dubey, N.K.; Mishra, V.K.; Dubey, R.; Deng, Y.-H.; Tsai, F.-C.; Deng, W.-P. Revisiting the advances in isolation, characterization and secretome of adipose-derived stromal/stem cells. *Int. J. Mol. Sci.* **2018**, *19*, 2200. [CrossRef]
37. Sasaki, A.; Mizuno, M.; Mochizuki, M.; Sekiya, I. Mesenchymal stem cells for cartilage regeneration in dogs. *World J. Stem Cells* **2019**, *11*, 254. [CrossRef]
38. Hagood, J.S. Thy-1 as an integrator of diverse extracellular signals. *Front. Cell Dev. Biol.* **2019**, *7*, 26. [CrossRef] [PubMed]
39. Harjunpää, H.; Asens, M.L.; Guenther, C.; Fagerholm, S.C. Cell adhesion molecules and their roles and regulation in the immune and tumor microenvironment. *Front. Immunol.* **2019**, *10*, 1078. [CrossRef]
40. Li, J.; Zanata, F.; Curley, J.L.; Martin, E.C.; Wu, X.; Dietrich, M.; Devireddy, R.V.; Wade, J.W.; Gimble, J.M. The relative functionality of freshly isolated and cryopreserved human adipose-derived stromal/stem cells. *Cells Tissues Organs* **2016**, *201*, 436–444.
41. Sun, Y.; Chen, S.; Zhang, X.; Pei, M. Significance of cellular cross-talk in stromal vascular fraction of adipose tissue in neovascularization. *Arter. Thromb. Vasc. Biol.* **2019**, *39*, 1034–1044. [CrossRef] [PubMed]
42. Duan, W.; Lopez, M.J. Effects of cryopreservation on canine multipotent stromal cells from subcutaneous and infrapatellar adipose tissue. *Stem Cell Rev. Rep.* **2016**, *12*, 257–268. [CrossRef] [PubMed]
43. Reich, C.M.; Raabe, O.; Wenisch, S.; Bridger, P.S.; Kramer, M.; Arnhold, S. Isolation, culture and chondrogenic differentiation of canine adipose tissue and bone marrow-derived mesenchymal stem cells–A comparative study. *Vet. Res. Commun.* **2012**, *36*, 139–148. [CrossRef] [PubMed]

animals

MDPI

Article

Effect of Scrapie Prion Infection in Ovine Bone Marrow-Derived Mesenchymal Stem Cells and Ovine Mesenchymal Stem Cell-Derived Neurons

Laura García-Mendívil [1,2], Diego R. Mediano [1], Adelaida Hernaiz [1], David Sanz-Rubio [1,3], Francisco J. Vázquez [1,4], Belén Marín [5], Óscar López-Pérez [1,5,6], Alicia Otero [5], Juan J. Badiola [5], Pilar Zaragoza [1,6], Laura Ordovás [2], Rosa Bolea [5] and Inmaculada Martín-Burriel [1,5,6,*]

1 Laboratorio de Genética Bioquímica (LAGENBIO), Instituto Agroalimentario de Aragón (IA2), Instituto de Investigación Sanitaria de Aragón (IISAragón), Universidad de Zaragoza, Miguel Servet 177, 50013 Zaragoza, Spain; lgmendivil@unizar.es (L.G.-M.); drmediano@gmail.com (D.R.M.); ahernaiz@unizar.es (A.H.); dsanz@iisaragon.es (D.S.-R.); pvazquez@unizar.es (F.J.V.); oscar.lopez@irta.cat (Ó.L.-P.); pilarzar@unizar.es (P.Z.)
2 Biomedical Signal Interpretation and Computational Simulation (BSICoS), Institute of Engineering Research (I3A), University of Zaragoza & Instituto de Investigación Sanitaria (IIS), 50018 Zaragoza, Spain; lordovas@unizar.es
3 Translational Research Unit, Instituto de Investigación Sanitaria de Aragón (IISAragón), Hospital Universitario Miguel Servet, 50009 Zaragoza, Spain
4 Departamento de Patología Animal, Facultad de Veterinaria, Universidad de Zaragoza, Miguel Servet 177, 50013 Zaragoza, Spain
5 Centro de Investigación en Encefalopatías y Enfermedades Transmisibles Emergentes, Instituto Agroalimentario de Aragón (IA2), Instituto de Investigación Sanitaria de Aragón (IISAragón), Universidad de Zaragoza, Miguel Servet 177, 50013 Zaragoza, Spain; belenm@unizar.es (B.M.); aliciaogar@unizar.es (A.O.); badiola@unizar.es (J.J.B.); rbolea@unizar.es (R.B.)
6 Centro de Investigación Biomédica en Red de Enfermedades Neurodegenerativas (CIBERNED), Instituto Carlos III, 28031 Madrid, Spain
* Correspondence: minma@unizar.es; Tel.: +34-976-761-662

Citation: García-Mendívil, L.; Mediano, D.R.; Hernaiz, A.; Sanz-Rubio, D.; Vázquez, F.J.; Marín, B.; López-Pérez, Ó.; Otero, A.; Badiola, J.J.; Zaragoza, P.; et al. Effect of Scrapie Prion Infection in Ovine Bone Marrow-Derived Mesenchymal Stem Cells and Ovine Mesenchymal Stem Cell-Derived Neurons. *Animals* **2021**, *11*, 1137. https://doi.org/10.3390/ani11041137

Academic Editors: Eleonora Iacono and Barbara Merlo

Received: 2 March 2021
Accepted: 12 April 2021
Published: 15 April 2021

Publisher's Note: MDPI stays neutral with regard to jurisdictional claims in published maps and institutional affiliations.

Simple Summary: Prion diseases are neurodegenerative disorders affecting humans and animals. The development of in vitro cellular models from naturally susceptible species like humans or ruminants can potentially make a great contribution to the study of many aspects of these diseases, including the ability of prions to infect and replicate in cells and therapeutics. Our study shows for the first time how ovine mesenchymal stem cells derived from bone marrow and their neural-like progeny are able to react to scrapie prion infection in vitro and assesses the effects of this infection on cell viability and proliferation. Finally, we observe that the differentiation of ovine mesenchymal stem cells into neuron-like cells makes them more permissive to prion infection.

Abstract: Scrapie is a prion disease affecting sheep and goats and it is considered a prototype of transmissible spongiform encephalopathies (TSEs). Mesenchymal stem cells (MSCs) have been proposed as candidates for developing in vitro models of prion diseases. Murine MSCs are able to propagate prions after previous mouse-adaptation of prion strains and, although ovine MSCs express the cellular prion protein (PrP^{C}), their susceptibility to prion infection has never been investigated. Here, we analyze the potential of ovine bone marrow-derived MSCs (oBM-MSCs), in growth and neurogenic conditions, to be infected by natural scrapie and propagate prion particles (PrP^{Sc}) in vitro, as well as the effect of this infection on cell viability and proliferation. Cultures were kept for 48–72 h in contact with homogenates of central nervous system (CNS) samples from scrapie or control sheep. In growth conditions, oBM-MSCs initially maintained detectable levels of PrP^{Sc} post-inoculation, as determined by Western blotting and ELISA. However, the PrP^{Sc} signal weakened and was lost over time. oBM-MSCs infected with scrapie displayed lower cell doubling and higher doubling times than those infected with control inocula. On the other hand, in neurogenic conditions, oBM-MSCs not only maintained detectable levels of PrP^{Sc} post-inoculation, as determined by ELISA, but this PrP^{Sc} signal also increased progressively over time. Finally, inoculation with CNS extracts seems to induce the proliferation of oBM-MSCs in both growth and neurogenic conditions. Our results suggest that oBM-MSCs respond to prion infection by decreasing their proliferation capacity and thus might not be permissive to prion replication, whereas ovine MSC-derived neuron-like cells seem to maintain and replicate PrP^{Sc}.

Keywords: scrapie; prion; sheep; infection; mesenchymal stem cell; in vitro model

1. Introduction

Transmissible spongiform encephalopathies (TSEs) or prion diseases are fatal neurodegenerative disorders that affect humans and animals [1]. These diseases are caused by the conformational conversion of the cellular prion protein (PrP^C) to an infectious isoform that is partially resistant to proteases and prone to forming aggregates called PrP^{Sc} [2]. The accumulation of this isoform in the central nervous system (CNS) causes spongiform neuronal degeneration, activation of glial cells and neuronal loss [3]. Scrapie, which affects sheep and goats, was the first reported TSE [4] and it is considered the prototype of these diseases [5].

Cell culture systems are useful tools to study prion protein propagation in TSEs and to identify new prion therapeutics [6]. However, only a few cell lines can be infected and display PrP^{Sc} accumulation and/or infectious capacity [7]. In most cases, murine cell lines are used, requiring a previous mouse-adaptation of the prion strain to eliminate the problem of the species barrier [8].

Mesenchymal stem cells (MSCs) are fibroblast-like cells characterized by their capacity for both self-renewal and differentiation in mesodermal tissues (osteoblasts, adipocytes, chondrocytes and myocytes) [9]. These cells can also transdifferentiate in vitro into neuron-like cells [10,11] and undifferentiated cells expressing PrP^C [12], which seems to play a key role in the neuronal differentiation process of MSCs [13–15].

Murine compact bone-derived MSCs (CB-MSCs) are able to migrate to brain extracts from prion-infected mice in vitro and significantly prolong the survival of mice infected with the Chandler prion strain when injected in vivo [16]. Furthermore, murine bone marrow-derived mesenchymal stem cells (BM-MSCs) can be infected with a Gerstmann–Sträussler–Scheinker strain adapted in mice ex vivo [17] and maintain the infectivity along passages. The susceptibility of these cells to prion infection makes them good candidates for use in developing in vitro models for prion research [18]. Therefore, the development of in vitro models from naturally prion-susceptible species like humans or ruminants, which would avoid the adaptation process, would be very useful for cutting-edge prion research. Although in recent studies, human cerebral organoids [19] and astrocytes [20], both derived from human induced pluripotent stem cells (iPSCs), have been described to maintain and propagate prion infectivity in vitro, in domestic species like sheep, the reprogramming of somatic cells to iPSCs might require adjustments of standard protocols.

We have previously described the isolation of ovine MSCs from peripheral blood (oPB-MSCs), which express PrP^C at the transcript level [21]. Our group also reported the presence of PrP^C in ovine bone marrow-derived MSCs (oBM-MSCs) at both transcript and protein levels [18]. However, in contrast to BM-MSCs obtained from individuals with sporadic Creutzfeldt-Jakob disease (CJD), who are positive to PrP^{Sc} [12], the pathogenic prion protein was not detected in oBM-MSCs isolated from scrapie sheep [18]. In addition to the lack of PrP^{Sc}, these cells displayed diminished proliferation potential compared to oBM-MSCs derived from healthy sheep. To the best of our knowledge, the susceptibility of oBM-MSCs to scrapie infection in vitro and their potential to replicate prions have never been investigated. The aim of the present study was to assess the susceptibility of oBM-MSCs and their derivative neuron-like cells to scrapie prion infection in vitro, their potential to replicate prions and the effects of this infection on cell viability and proliferation.

2. Materials and Methods

2.1. Animals and Sample Collection

Bone marrow samples were obtained from 11 adult female (n = 7) and male (n = 4) sheep, aged from 1 to 7 years and carrying different genotypes for the *PRNP* gene (Table 1). After animal sedation (Xylazine) and local anesthesia (Lidocaine), bone marrow aspirates were harvested from the humeral head using a 13 G Jamshidi needle and 10-mL syringes previously loaded with 0.5 mL of sodium heparin. All procedures were carried out under Project Licence PI06/12, approved by the Ethical Committee for Animal Experiments from the University of Zaragoza. The care and use of animals were performed in accordance with the Spanish Policy for Animal Protection, RD53/2013, which meets European Union Directive 2010/63 on the protection of animals used for experimental and other scientific purposes.

Table 1. Characteristics of the animals selected to obtain bone marrow mesenchymal stem cells. The different assays in which the ovine bone marrow-derived mesenchymal stem cells (oBM-MSCs) were used, are also shown—Western blotting assay (WB), proliferation assay (PA), cell viability assay (MTT) and ELISA.

Sheep	Genotype	Sex	Age (Years)	Scrapie Status	Breed	Assay
BMO1	ARQ/ARQ	Female	7	Exposed, not detected	Rasa Aragonesa	WB, PA
BMO2	ARQ/ARQ	Female	4	Exposed, not detected	Rasa Aragonesa	WB, PA
BMO3	ARQ/ARQ	Female	4	Exposed, not detected	Rasa Aragonesa	WB, PA
BMO4	ARQ/ARQ	Male	1	Exposed, not detected	Rasa Aragonesa	MTT, ELISA
BMO5	ARQ/ARQ	Male	2	Preclinical	Crossbreed	MTT, ELISA
BMO6	ARQ/ARQ	Female	6	Exposed, not detected	Ojinegra	MTT, ELISA
BMO7	ARQ/ARQ	Male	3	Exposed, not detected	Crossbreed	MTT, ELISA
BMO8	ARQ/ARQ	Female	7	Exposed, not detected	Rasa Aragonesa	MTT
BMO9	ARR/ARQ	Female	5	Exposed, not detected	Rasa Aragonesa	MTT
BMO10	ARQ/VRQ	Female	4	Exposed, not detected	Crossbreed	MTT
BMO11	ARQ/VRQ	Male	2	Preclinical	Crossbreed	MTT, ELISA

The animals used in this study were maintained in an experimental flock in which the prevalence of scrapie was high. Although none of the animals displayed clinical signs compatible with scrapie, an in vivo test for PrP^{Sc} determination using third-eyelid biopsies was performed as previously described [22,23] to identify any scrapie-infected preclinical sheep. Two males were positive to scrapie but their cultures were maintained in the study to evaluate if these cultures could react differently to those obtained from negative sheep, although in previous studies infectivity was not detected in oBM-MSCs derived from scrapie sheep [18]. Negative animals were those that did not show scrapie compatible symptoms and were negative to PrP^{Sc} based on a third-eyelid biopsy.

2.2. Ovine Mesenchymal Stem Cell Isolation and Culture

MSC isolation from bone marrow aspirates (3–5 mL) was performed following the previously described protocol [18,21,24]. This protocol is based on the separation of the mononuclear fraction after density gradient centrifugation in Lymphoprep (Atom) and further isolation thanks to the ability of MSCs to adhere to plastic. After isolation, cells were expanded up to passage 3 in basal medium, consisting of low glucose Dulbecco's modified Eagle's medium (DMEM, Sigma-Aldrich, St. Louis, MO, USA) supplemented with 10% fetal bovine serum (FBS), 1% L-glutamine (Sigma-Aldrich) and 1% streptomycin/penicillin (Sigma-Aldrich).

In addition to plastic-adherence in standard culture conditions, the minimal criteria to define MSCs are the expression of certain cell surface markers and the ability to differentiate into adipocytes, osteoblasts and chondroblasts in vitro [25]. The ability to differentiate to mesodermal lineages and the expression of mesenchymal and hematopoietic markers of these cultures have been evaluated previously [18]. After characterization, the expression of PrP^C in oBM-MSCs was confirmed by RT-qPCR and dot blotting [18].

2.3. Neurogenic Differentiation

To study whether neural differentiation increased the susceptibility to prion infection, oBM-MSC cultures were seeded at 1500 cells/cm^2 and differentiated into neuron-like cells using HyClone neurogenic medium (Thermo Scientific, Waltham, MA, USA) according to the manufacturer's instructions. The differentiation process lasted three days. Neural differentiation was monitored and confirmed by observing the cultures through an inverted optical microscope. The formation of neuron-like cells was seen within 24 h, peaking at 72 h (Figure 1a,b), as previously described [18].

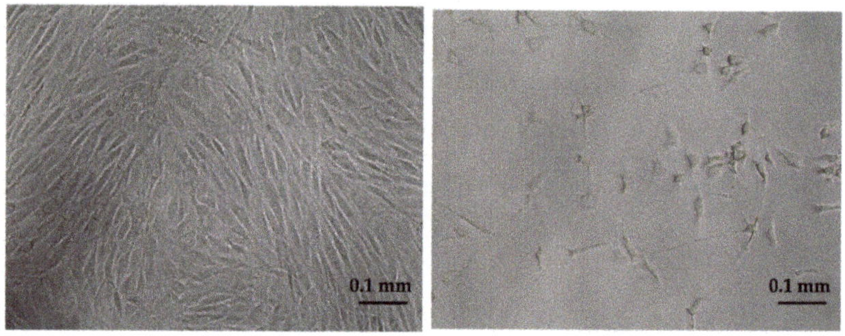

(a) (b)

Figure 1. oBM-MSC differentiation into neuron-like cells 3 days after neurogenic induction with HyClone neurogenic medium: (**a**) oBM-MSCs in growth conditions and (**b**) oBM-MSCs in neurogenic differentiation.

2.4. Scrapie Inocula and Infection of oBM-MSC and Neuron-Like Cultures

Inocula were prepared using CNS samples from one healthy (negative controls) and one classical scrapie-infected sheep carrying the ARQ/ARQ genotype and preserved at the tissue bank of the Center of Encephalopathies and Emerging Transmissible Diseases (CEETE; University of Zaragoza). The presence/absence of PrP^{Sc} in the tissues was confirmed following protocols reported in other works [26], using two rapid diagnostic tests (Prionics-Check Western blotting; ThermoFisher Scientific and Idexx HerdChek; IDEXX, Westbrook, ME, USA) and confirmation by immunohistochemical examination of CNS tissue. CNS samples were homogenized and diluted 1:10 (g/mL) in physiological saline solution (Braun). Afterwards, samples were treated at 70 °C for 10 min before adding streptomycin sulphate (100 µg/mL) and benzylpenicillin (100 µg/mL). In order to check the safety of the inocula once generated, samples were incubated in blood agar plates, and the absence of any bacterial growth was confirmed.

To determine the effect of prion infection on the proliferation potential and the ability of prion replication, oBM-MSCs cultures were seeded at 5000 cells/cm^2 for proliferation conditions and at 1500 cells/cm^2 for neurogenic conditions. In both cases, three groups were established: positive, negative and control cultures. Positive cultures were infected with inocula from a scrapie-infected sheep, negative cultures with inocula from a healthy sheep and control cultures were kept in standard conditions. After adhesion for 24 h, basal media was substituted by inocula diluted 1:10 in DMEM media (10% FBS, 1% L-glutamine and 1% streptomycin/penicillin) for the oBM-MSC cultures and in HyClone media for the oBM-MSC cultures in neurogenic differentiation. Cells were maintained in this medium for 48 h to analyze the proliferation potential and cell viability and for 72 h for the MTT/ELISA assays. Afterwards, the medium was changed twice a week.

2.5. Proliferation Potential and Cell Viability

To determine the effect of prion infection on oBM-MSC proliferation potential, cultures from three different donors (biological replicates) were seeded in 6-well plates at

5000 cells/cm^2, inoculated with scrapie and control inocula and maintained until passage 3 post-infection; every passage was performed at around 80% confluence. Adherent cells were counted every passage and the cell doubling number (CD) and cell doubling time (DT), used to determine the time it takes for a population of cells to double in size, were calculated as previously described [21,24]. The results were evaluated using the paired Student's *t*-test.

To assess early prion toxicity, cell viability was also evaluated using MTT in oBM-MSC from 8 donors at 3, 7 and 10 days post-inoculation (dpi), seeding 4 technical replicates for each culture. oBM-MSC cultures were seeded in 96-well plates at 5000 cells/cm^2 in growth conditions and at 1500 cells/cm^2 in neurogenic differentiation conditions. Briefly, the MTT assay was performed by adding 25 µL of MTT solution (2 mg/mL) per well. Then, the plates were incubated at 37 °C for 4 h. Afterwards, the content of each well was removed and was substituted with 150 µL of HCl solution (HCl 40 mM in isopropanol) per well. Plates were then incubated for 1 h at room temperature protected from light. The absorbance was measured at 570 nm in an Infinite F200 microplate reader (Tecan Ibérica Instrumentación, Barcelona, Spain). A calibration curve was prepared with different amounts of cells. Since oBM-MSCs in growth conditions are seeded in a higher density than the ones in neurogenic differentiation conditions, two calibration curves were prepared: a more concentrated one to compare oBM-MSCs in growth conditions (Figure S1a) and a more diluted one to compare cultures in neurogenic conditions (Figure S1b). In both cases, the calibration curve enabled us to establish the relationship between absorbance and the amount of cultured cells. The toxicity of the prion was studied in three conditions (inoculated with scrapie-positive brain homogenates, negative brain inoculum and non-inoculated controls) and at three different stages (3, 7 and 10 dpi). As cells were kept in contact with the inoculum for 72 h, the stage 3 dpi corresponds to the moment just after inoculum removal. The moment of the infection with the inocula was considered as day 0. The normality of the results was evaluated with Shapiro–Wilk and D'Agostino–Pearson tests. Differences in cell viability and proliferation were evaluated with Student's *t*-test. Statistical significance was defined as $p < 0.05$.

2.6. PrP^{Sc} Detection

Cells from the three biological replicates analyzed in the proliferation assay were used to evaluate if PrP^{Sc} was increased or maintained along the passages in MSC cultures infected with scrapie and maintained under grown conditions. Approximately 10^6 cells of passages 1, 2 and 3 post-infection were frozen at −80 °C for further PrP^{Sc} determination by Western blotting. Pellets of frozen cells were homogenized in 100 µL of PBS. Afterwards, samples were analyzed using the BSE Scrapie Discriminatory Kit (Bio-Rad Laboratories, Hercules, CA, USA) and treated following the manufacturer's recommendations. Electrophoresis was developed in 12% SDS-PAGE gels. Protein was then transferred to a 0.20-µm nitrocellulose membrane (Bio-Rad). CDP-Star substrate (ThermoFisher Scientific, Westbrook, ME, USA) was used to determine chemiluminescence in a Versa-Doc Imaging System (Bio-Rad Laboratories). Chemiluminescence signals were evaluated using ImageJ 1.4.3.67 (Psion Image), as described previously [27].

Neurogenic differentiation of MSCs requires seeding cells at low density and differentiated cells cannot be maintained along passages. To test the ability of these cells to replicate PrP^{Sc}, we quantified the amount of the pathogenic protein soon after prion infection at three different stages (3 dpi, which corresponds to inoculum removal, 7 and 10 dpi) in oBM-MSCs in growth and neurogenic differentiation conditions. We used a more sensitive test, the ELISA kit EEB-Scrapie HerdCheck kit (IDEXX), following the manufacturer's recommendations. oBM-MSCs cultures from 5 donors were seeded in 6-well plates and the retrieval of the cells was performed by means of trypsinization and subsequent centrifugation. To quantify the PrP^{Sc} detection range of the ELISA kit, a calibration curve was performed using different concentrations of scrapie inocula (Figure S2). PrP^{Sc} was detected in all inoculum concentrations, showing that this kit is suitable to detect PrP^{Sc} in oBM-MSC cultures, as the amount of inocula used in oBM-MSC infection was higher than the most diluted concentration of the calibration curve. The inoculum used in the calibration curve was the same used to infect oBM-MSCs in growth and neurogenic conditions. For infection, a volume of 100 μL of scrapie inoculum per well was employed, which would correspond to >3 units of absorbance.

3. Results

3.1. Proliferation Potential of Infected oBM-MSC

The effect of scrapie infection on the proliferation capacity was analyzed in oBM-MSCs, calculating the CD and DT. Significant differences between cultures infected with scrapie and control inocula were found for both CD and DT at the first passage post-infection. CD was higher and DT was lower in the cultures treated with control inocula compared to those inoculated with scrapie brain cells (Table 2).

Table 2. Cell doubling number (CD) and cell doubling time (DT) of oBM-MSCs from 3 donors through passages 1 to 3 post-inoculation with 1% brain homogenates obtained from healthy and scrapie sheep and the average value for the three passages (Av).

Passage		Inocula	
		Healthy	Scrapie
1	CD	3.150 ± 0.286 *	2.949 ± 0.219 *
	DT (days)	1.714 ± 0.355 **	1.825 ± 0.343 **
2	CD	3.22 ± 0.651	2.870 ± 0.531
	DT (days)	2.054 ± 0.653	2.291 ± 0.681
3	CD	1.93 ± 0.390	1.807 ± 0.027
	DT (days)	2.116 ± 0.428	2.214 ± 0.033
Av	CD	2.871 ± 0.711 *	2.634 ± 0.597 *
	DT (days)	1.942 ± 0.469	2.097 ± 0.469

Significant differences were calculated using Student's *t*-test (* $p < 0.05$, ** $p < 0.01$).

3.2. Cell Viability of Infected Cultures

The effect of prion infection on cell viability was studied in three conditions (scrapie positive inoculum, healthy/negative inoculum and control without inoculum) and at three different stages (3, 7 and 10 dpi) in oBM-MSCs in growth conditions and in neurogenic differentiation.

Proliferation was evidenced in oBM-MSC cultures maintained in growth medium under the three conditions. Inoculated cultures displayed higher number of cells than controls at the three stages (3, 7 and 10 dpi). Proliferation was significantly lower in scrapie infected cells than in cultures treated with negative inocula at 3 dpi but, in subsequent stages (7 and 10 dpi), the positive cultures displayed significantly more cells than the negative cultures (Figure 2a).

In neurogenic differentiation conditions, the number of cells also increased over time in the three conditions, whereas inoculated cultures showed a higher growth than controls. Comparing between the inoculated cultures, the number of cells was significantly higher in cultures that were in contact with negative inoculum than the ones infected with scrapie, and this difference was statistically significant at 10 dpi (Figure 2b).

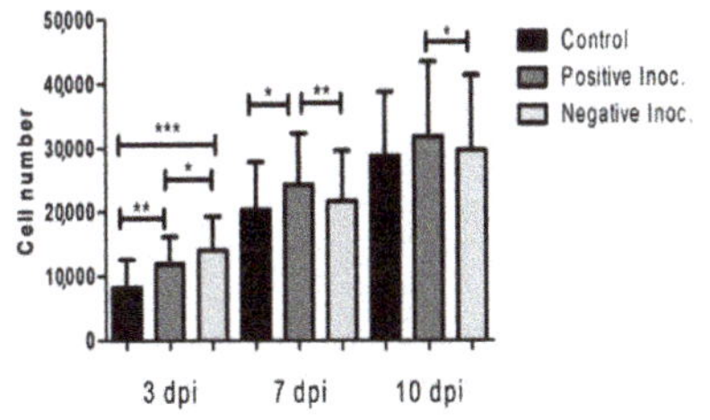

(a)

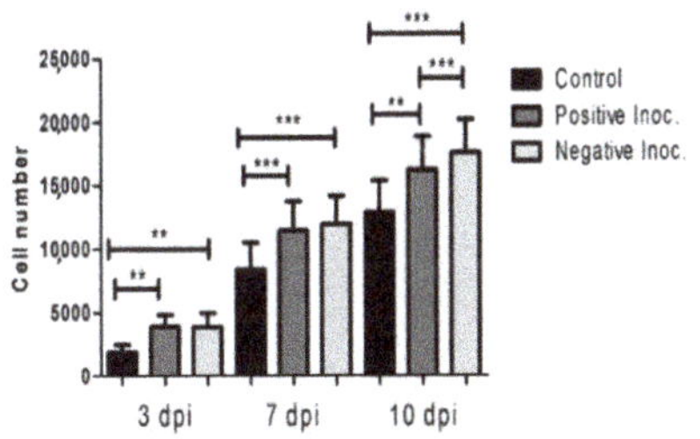

(b)

Figure 2. Cell viability study by MTT in infected oBM-MSC cultures in growth conditions (**a**) and neurogenic differentiation (**b**) 3, 7 and 10 days post-inoculation (dpi). oBM-MSCs were from 8 different donors and 4 technical replicates per culture were seeded. Significant differences were calculated using the Student *t*-test (* $p < 0.05$, ** $p < 0.01$, *** $p < 0.001$).

3.3. PrP^{Sc} Detection in Infected oBM-MSCs, Analyzed by Western Blotting

After inoculation of oBM-MSCs in growth conditions, surviving cells retained their ability to proliferate and were expanded until passage 3 post-infection. Western blotting analysis revealed the presence of PrP^{Sc} in the cultures during these three passages although the intensity of bands decreased with the number of passages (Figures 3 and S3) and was lost in further subcultures.

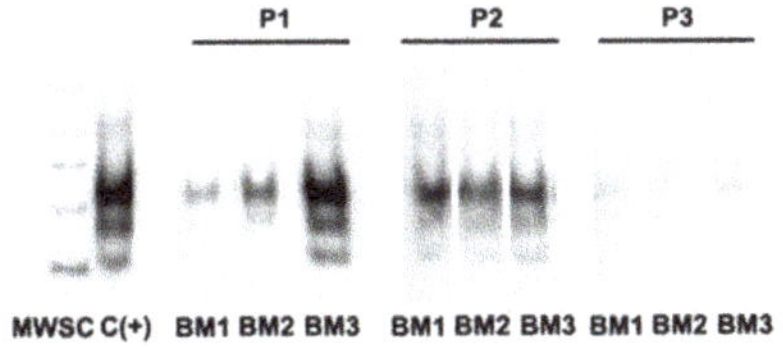

Figure 3. Determination of PrP^{Sc} by Western blotting in oBM-MSCs (BM1, BM2, BM3) infected with scrapie inocula at passages 1 to 3 (P1, P2, P3). MWSC = molecular weight marker; C (+) = positive control.

3.4. Prion Detection by ELISA in oBM-MSC and Neuron-Like Cultures Infected with Scrapie

To test the ability of MSC-derived neuron-like cells to replicate prions, the presence of PrP^{Sc} was studied by means of ELISA immediately after infection at three different stages (3, 7 and 10 dpi) in oBM-MSCs infected with positive inocula in growth and neurogenic differentiation conditions.

In oBM-MSCs maintained under growth conditions, a decrease in ELISA absorbance was observed, which could be associated with a loss of the PrP^{Sc} signal. Significant differences were found between 3 dpi and 7 dpi ($p < 0.05$) and 3 dpi and 10 dpi ($p < 0.01$) (Figure 4a).

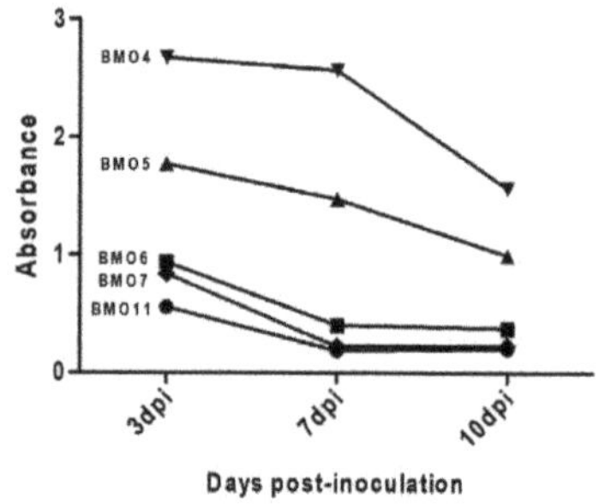

(a)

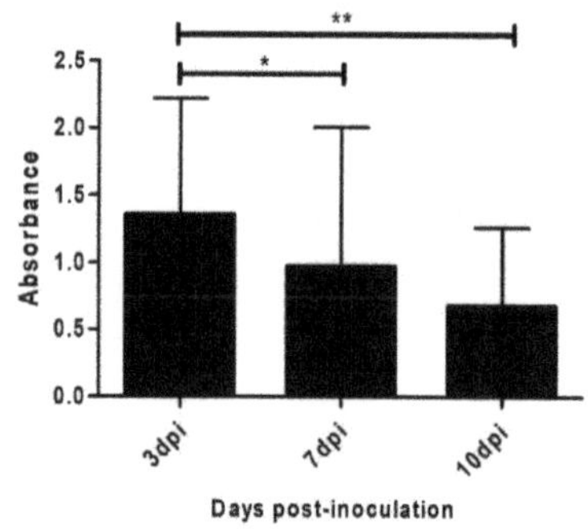

(b)

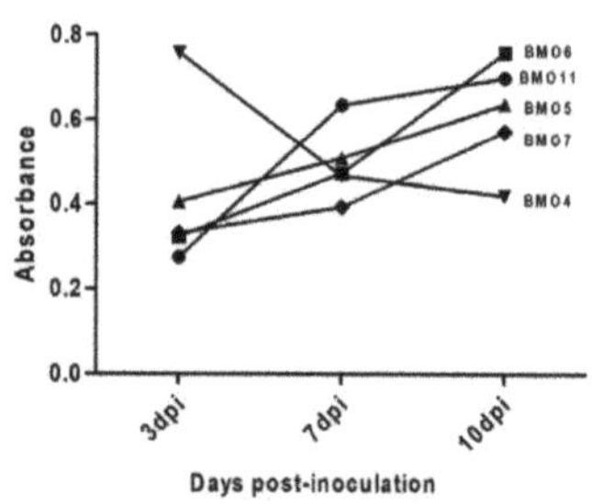

(c)

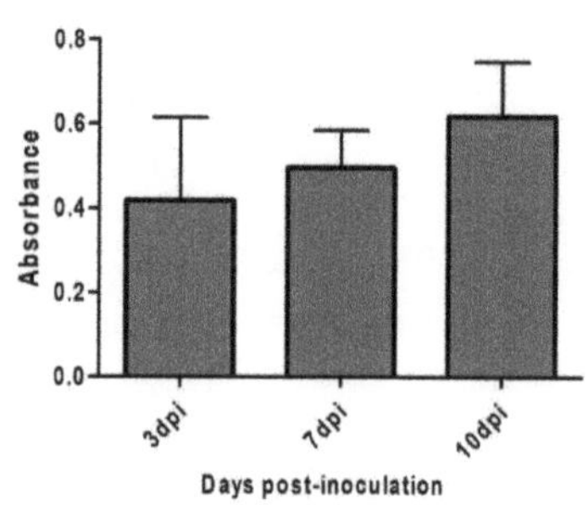

(d)

Figure 4. PrP^{Sc} detection via ELISA in infected oBM-MSC cultures from 5 donors in growth (**a**,**b**) and neurogenic differentiation (**c**,**d**) conditions 3, 7 and 10 days post-inoculation (dpi). Significant differences were calculated using Student's *t*-test (* $p < 0.05$, ** $p < 0.01$).

In contrast, the majority (four out of five) of neuron-like differentiated cultures showed an increase in absorbance over time, which could be associated with a progressive increase in the PrP^{Sc} signal. Even though most of the cultures had an increasing signal pattern, as one of them (BMO4) displayed decreased absorbance over time, no significant differences were found between any of the three stages (Figure 4b).

Although we used the same inoculum in all cultures, the initial amount of PrP^{Sc} was different in each culture at 3 dpi, suggesting a heterogeneity in the ability to retain prions. Higher cellular density under growth conditions would explain the higher absorbance observed in this condition, compared to neurogenic conditions.

4. Discussion

Prion diseases are fatal neurodegenerative disorders affecting humans and animals. Over the years, a substantial effort has been made to develop in vitro models for the study of these pathologies. Most of the cellular models are based on the culturing of murine cell lines [8] and require a previous adaptation of the strain to mice, due to the well-known phenomenon of the species barrier. Therefore, in vitro models with a natural host background would be very useful tools for research into many prion topics, e.g., prion replication, toxicity, genetic susceptibility, differences in strain susceptibility, early mechanisms of infection and new treatment testing.

MSCs can be easily collected from several accessible adult tissues like bone marrow or peripheral blood [8,28] and they show the ability to transdifferentiate into neuronal elements in vitro [10,29]. Several works have described the ability of murine stromal cells to propagate prion infectivity [12,17,30,31] and expanded MSCs obtained from sporadic Creutzfeldt-Jakob disease CJD patients have been shown to be positive for PrP^{Sc} [12]. The infectivity of MSCs obtained from sick individuals does not seem to be a pan-species characteristic, as oBM-MSCs from scrapie infected sheep did not show PrP^{Sc} [18]. Although MSCs derived from human, cattle and sheep express PrP^{C} [12,18] to the best of our knowledge, the potential of MSCs derived from these naturally susceptible species to propagate prion infection in vitro has never been investigated. In the present work, we infected ovine bone marrow-derived MSCs and their neuron-like derivatives with scrapie-infected sheep isolates to study the response of these cells to prion infection during a certain period of time.

MSCs can migrate to prion-affected neurological tissues as a response to secreting trophic factors that activate endogenous restorative reactions in the injured brain [32–34]. In our study, cell viability was higher in both oBM-MSCs and neural-differentiated cultures after inoculation compared with non-inoculated control cultures in the three monitored stages (3, 7 and 10 dpi), which suggests that brain inocula, independently of their origin, may contain factors that stimulate oBM-MSC proliferation.

Murine stromal cells are able to propagate prions for many passages [17,30]. On the contrary, oBM-MSCs do not seem to be permissive to PrP^{Sc} infection. The loss of PrP^{Sc} signal over time detected by ELISA in oBM-MSC cultures soon after infection with positive inocula suggests that these cells, if infected, are unable to replicate the prions, unlike what happens in mice. Similarly, Western blotting revealed the presence of PrP^{Sc} in scrapie-infected oBM-MSC cultures three passages after inoculation, but the presence of the pathologic protein seemed also to be weakened between passages 2 and 3, suggesting that PrP^{Sc} may be taken up by oBM-MSCs without leading to a successful prion infection. In some works, murine BM-MSCs infected with prions in vitro showed few or no PrP^{Sc} production during the first 10 or even 50 passages [12,31] and stable and detectable (by Western blotting) production of PrP^{Sc} afterwards. We could not explore this possibility because, contrary to murine cells, MSCs obtained from humans or unconventional model organisms including sheep are able to be maintained in culture for far fewer passages [35,36].

Moreover, after inoculation with scrapie, oBM-MSC cultures displayed a high proliferation rate, with an average doubling time during the three passages that was lower than the DT described previously for BM-MSC cultures derived from scrapie and healthy sheep [18]. Cell division modulates prion accumulation in cultured cells [37] and direct proximity between donor and recipient cells increases the infection in other cell culture models [38]. The high proliferation rate observed in oBM-MSCs could help to avoid the transmission of PrP^{Sc} from infected cells to non-infected ones because the cells are not in contact for a long enough time. Therefore, only the cells infected during the inoculation process and their daughters would show infection and this would be diluted in successive passages. Changes in culture conditions focused on slowing down the proliferation rate or increasing their contact in spheroid cultures could facilitate the propagation.

On the other hand, we cannot discard the possibility that infection with scrapie-infected brain cells could be toxic for the oBM-MSC cultures. Even though no toxicity was observed in murine MSCs infected with the CJD agent [12], we have to take into account that our cells come from a naturally susceptible host. In our study, CD was significantly higher in cells infected with healthy brain extracts and, accordingly, DT was higher in cells infected with scrapie inocula. Therefore, those cells exposed to scrapie prions showed lower proliferation potential, similar to the findings observed in MSCs obtained from scrapie sheep [18]. This could be a consequence of the loss of infected cells due to prion toxicity. In the MTT assay, the effect of prion infection on cell viability was evaluated during the first passage after prion infection. Toxicity seems to be an early effect of prion infection, as 3 days after inoculation, viability was lower in scrapie cultures than in healthy infected cultures. In contrast with the first assay, at the end of this passage (10 dpi), the number of cells in scrapie-infected cultures was higher than that in the cultures inoculated with control brain cells. This could be a consequence of differences in the inocula, as brain tissues used in the different experiments were different and could have contained different amounts of PrP^{Sc} and therefore exhibited different degrees of toxicity. Nevertheless, throughout all passages, CD and DT differences were lower and, similarly, the number of cells in scrapie-infected cultures increased after early toxicity. In both cases, this increase in proliferation and viability was accompanied by the loss of PrP^{Sc} detection, which might indicate a recovery of the cell culture conditions after the elimination of PrP^{Sc} infected cells, increasing the proportion of non-infected cells, which display higher proliferation potential.

Regarding the differentiation of oBM-MSC cultures into neuron-like cells, although certain toxicity was also observed 3 days after prion infection, four out of the five cultures analyzed seemed to be infected and possibly displayed the ability to replicate the pathological prion protein. Although we did not obtain statistical support for this observation, the ELISA assay showed that these cells maintained the PrP^{Sc} signal and this signal increased progressively over time. Similarly, astrocytes derived from human induced pluripotent stem cells are capable of replicating prions from brain samples of CJD patients, generating prion infectivity in vitro [20]. Taking this into account and knowing that cells from the central nervous system are the target of the pathological prion protein, oBM-MSC-derived neuron-like cells may have a greater ability to capture and replicate PrP^{Sc} than oBM-MSCs in growth conditions. The lack of statistical significance in our results was due to the existence of variability in prion replication, as one culture (BMO4) failed to replicate prions. This culture displayed a *PRNP* genotype identical to other three cultures (ARQ/ARQ), suggesting that, in addition to the *PRNP* genotype, other factors influence prion replication. Despite all the cultures being infected with the same amount of inoculum, BMO4 was the one that showed the highest absorbance for PrP^{Sc} under both growth and differentiation conditions. Differences in the ability to access prions could explain differences in toxicity and prion replication.

The observed differences in undifferentiated and differentiated oBM-MSCs suggest that the latter possess a competence for infection that it is not present at the MSC stage, even though they share a genetic background for each given animal. This system, using oBM-MSC-neuron-like derivates, could serve for the investigation (in an isogenic context) of the molecular trigger that sustains scrapie infection in vitro, specifically in the neural lineage. In addition, differences between cultures harboring the same *PRNP* genotype could help in the identification of other factors related to prion susceptibility.

5. Conclusions

This work describes for the first time the infection with scrapie agents of bone marrow-derived MSCs obtained from sheep, which is a natural host of prion diseases. Culturing ovine MSCs with CNS extracts in growth and neurogenic conditions induced cell proliferation, although some toxicity was observed in scrapie-infected cultures. Inoculated oBM-MSCs in growth conditions were not permissive to prion infection, whereas most cultures under neurogenic differentiation conditions seemed to retain and replicate the pathological prion protein. oBM-MSC-derived neuron-like cells could be a good candidate for developing in vitro studies in species for which iPSC reprogramming is not standardized, like sheep. Further studies focusing on elucidating the molecular mechanisms implicated in retaining prion infectivity and inducing prion toxicity in mesenchymal stem cells and MSC-derived neuron-like cells are warranted.

Supplementary Materials: The following are available online at https://www.mdpi.com/article/10.3390/ani11041137/s1, Figure S1: Calibration curves used in the MTT assay: (a) calibration curve used to establish the relationship between absorbance and the amount of MSCs in growth conditions ($r^2 = 0.96$) and (b) calibration curve used to establish the relationship between absorbance and the amount of MSCs in neurogenic conditions ($r^2 = 0.98$). Figure S2: Calibration logarithmic curve used to evaluate the sensitivity of PrPSc detection of the EEB-Scrapie HerdCheck kit, where $y = 0.626\ln(x) + 0.4023$ and $r^2 = 0.9911$. Figure S3: Full Western blotting membranes used for PrP^{Sc} determination in oBM-MSCs (BM1, BM2, BM3) infected with scrapie inocula at passages 1 to 3 (P1, P2, P3). MWSC = molecular weight marker; C (+) = positive control.

Author Contributions: Conceptualization, I.M.-B.; methodology, I.M.-B.; formal analysis, L.G.-M. and D.R.M.; investigation, L.G.-M., D.R.M., A.H., D.S.-R., Ó.L.-P. and A.O.; animal handling and surgical treatment, B.M. and F.J.V.; writing—original draft preparation, A.H. and L.G.-M.; writing—review and editing, I.M.-B. and L.O.; visualization, A.H.; supervision, I.M.-B., R.B. and L.O.; project administration, I.M.-B., P.Z., R.B. and J.J.B.; funding acquisition, I.M.-B., R.B., P.Z. and J.J.B. All authors have read and agreed to the published version of the manuscript.

Funding: This work was performed as part of the RTI2018-098711-B-I00 project financed by Ministerio de Ciencia, Innovación y Universidades, Agencia Estatal de Investigación and Fondo Europeo de Desarrollo Regional and by REDPRION, 65% co-financed by the European Regional Development Fund (ERDF) through the Interreg V-A Spain-France-Andorra program (POCTEFA 2014-2020). POCTEFA aims to reinforce the economic and social integration of the French–Spanish–Andorran border. Its support is focused on developing economic, social and environmental cross-border activities through joint strategies, favoring sustainable territorial development.

Institutional Review Board Statement: The study was conducted under Project License PI06/12 approved by the Ethical Committee for Animal Experiments from the University of Zaragoza. The care and use of animals were performed in accordance with the Spanish Policy for Animal Protection RD53/2013, which meets the European Union Directive 2010/63 on the protection of animals used for experimental and other scientific purposes.

Data Availability Statement: The raw data of the results presented in this study are available on request from the corresponding author.

Acknowledgments: We thank the staff of the Large Animal section of the Veterinary Hospital of the University of Zaragoza, co-financed by the Cátedra Ruralia-Bantierra, for their assistance in sheep management and sedation for bone marrow sampling. This publication uses extracts from Diego Rubén Mediano Martín-Maestro's PhD thesis [39] and the University of Zaragoza does not reserve any right of exclusive publication.

Conflicts of Interest: The authors declare no conflict of interest.

References

1. Prusiner, S.B. The prion diseases. *Brain Pathol.* **1998**, *8*, 499–513. [CrossRef]
2. Prusiner, S.B. Novel proteinaceous infectious particles cause scrapie. *Science* **1982**, *216*, 136–144. [CrossRef]
3. Bell, J.E.; Ironside, J.W. Neuropathology of spongiform encephalopathies in humans. *Br. Med. Bull.* **1993**, *49*, 738–777. [CrossRef] [PubMed]

4. Pattison, I.H.; Jones, K.M. The astrocytic reaction in experimental scrapie in the rat. *Res. Vet. Sci.* **1967**, *8*, 160–165. [CrossRef]
5. Zabel, M.D.; Reid, C. A brief history of prions. *Pathog. Dis.* **2015**, *73*. [CrossRef] [PubMed]
6. McMahon, H.E. Cell culture methods for screening of prion therapeutics. In *Methods in Molecular Biology*; Humana Press Inc.: Totowa, NJ, USA, 2017; Volume 1658, pp. 295–304.
7. Bedecs, K. Cell culture models to unravel prion protein function and aberrancies in prion diseases. *Methods Mol. Biol.* **2008**, *459*, 1–20. [CrossRef] [PubMed]
8. Solassol, J.; Crozet, C.; Lehmann, S. Prion propagation in cultured cells. *Br. Med. Bull.* **2003**, *66*, 87–97. [CrossRef] [PubMed]
9. Pittenger, M.F.; Mackay, A.M.; Beck, S.C.; Jaiswal, R.K.; Douglas, R.; Mosca, J.D.; Moorman, M.A.; Simonetti, D.W.; Craig, S.; Marshak, D.R. Multilineage potential of adult human mesenchymal stem cells. *Science* **1999**, *284*, 143–147. [CrossRef]
10. Woodbury, D.; Reynolds, K.; Black, I.B. Adult bone marrow stromal stem cells express germline, ectodermal, endodermal, and mesodermal genes prior to neurogenesis. *J. Neurosci. Res.* **2002**, *69*, 908–917. [CrossRef] [PubMed]
11. Zhao, L.R.; Duan, W.M.; Reyes, M.; Keene, C.D.; Verfaillie, C.M.; Low, W.C. Human bone marrow stem cells exhibit neural phenotypes and ameliorate neurological deficits after grafting into the ischemic brain of rats. *Exp. Neurol.* **2002**, *174*, 11–20. [CrossRef]
12. Takakura, Y.; Yamaguchi, N.; Nakagaki, T.; Satoh, K.; Kira, J.; Nishida, N. Bone marrow stroma cells are susceptible to prion infection. *Biochem. Biophys. Res. Commun.* **2008**, *377*, 957–961. [CrossRef]
13. Shi, F.; Yang, Y.; Wang, T.; Kouadir, M.; Zhao, D.; Hu, S. Cellular Prion Protein Promotes Neuronal Differentiation of Adipose-Derived Stem Cells by Upregulating miRNA-124. *J. Mol. Neurosci.* **2016**, *59*, 48–55. [CrossRef]
14. Martellucci, S.; Santacroce, C.; Manganelli, V.; Santilli, F.; Piccoli, L.; Cassetta, M.; Misasi, R.; Sorice, M.; Mattei, V. Isolation, propagation, and prion protein expression during neuronal differentiation of human dental pulp stem cells. *J. Vis. Exp.* **2019**, *2019*, e59282. [CrossRef]
15. Martellucci, S.; Santacroce, C.; Santilli, F.; Piccoli, L.; Monache, S.D.; Angelucci, A.; Misasi, R.; Sorice, M.; Mattei, V. Cellular and molecular mechanisms mediated by recPrP C involved in the neuronal differentiation process of mesenchymal stem cells. *Int. J. Mol. Sci.* **2019**, *20*, 345. [CrossRef] [PubMed]
16. Shan, Z.; Hirai, Y.; Nakayama, M.; Hayashi, R.; Yamasaki, T.; Hasebe, R.; Song, C.H.; Horiuchi, M. Therapeutic effect of autologous compact bone-derived mesenchymal stem cell transplantation on prion disease. *J. Gen. Virol.* **2017**, *98*, 2615–2627. [CrossRef] [PubMed]
17. Akimov, S.; Vasilyeva, I.; Yakovleva, O.; McKenzie, C.; Cervenakova, L. Murine bone marrow stromal cell culture with features of mesenchymal stem cells susceptible to mouse-adapted human TSE agent, Fukuoka-1. *Folia Neuropathol.* **2009**, *47*, 205–214. [PubMed]
18. Mediano, D.R.; Sanz-Rubio, D.; Bolea, R.; Marín, B.; Vázquez, F.J.; Remacha, A.R.; López-Pérez, Ó.; Fernández-Borges, N.; Castilla, J.; Zaragoza, P.; et al. Characterization of mesenchymal stem cells in sheep naturally infected with scrapie. *J. Gen. Virol.* **2015**, *96*, 3715–3726. [CrossRef] [PubMed]
19. Groveman, B.R.; Foliaki, S.T.; Orru, C.D.; Zanusso, G.; Carroll, J.A.; Race, B.; Haigh, C.L. Sporadic Creutzfeldt-Jakob disease prion infection of human cerebral organoids. *Acta Neuropathol. Commun.* **2019**, *7*, 90. [CrossRef]
20. Krejciova, Z.; Alibhai, J.; Zhao, C.; Krencik, R.; Rzechorzek, N.M.; Ullian, E.M.; Manson, J.; Ironside, J.W.; Head, M.W.; Chandran, S. Human stem cell-derived astrocytes replicate human prions in a PRNP genotype-dependent manner. *J. Exp. Med.* **2017**, *214*, 3481–3495. [CrossRef]
21. Lyahyai, J.; Mediano, D.R.; Ranera, B.; Sanz, A.; Remacha, A.R.; Bolea, R.; Zaragoza, P.; Rodellar, C.; Martín-Burriel, I. Isolation and characterization of ovine mesenchymal stem cells derived from peripheral blood. *BMC Vet. Res.* **2012**, *8*, 169. [CrossRef]
22. Hortells, P.; Monzón, M.; Monleón, E.; Acín, C.; Vargas, A.; Bolea, R.; Luján, L.; Badiola, J.J. Pathological findings in retina and visual pathways associated to natural Scrapie in sheep. *Brain Res.* **2006**, *1108*, 188–194. [CrossRef]
23. O'Rourke, K.I.; Baszler, T.V.; Besser, T.E.; Miller, J.M.; Cutlip, R.C.; Wells, G.A.H.; Ryder, S.J.; Parish, S.M.; Hamir, A.N.; Cockett, N.E.; et al. Preclinical diagnosis of scrapie by immunohistochemistry of third eyelid lymphoid tissue. *J. Clin. Microbiol.* **2000**, *38*, 3254–3259. [CrossRef] [PubMed]
24. Ranera, B.; Ordovás, L.; Lyahyai, J.; Bernal, M.L.; Fernandes, F.; Remacha, A.R.; Romero, A.; Vázquez, F.J.; Osta, R.; Cons, C.; et al. Comparative study of equine bone marrow and adipose tissue-derived mesenchymal stromal cells. *Equine Vet. J.* **2012**, *44*, 33–42. [CrossRef] [PubMed]
25. Dominici, M.; Le Blanc, K.; Mueller, I.; Slaper-Cortenbach, I.; Marini, F.C.; Krause, D.S.; Deans, R.J.; Keating, A.; Prockop, D.J.; Horwitz, E.M. Minimal criteria for defining multipotent mesenchymal stromal cells. The International Society for Cellular Therapy position statement. *Cytotherapy* **2006**, *8*, 315–317. [CrossRef]
26. Bolea, R.; Monleón, E.; Schiller, I.; Raeber, A.J.; Acín, C.; Monzón, M.; Martín-Burriel, I.; Struckmeyer, T.; Oesch, B.; Badiola, J.J. Comparison of immunohistochemistry and two rapid tests for detection of abnormal prion protein in different brain regions of sheep with typical scrapie. *J. Vet. Diagn. Investig.* **2005**, *17*, 467–469. [CrossRef]
27. Filali, H.; Vidal, E.; Bolea, R.; Márquez, M.; Marco, P.; Vargas, A.; Pumarola, M.; Martin-Burriel, I.; Badiola, J.J. Gene and protein patterns of potential prion-related markers in the central nervous system of clinical and preclinical infected sheep. *Vet. Res.* **2013**, *44*, 14. [CrossRef] [PubMed]
28. Zvaifler, N.J.; Marinova-Mutafchieva, L.; Adams, G.; Edwards, C.J.; Moss, J.; Burger, J.A.; Maini, R.N. Mesenchymal precursor cells in the blood of normal individuals. *Arthritis Res.* **2000**, *2*, 477–488. [CrossRef]

29. Sanchez-Ramos, J.; Song, S.; Cardozo-Pelaez, F.; Hazzi, C.; Stedeford, T.; Willing, A.; Freeman, T.B.; Saporta, S.; Janssen, W.; Patel, N.; et al. Adult bone marrow stromal cells differentiate into neural cells in vitro. *Exp. Neurol.* **2000**, *164*, 247–256. [CrossRef]
30. Akimov, S.; Yakovleva, O.; Vasilyeva, I.; McKenzie, C.; Cervenakova, L. Persistent Propagation of Variant Creutzfeldt-Jakob Disease Agent in Murine Spleen Stromal Cell Culture with Features of Mesenchymal Stem Cells. *J. Virol.* **2008**, *82*, 10959–10962. [CrossRef] [PubMed]
31. Cervenakova, L.; Akimov, S.; Vasilyeva, I.; Yakovleva, O.; McKenzie, C.; Cervenak, J.; Piccardo, P.; Asher, D.M. Fukuoka-1 strain of transmissible spongiform encephalopathy agent infects murine bone marrow-derived cells with features of mesenchymal stem cells. *Transfusion* **2011**, *51*, 1755–1768. [CrossRef]
32. Song, C.-H.; Honmou, O.; Ohsawa, N.; Nakamura, K.; Hamada, H.; Furuoka, H.; Hasebe, R.; Horiuchi, M. Effect of Transplantation of Bone Marrow-Derived Mesenchymal Stem Cells on Mice Infected with Prions. *J. Virol.* **2009**, *83*, 5918–5927. [CrossRef] [PubMed]
33. Chopp, M.; Li, Y. Treatment of neural injury with marrow stromal cells. *Lancet Neurol.* **2002**, *1*, 92–100. [CrossRef]
34. Song, C.-H.; Honmou, O.; Furuoka, H.; Horiuchi, M. Identification of Chemoattractive Factors Involved in the Migration of Bone Marrow-Derived Mesenchymal Stem Cells to Brain Lesions Caused by Prions. *J. Virol.* **2011**, *85*, 11069–11078. [CrossRef] [PubMed]
35. Bonab, M.M.; Alimoghaddam, K.; Talebian, F.; Ghaffari, S.H.; Ghavamzadeh, A.; Nikbin, B. Aging of mesenchymal stem cell in vitro. *BMC Cell Biol.* **2006**, *7*, 14. [CrossRef]
36. Calloni, R.; Viegas, G.S.; Türck, P.; Bonatto, D.; Henriques, J.A.P. Mesenchymal stromal cells from unconventional model organisms. *Cytotherapy* **2014**, *16*, 3–16. [CrossRef] [PubMed]
37. Ghaemmaghami, S.; Phuan, P.W.; Perkins, B.; Ullman, J.; May, B.C.H.; Cohen, F.E.; Prusiner, S.B. Cell division modulates prion accumulation in cultured cells. *Proc. Natl. Acad. Sci. USA* **2007**, *104*, 17971–17976. [CrossRef] [PubMed]
38. Kanu, N.; Imokawa, Y.; Drechsel, D.N.; Williamson, R.A.; Birkett, C.R.; Bostock, C.J.; Brockes, J.P. Transfer of scrapie prion infectivity by cell contact in culture. *Curr. Biol.* **2002**, *12*, 523–530. [CrossRef]
39. Mediano, D.R. *Caracterización de Células Madre Mesenquimales Ovinas Infectadas Por Scrapie*; University of Zaragoza: Zaragoza, Spain, 2016. (In Spanish)

animals

MDPI

Review

The Usefulness of Mesenchymal Stem Cells beyond the Musculoskeletal System in Horses

Alina Cequier [1,†], Carmen Sanz [2,†], Clementina Rodellar [1] and Laura Barrachina [1,2,*]

1. Laboratorio de Genética Bioquímica LAGENBIO—Instituto de Investigación Sanitaria de Aragón (IIS)—Instituto Agroalimentario de Aragón (IA2), Universidad de Zaragoza, C/Miguel Servet, 177, 50013 Zaragoza, Spain; alinacs@unizar.es (A.C.); rodellar@unizar.es (C.R.)
2. Servicio de Cirugía y Medicina Equina, Hospital Veterinario, Universidad de Zaragoza, C/Miguel Servet, 177, 50013 Zaragoza, Spain; cap141297@gmail.com

* Correspondence: lbarrach@unizar.es

† These authors equally contributed to this article.

Simple Summary: The main target of mesenchymal stem cell therapy in horses has long been the locomotor system, because these athletic animals commonly suffer from tendon and joint lesions. Originally, mesenchymal stem cells were thought to act by just differentiating into the cells of the injured tissue. However, these cells are also able to regulate and stimulate the body's own repair mechanisms, opening the door to many applications in inflammatory and immune-mediated disorders in both animals and humans. In horses, beyond their traditional application in the musculoskeletal system, these cells have been studied for ophthalmologic pathologies such as corneal ulcers or immune-mediated processes, and for reproductive disorders such as endometritis/endometrosis. Their potential has been explored for equine pathologies very similar to those affecting people, such as asthma, metabolic syndrome, aberrant wound healing, or endotoxemia, as well as for equine-specific pathologies such as laminitis. Current evidence is still preliminary, and further research is needed to clarify different aspects, although research performed so far shows the promising potential of mesenchymal stem cells to treat a wide variety of equine pathologies, some of which are analogous to human disorders. Therefore, advancements in this path will be beneficial for both animals and people.

Abstract: The differentiation ability of mesenchymal stem cells (MSCs) initially raised interest for treating musculoskeletal injuries in horses, but MSC paracrine activity has widened their scope for inflammatory and immune-mediated pathologies in both equine and human medicine. Furthermore, the similar etiopathogenesis of some diseases in both species has advanced the concept of "One Medicine, One Health". This article reviews the current knowledge on the use of MSCs for equine pathologies beyond the locomotor system, highlighting the value of the horse as translational model. Ophthalmologic and reproductive disorders are among the most studied for MSC application. Equine asthma, equine metabolic syndrome, and endotoxemia have been less explored but offer an interesting scenario for human translation. The use of MSCs in wounds also provides a potential model for humans because of the healing particularities in both species. High-burden equine-specific pathologies such as laminitis have been suggested to benefit from MSC-therapy, and MSC application in challenging disorders such as neurologic conditions has been proposed. The available data are preliminary, however, and require further development to translate results into the clinic. Nevertheless, current evidence indicates a significant potential of equine MSCs to enlarge their range of application, with particular interest in pathologies analogous to human conditions.

Keywords: One Medicine; mesenchymal stem cells; immune-mediated disorders; ophthalmology; reproduction; equine asthma; equine metabolic syndrome; wounds; laminitis; neurologic disorders

Citation: Cequier, A.; Sanz, C.; Rodellar, C.; Barrachina, L. The Usefulness of Mesenchymal Stem Cells beyond the Musculoskeletal System in Horses. *Animals* **2021**, *11*, 931. https://doi.org/10.3390/ani11040931

Academic Editors: Eleonora Iacono and Barbara Merlo

Received: 9 March 2021
Accepted: 22 March 2021
Published: 25 March 2021

1. Introduction

The equine industry is a sector of great economic importance, mainly due to the sporting dedication of the horse. Thus, musculoskeletal conditions have a great relevance in these animals. The limited healing capacity of ligaments, tendons, and cartilage have led to the interest in advanced biological therapies [1] aimed at restoring the structure and functions of tissues or organs using the own mechanisms of the body [2], of which mesenchymal stem cells (MSCs) are of particular relevance.

Initially, MSC therapeutics were predicated to be based on their ability to differentiate into cells of the appropriate tissue type, and thus to directly stimulate regeneration of the damaged structures. However, it has been shown that MSCs exert their effects mainly through the secretion of a wide range of bioactive molecules [1], which significantly increases the scope for their therapeutic applications. Actually, the main focus on musculoskeletal application in equine practice is in contrast with human medicine, where MSC therapies are primarily focused on immune-mediated, inflammatory, and ischemic diseases [3].

Even though musculoskeletal pathologies are among the most frequent in horses, the alteration of other systems is also of great relevance provided their economic and welfare impact [4]. The treatment of some of these pathologies presents important limitations owing to different factors, such as horse anatomy and healing physiology in the case of wounds [5], or intrinsic characteristics of immune-based pathologies such as asthma, where only palliative treatments are available [6]. Thus, the range of therapeutic mechanisms elicited by equine MSCs could extend their application beyond the locomotor system. Moreover, numerous naturally occurring diseases in both animals and humans (developmental, infectious, autoimmune, or allergic) have similar pathophysiological bases [7].

This review aims at revising the current knowledge on the use of MSCs for equine pathologies beyond the locomotor system, including ophthalmic and reproductive pathologies, equine metabolic syndrome, equine asthma, wounds, laminitis, neurological disorders, and systemic inflammatory response syndrome, highlighting the value of the horse as a translational model for developing novel treatments that could benefit both animals and humans.

1.1. "One Medicine, One Health": The Role of the Horse

In comparison with small animals such as rodents, large species such as horses better resemble the anatomy and physiology of humans, and their greater lifespan allows for longer-term follow-up. Several organizations such as the European Medicines Agency, the U.S. Food and Drug Administration, and the International Society for Stem Cell Research are recommending the use of large animal models to evaluate the efficacy, durability, dose response, and safety of advanced therapeutic medicinal products [8]. Horses are already recognized as models for several human diseases, including metabolic syndrome [9], asthma [10], musculoskeletal diseases [11,12], melanoma [13], or autoimmune uveitis [14]. Moreover, other equine conditions may also serve as models, including infectious diseases [15], fertility disorders [16], and even depression and mental conditions [17,18].

All of this led to the development of the "One Health, One Medicine" approach, which, according to the World Health Organization, promotes a vision of the health of humans, animals and the environment as a coherent system, and presumes that diseases in humans and animals require analogous therapeutic approaches [19]. Hence, progress in animal models is mutually beneficial for animals, researchers, and human and veterinary patients. Importantly, the rapid advancement of regenerative medicine in the horse makes this species particularly relevant for translational research [20].

1.2. Properties of Mesenchymal Stem Cells

Mesenchymal stem cells are adult and multipotent stem cells of mesodermal origin, which have raised interest in the field of regenerative medicine due to their unique biological properties [21]. Equine MSC characterization is based on the recommendations of the

International Society for Cell Therapy (ISCT) set for human MSCs, which include cellular plastic adherence, expression of the surface markers cluster of differentiation (CD)90, CD105 and CD73, and the lack of expression of CD45, CD34, CD14 or CD11b, CD79a or CD19 and human leukocyte antigen (HLA) class II. In addition, these cells must be able to differentiate at least into osteoblasts, adipocytes and chondrocytes in vitro [22]. Equine MSCs meet the set criteria of plastic adherence and multipotency, but variably express different surface markers depending on tissue source [23,24].

In equine medicine, MSCs are often the choice for advanced therapies, due to their ease of isolation and culture, their multipotency, and their ability to migrate to damaged tissues [25]. The mechanisms of action through which MSCs exert their effects have not been fully characterized [21]. Initially, it was thought that MSCs differentiated directly into cells of the affected tissue or enhanced the activity of resident cells [3], although it has been reported that MSCs can act indirectly by secreting immunomodulatory and bioactive factors [22–24,26,27].

These paracrine effects can be divided into immunomodulatory, anti-scarring, chemoattractant and trophic effects, which can be further subdivided into anti-apoptotic, supportive (stimulation of mitosis, proliferation and differentiation of precursors) and angiogenic. The number of molecules known to mediate the paracrine action of MSCs increases every day [28]; thus indicating that there is a substantial potential to harness these properties to treat several medical conditions in horses.

For example, MSCs secrete different chemokines that recruit and regulate the function of several cell types, as well as growth factors such as vascular endothelial growth factor (VEGF), hepatocyte growth factor (HGF) or the leukemia inhibitory factor (LIF) that promote survival and stimulate the proliferation of resident cells. Some of these growth factors, such as VEGF, also elicit pro-angiogenic effects of great importance, because the restoration of blood supply is essential for the recovery of damaged tissues [25].

In addition, MSCs secrete different cytokines and mediators such as interleukin (IL)-6 and 10, prostaglandin E-2 (PGE-2), transforming growth factor (TGF-β), or nitric oxide (NO), which elicit immunomodulatory actions such as inhibiting the proliferation of T lymphocytes [29], prevent lysis mediated by cytotoxic T cells [30], suppress the activation of natural killer (NK) cells [31] and macrophages [32], or modulate B cell proliferation [33].

2. Applications of Equine MSCs in Ophthalmology

The corneal epithelium contains limbal stem cells (LSCs), whose deficiency may greatly affect corneal transparency and integrity of the ocular surface [34]. Mesenchymal stem cells can differentiate into corneal epithelial cells both in vitro and in vivo in rabbits [35]. In addition, MSC paracrine activity can reduce oxidative stress and inhibit the release of pro-inflammatory cytokines, thereby reducing corneal inflammation and neovascularization [36,37]. The in vivo ophthalmologic studies presented here are summarized in Table 1.

2.1. Corneal Ulcers and Ulcerative Keratitis

Horses are more prone to corneal damage than other species because of their large-sized eyes placed laterally and prominently [38], and the active movements of their head which favors exposure to bacterial or fungal contamination [39]. Some corneal ulcers can be very severe due to protease activity and can lead to vision loss. The therapeutic strategy should be aimed at eradicating the infection and limiting cornea destruction, in order to control pain and minimize scar formation [40]. Current therapies include topical antibiotics, anti-proteases, and mydriatic or cycloplegic drugs [41], but the typically associated pain makes repeated local administration difficult.

The wound healing capacity of equine autologous bone marrow-derived MSCs (BM-MSCs) and their supernatant (MSC-Sp) was evaluated in vitro using a scratch assay in corneal fibroblasts. The significant decrease in the scratch area after exposure to either MSCs or MSC-Sp suggested their potential to improve corneal healing. Interestingly, the use of MSC-Sp may provide a stem cell-derived but cell-free product that could be more easily stored and applied [42].

Regarding in vivo studies, in one case of bacterial ulcerative keratitis unresponsive to conventional therapies, a single dose of autologous peripheral blood-derived stem cells (PB-SCs) was applied both systemically (intravenously (IV), jugular vein) and in transverse facial artery. In addition, local application was performed three times a day for seven days. Three months later, the ulcer was almost unnoticed and the clinical signs of inflammation, pain and irritation disappeared [43]. Similar results were observed in another case of bacterial ulcerative keratitis and in three cases of corneal ulcers treated with IV and local administration of autologous PB-SCs [40].

2.2. Equine Recurrent Uveitis

Equine recurrent uveitis (ERU) is a spontaneous and immune-mediated disorder characterized by recurrent episodes of intraocular inflammation separated by periods of remission [44]. The exact pathophysiology of ERU is not clear, although it is thought that *Leptospira interrogans* can be implicated by initiating an infection that leads to ocular immune privilege breakdown [45]. The subsequent immune response involves cytokines and chemokines that activate helper T cells (Th), and Th17-associated cytokines seem to play a role [46]. Currently, there is no cure, and treatment is focused on preserving vision, alleviating pain, and limiting the recurrence of episodes by using mydriatics and anti-inflammatory drugs. The end stage of the disease in the majority of affected horses is blindness [44].

Mesenchymal stem cells are effective at reducing immune cell activation in vitro in many species, making them a potential therapeutic option for ERU [47]. It has been shown in several mammals, such as cats, dogs and horses, that MSCs can induce a switch from pro-inflammatory to regulatory T cell subsets when applied in immune diseases [47–49]. The same has been suggested for ERU in an in vitro assay in which lymphocytes from ERU-affected horses were co-cultured with adipose-derived MSCs (AT-MSCs) [50]. However, to the best of the authors' knowledge, there is no literature on controlled equine clinical studies employing MSCs in ERU except for a brief mention in the study of Saldinger et al., which stated satisfactory treatment of three cases (unpublished data).

Interestingly, ERU closely resembles human autoimmune uveitis regarding clinical and immune pathological features, including the same autoantigens involved and the remitting-relapsing onset of the disease. Therefore, ERU has been suggested as a reliable spontaneous model to study the histopathological changes and the inflammatory processes in uveitis [14,51]; thus, therapeutic advancements in ERU and human uveitis can be mutually beneficial.

2.3. Equine Immune Mediated Keratitis

Equine immune mediated keratitis (IMMK) is a generic term used to describe a heterogeneous group of chronic, non-ulcerative corneal opacities accompanied by intraocular inflammation. The etiopathogenesis is unknown, but seems to be related with dysregulated immune responses involving a complex cytokine cascade and amplified pro-inflammatory response [41]. Long-term topical anti-inflammatory and immunosuppressive therapy is the main management protocol, although it does not offer a definitive solution and can be challenging to maintain with low owner compliance or poor response to treatment [52].

Table 1. Ophthalmology in vivo studies using stem cells in horses. Peripheral blood stem cells (PB-SCs), phosphate-buffered saline (PBS), equine recurrent uveitis (ERU), bone marrow-derived mesenchymal stem cells (BM-MSCs), immune-mediated keratitis (IMMK), intravenous (IV).

Pathology	Study	Type of MSCs	Experimental Design	Administration Regime	Outcome	Considerations
Corneal ulcers	Spaas et al., 2011 [43]	Autologous PB-SCs	Naturally occurring pathology (one case): 20-year-old gelding, with bacterial (*Pseudomonas aeruginosa*) ulcerative keratitis, resistant for 6 months against conventional therapies and surgical intervention.	A one-time injection in the jugular vein (125 × 10^3 PB-SCs in 5 mL PBS) and in transverse facial artery (125 × 10^3 PB-MSCs in 5 mL PBS) + local application (eye drops formulation, 500 × 10^3 PB-SCs in 5 mL PBS) 3 times/day, 7 days.	• 2 weeks: inflammation and tear flow decreased; ulcer size reduced • 3 months: eye ulcer almost invisible and inflammation, pain and irritation disappeared. Better general condition.	Single case with no control Cells used not expanded and not fully characterized as MSCs After 7 days, eye drop application was stopped because the cell suspension appeared cloudy, probably because of cell death. Storage conditions of drop bottle between administrations not specified.
	Marfe et al., 2012 [40]	Autologous PB-SCs	Naturally occurring pathology (4 cases) • Case 1: 20-year-old gelding with bacterial (*Pseudomonas aeruginosa*) ulcerative keratitis resistant for 6 months against conventional therapies and surgical intervention • Case 2: 7-year-old mare with a corneal ulcer treated for 2 weeks and IMKK treated for a year • Case 3: 12-year-old gelding with traumatic corneal ulcer treated for 6 months • Case 4: 13-year-old gelding with ERU-derived corneal ulcer treated for 2 weeks	1–2 systemic administration (IV) + local instillation 2–3/day for 2 weeks.	Case 1: • 2 weeks: inflammation and lacrimation decreased; ulcer size reduced and stable. • 3 months: eye ulcer reduced, inflammation stable, pain and irritation disappeared. • >3 months: eye ulcer disappeared Case 2: • 1 month: no signs of relapse Case 3: • 2 weeks: ulcer significantly reduced • 1 month: deposit of melanin (scarring effects) Case 4: • 2 months: corneal ulcer completely disappeared	Low number of cases and with corneal ulcers of different origins Absence of control group, but animals unresponsive to previous treatments Cells used not expanded and not fully characterized as MSCs Number of PB-SCs administrated not stated Administration route and protocol unclear at some points

Table 1. *Cont.*

Pathology	Study	Type of MSCs	Experimental Design	Administration Regime	Outcome	Considerations
IMMK	Marfe et al., 2012 [40]	Autologous PB-SCs	Naturally occurring pathology (one case): 7-year-old mare, poorly responsive to traditional medical treatments for a year (Case 2 above)	2 systemic administration (IV) + local instillation 2/day for 2 weeks	• 1 month: no signs of relapse • 3 months: complete recovery of clinical signs	A single case with no control group Number of PB-SCs administered not stated
	Davis et al., 2019 [53]	Autologous BM-MSCs	Naturally occurring pathology (4 cases): unilateral IMMK poorly responsive to traditional medical treatments • Case 1: 9-year-old gelding with midstromal keratitis • Case 2: 12year-old mare with superficial to midstromal keratitis • Case 3: 9-year-old gelding with midstromal keratitis • Case 4: 10-year-old gelding with midstromal keratitis	Subconjunctival injection (15×10^6 MSCs in 1 mL PBS) every 3–4 weeks for 1–5 injections • Case 1: 1 single injection • Case 2: 3 injections every 3–40 weeks • Case 3: 3 injections every 3–6 weeks • Case 4: 3 injections every 3 weeks Cases 2 and 3 received 1 and 2 additional injections, respectively	• 3 weeks: improvement of clinical signs (decreased fibrosis/opacity, irregularity, and vascularization). • No relapse for average 1 year. • Case 1, 2 and 3: Resolution of fibrosis and neovascularization • Case 4: Enucleation due to disease worsening and discomfort	Low number of cases with no control group, but selected upon disease similarities and unresponsiveness to medical treatment. Variability in the administration protocol Additional topical treatments in some case in conjunction with the BM-MSCs (cyclosporine, bromfenac, diclofenac or flunixin meglumine)

Due to the immunomodulatory action of MSCs, these have been suggested as potential therapeutic tools. One case of IMMK was reported to have a positive response after two intravenous injections of autologous PB-SCs and local instillation, without signs of relapse [40]. In another study involving four horses with IMMK unresponsive to conventional treatment, subconjunctival injection of autologous BM-MSCs (1–5 injections every 3–4 weeks) was tested along with usual treatment. Three horses had a positive clinical response, as demonstrated by decreased corneal opacity, diminished neovascularization, and improvement in surface irregularity [53].

The limitations of most in vivo ophthalmologic studies include a small sample and variability in patient selection, along with lack of a control group. Nevertheless, all cases were poorly responsive to medical management and showed improvement after MSC therapy. The administration protocol, including route, number of applications, and duration of the treatment, as well as medical management before and during MSC therapy, varied significantly. Regarding the route of administration, the subconjunctival space is relatively common for ocular medications and provides high drug concentration to the cornea and anterior structures for prolonged periods [54]; therefore, it could represent a suitable approach for local MSC administration that avoids repeated manipulation of the painful eye for local instillation, which also poses the problem of prolonged storage [43]. Other delivery systems include local intra-arterial injection in order to reach the internal structures of the eye, and contact lenses seeded with MSCs have been used in other models [55]. Further research based on controlled studies and standardized protocols supported by distribution assays is needed to demonstrate the benefits of such treatments and implement its use in clinical practice.

3. Applications of Equine MSCs in Reproduction

Due to the MSC regenerative and immunomodulatory properties, these might be used to treat damaged reproductive tissues, as has been proposed in different animal models, such as rats [56]. Table 2 summarizes the in vivo reproduction-related studies discussed below.

3.1. *Endometritis*

Endometritis is the infection and/or inflammation of the endometrium, and constitutes the main cause of subfertility and is the third most common disease affecting horses [57]. Repeated inflammation of the endometrium can lead to endometrosis, a chronic state characterized by fibrosis with glandular alterations [58]. Both persistent breeding-induced endometritis (PBIE) and endometrosis are of pivotal importance for reproductive health in mares.

3.1.1. Equine Persistent Breeding-Induced Endometritis

Persistent breeding-induced endometritis is an acute inflammatory response of the endometrium to semen, linked to an incapability of the uterus to remove bacteria, spermatozoa, and inflammatory exudate post-breeding. This disease affects broodmares of all breeds, leading to reproductive inefficiency and important economic losses [59]. Traditional therapeutic modalities lack sufficient effectivity, which along with increasing antibiotic-resistances, has led to the development of alternative therapies such as MSCs [60], because of their engraftment and immunomodulatory abilities.

Studies in healthy mares have provided a first insight into the effects of MSCs on reproductive tissues. Autologous MSCs isolated from endometrial biopsies harvested from healthy mares were infused into both uterine horns during the early diestrus to avoid ready elimination from the uterine lumen. Endometrial mesenchymal stem cells (eMSCs) were detected in the uterine lumen up to 24 h after infusion, but they did not engraft into the endometrium. Moreover, eMSCs effectively attenuated the inflammatory response produced by the uterine infusion itself [61]. Another study showed that allogenic BM-MSCs infused into the uterus 24 h before insemination modulated the uterine inflammatory response in healthy mares [62]. Subsequently, PBIE was induced in nine healthy mares which received intrauterine instillation of allogeneic AT-MSCs or eMSCs. The MSC administration significantly reduced inflammation regardless of the origin of the cells, but their engraftment after one month was limited, suggesting that their function in this context is preferentially exerted by paracrine mechanisms [63].

3.1.2. Endometrosis

Endometrosis is characterized by periglandular and/or stromal endometrial fibrosis, including glandular alterations within fibrotic foci. The etiology is not fully elucidated, but it seems to be age-related [64]. The endometrial changes caused by this disease are irreversible, and therapeutic management is challenging [65].

In an ex vivo model, endometrial biopsies from healthy and pathological mares were exposed to allogeneic AT-MSCs, which were able to infiltrate both the periglandular space and single glands of endometrosis-affected tissue [66]. In an in vivo study, autologous BM-MSCs were injected directly into the uterus using a catheter in mares with subfertility history and degenerated endometrium to assess safety in pathological conditions, observing no clinical alterations, histological changes, or endometrial edema [67]. These studies suggest the feasibility and safety of endometrial injection of MSCs as a new therapeutic approach for uterine disorders.

Another study reported that allogeneic AT-MSCs infused in the uterine lumen of six endometrosis-affected mares effectively homed into both the glandular and non-glandular endometrium when delivered by a simple technique, similar to that used for artificial insemination [68]. A continuation of this study showed histological improvement of endometrosis after infusing allogeneic AT-MSCs. The authors suggested that multiple mechanisms, including homing to fibrotic areas and increased epithelial cell proliferation, might mediate the anti-scarring effects observed [69]. Furthermore, amniotic membrane-derived MSCs may improve the endometrial cell replenishment when scarcity or low proliferation of endometrial cells is associated with pregnancy failure [70].

Compared to BM-MSCs, eMSCs have a higher ability to differentiate into smooth muscle [71], and display robust immunomodulatory [72] and migratory properties [73]. Moreover, eMSCs can be collected from simple endometrial biopsies, which can be safely and repeatedly obtained, thus suggesting a novel source of therapeutic cells for inflammatory conditions of the uterus [74].

3.2. Ovarian Diseases

Intra-ovarian injection of allogeneic BM-MSCs to treat ovarian dysfunction in old mares has been suggested in one study. Although there were no side effects, MSC injection was not associated with significant changes in follicle number, oocyte recovery and maturation rate, or blastocyst rate [75]. This contrasts with MSC beneficial effects observed in chemotherapy-damaged ovaries in other species [76,77], but these data should be extended to be conclusive, because to the best of the authors' knowledge, there are no further studies exploring this possibility.

3.3. Testicular Diseases

Testicular disorders are characterized by altered or suppressed activity leading to important reproductive issues in stallions and consequent economic impacts [78]. Immunomodulatory, anti-inflammatory, and anti-apoptotic effects of MSCs, as well as their ability to migrate to injured tissues, support their potential to treat testicular disorders, as suggested in an induced testicular torsion model in rats [79]. One in vivo study evaluated the intratesticular injection of allogeneic BM-MSCs in healthy stallions. The absence of clinical abnormalities and altered semen parameters suggested that this is a safe procedure [80], although further studies are required to test their therapeutic potential.

Table 2. In vivo studies using mesenchymal stem cells in equine reproduction. Endometrial mesenchymal stromal cells (eMSCs), phosphate-buffered saline (PBS), polymorphonuclear neutrophils (PMNs), bone marrow mesenchymal stem cells (BM-MSCs), anti-Müllerian hormone (AMH), autologous conditioned serum (ACS), lactate ringer (LR), intrauterine (IU), post-breeding induced endometritis (PBIE), adipose tissue MSCs (AT-MSCs), artificial insemination (AI), fetal bovine serum (FBS), dimethyl sulfoxide (DMSO), interleukin 6 (IL-6), tumor necrosis factor alpha (TNFα), intracytoplasmic sperm injection (ICSI).

Pathology	Study	Type of MSCs	Experimental Design	Administration Regime	Outcome	Considerations
Safety and distribution assays in healthy mares	Rink et al., 2018 [61]	Autologous eMSCs	Intra-uterine administration of labelled eMSCs and follow-up: • Uterine cytology • Tracking of eMSCs in peripheral blood • Tracking of eMSCs in uterus (lumen and endometrial biopsies) 7 healthy cycling mares, 4–7 years old.Four of these mares were used as controls, two estrous cycles before the eMSC infusion cycle.	15×10^6 eMSCs in 1 mL or PBS infused into each uterine horn during early diestrus (day 4 after ovulation).	Mild inflammatory reaction after infusion was attenuated by eMSCs (percentage of PMNs lower in eMSC than PBS-infused mares at 6h) eMSCs detected in the uterine horn lumen for up to 24 h after infusion but did not migrate into healthy endometrium. eMSCs were not found in the peripheral blood at 6, 12, and 24 h after application.	eMSCs intra-uterine administration is safe and cells persist in the uterine lumen for up to 24 h after infusion, but do not engraft into healthy endometrium at that time.
PBIE	Ferris et al., 2014 [62]	Allogeneic BM-MSCs	Evaluate the ability of ACS, BM-MSCs or dexamethasone to modulate the inflammatory response to spermatozoa after breeding (24 h) 12 healthy mares Experiment 1: Crossover study Experiment 2: Two-way crossover study Administration of treatments, sperm challenge and follow-up: • Ultrasonographic evaluations • Lavages: PMN and cytokine analysis	Experiment 1: 6 mares treated with an IU infusion of: • 20 mL of ACS in 20 mL PBS • 20 mg of dexamethasone QS in 20 mL PBS • Control: 20 mL PBS One treatment per estrous in the same mares (every 3 weeks). Experiment 2: 6 mares treated with an IU infusion of: • 20×10^6 allogeneic BM-MSCs in 20 mL LR • 20 mL PBS One treatment per estrous in the same mares (every 3 weeks).	BM-MSC and ACS were able to modulate the uterine inflammatory response to spermatozoa in normal mares Decreased neutrophil migration into the uterine lumen in response to insemination after BM-MSC treatment may be due to increased anti-inflammatory cytokine IL-1Ra and reduced proinflammatory mediator IL-1	Healthy mares (proof of concept for PBIE-affected mares). Same mares used for different treatments. Age of mares not stated.

Table 2. *Cont.*

Pathology	Study	Type of MSCs	Experimental Design	Administration Regime	Outcome	Considerations
PBIE	Navarrete et al., 2020 [63]	Allogeneic AT-MSCs and eMSCs	Evaluate anti-inflammatory and engraftment properties of AT-MSCs and eMSCs from the same donors in vivo in mares with induced PBIE. 9 healthy mares with induced PBIE. Follow-up: • Lavages: cytokine and gene expression analysis • Uterine biopsies	2×10^7 AT-MSCs ($n = 3$) or eMSCs ($n = 3$) in 20 mL of NaCl 0.9% Control group ($n = 3$): 20 mL NaCl 0.9%	Both MSC types significantly reduced inflammation and showed limited engraftment, detectable after one month of infusion Decrease in IL-6 and TNFα in both MSC-treated groups over control.	Age of mares not stated. Virgin mares with no previously reported PBIE (possibility of natural PBIE resistance)
Endometrosis	Alvarenga et al., 2016 [67]	Autologous BM-MSCs	Evaluate the feasibility and safety of MSC endometrial injections 16 mares (15–24 years) with reproductive history of subfertility and endometrial degeneration Control: baseline values before endometrial injection Follow-up: • Ultrasonographic evaluations • Uterine biopsies	12 endometrial injections of 1×10^6 MSCs in 0.5 mL PBS, each one at 12 different sites, 1 cm apart from one uterine horn to another (12×10^6 MSCs in total)	Neither clinical alteration nor intrauterine fluid and endometrial edema were observed after MSC administration. No histological worsening The results suggest that the procedure is safe	Proof of concept for safety. Therapeutic effects not thoroughly assessed.
	Mambelli et al., 2013 [68] Mambelli et al., 2014 [69]	Allogeneic AT-MSCs	Evaluate the feasibility of an MSC delivery system for endometrosis-affected mares. 6 mares (6–21 years) with varying degrees of naturally occuring endometrosis Follow-up: • Uterine biopsies 7, 21 and 60 days	20×10^6 AT-MSCs in 20 mL NaCl 0.9% inoculated into uterus using a technique similar to AI Control ($n = 2$): 20 mL NaCl 0.9%	Migration of AT-MSCs to the uterine body and both horns. Engraftment in both glandular and periglandular spaces in three mares. AT-MSCs beneficially modulated the expression pattern of secretory proteins and promoted proliferation of glandular epithelial cells.	Small control group Different degrees of endometrosis among recruited mares

Table 2. *Cont.*

Pathology	Study	Type of MSCs	Experimental Design	Administration Regime	Outcome	Considerations
Ovarian and testicular diseases	Grady et al., 2019 [75]	Allogeneic BM-MSCs	Determine if intra-ovarian injection of BM-MSCs improves or restores ovarian function in aged mares 8 aged mares (20–29 years old) and 6 young mares (7–12 years old) Assessment (aged and young mares): • Antral follicle count • Serum AMH • Assessment (aged mares only): • Oocyte meiotic and developmental competence • Gross and histological ovarian assessment • Gene expression in ovarian tissue as assessed by RNA sequencing.	2 intraovarian injections of 10 × 10^6 BM-MSCs (different donors) in 1mL (95%FBS and 5%DMSO) in four different locations per ovary Aged mares: First injection • Group 1 ($n = 3$): BM-MSCs from donor A • Group 2 ($n = 3$): BM-MSCs from donor B • Group 3: Control ($n = 2$): 95%FBS and 5%DMSO Second injection: at 6 weeks from different donor Young mares: • Group 1 ($n = 2$): BM-MSCs from donor A • Group 2 ($n = 2$): BM-MSCs from donor B • Group 3 ($n = 2$): Control (95% FBS and 5% DMSO)	No adverse events after intra-ovarian injections were observed Oocyte recovery on follicle aspiration, oocyte maturation, and blastocyst development rates after ICSI remained unchanged in the MSC-treated aged mares, suggesting no detriment of ovarian function but neither a beneficial effect. Injection of BM-MSCs not associated with significant changes in follicle number in young mares. No significant changes in peripheral AMH concentrations in aged and young mares observed, indicating a lack of effect on growing follicles.	BM-MSCs administered immediately after thawing (viability not stated) Small size of groups

Table 2. *Cont.*

Pathology	Study	Type of MSCs	Experimental Design	Administration Regime	Outcome	Considerations
Ovarian and testicular diseases	Papa et al., 2020 [80]	Allogeneic BM-MSCs	Evaluate the effect of intratesticular injection of BM-MSCs in healthy stallions, and its outcome on seminal parameters and fertility Experiment 1: 24 stallions (3–4 years) Experiment 2: 3 stallions (3–10 years) Assessment: Experiment 1: • Testicular morphometry, thermography • Testosterone concentrations • Ultrasonography • Histopathology (after orchiectomy) Experiment 2: • Sperm collection: concentration, kinetics, plasma membrane integrity • Insemination of six healthy mares (8 ± 3.5 years old)	Experiment 1: • Intratesticular injection 10 × 10^6 BM-MSCs in PBS in 5 different points (1 mL in each point) (n = 12) • Same injections with 5 mL PBS (n = 12) Experiment 2 (n = 3): • Intratesticular injection 20 × 10^6 BM-MSCs in PBS in 5 different points (1 mL in each point)	Experiment 1: No signs of inflammation. No differences on testicular volume, parenchyma echogenicity, testicular blood flow and serum testosterone levels between treated and control at 24 h and 28 h. Experiment 2: no physical alterations or changes in sperm parameters. Satisfactory fertility rate (83%; 5/6) after AI. The results suggest that MSC intra-testicular administration would be safe and would not affect testicular function.	No sperm parameters evaluated after BM-MSCs injection in experiment 1 Absence of control group in experiment 2 Small size of group in experiment 2

4. Application of Equine MSCs in Metabolic Disorders

Equine Metabolic Syndrome

Equine metabolic syndrome (EMS) is an endocrine disorder characterized by pathological obesity, insulin dysregulation, altered hepatic function, and predisposition to developing laminitis. Obesity is the main risk factor for EMS, because adipose tissue acts as an important secretory organ producing different molecules such as pro-inflammatory cytokines with associated adverse local and systemic effects [81]. Management of EMS should include a well-balanced diet and physical activity. Some drugs such as metformin can help regulate this disorder, although there are no treatments available that can effectively resolve the EMS [82].

Therefore, MSCs have been proposed as a therapeutic strategy in different metabolic syndromes in several species, such as rodents [83–85], dogs [86] and monkeys [87]. Importantly, great attention has been paid to MSC treatment of human diabetes type 2 [88,89]. Thus, the One Medicine concept applies in both directions; knowledge from human studies is also highly valuable for veterinary medicine.

Adipose-derived MSCs from EMS-affected horses display senescent phenotype, increased apoptosis, and reduced viability and differentiation capacity, which would make autologous therapy suboptimal [90–92]. In a recent study of the same group, autologous AT-MSCs were exposed in vitro to pharmacotherapy with 5-azacytidine (AZA) and resveratrol (RES) before their clinical application in order to reverse the aged phenotype of these cells. These rejuvenated autologous AT-MSCs showed in vivo potential to improve liver metabolism in one EMS-diagnosed horse, as demonstrated by a decrease in specific liver enzymes. Nonetheless, MSC therapy was combined with conventional EMS management, which could have also contributed to this improvement [93].

5. Application of Equine MSCs in the Respiratory System

Equine Asthma

Equine asthma syndrome encompasses a spectrum of inflammatory airway diseases characterized by chronic respiratory signs ranging in severity that can significantly affect athletic performance [6]. Although its exact etiopathogenesis remains incompletely defined, immune-mediated responses are involved and lead to excessive mucus production, neutrophilic accumulation, bronchial hyperreactivity, and bronchospasm [94]. Treatment is limited to environmental management, anti-inflammatory drugs, and bronchodilator therapy. Nevertheless, the use of inhaled corticosteroids may be contraindicated in some cases, and may also be cost-prohibitive and inconvenient for some owners [6].

Studies in murine models of induced asthma have shown that MSCs may be beneficial for managing this process [32,95,96]. To date, there are no published reports on the use of MSCs in vivo to treat equine asthma. However, the use of cell derivatives has been tested. An in vitro assay showed the ability of conditioned medium and microvesicles from equine amniotic mesenchymal cells to modulate lipopolysaccharide (LPS)-stimulated equine alveolar macrophages, suggesting that these products can play a role in treating inflammatory diseases of the lung [97]. In vivo, the intra-tracheal instillation of autologous bone marrow-derived mononuclear cells in eight asthma-affected horses showed that the procedure is safe and improved clinical signs, suggesting amelioration of the asthmatic inflammatory response [98].

Moreover, equine and human asthma syndromes share several features, with the difference that the process is dominated by neutrophils in horses and by eosinophils in humans [6]. The similarities in the airway remodeling processes make equine asthma an ideal model to study the cellular and molecular pathways associated with the asthmatic airway response and its reversibility [10].

6. Application of Equine MSCs in Disorders of the Integumentary System

The main findings of in vivo studies using MSCs for integumentary-related disorders are highlighted in Table 3.

6.1. Wounds

Due to their "flight" instinct in response to frights, horses are particularly susceptible to trauma [99]. Actually, wound injuries are the most common medical condition affecting horses [100,101]. Traumatic wounds commonly occur in the distal limb, where healing is often delayed due to high skin tension, minimal vascularization, and a low-grade inflammatory response. This aberrant inflammation leads to a fibroproliferative response, resulting in a dysplastic healing with exuberant granulation tissue [102]. These particularities make successful wound management difficult, to which MSCs could contribute owing to their varied paracrine activities.

In vitro, equine MSCs promote dermal wound repair through the mobilization of dermal fibroblasts and by increasing the expression of genes involved in wound healing [103]. In an equine distal limb wound model using six horses, allogeneic umbilical cord blood-derived MSCs (UCB-MSCs) were applied after culture in normoxic or hypoxic conditions, either directly injected into wound margins or topically applied embedded in an autologous fibrin gel. Early healing was enhanced, with histology suggesting a pro-healing rather than a pro-inflammatory scenario and smaller sizes in MSC-treated wounds, with no additional advantage of hypoxic preconditioning [104]. Systemic administration of allogeneic UCB-MSCs resulted in engraftment into induced wounds on the forelimb and thorax at early stages, without clinically adverse reactions [105]. Autologous PB-SCs injected both locally and systemically in four naturally occurring chronic dermal wounds in the metatarsus, unresponsive to conventional treatments, showed positive outcomes with granulation tissue formed within four weeks, and no adverse effects were noted [106]. These results suggest MSCs as a promising tool to improve and accelerate the healing of chronic wounds.

Equine distal limb wounds display similarities to human wounds, with keloid formation in the latter being similar to exuberant granulation tissue formation in horses. Humans and equines are the only two species that spontaneously develop these fibroproliferative disorders, making the horse a suitable model for studying the pathogenesis and treatment of keloids and hypertrophic scars [99]. In both species, chronic wound management is a growing problem leading to a significant economic impact; thus, there are demands for advanced therapeutic options able to decrease healing time and minimize complications.

6.2. Decubitus Ulcers

Decubitus ulcers result from pressure, shear, and/or friction when the patient is lifted or put in decubitus, leading to tissue ischemia, cell death and necrosis [107]. Thus, the pathogenesis of this injury differs from that in traumatic wounds. Neonatal foals are prone to decubitus ulcers due to their thin skin and possible concomitant disorders, which force them into prolonged decubitus [108]. Mesenchymal stem cells play a key role in skin homeostasis and repair by promoting cell differentiation, immunomodulation, and the secretion of growth factors to drive re-epithelialization and neovascularization [109].

In a septic neonatal foal, local implantation of amniotic fluid-derived MSCs (AF-MSCs) along with platelet-rich plasma (PRP) gel effectively improved decubital ulcers earlier than in those ulcers treated with *Aloe vera* gel only [110]. In another study, several deep sore wounds, presented concomitantly in five septic foals, were treated by either allogeneic AF-MSCs embedded in a carboxymethylcellulose scaffold or by a commercial preparation of formosulfathiazole, the former leading to faster healing [111]. Additionally, the effectiveness of local application of allogeneic Wharton's jelly MSCs (WJ-MSCS) was reported in a six-month-old filly with a large non-healing skin wound in the hock [112]. Therefore, MSCs may be considered for healing skin ulcers in foals, with no side effects noticed, although data are limited to a few reports including low numbers of animals.

6.3. Laminitis

Laminitis is included as part of the integumentary system provided the distal third phalanx is enveloped by a specialized epithelial tissue, the corium, attaching the bone to the hoof. However, it should be noted that this is a complex and multifactorial disease [113]. This disorder affecting the hooves of ungulates is of particular importance in equids, carrying a poor prognosis, with a strong economic impact and constituting a serious issue for animal welfare that may eventually require euthanasia [114]. A great variety of initiating causes can lead to the onset of the disease and can be influenced by different body systems, including inflammation, metabolic disorders, and endothelial or vascular dysfunction [113]. Whatever the triggering factors are, the release of pro-inflammatory mediators and the activation of metalloproteinases leads to the degradation of the basement membrane that may result in a complete mechanical collapse of the foot [115].

The largely incomplete knowledge of the pathogenic mechanisms involved in laminitis makes both prevention and treatment difficult [116]. Laminitis has been associated with the loss of p63+ epidermal stem cells in the hoof lamellae, suggesting limited proliferative potential of the laminitic hoof epithelium [117]. Therapy with MSCs is a promising tool to improve cell proliferation and tissue quality, as well as contribute to vascular stabilization and control the pro-inflammatory environment.

In clinical studies on chronic refractory laminitis, allogeneic UCB-MSCs were delivered directly to the affected foot via regional perfusion (digital vein). The outcome suggested a positive MSC effect on the prognosis of animals treated early, as reflected by the evolution with decreased radiologic distance between the bone and hoof wall [114,118]. Allogeneic BM-MSCs provided better results when digitally perfused compared to epidermally injected [119]. Another study on chronic laminitis administered allogeneic and autologous AT-MSCs suspended in autologous PRP by distal digital venous injection. Improvement of both hoof quality and animal mobility was reported, improving the life quality of treated horses [120].

These results encourage further exploring the use of MSCs to treat chronic laminitis, but it is important to highlight that MSC administration does not eliminate the need for routine management and appropriate hoof support.

Table 3. In vivo studies using mesenchymal stem cells for integumentary system disorders. Umbilical cord-derived stem cells (UCB-MSCs), transforming growth factor beta (TGF-β), cyclooxygenase-2 (COX-2), intravenous (IV), HypoThermosol FRS (HTS-FRS), peripheral blood stem cells (PB-SCs), phosphate-buffered saline (PBS), amniotic fluid MSCs (AF-MSCs), platelet-rich plasma (PRP), carboxymethylcellulose (CMC), Wharton's jelly MSCs (WJ-MSCs), bone marrow mesenchymal stem cells (BM-MSCs), adipose tissue MSCs (AT-MSCs).

Pathology	Study	Type MSCs	Experimental Design	Administration Regime	Outcome	Considerations
Traumatic wounds	Textor et al., 2017 [104]	Allogeneic UCB-MSCs	Six horses (3 mares, 3 geldings; 5–19 years) Induced model: 3 full-thickness cutaneous wounds surgically created on each distal forelimb Assessment and follow-up (6 weeks): • Wound surface area • Thermography • Gene expression • Histologic scoring	10–20 $\times$ 10^6 normoxic UCB-MSCs, hypoxic UCB-MSCs or control in 1 mL NaCl 0.9% were injected into wound margins or topically applied embedded in an autologous fibrin gel, 1 day after wound creation	MSC administration by either delivery method was safe and improved histologic outcomes and reduced wound area over control MSC-injected wounds were consistently smaller than gel-treated or control wounds. Hypoxic pre-conditioned MSCs did not provide a substantial advantage	Treatments applied very early and in aseptically created wounds (proof of concept for application in clinical situation)
	Mund et al., 2020 [105]	Allogeneic UCB-MSCs	Two 7-year-old mares Determine adverse reactions after IV MSC administration and assess their engraftment potential into wounds Induced model: standardized cutaneous wounds surgically created on the left lateral third metacarpus (7 wounds) and left hemi-thorax (7 wounds)	1.02 $\times$ 10^8 UCB-MSCs (fluorescently labelled) in 60 mL HTS-FRS via IV administration Control: 2 wounds were left to heal by second intention (in the same horses)	No clinically adverse effects (largest recorded dose of IV UCB-MSCs) UCB-MSCs preferentially engrafted into wounds during the acute and early remodeling phases.Results suggest no difference in homing potential between limb and thoracic wounds	Low number of animals Biopsy collection at the control sites created inflammation that may have influenced homing to the sequential control site (untreated control not included)

Table 3. *Cont.*

Pathology	Study	Type MSCs	Experimental Design	Administration Regime	Outcome	Considerations
Traumatic wounds	Spaas et al., 2013 [106]	Autologous PB-SCs	4 horses with naturally occurring traumatic wounds unresponsive to conventional therapies for at least 3 months: • Case 1 (11-year-old mare): dorsal surface of the metatarsal bone infected with *Clostridium* spp. • Case 2 (16-year-old gelding): plantar surface of the metatarsal bone infected (presence of pus) • Case 3 (26-year-old gelding): deep wound with bone exposition at the medial surface of the tibia • Case 4 (26-year-old gelding): wound presenting non-neoplastic exuberant granulation tissue on the plantar surface of the metatarsal bone. Nodular proliferative lesions recurred after resection and tended to ulcerate.	5×10^5 PB-SCs in 2 mL PBS were locally (intradermally) injected into 5–6 different locations at the wound's edges and 1.25×10^5 PB-SCs via IV administration	Granulation tissue began forming within 4 weeks of the PB-SC therapy in cases 1, 2 and 3. Crust formation was achieved within 2 months. In case 4, the granulation tissue could be easily removed without recurrence of the wound. 1 year: no wound recurrence or other adverse effects	Low number of cases of varying presentation with no control group, but unresponsive to previous treatments. Cells used not expanded and not fully characterized as MSCs. In cases with bacterial infection antibiotic, administration was continued.Outcome of each individual case is not deeply explained

Table 3. *Cont.*

Pathology	Study	Type MSCs	Experimental Design	Administration Regime	Outcome	Considerations
Decubitus ulcers	Iacono et al., 2012 [110]	Allogeneic AF-MSCs and allogeneic PRP gel	One septic neonatal foal with severe ulcers in fetlocks, carpus and right stifle	• Aloe gel every 48 h: Left hock and carpus • 10 mL PRP gel: right carpus • 5×10^6 AF-MSCs + 10 mL PRP gel (twice a week for 2 weeks): right hock	None of the wounds treated developed exuberant scar tissue Healing was faster using AF-MSCs + PRP. The ulcer treated this way resulted in a linear scar, while the other lesions produced star scars 7 months: ulcer treated with aloe gel was not fully healed	One single case Different wound locations Concomitant disease Oral antibiotic therapy ongoing during ulcers healing
	Iacono et al., 2016 [111]	Allogeneic AF-MSCs	5 hospitalized neonatal foals (10–15 days old) with a total of 9 pressure sores on the carpus (4), fetlock (2), and hock (3). Sores were divided into group 1 ($n = 6$) and group 2 ($n = 3$)	• Group 1: 5×10^5 AF-MSCs in CMC gel applied twice a week for 2 weeks • Group 2: formosulfathiazole ointment every 48 h	Sores treated with AF-MSCs in CMC gel healed quicker 30 days: no further treatments were needed	Low number of cases with no substance vehicle (CMC gel) treated control. Variable presentation of treated sores
	Lanci et al., 2019 [112]	Allogeneic WJ-MSCs	One 6-month-old filly hospitalized by the re-injury of a pressure wound on the left hock	5×10^6 WJ-MSCs in a CMC gel were applied every 4 days for 4 times Four days after the last application, no further bandages were applied and the wound was daily cleaned and treated with hydrotherapy (cold tap water 10 min/day)	No side effects and fast wound regression No evident exuberant scar. The hair grew completely without changing color80% regression rate between 8 days and 39 days No relapse	One single case with no control

Table 3. *Cont.*

Pathology	Study	Type MSCs	Experimental Design	Administration Regime	Outcome	Considerations
Laminitis	Morrison, 2011 [118]	Allogeneic UCB-MSCs	12 horses with naturally occurring chronic laminitis unresponsive to other treatments	20–25 × 10^6 UCB-MSCs in NaCl 0.9% infused by regional perfusion (digital vein) every 3–4 weeks (3 infusions in total per affected foot)	83% of horses with positive evolution by the time of publication.	Routine treatments for laminitis continued Long-term success rates still need to be determined Only clinical follow-up Absence of control group No age or breed information stated
	Dryden et al., 2013 [114]	Allogeneic UCB-MSCs and autologous BM-MSCs	30 horses with naturally occurring chronic laminitis	20–30 × 10^6 allogeneic UCB-MSCs in NaCl 0.9% infused by regional perfusion (digital vein) and subsequent injections with either 20–30 × 10^6 autologous BM-MSCs or allogeneic UCB-MSCs (4 infusions in total at 1 month intervals)	21 patients (70%): successful outcome. Decreased radiologic distance between the bone and hoof wall Cases receiving MSCs < 30 days after the onset of laminitis: success rate was 100%. Cases receiving MSCs > 90 days after the onset of laminitis: success rate was 50%. Results suggest improved prognosis in cases treated early.	Routine treatments for laminitis continued Absence of control group Only clinical and radiologic follow-up Variation in therapeutic regime among cases. No age or breed information stated
	Angelone et al., 2017 [120]	Allogeneic AT-MSCs and autologous AT-MSCs	9 horses (5 mares, 4 geldings; 10–21 years) with severe naturally occurring laminitis unresponsive to conventional therapies	15 × 10^6 allogeneic AT-MSCs in 15 mL autologous PRP infused by regional perfusion (digital vein) and subsequent injections with autologous AT-MSCs (3 infusions in total at 1 month intervals)	Clinical and radiologic signs improved All the animals returned to activity at six months from the first treatment. 12 months: 7 horses still performing activity 24 months: • 2 animals relapsed and were euthanized • 3 horses deceased (unrelated to laminitis)	Absence of control group Horses enrolled presented different laminitis stages. Only clinical and radiologic follow-up

7. Application of Equine MSCs in Neurological Disorders

Neurological disorders affecting the brain and spinal cord can represent a therapeutic challenge, and many horses can have sequelae even after recommended treatment [121]. The neuroprotective effects of MSCs have been described in other species and involve anti-inflammatory, immunomodulatory, pro-angiogenic and trophic mechanisms [122,123], which could ameliorate the symptoms of several neurodegenerative disorders [124]. Furthermore, MSCs can trans-differentiate in vitro into neuronal lineages [125]. In vivo related studies in horses are summarized in Table 4.

7.1. Peripheral Nerve Injury

Horses suffer injury to peripheral nerves from trauma, metabolic and genetic disorders, toxins or degenerative and infectious diseases [126]. The degree of restoration of nerve function depends on the severity and chronicity of the damage, with the worst prognosis in cases where the nerve is transected [127]. There are few surgical techniques for repairing nerves, and clinical results are often poor [128]. Consequences frequently include poor performance, disability, or even death [129,130].

In an in vivo model of acute peripheral nerve injury using three horses, allogeneic BM-MSCs were implanted into the fascia surrounding the *ramus communicans* of one forelimb after a portion was transected. No evidence of nerve regeneration was observed, neither were histological differences between MSC-treated and control injuries found 45 days later [127]. Nevertheless, improvement of nerve regeneration after MSC treatment has been observed in other large animals such as sheep [131].

7.2. Wobblers Syndrome

Wobblers syndrome, also known as cervical vertebral stenotic myelopathy (CVSM) or incoordination syndrome, is characterized by ataxia and weakness, caused by the narrowing of the cervical vertebral canal and/or compression of the spinal cord. Medical treatment commonly includes steroidal and nonsteroidal anti-inflammatory drugs, along with diet and exercise restrictions. Surgery can be considered in some cases, but this option is expensive and involves significant risks [132].

An intrathecal injection could extensively deliver cells through the cerebrospinal fluid to reach the equine central nervous system. To evaluate the feasibility and safety of intrathecal transplantation of cells, autologous BM-MSCs were administered to healthy horses and no clinical alterations were developed [133]. A posterior study aimed at determining the safety of a relatively high dose of intrathecal allogeneic AT-MSCs in both healthy and CVSM-affected horses. Neurological status was not altered regardless of atlanto-occipital or lumbosacral administration. Atlanto-occipital injection is apparently distributed more efficiently through the subarachnoid space, suggesting that this approach might be more suitable for cranial spinal cord lesions. As for diseased horses, MSCs could not be found at 15 days after injection at the site of injury, so either cells did not reach the lesion site or did not persist at that time [134].

7.3. Laryngeal Hemiplegia or Left Recurrent Laryngeal Neuropathy

Recurrent laryngeal neuropathy is characterized by varying degrees of arytenoid paralysis, constituting a highly prevalent pathology of the upper airway in horses [135]. Affected horses emit abnormal respiratory sounds and may present exercise intolerance in severe cases. Although the overall success rate of laryngoplasty with or without ventriculo-cordectomy is elevated, post-operative complications such as a gradual loss of abduction are very common [136].

Table 4. In vivo studies using mesenchymal stem cells in equine neurological disorders. Bone marrow mesenchymal stem cells (BM-MSCs), adipose tissue MSCs (AT-MSCs), cervical vertebral compressive myelopathy (CVCM), 99m technetium-hexamethyl-propylene-amine-oxyme (99mTc-HMPAO), atlanto-occipital (AO), lumbosacral (LS), recurrent laryngeal neuropathy (RLN), muscle-derived MSCs (M-MSCs).

Pathology	Study	Type MSCs	Experimental Design	Administration Regime	Outcome	Considerations
Peripheral nerve injury	Villagrán et al., 2016 [127]	Allogeneic BM-MSCs	Induced model: 3 healthy mares (9–13 years old) with surgically created 15-mm longitudinal incision over the *ramus communicans*	10×10^6 BM-MSCs in 1 mL NaCl 0.9% or 1 mL NaCl 0.9% (control) instilled into the fascia surrounding the medial and lateral stumps Stumps of *ramus communicans* of each fore limb were harvested 45 days after treatment or control administration	No evidence of nerve regeneration No histological differences between MSC-treated and control nerve stumps No histological evidence of BM-MSCs or primitive cells (e.g., neural or Schwann-cell progenitors)	Small size Immediate treatment of aseptically created injury (proof of concept) Poor vasculature of the anatomical region may have influenced the outcome
CVCM	Barberini et al., 2018 [134]	Allogeneic AT-MSCs	Distribution and safety assessment in 6 healthy mares (6–21 years) and 3 diseased horses presenting moderate to severe neurological signs (presumedly CVCM) Assessment: • Scintigraphy • Detection of anti-AT-MSC alloantibody • Necropsy (diseased horses)	Healthy horses: 100×10^6 AT-MSCs in 5 mL NaCl 0.9% were injected either AO (n = 3) or LS (n = 3). One horse in each group received 99 mTc-HMPAO-labeled AT-MSCs. One additional horse was injected with free label as a control. Diseased horses: 100×10^6 AT-MSCs via AO	Healthy horses: AO and LS intrathecal injection of relatively high doses of AT-MSCs was well tolerated AT-MSCs apparently distributed more efficiently through the subarachnoid space after AO injection (suggested as preferred route to deliver MSCs to the cervical area) Diseased horses: AT-MSCs not found at 15 days after injection at the site of injury (either did not have time to reach the lesion site or did not survive long enough) No horses developed detectable anti-AT-MSC alloantibodies.	Low number of horses per group. Tracking and control only in one horse
RLN	Sandersen et al., 2018 [137]	AutologousM-MSCs	5 healthy mares (ages 10–22) Evaluate the feasibility and safety of administering MSCs by a peri-neural injection to the left recurrent laryngeal nerve in healthy horses by using an electrical nerve stimulator	10×10^6 M-MSCs in 1 mL cryopreservation medium directly administered into recurrent laryngeal nerve by a nerve stimulator-guided injection	Feasibility and safety of the procedure suggested by absence of functional changes upon endoscopic evaluation up to 28 days. No signs of adverse events in four out of the five mares up to 1 year after the injection	No control group M-MSCs administered immediately after thawing Composition of cryopreservation (administration) media not mentioned

To explore the feasibility of using MSCs to promote nerve function restoration, a nerve-stimulator-guided injection of muscle-derived autologous MSCs near the left recurrent laryngeal nerve was performed in five healthy horses. Laryngeal function was not affected, thus suggesting that this delivery technique is safe [137]. These findings would facilitate future studies assessing MSC effectiveness to treat this pathology, which has not yet been tested, to the best of the authors' knowledge.

8. Application of Equine MSCs in Endotoxemia

Systemic inflammatory response syndrome (SIRS) is characterized by an exaggerated inflammatory response to an aggression, which causes a series of unspecific clinical signs that can seriously compromise the patient's life. This process is of great importance in horses due to the particularities of their inflammatory response and the high incidence of pathologies such as acute abdominal syndrome, pneumonia, and metritis, which can provoke an endotoxemia or sepsis and subsequent SIRS development. These processes can lead to complications such as disseminated intravascular coagulation, vascular endothelial damage, laminitis, and multiple organ dysfunction syndrome [138]. Therefore, SIRS is associated with a significantly higher risk of death in horses presenting acute colic [139] or other disorders. There are no specific treatments for endotoxemia and SIRS other than controlling the primary cause and providing supportive therapy, mostly based on fluids and antiendotoxics [138]. The complexity and poor prognoses of these processes and the lack of effective treatments have led to the interest in using MSCs because of their immunomodulatory capacity. Actually, there are a number of studies in rodent models of endotoxemia showing promising results of the administration of MSCs, including decreased levels of circulating proinflammatory cytokines and increased survival rates [140].

In horses, and to the best of the authors' knowledge, there is only one study reporting the effects of administering MSCs in an experimental model of endotoxemia [141]. In this study, lipopolysaccharide was IV injected into six horses, and three of them immediately received 100×10^6 allogeneic BM-MSCs IV. Serial clinicopathological assessment and determination of pro-inflammatory cytokine production did not show significant differences between treated and control animals, but adverse reactions after MSC infusion were not observed either, thus suggesting that the procedure is safe; however, further studies with higher numbers of animals and studying variables such as the dosage and moment of administration are needed to elucidate the potential benefit of this therapy. Importantly, endotoxemia, sepsis, and SIRS also affect a high percentage of human patients in intensive care units, and the horse has been proposed as a model for understanding human innate immunity and shedding light onto pathology processes and therapeutics [142].

9. Conclusions and Future Perspectives

The unique properties of MSCs hold a great potential for the treatment of different equine diseases other than musculoskeletal injuries, for which current pharmacologic or surgical approaches often do not provide satisfactory results.

Several studies have explored the use of MSCs for different equine diseases beyond the locomotor system. Their application for ophthalmologic and reproductive disorders has been particularly investigated, and the treatment of wounds, asthma, and SIRS holds a special interest to develop One Medicine approaches. However, the variability among clinical case conditions, source of the MSCs used, cell isolation and culture techniques, and therapeutic protocols (MSC dose, route, number, and frequency of administrations), as well as the low sample size and lack of control groups in some of the studies, limit extracting definitive conclusions. Nevertheless, case reports are highly valuable as a proof of concept, indicating that there is potential for in further investigations. In addition, in vitro studies provide interesting preliminary insight into MSC mechanisms for each pathology, and in vivo studies in healthy animals enable safety and feasibility assessments.

More in-depth research is needed to test the safety and efficacy of these novel treatments, and future clinical trials would include a larger number of similar cases and standardized measurements of the outcomes, in order to establish specific therapeutic protocols. Furthermore, more in vitro and experimental work is needed to understand the pathways through which MSCs elicit their effects, in order to achieve their highest therapeutic potential. Therefore, researchers and clinicians should work together to develop evidence-based treatments and exploit the MSC potential by extending their use to different pathologies in both equine and human patients.

Author Contributions: Conceptualization, L.B.; methodology, A.C. and C.S.; writing—original draft preparation, A.C and C.S.; writing—review and editing, L.B. and C.R.; supervision, L.B and C.R; funding acquisition, C.R. All authors have read and agreed to the published version of the manuscript.

Funding: This research was funded by Ministerio de Industria, Economía y Competitividad, Spain; grant number AGL2017-84411-P. Alina Cequier was supported by a PhD fellowship from the Gobierno de Aragón and co-funded by European Social Fund.

Institutional Review Board Statement: Not applicable.

Data Availability Statement: No new data were created or analyzed in this study. Data sharing is not applicable to this article.

Conflicts of Interest: The authors declare no conflict of interest.

References

1. Stewart, M.C.; Stewart, A.A. Mesenchymal stem cells: Characteristics, sources, and mechanisms of action. *Vet. Clin. N. Am. Equine Pract.* **2011**, *27*, 243–261. [CrossRef] [PubMed]
2. Mao, A.S.; Mooney, D.J. Regenerative medicine: Current therapies and future directions. *Proc. Natl. Acad. Sci. USA* **2015**, *112*, 14452–14459. [CrossRef] [PubMed]
3. Borjesson, D.L.; Peroni, J.F. The regenerative medicine laboratory: Facilitating stem cell therapy for equine disease. *Clin. Lab. Med.* **2011**, *31*, 109–123. [CrossRef] [PubMed]
4. Horseman, S.V.; Buller, H.; Mullan, S.; Whay, H.R. Current welfare problems facing horses in Great Britain as identified by equine stakeholders. *PLoS ONE* **2016**, *11*, e0160269. [CrossRef]
5. Theoret, C.L.; Olutoye, O.O.; Parnell, L.K.S.; Hicks, J. Equine exuberant granulation tissue and human keloids: A comparative histopathologic study. *Vet. Surg.* **2013**, *42*, 783–789. [CrossRef] [PubMed]
6. Couëtil, L.L.; Cardwell, J.M.; Gerber, V.; Lavoie, J.-P.; Léguillette, R.; Richard, E.A. Inflammatory airway disease of horses—Revised consensus statement. *J. Vet. Intern. Med.* **2016**, *30*, 503–515. [CrossRef]
7. Gershwin, L. Veterinary autoimmunity: Autoimmune diseases in domestic animals. *Ann. N. Y. Acad. Sci.* **2007**, *1109*, 109–116. [CrossRef]
8. Hurtig, M.B.; Buschmann, M.D.; Fortier, L.A.; Hoemann, C.D.; Hunziker, E.B.; Jurvelin, J.S.; Mainil-Varlet, P.; McIlwraith, C.W.; Sah, R.L.; Whiteside, R.A. Preclinical studies for cartilage repair: Recommendations from the International Cartilage Repair Society. *Cartilage* **2011**, *2*, 137–152. [CrossRef]
9. Al Naem, M.; Bourebaba, L.; Kucharczyk, K.; Röcken, M.; Marycz, K. Therapeutic mesenchymal stromal stem cells: Isolation, characterization and role in equine regenerative medicine and metabolic disorders. *Stem Cell Rev. Rep.* **2019**, *16*, 301–322. [CrossRef]
10. Bullone, M.; Lavoie, J.-P. Asthma 'of horses and men'—How can equine heaves help us better understand human asthma immunopathology and its functional consequences? *Mol. Immunol.* **2015**, *66*, 97–105. [CrossRef]
11. Smith, R.K.W.; Garvican, E.R.; Fortier, L.A. The current "state of play" of regenerative medicine in horses: What the horse can tell the human. *Regen. Med.* **2014**, *9*, 673–685. [CrossRef] [PubMed]
12. Colbath, A.C.; Frisbie, D.D.; Dow, S.W.; Kisiday, J.D.; McIlwraith, C.W.; Goodrich, L.R. Equine models for the investigation of mesenchymal stem cell therapies in orthopaedic disease. *Oper. Tech. Sports Med.* **2017**, *25*, 41–49. [CrossRef]
13. van der Weyden, L.; Brenn, T.; Patton, E.E.; Wood, G.A.; Adams, D.J. Spontaneously occurring melanoma in animals and their relevance to human melanoma. *J. Pathol.* **2020**, *252*, 4–21. [CrossRef]
14. Witkowski, L.; Cywinska, A.; Paschalis-Trela, K.; Crisman, M.; Kita, J. Multiple etiologies of equine recurrent uveitis—A natural model for human autoimmune uveitis: A brief review. *Comp. Immunol. Microbiol. Infect. Dis.* **2016**, *44*, 14–20. [CrossRef]
15. Horohov, D.W. The equine immune responses to infectious and allergic disease: A model for humans? *Mol. Immunol.* **2015**, *66*, 89–96. [CrossRef]
16. Carnevale, E.M. The mare model for follicular maturation and reproductive aging in the woman. *Theriogenology* **2008**, *69*, 23–30. [CrossRef]
17. Fureix, C.; Jego, P.; Henry, S.; Lansade, L.; Hausberger, M. Towards an ethological animal model of depression? A study on horses. *PLoS ONE* **2012**, *7*, e39280. [CrossRef] [PubMed]

18. Danek, M.; Danek, J.; Araszkiewicz, A. Large animal as a potential models of humans mental and behavioral disorders. *Psychiatr. Pol.* **2017**, *51*, 1009–1027. [CrossRef] [PubMed]
19. Fortier, L.A.; Goodrich, L.R.; Ribitsch, I.; Schnabel, L.V.; Shepard, D.O.; Watts, A.E.; Smith, R.K.W. One health in regenerative medicine: Report on the second Havemeyer symposium on regenerative medicine in horses. *Regen. Med.* **2020**, *15*, 1775–1787. [CrossRef] [PubMed]
20. Ribitsch, I.; Baptista, P.M.; Lange-Consiglio, A.; Melotti, L.; Patruno, M.; Jenner, F.; Schnabl-Feichter, E.; Dutton, L.C.; Connolly, D.J.; van Steenbeek, F.G.; et al. Large animal models in regenerative medicine and tissue engineering: To do or not to do. *Front. Bioeng. Biotechnol.* **2020**, *8*. [CrossRef] [PubMed]
21. Wang, S.; Zhao, R.C. A historical overview and concepts of mesenchymal stem cells. In *Essentials of Mesenchymal Stem Cell Biology and Its Clinical Translation*; Zhao, R.C., Ed.; Springer: Heidelberg, Germany, 2013; pp. 3–15. ISBN 9789400767164.
22. Dominici, M.; Le Blanc, K.; Mueller, I.; Slaper-Cortenbach, I.; Marini, F.C.; Krause, D.S.; Deans, R.J.; Keating, A.; Prockop, D.J.; Horwitz, E.M. Minimal criteria for defining multipotent mesenchymal stromal cells. The International Society for Cellular Therapy position statement. *Cytotherapy* **2006**, *8*, 315–317. [CrossRef]
23. Radcliffe, C.H.; Flaminio, M.J.B.F.; Fortier, L.A. Temporal analysis of equine bone marrow aspirate during establishment of putative mesenchymal progenitor cell populations. *Stem Cells Dev.* **2010**, *19*, 269–282. [CrossRef]
24. Ranera, B.; Lyahyai, J.; Romero, A.; Vázquez, F.J.; Remacha, A.R.; Bernal, M.L.; Zaragoza, P.; Rodellar, C.; Martín-Burriel, I. Immunophenotype and gene expression profiles of cell surface markers of mesenchymal stem cells derived from equine bone marrow and adipose tissue. *Vet. Immunol. Immunopathol.* **2011**, *144*, 147–154. [CrossRef] [PubMed]
25. da Silva Meirelles, L.; Fontes, A.M.; Covas, D.T.; Caplan, A.I. Mechanisms involved in the therapeutic properties of mesenchymal stem cells. *Cytokine Growth Factor Rev.* **2009**, *20*, 419–427. [CrossRef] [PubMed]
26. Gugjoo, M.B.; Amarpal; Makhdoomi, D.M.; Sharma, G.T. Equine mesenchymal stem cells: Properties, sources, characterization, and potential therapeutic applications. *J. Equine Vet. Sci.* **2019**, *72*, 16–27. [CrossRef]
27. Fortier, L.A.; Travis, A.J. Stem cells in veterinary medicine. *Stem Cell Res. Ther.* **2011**, *2*, 1–9. [CrossRef] [PubMed]
28. Gnecchi, M.; Danieli, P.; Malpasso, G.; Ciuffreda, M.C. Paracrine mechanisms of mesenchymal stem cells in tissue repair. *Methods Mol. Biol.* **2016**, *1416*, 123–146. [CrossRef] [PubMed]
29. Martinet, L.; Fleury-Cappellesso, S.; Gadelorge, M.; Dietrich, G.; Bourin, P.; Fournie, J.J.; Poupot, R. A regulatory cross-talk between Vγ9Vδ2 T lymphocytes and mesenchymal stem cells. *Eur. J. Immunol.* **2009**, *39*, 752–762. [CrossRef]
30. Rasmusson, I.; Uhlin, M.; Le Blanc, K.; Levitsky, V. Mesenchymal stem cells fail to trigger effector functions of cytotoxic T lymphocytes. *J. Leukoc. Biol.* **2007**, *82*, 887–893. [CrossRef]
31. Sotiropoulou, P.A.; Perez, S.A.; Gritzapis, A.D.; Baxevanis, C.N.; Papamichail, M. Interactions between human mesenchymal stem cells and natural killer cells. *Stem Cells* **2006**, *24*, 74–85. [CrossRef]
32. Nemeth, K.; Keane-Myers, A.; Brown, J.M.; Metcalfe, D.D.; Gorham, J.D.; Bundoc, V.G.; Hodges, M.G.; Jelinek, I.; Madala, S.; Karpati, S.; et al. Bone marrow stromal cells use TGF-β to suppress allergic responses in a mouse model of ragweed-induced asthma. *Proc. Natl. Acad. Sci. USA* **2010**, *107*, 5652–5657. [CrossRef] [PubMed]
33. Asari, S.; Itakura, S.; Ferreri, K.; Liu, C.; Kuroda, Y.; Kandeel, F.; Mullen, Y. Mesenchymal stem cells suppress B-cell terminal differentiation. *Exp. Hematol.* **2009**, *37*, 604–615. [CrossRef] [PubMed]
34. Jiang, T.S.; Cai, L.; Ji, W.Y.; Hui, Y.N.; Wang, Y.S.; Hu, D.; Zhu, J. Reconstruction of the corneal epithelium with induced marrow mesenchymal stem cells in rats. *Mol. Vis.* **2010**, *16*, 1304–1316. [PubMed]
35. Gu, S.; Xing, C.; Han, J.; Mark, O.M.T.; Hong, J. Differentiation of rabbit bone marrow mesenchymalstem cells into corneal epithelial cells in vivo and ex vivo. *Mol. Vis.* **2009**, *15*, 99–107. [PubMed]
36. Cejka, C.; Holan, V.; Trosan, P.; Zajicova, A.; Javorkova, E.; Cejkova, J. The favorable effect of mesenchymal stem cell treatment on the antioxidant protective mechanism in the corneal epithelium and renewal of corneal optical properties changed after Alkali Burns. *Oxid. Med. Cell. Longev.* **2016**, *2016*, 1–12. [CrossRef] [PubMed]
37. Cejkova, J.; Trosan, P.; Cejka, C.; Lencova, A.; Zajicova, A.; Javorkova, E.; Kubinova, S.; Sykova, E.; Holan, V. Suppression of alkali-induced oxidative injury in the cornea by mesenchymal stem cells growing on nanofiber scaffolds and transferred onto the damaged corneal surface. *Exp. Eye Res.* **2013**, *116*, 312–323. [CrossRef]
38. Clode, A.B. Diseases and surgery of the cornea. In *Equine Ophthalmology*; Gilger, B.C., Ed.; John Wiley & Sons, Inc.: Hoboken, NJ, USA, 2011; pp. 181–266. ISBN 9781437708462.
39. Greenberg, S. Ulcerative keratitis and keratomalacia in horses: Medical management and surgical considerations. *Equine Vet. Educ.* **2017**, *31*, 179–182. [CrossRef]
40. Marfe, G.; Massaro-Giordano, M.; Ranalli, M.; Cozzoli, E.; Di Stefano, C.; Malafoglia, V.; Polettini, M.; Gambacurta, A. Blood derived stem cells: An ameliorative therapy in veterinary ophthalmology. *J. Cell. Physiol.* **2012**, *227*, 1250–1256. [CrossRef]
41. Brooks, D.E.; Matthews, A.; Clode, A.B. Diseases of the cornea. In *Equine Ophthalmology*; Gilger, B.C., Ed.; John Wiley & Sons, Inc.: Hoboken, NJ, USA, 2016; pp. 252–368. ISBN 9781119047919.
42. Sherman, A.B.; Gilger, B.C.; Berglund, A.K.; Schnabel, L.V. Effect of bone marrow-derived mesenchymal stem cells and stem cell supernatant on equine corneal wound healing in vitro. *Stem Cell Res. Ther.* **2017**, *8*, 120. [CrossRef]
43. Spaas, J.H.; Gambacurta, A.; Polettini, M.; Broeckx, S.; Van Hoeck, F.; De Schauwer, C.; Van De Walle, G.R.; Van Soom, A. Purification and expansion of stem cells from equine peripheral blood, with clinical applications. *Vlaams Diergeneeskd. Tijdschr.* **2011**, *80*, 129–135.

44. Gilger, B.C.; Hollingsworth, S.R. Diseases of the uvea, uveitis, and recurrent uveitis. In *Equine Ophthalmology*; Gilger, B., Ed.; John Wiley & Sons, Inc.: Hoboken, NJ, USA, 2016; pp. 369–415. ISBN 9781119047919.
45. Sauvage, A.C.; Monclin, S.J.; Elansary, M.; Hansen, P.; Grauwels, M.F. Detection of intraocular Leptospira spp. by real-time polymerase chain reaction in horses with recurrent uveitis in Belgium. *Equine Vet. J.* **2019**, *51*, 299–303. [CrossRef]
46. Regan, D.P.; Aarnio, M.C.; Davis, W.S.; Carmichael, K.P.; Vandenplas, M.L.; Lauderdale, J.D.; Moore, P.A. Characterization of cytokines associated with Th17 cells in the eyes of horses with recurrent uveitis. *Vet. Ophthalmol.* **2012**, *15*, 145–152. [CrossRef]
47. Kol, A.; Walker, N.J.; Nordstrom, M.; Borjesson, D.L. Th17 pathway as a target for multipotent stromal cell therapy in dogs: Implications for translational research. *PLoS ONE* **2016**, *11*, e0148568. [CrossRef] [PubMed]
48. Arzi, B.; Mills-Ko, E.; Verstraete, F.J.M.; Kol, A.; Walker, N.J.; Badgley, M.R.; Fazel, N.; Murphy, W.J.; Vapniarsky, N.; Borjesson, D.L. Therapeutic efficacy of fresh, autologous mesenchymal stem cells for severe refractory gingivostomatitis in cats. *Stem Cells Transl. Med.* **2016**, *5*, 75–86. [CrossRef]
49. Holt, D.D.C.; Wood, J.A.; Granick, J.L.; Walker, N.J.; Clark, K.C.; Borjesson, D.L. Equine mesenchymal stem cells inhibit T cell proliferation through different mechanisms depending on tissue source. *Stem Cells Dev.* **2014**, *23*, 1258–1265. [CrossRef] [PubMed]
50. Saldinger, L.K.; Nelson, S.G.; Bellone, R.R.; Lassaline, M.; Mack, M.; Walker, N.J.; Borjesson, D.L. Horses with equine recurrent uveitis have an activated CD4+ T-cell phenotype that can be modulated by mesenchymal stem cells in vitro. *Vet. Ophthalmol.* **2020**, *23*, 160–170. [CrossRef] [PubMed]
51. Malalana, F.; Stylianides, A.; McGowan, C. Equine recurrent uveitis: Human and equine perspectives. *Vet. J.* **2015**, *206*, 22–29. [CrossRef]
52. Matthews, A.; Gilger, B. Equine immune-mediated keratopathies. *Equine Vet. J.* **2010**, *42*, 31–37. [CrossRef]
53. Davis, A.B.; Schnabel, L.V.; Gilger, B.C. Subconjunctival bone marrow-derived mesenchymal stem cell therapy as a novel treatment alternative for equine immune-mediated keratitis: A case series. *Vet. Ophthalmol.* **2019**, *22*, 674–682. [CrossRef] [PubMed]
54. Dwyer, A.E. Practical general field ophthalmology. In *Equine Ophthalmology*; Gilger, B.C., Ed.; John Wiley & Sons, Inc.: Hoboken, NJ, USA, 2016; pp. 71–111. ISBN 9781119047919.
55. Lin, H.-B.; Li, W.-L.; Zhuo, X.; He, A.-Q. Application effect of soft corneal contact lens on corneal limbal stem cell transplantation in elderly patients with pterygium. *Int. J. Ophthalmol.* **2019**, *19*, 329–331.
56. Zhao, J.; Zhang, Q.; Wang, Y.; Li, Y. Uterine infusion with bone marrow mesenchymal stem cells improves endometrium thickness in a rat model of thin endometrium. *Reprod. Sci.* **2015**, *22*, 181–188. [CrossRef]
57. Canisso, I.F.; Segabinazzi, L.G.T.M.; Fedorka, C.E. Persistent breeding-induced endometritis in mares—A multifaceted challenge: From clinical aspects to immunopathogenesis and pathobiology. *Int. J. Mol. Sci.* **2020**, *21*, 1432. [CrossRef] [PubMed]
58. LeBlanc, M.M.; Causey, R.C. Clinical and subclinical endometritis in the mare: Both threats to fertility. *Reprod. Domest. Anim.* **2009**, *44*, 10–22. [CrossRef] [PubMed]
59. Woodward, E.M.; Troedsson, M.H. Equine breeding-induced endometritis: A review. *J. Equine Vet. Sci.* **2013**, *33*, 673–682. [CrossRef]
60. Canisso, I.F.; Stewart, J.; da Silva, M.A.C. Endometritis: Managing persistent post-breeding endometritis. *Vet. Clin. N. Am. Equine Pract.* **2016**, *32*, 465–480. [CrossRef] [PubMed]
61. Rink, B.E.; Beyer, T.; French, H.M.; Watson, E.; Aurich, C.; Donadeu, F.X. The fate of autologous endometrial mesenchymal stromal cells after application in the healthy equine uterus. *Stem Cells Dev.* **2018**, *27*, 1046–1052. [CrossRef]
62. Ferris, R.A.; Frisbie, D.D.; McCue, P.M. Use of mesenchymal stem cells or autologous conditioned serum to modulate the inflammatory response to spermatozoa in mares. *Theriogenology* **2014**, *82*, 36–42. [CrossRef]
63. Navarrete, F.; Saravia, F.; Cisterna, G.; Rojas, F.; Silva, P.P.; Rodríguez-Alvarez, L.; Rojas, D.; Cabezas, J.; Mançanares, A.C.F.; Castro, F.O. Assessment of the anti-inflammatory and engraftment potential of horse endometrial and adipose mesenchymal stem cells in an in vivo model of post breeding induced endometritis. *Theriogenology* **2020**, *155*, 33–42. [CrossRef]
64. Hoffmann, C.; Ellenberger, C.; Mattos, R.C.; Aupperle, H.; Dhein, S.; Stief, B.; Schoon, H.-A. The equine endometrosis: New insights into the pathogenesis. *Anim. Reprod. Sci.* **2009**, *111*, 261–278. [CrossRef]
65. Buczkowska, J.; Kozdrowski, R.; Nowak, M.; Raś, A.; Mrowiec, J. Endometrosis—Significance for horse reproduction, pathogenesis, diagnosis, and proposed therapeutic methods. *Pol. J. Vet. Sci.* **2014**, *17*, 547–554. [CrossRef]
66. Falomo, M.E.; Ferroni, L.; Tocco, I.; Gardin, C.; Zavan, B. Immunomodulatory role of adipose-derived stem cells on equine endometriosis. *Biomed Res. Int.* **2015**, *2015*, 1–6. [CrossRef] [PubMed]
67. Alvarenga, M.A.; do Carmo, M.T.; Segabinazzi, L.G.; Guastali, M.D.; Maia, L.; Landim-Alvarenga, F.C. Feasibility and safety of endometrial injection of autologous bone marrow mesenchymal stem cells in mares. *J. Equine Vet. Sci.* **2016**, *42*, 12–18. [CrossRef]
68. Mambelli, L.I.; Winter, G.H.Z.; Kerkis, A.; Malschitzky, E.; Mattos, R.C.; Kerkis, I. A novel strategy of mesenchymal stem cells delivery in the uterus of mares with endometrosis. *Theriogenology* **2013**, *79*, 744–750. [CrossRef]
69. Mambelli, L.I.; Mattos, R.C.; Winter, G.H.Z.; Madeiro, D.S.; Morais, B.P.; Malschitzky, E.; Miglino, M.A.; Kerkis, A.; Kerkis, I. Changes in expression pattern of selected endometrial proteins following mesenchymal stem cells infusion in mares with endometrosis. *PLoS ONE* **2014**, *9*, e97889. [CrossRef] [PubMed]
70. Corradetti, B.; Correani, A.; Romaldini, A.; Marini, M.G.; Bizzaro, D.; Perrini, C.; Cremonesi, F.; Lange-Consiglio, A. Amniotic membrane-derived mesenchymal cells and their conditioned media: Potential candidates for uterine regenerative therapy in the horse. *PLoS ONE* **2014**, *9*, e111324. [CrossRef] [PubMed]

71. Rink, B.E.; Amilon, K.R.; Esteves, C.L.; French, H.M.; Watson, E.; Aurich, C.; Donadeu, F.X. Isolation and characterization of equine endometrial mesenchymal stromal cells. *Stem Cell Res. Ther.* **2017**, *8*. [CrossRef]
72. Cortés-Araya, Y.; Amilon, K.; Rink, B.E.; Black, G.; Lisowski, Z.; Donadeu, F.X.; Esteves, C.L. Comparison of antibacterial and immunological properties of mesenchymal stem/stromal cells from equine bone marrow, endometrium, and adipose tissue. *Stem Cells Dev.* **2018**, *27*, 1518–1525. [CrossRef]
73. Cabezas, J.; Rojas, D.; Navarrete, F.; Ortiz, R.; Rivera, G.; Saravia, F.; Rodriguez-Alvarez, L.; Castro, F.O. Equine mesenchymal stem cells derived from endometrial or adipose tissue share significant biological properties, but have distinctive pattern of surface markers and migration. *Theriogenology* **2018**, *106*, 93–102. [CrossRef]
74. Snider, T.A.; Sepoy, C.; Holyoak, G.R. Equine endometrial biopsy reviewed: Observation, interpretation and application of histopathologic data. *Theriogenology* **2011**, *75*, 1567–1581. [CrossRef]
75. Grady, S.T.; Watts, A.E.; Thompson, J.A.; Penedo, M.C.T.; Konganti, K.; Hinrichs, K. Effect of intra-ovarian injection of mesenchymal stem cells in aged mares. *J. Assist. Reprod. Genet.* **2019**, *36*, 543–556. [CrossRef]
76. Takehara, Y.; Yabuuchi, A.; Ezoe, K.; Kuroda, T.; Yamadera, R.; Sano, C.; Murata, N.; Aida, T.; Nakama, K.; Aono, F.; et al. The restorative effects of adipose-derived mesenchymal stem cells on damaged ovarian function. *Lab. Investig.* **2013**, *93*, 181–193. [CrossRef] [PubMed]
77. Wang, S.; Yu, L.; Sun, M.; Mu, S.; Wang, C.; Wang, D.; Yao, Y. The therapeutic potential of bone marrow mesenchymal stem cells in premature ovarian failure. *Stem Cell Res. Ther.* **2018**, *9*. [CrossRef]
78. Oristaglio Turner, R.M.O. Pathogenesis, diagnosis, and management of testicular degeneration in stallions. *Clin. Tech. Equine Pract.* **2007**, *6*, 278–284. [CrossRef]
79. Hsiao, C.; Ji, A.; Chang, C.; Cheng, C.; Lee, L.; Ho, J. Local injection of mesenchymal stem cells protects testicular torsion-induced germ cell injury. *Stem Cell Res. Ther.* **2015**, *6*, 113. [CrossRef] [PubMed]
80. de Papa, P.M.; Guasti, P.N.; De Vita, B.; Nakazato, N.G.; Maia, L.; de Freitas Dell'Aqua, C.P.; da Scheeren, V.F.C.; Segabinazzi, L.G.T.M.; de Andrade Junior, L.R.P.; Silva, L.F.M.C.; et al. Clinical safety of intratesticular transplantation of allogeneic bone marrow multipotent stromal cells in stallions. *Reprod. Domest. Anim.* **2020**, *55*, 429–437. [CrossRef] [PubMed]
81. Frank, N.; Geor, R.J.; Bailey, S.R.; Durham, A.E.; Johnson, P.J. Equine metabolic syndrome. *J. Vet. Intern. Med.* **2010**, *24*, 467–475. [CrossRef]
82. Marycz, K.; Michalak, I.; Kornicka, K. Advanced nutritional and stem cells approaches to prevent equine metabolic syndrome. *Res. Vet. Sci.* **2018**, *118*, 115–125. [CrossRef]
83. Fiorina, P.; Jurewicz, M.; Augello, A.; Vergani, A.; Dada, S.; La Rosa, S.; Selig, M.; Godwin, J.; Law, K.; Placidi, C.; et al. Immunomodulatory function of bone marrow-derived mesenchymal stem cells in experimental autoimmune type 1 diabetes. *J. Immunol.* **2009**, *183*, 993–1004. [CrossRef] [PubMed]
84. Lee, R.H.; Seo, M.J.; Reger, R.L.; Spees, J.L.; Pulin, A.A.; Olson, S.D.; Prockop, D.J. Multipotent stromal cells from human marrow home to and promote repair of pancreatic islets and renal glomeruli in diabetic NOD/scid mice. *Proc. Natl. Acad. Sci. USA* **2006**, *103*, 17438–17443. [CrossRef]
85. Lin, G.; Wang, G.; Liu, G.; Yang, L.-J.; Chang, L.-J.; Lue, T.F.; Lin, C.-S. Treatment of type 1 diabetes with adipose tissue-derived stem cells expressing pancreatic duodenal homeobox 1. *Stem Cells Dev.* **2009**, *18*, 1399–1406. [CrossRef]
86. Zhu, S.; Lu, Y.; Zhu, J.; Xu, J.; Huang, H.; Zhu, M.; Chen, Y.; Zhou, Y.; Fan, X.; Wang, Z. Effects of intrahepatic bone-derived mesenchymal stem cells autotransplantation on the diabetic Beagle dogs. *J. Surg. Res.* **2011**, *168*, 213–223. [CrossRef]
87. Berman, D.M.; Willman, M.A.; Han, D.; Kleiner, G.; Kenyon, N.M.; Cabrera, O.; Karl, J.A.; Wiseman, R.W.; Connor, D.H.; Bartholomew, A.M.; et al. Mesenchymal stem cells enhance allogeneic islet mgraftment in nonhuman primates. *Diabetes* **2010**, *59*, 2558–2568. [CrossRef] [PubMed]
88. Bhansali, A.; Upreti, V.; Khandelwal, N.; Marwaha, N.; Gupta, V.; Sachdeva, N.; Sharma, R.R.; Saluja, K.; Dutta, P.; Walia, R.; et al. Efficacy of autologous bone marrow-derived stem cell transplantation in patients with type 2 diabetes mellitus. *Stem Cells Dev.* **2009**, *18*, 1407–1416. [CrossRef] [PubMed]
89. Pileggi, A. Mesenchymal stem cells for the treatment of diabetes. *Diabetes* **2012**, *61*, 1355–1356. [CrossRef]
90. Marycz, K.; Kornicka, K.; Markedziak, M.; Golonka, P.; Nicpoń, J. Equine metabolic syndrome impairs adipose stem cells osteogenic differentiation by predominance of autophagy over selective mitophagy. *J. Cell. Mol. Med.* **2016**, *20*, 2384–2404. [CrossRef]
91. Marycz, K.; Kornicka, K.; Grzesiak, J.; Śmieszek, A.; Szłapka, J. Macroautophagy and selective mitophagy ameliorate chondrogenic differentiation potential in adipose stem cells of equine metabolic syndrome: New findings in the field of progenitor cells differentiation. *Oxid. Med. Cell. Longev.* **2016**, *2016*, 1–18. [CrossRef]
92. Marycz, K.; Kornicka, K.; Basinska, K.; Czyrek, A. Equine metabolic syndrome affects viability, senescence, and stress factors of equine adipose-derived mesenchymal stromal stem cells: New insight into EqASCs isolated from EMS horses in the context of their aging. *Oxid. Med. Cell. Longev.* **2016**, *2016*. [CrossRef]
93. Marycz, K.; Szłapka-Kosarzewska, J.; Geburek, F.; Kornicka-Garbowska, K. Systemic sdministration of rejuvenated adipose-derived mesenchymal stem cells improves liver metabolism in equine metabolic syndrome (EMS)—New approach in veterinary regenerative medicine. *Stem Cell Rev. Rep.* **2019**, *15*, 842–850. [CrossRef]
94. Bond, S.; Léguillette, R.; Richard, E.A.; Couetil, L.; Lavoie, J.-P.; Martin, J.G.; Pirie, R.S. Equine asthma: Integrative biologic relevance of a recently proposed nomenclature. *J. Vet. Intern. Med.* **2018**, *32*, 2088–2098. [CrossRef]

95. Urbanek, K.; De Angelis, A.; Spaziano, G.; Piegari, E.; Matteis, M.; Cappetta, D.; Esposito, G.; Russo, R.; Tartaglione, G.; De Palma, R.; et al. Intratracheal administration of mesenchymal stem cells modulates tachykinin system, suppresses airway remodeling and reduces airway hyperresponsiveness in an animal model. *PLoS ONE* **2016**, *11*, e0158746. [CrossRef] [PubMed]
96. Bonfield, T.L.; Koloze, M.; Lennon, D.P.; Zuchowski, B.; Yang, S.E.; Caplan, A.I. Human mesenchymal stem cells suppress chronic airway inflammation in the murine ovalbumin asthma model. *Am. J. Physiol. Lung Cell. Mol. Physiol.* **2010**, *299*, L760–L770. [CrossRef] [PubMed]
97. Zucca, E.; Corsini, E.; Galbiati, V.; Lange-Consiglio, A.; Ferrucci, F. Evaluation of amniotic mesenchymal cell derivatives on cytokine production in equine alveolar macrophages: An in vitro approach to lung inflammation. *Stem Cell Res. Ther.* **2016**, *7*, 137. [CrossRef] [PubMed]
98. Barussi, F.C.M.; Bastos, F.Z.; Leite, L.M.B.; Fragoso, F.Y.I.; Senegaglia, A.C.; Brofman, P.R.S.; Nishiyama, A.; Pimpão, C.T.; Michelotto, P.V. Intratracheal therapy with autologous bone marrow-derived mononuclear cells reduces airway inflammation in horses with recurrent airway obstruction. *Respir. Physiol. Neurobiol.* **2016**, *232*, 35–42. [CrossRef] [PubMed]
99. Theoret, C.L.; Wilmink, J.M. Aberrant wound healing in the horse: Naturally occurring conditions reminiscent of those observed in man. *Wound Repair. Regen.* **2013**, *21*, 365–371. [CrossRef] [PubMed]
100. Sole, A.; Bolwell, C.; Riley, C.; Theoret, C. Descriptive survey of wounds in horses presented to Australian veterinarians. *Austral Equine Vet.* **2015**, *34*, 68–74.
101. Theoret, C.L.; Bolwell, C.F.; Riley, C.B. A cross-sectional survey on wounds in horses in New Zealand. *N. Z. Vet. J.* **2016**, *64*, 90–94. [CrossRef]
102. Deschene, K.; Céleste, C.; Boerboom, D.; Theoret, C.L. Hypoxia regulates the expression of extracellular matrix associated proteins in equine dermal fibroblasts via HIF1. *J. Dermatol. Sci.* **2012**, *65*, 12–18. [CrossRef] [PubMed]
103. Bussche, L.; Harman, R.M.; Syracuse, B.A.; Plante, E.L.; Lu, Y.-C.; Curtis, T.M.; Ma, M.; de Walle, G.R. Van Microencapsulated equine mesenchymal stromal cells promote cutaneous wound healing in vitro. *Stem Cell Res. Ther.* **2015**, *6*, 1–15. [CrossRef] [PubMed]
104. Textor, J.A.; Clark, K.C.; Walker, N.J.; Aristizobal, F.A.; Kol, A.; LeJeune, S.S.; Bledsoe, A.; Davidyan, A.; Gray, S.N.; Bohannon-Worsley, L.K.; et al. Allogeneic stem cells alter gene expression and improve healing of distal limb wounds in horses. *Stem Cells Transl. Med.* **2018**, *7*, 98–108. [CrossRef] [PubMed]
105. Mund, S.J.K.; Kawamura, E.; Awang-Junaidi, A.H.; Campbell, J.; Wobeser, B.; MacPhee, D.J.; Honaramooz, A.; Barber, S. Homing and engraftment of intravenously administered equine cord blood-derived multipotent mesenchymal stromal cells to surgically created cutaneous wound in horses: A pilot project. *Cells* **2020**, *9*, 1162. [CrossRef] [PubMed]
106. Spaas, J.H.; Broeckx, S.; Van De Walle, G.R.; Polettini, M. The effects of equine peripheral blood stem cells on cutaneous wound healing: A clinical evaluation in four horses. *Clin. Exp. Dermatol.* **2013**, *38*, 280–284. [CrossRef]
107. Agrawal, K.; Chauhan, N. Pressure ulcers: Back to the basics. *Indian J. Plast. Surg.* **2012**, *45*, 244–254. [CrossRef] [PubMed]
108. Knottenbelt, D. *Pascoe's Principles and Practice of Equine Dermatology*, 2nd ed.; Saunders Ltd.: Philadelphia, PA, USA, 2009; ISBN 9780702042515.
109. Chen, M.; Przyborowski, M.; Berthiaume, F. Stem cells for skin tissue engineering and wound healing. *Crit. Rev. Biomed Eng.* **2009**, *37*, 399–421. [CrossRef]
110. Iacono, E.; Merlo, B.; Pirrone, A.; Antonelli, C.; Brunori, L.; Romagnoli, N.; Castagnetti, C. Effects of mesenchymal stem cells isolated from amniotic fluid and platelet-rich plasma gel on severe decubitus ulcers in a septic neonatal foal. *Res. Vet. Sci.* **2012**, *93*, 1439–1440. [CrossRef] [PubMed]
111. Iacono, E.; Lanci, A.; Merlo, B.; Ricci, F.; Pirrone, A.; Antonelli, C.; Mariella, J.; Castagnetti, C. Effects of amniotic fluid mesenchymal stem cells in carboxymethyl cellulose gel on healing of spontaneous pressure sores: Clinical outcome in seven hospitalized neonatal foals. *Turkish J. Biol.* **2016**, *40*, 484–492. [CrossRef]
112. Lanci, A.; Merlo, B.; Mariella, J.; Castagnetti, C.; Iacono, E. Heterologous Wharton's Jelly derived mesenchymal stem cells application on a large chronic skin wound in a 6-month-old filly. *Front. Vet. Sci.* **2019**, *6*, 9. [CrossRef] [PubMed]
113. Katz, L.M.; Bailey, S.R. A review of recent advances and current hypotheses on the pathogenesis of acute laminitis. *Equine Vet. J.* **2012**, *44*, 752–761. [CrossRef] [PubMed]
114. Dryden, V.C.; Morrison, S.; Bras, R.; Morrell, S.A. Using stem cells in clinical cases. *J. Equine Vet. Sci.* **2013**, *33*, 872–873. [CrossRef]
115. Engiles, J.B. Pathology of the distal phalanx in equine laminitis: More than just skin deep. *Vet. Clin. N. Am. Equine Pract.* **2010**, *26*, 155–165. [CrossRef]
116. Pollitt, C.C. The anatomy and physiology of the suspensory apparatus of the distal phalanx. *Vet. Clin. N. Am. Equine Pract.* **2010**, *26*, 29–49. [CrossRef]
117. Carter, R.A.; Engiles, J.B.; Megee, S.O.; Senoo, M.; Galantino-Homer, H.L. Decreased expression of p63, a regulator of epidermal stem cells, in the chronic laminitic equine hoof. *Equine Vet. J.* **2011**, *43*, 543–551. [CrossRef]
118. Morrison, S. Successful use of allogenic umbilical cord-derived stem cells in nonresponsive chronic laminitic cases. *J. Equine Vet. Sci.* **2011**, *31*, 603. [CrossRef]
119. Sole, A.; Spriet, M.; Galuppo, L.D.; Padgett, K.A.; Borjesson, D.L.; Wisner, E.R.; Brosnan, R.J.; Vidal, M.A. Scintigraphic evaluation of intra-arterial and intravenous regional limb perfusion of allogeneic bone marrow-derived mesenchymal stem cells in the normal equine distal limb using 99mTc-HMPAO. *Equine Vet. J.* **2012**, *44*, 594–599. [CrossRef] [PubMed]

120. Angelone, M.; Conti, V.; Biacca, C.; Battaglia, B.; Pecorari, L.; Piana, F.; Gnudi, G.; Leonardi, F.; Ramoni, R.; Basini, G.; et al. The contribution of adipose tissue-derived mesenchymal stem cells and platelet-rich plasma to the treatment of chronic equine laminitis: A proof of concept. *Int. J. Mol. Sci.* **2017**, *18*, 2122. [CrossRef] [PubMed]
121. Cole, C.; Bentz, B. Treatment of equine nervous system disorders. In *Equine Pharmacology*; Maxwell, L., Ed.; John Wiley & Sons, Inc.: Hoboken, NJ, USA, 2014; pp. 192–217. ISBN 9780813822624.
122. Schaakxs, D.; Kalbermatten, D.F.; Raffoul, W.; Wiberg, M.; Kingham, P.J. Regenerative cell injection in denervated muscle reduces atrophy and enhances recovery following nerve repair. *Muscle Nerve* **2013**, *47*, 691–701. [CrossRef]
123. Wright, K.T.; El Masri, W.; Osman, A.; Chowdhury, J.; Johnson, W.E.B. Concise Review: Bone marrow for the treatment of spinal cord injury: Mechanisms and clinical applications. *Stem Cells* **2011**, *29*, 169–178. [CrossRef]
124. Sadan, O.; Melamed, E.; Offen, D. Bone-marrow-derived mesenchymal stem cell therapy for neurodegenerative diseases. *Expert Opin. Biol. Ther.* **2009**, *9*, 1487–1497. [CrossRef]
125. Jamnig, A.; Lepperdinger, G. From tendon to nerve: An MSC for all seasons. *Can. J. Physiol. Pharmacol.* **2012**, *90*, 295–306. [CrossRef]
126. James, F.M.; Engiles, J.B.; Beech, J. Meningitis, cranial neuritis, and radiculoneuritis associated with Borrelia burgdorferi infection in a horse. *J. Am. Vet. Med. Assoc.* **2010**, *237*, 1180–1185. [CrossRef]
127. Villagrán, C.C.; Schumacher, J.; Donnell, R.; Dhar, M.S. A novel model for acute peripheral nerve injury in the horse and evaluation of the effect of mesenchymal stromal cells applied in situ on nerve regeneration: A preliminary study. *Front. Vet. Sci.* **2016**, *3*, 80. [CrossRef]
128. Kuffler, D.P. An assessment of current techniques for inducing axon regeneration and neurological recovery following peripheral nerve trauma. *Prog. Neurobiol.* **2014**, *116*, 1–12. [CrossRef]
129. Villagrán, C.C.; Amelse, L.; Neilsen, N.; Dunlap, J.; Dhar, M. Differentiation of equine mesenchymal stromal cells into cells of neural lineage: Potential for clinical applications. *Stem Cells Int.* **2014**, *2014*, 13–18. [CrossRef]
130. Reed, S.M. Medical aspects of traumatic brain injury. In *Equine Fracture Repair*; Nixon, A.J., Ed.; John Wiley & Sons, Ltd.: Hoboken, NJ, USA, 2019; pp. 800–803. ISBN 9781119108757.
131. Casañas, J.; de la Torre, J.; Soler, F.; Garcia, F.; Rodellar, C.; Pumarola, M.; Climent, J.; Soler, R.; Orozco, L. Peripheral nerve regeneration after experimental section in ovine radial and tibial nerves using synthetic nerve grafts, including expanded bone marrow mesenchymal cells: Morphological and neurophysiological results. *Injury* **2014**, *45*, S2–S6. [CrossRef]
132. Johnson, A.; Reed, S. Cervical vertebral stenotic myelopathy. In *Equine Neurology*; Reed, S., Furr, M., Eds.; John Wiley & Sons, Inc.: Hoboken, NJ, USA, 2015; pp. 349–367. ISBN 9781118993712.
133. Maia, L.; da Cruz Landim-Alvarenga, F.; Taffarel, M.O.; de Moraes, C.N.; Machado, G.F.; Melo, G.D.; Amorim, R.M. Feasibility and safety of intrathecal transplantation of autologous bone marrow mesenchymal stem cells in horses. *BMC Vet. Res.* **2015**, *11*, 63. [CrossRef] [PubMed]
134. Barberini, D.J.; Aleman, M.; Aristizabal, F.; Spriet, M.; Clark, K.C.; Walker, N.J.; Galuppo, L.D.; Amorim, R.M.; Woolard, K.D.; Borjesson, D.L. Safety and tracking of intrathecal allogeneic mesenchymal stem cell transplantation in healthy and diseased horses. *Stem Cell Res. Ther.* **2018**, *9*, 96. [CrossRef] [PubMed]
135. Draper, A.C.E.; Piercy, R.J. Pathological classification of equine recurrent laryngeal neuropathy. *J. Vet. Intern. Med.* **2018**, *32*, 1397–1409. [CrossRef] [PubMed]
136. Biasutti, S.; Dart, A.J.; Jeffcott, L.B. A review of recent developments in the clinical application of prosthetic laryngoplasty for recurrent laryngeal neuropathy: Indications, complications and outcome. *Equine Vet. Educ.* **2016**, *29*, 337–345. [CrossRef]
137. Sandersen, C.; Ceusters, J.; Fourez, A.; Tosi, I.; Graide, H.; Lejeune, J.P.; Serteyn, D. Nerve stimulator-guided injection of autologous stem cells near the equine left recurrent laryngeal nerve. *J. Vis. Exp.* **2018**, *2018*, 58023. [CrossRef]
138. Sheats, M.K. A comparative review of equine SIRS, sepsis, and neutrophils. *Front. Vet. Sci.* **2019**, *6*, 69. [CrossRef]
139. Roy, M.F.; Kwong, G.P.S.; Lambert, J.; Massie, S.; Lockhart, S. Prognostic value and development of a scoring system in horses with systemic inflammatory response syndrome. *J. Vet. Intern. Med.* **2017**, *31*, 582–592. [CrossRef]
140. Lalu, M.M.; Sullivan, K.J.; Mei, S.H.J.; Moher, D.; Straus, A.; Fergusson, D.A.; Stewart, D.J.; Jazi, M.; Macleod, M.; Winston, B.; et al. Evaluating mesenchymal stem cell therapy for sepsis with preclinical meta-analyses prior to initiating a first-in-human trial. *Elife* **2016**, *5*, 1–28. [CrossRef] [PubMed]
141. Kilcoyne, I.; Nieto, J.E.; Watson, J.L.; Galuppo, L.D.; Borjesson, D.L. Do allogeneic bone marrow derived mesenchymal stem cells diminish the inflammatory response to lipopolysaccharide infusion in horses? A pilot study. *Vet. Immunol. Immunopathol.* **2021**, *231*, 110146. [CrossRef] [PubMed]
142. Karagianni, A.E.; Lisowski, Z.M.; Hume, D.A.; Scott Pirie, R. The equine mononuclear phagocyte system: The relevance of the horse as a model for understanding human innate immunity. *Equine Vet. J.* **2021**, *53*, 231–249. [CrossRef] [PubMed]

animals

MDPI

Article

The Effect of a 7 Year-Long Cryopreservation on Stemness Features of Canine Adipose-Derived Mesenchymal Stem Cells (cAD-MSC)

Santina Di Bella *, Vincenza Cannella *, Francesco Mira, Patrizia Di Marco, Antonio Lastra, Francesca Gucciardi, Giuseppa Purpari and Annalisa Guercio

Instituto Zooprofilattico Sperimentale della Sicilia "A. Mirri", via G. Marinuzzi 3, 90129 Palermo, Italy; dottoremira@gmail.com (F.M.); patrizia.dimarco@izssicilia.it (P.D.M.); antonio.lastra@izssicilia.it (A.L.); francesca.gucciardi@izssicilia.it (F.G.); giuseppa.purpari@izssicilia.it (G.P.); annalisa.guercio@izssicilia.it (A.G.)
* Correspondence: santina.dibella@izssicilia.it (S.D.B.); vincenza.cannella@izssicilia.it (V.C.)

Simple Summary: The use of canine adipose tissue-derived mesenchymal stem cells represents a promising tool in the emerging field of autologous cell therapy in veterinary medicine. Cells need to be isolated and expanded in vitro in order to obtain the sufficient amount for clinical application, but long-term cultivation before therapeutic use is not recommended, since the cells may lose their stemness features. Alternatively, the cells can be cryopreserved and used as needed and in a short time after thawing. This study evaluated the effect of a 7 year-long cryopreservation using 10% dimethyl sulfoxide with different fetal bovine serum concentrations (from 10 to 90%) on different cell passages. The aim was to establish the most appropriate cell passage and serum percentage for the long-term cryopreservation of cells ensuring the maintenance of the stemness features. Cells were expanded in vitro from P0 to P1–P2 passages and subsequently frozen. This study demonstrated that a high percentage of serum (80%) is necessary to obtain optimal cryopreservation with 10% DMSO. Cells thawed at passages from P2 to P4, even after seven years, could be considered in the studies on therapeutic application and in the in vitro study, because they maintain stem potential after cryopreservation.

Abstract: Mesenchymal stem cells (MSCs) are used in therapy in animal models and veterinary medicine, due to their capacity of inducing tissue regeneration and immunomodulation. Their clinical application requires a ready off-the-shelf amount of viable therapeutics doses. For this purpose, it is useful to cryopreserve MSCs to gain a ready and controlled source of abundant autologous stem cells. We evaluated the effect of 7 years cryopreservation using 10% dimethyl sulfoxide (DMSO) with different fetal bovine serum (FBS) concentrations (from 10 to 90%) on different passages of MSCs isolated from canine adipose tissue (cAD-MSCs). The study aimed to evaluate the most adequate cell passage and FBS percentage for the long-term cryopreservation of cells by maintaining the stemness features. Phenotype morphology, cell viability, osteogenic and adipogenic differentiation potentials, proliferative potential and expression of pluripotency markers were analyzed in thawed cells and compared with fresh ones. We demonstrated that cells cryopreserved with at least 80% FBS maintain unaltered the stemness characteristics of the freshly isolated cells. In particular, cells of P0–P1 passages have to be expanded in vitro and subsequently cryopreserved and cells of P2–P4 passages should be considered in the studies on therapeutic application and in vitro study of cAD-MSCs.

Keywords: canine adipose-derived mesenchymal stem cells (cAD-MSCs); cryopreservation; dimethyl sulfoxide (DMSO); fetal bovine serum (FBS)

Citation: Di Bella, S.; Cannella, V.; Mira, F.; Di Marco, P.; Lastra, A.; Gucciardi, F.; Purpari, G.; Guercio, A. The Effect of a 7 Year-Long Cryopreservation on Stemness Features of Canine Adipose-Derived Mesenchymal Stem Cells (cAD-MSC). *Animals* **2021**, *11*, 1755. https://doi.org/10.3390/ani11061755

Academic Editors: Academic Editor(s): Eleonora Iacono and Barbara Merlo

Received: 15 April 2021
Accepted: 9 June 2021
Published: 11 June 2021

1. Introduction

Mesenchymal stem cells (MSCs) have attracted increasing attention due to their potential use in regenerative medicine and tissue engineering. These cells display a significant therapeutic plasticity as reflected by their advantageous characteristics: the ability to enhance tissue renovation, the immunomodulatory and anti-inflammatory effects [1]. Although MSCs can be easily isolated from several tissues, clinical use has favored adipose tissue because of its relative ease of stem cell recovery and the minimal donor-site morbidity [2]. In dogs, adipose tissue can be collected either by a simple adapted liposuction procedure, or through biopsies or in routine veterinary surgery procedures [3]. The visceral fat is particularly easy to collect because this species is subjected to a large number of ovariohysterectomies [4]. Interest in MSCs, both for regenerative and reparative therapies in dogs, is emerging as the current treatment options for several conditions that often do not result either in the desired clinical outcome or in the patients' return to normal function [2]. Moreover, canine MSCs have been evaluated in some experimental and preclinical studies on efficacy and safety testing of novel treatments for humans, since the dog is considered to be a superior model for humans than rodents [2].

The clinical application of MSCs requires on demand access to a ready off-the-shelf amount of viable therapeutics doses. Cell dosage varies widely among applications and it is not established for any single treatment. MSCs require extensive culture expansion due to low cell number, and genetic alterations and contamination risks increase with culture time. For this purpose, it is very useful to cryopreserve these cells in order to gain a ready and controlled source of abundant autologous stem cells that maintain unaltered characteristics of the freshly isolated cells by preserving their vitality and maintaining their pluripotent phenotype. Cell aliquots can be biobanked for later administration immediately upon revitalization or after short-term expansion. Cryopreservation also increases MSC availability as frozen cells can be delivered over long distances [5].

Most cryopreservation protocols for stem cells use dimethyl sulfoxide (DMSO) at a concentration of 10% (*v*/*v*), often combined with fetal bovine serum (FBS) in concentrations ranging from 10% to 90% (*v*/*v*) [4,6–8]. FBS is commonly added to freezing solutions for its benefits to stabilize the cell membrane and adjust cell osmotic pressure [9]. On the other hand, it always accompanied with risk of infections due to its xenobiotic origin [10].

In this study, the effect of a 7 year-long cryopreservation using 10% DMSO with different FBS concentrations (from 10 to 90%) was evaluated on different cell passages of MSCs isolated from adipose tissue of *Canis familiaris* (cAD-MSCs). Phenotype morphology, cell viability, osteogenic and adipogenic differentiation potential, proliferative potential and expression of pluripotency markers were analyzed and compared in both fresh and thawed cells.

The aim was to establish the most appropriate cell passage and FBS percentage for the long-term cryopreservation of cAD-MSCs ensuring the maintenance of the stemness features.

2. Materials and Methods

2.1. Cell Culture

Samples were collected from visceral adipose tissue of 10 female dogs. Each sample was weighted, cleaned of large blood vessels and chopped, washed with Hank's balanced salt solution (HBSS, Sigma–Aldrich®, Milan, Italy) and digested for 3–4 h at 37 °C with 0.2% collagenase type IA (GIBCO BRL/Life Technologies, Milan, Italy) prepared in sterile phosphate buffered saline (PBS) supplemented with 1% antibiotics (penicillin, streptomycin and amphotericin). The collagenase activity was neutralized by adding 10% FBS (EuroClone®, Milan, Italy). After centrifugation (300× *g* for 10 min) and washing of the pellet, cells were cultured in T25 flasks (Falcon, BD Bioscence, Basel, Switzerland), in non-inductive medium consisting of Dulbecco's modified Eagle's medium (D-MEM) low glucose (Sigma–Aldrich®, Milan, Italy) with 10% FBS and 1% penicillin, streptomycin and amphotericin. Cells were incubated at 37 °C in a humidified atmosphere containing 5% CO_2. After the overnight incubation, non-adherent cells were removed and fresh medium was added to the flasks.

The medium was renewed every 3 days. Adherent cells, grown to semiconfluency, were harvested, quantified and subcultured. A small volume of sterile and warm HBSS was added to the flasks for harvesting viable cAD-MSCs. HBSS was replaced with 500 μL of Trypsin/EDTA solution (0.5%) (Sigma–Aldrich®, Milan, Italy). Cells were resuspended in a culture medium, transferred from the flask to a sterile tube of 15 mL, and centrifuged at 300× *g* for 5 min. The supernatant was aspirated and the cells resuspended in a small volume of culture medium. Cells were counted using the hemocytometer (Cellometer Auto T4 EuroClone®, Milan, Italy). The primary cells cultured for 5–6 days were defined as passage 'P0'. At 80% confluence, the MSCs were split and expanded (P1); cell expansion was continued until passage 6 (P6).

2.2. Cryopreservation and Thawing

At each passage from P1, cells were collected and resuspended at density of 1×10^6 cells/cryovials in 1 mL of cryopreservation media composed of 10% DMSO (Sigma–Aldrich®, Milan, Italy) and increasing percentages (from 10% to 90%) of FBS. Cells were frozen by Mr. Frosty container (Thermo Fisher Scientific, Monza, Italy) decreasing −1 °C/min until −80 °C for 1 week and then they were transferred to liquid nitrogen tank for long time storage. After a 7 year-long cryopreservation, cells were thawed. cAD-MSCs were placed into a 37 °C water-bath for 1–2 min and washed in 90% D-MEM and 10% FBS to eliminate DMSO. Cells were counted and plated in T25 flasks with culture medium.

2.3. Cell Viability Analysis

Cryopreserved cAD-MSCs were thawed, the freezing solutions were removed and cell viability was assessed with a Trypan blue dye exclusion assay. A 1:2 dilution of the suspension was prepared using a 0.4% Trypan Blue solution (Sigma–Aldrich®, Milan, Italy). The number of the non-viable (stained) and viable (trypan blue excluded) cells were counted. Viability was expressed as the percentage of the number of the viable unstained cells obtained after thawing divided by the number of the viable cells before freezing.

The cell viability and proliferation rate for fresh and frozen-thawed cAD-MSCs were also evaluated by measuring 3-(4,5dimethylthiazol-2yl)-5-(3carboxymethoxyphenyl)-2-(4-sulfophenyl)-2H-tetrazolium inner salt (MTS), using a commercially available kit (Cell Titer 96 Aqueous One solution Cell proliferation Assay, Promega, Milan, Italy). Thawed cells were plated in a 96-well plate at a density of 1×10^4 cells/100 μL per well. The Cell Titer 96 Aqueous One Solution Reagent (20 μL) was added and the plate was cultured for 4 h at 37 °C in a humidified 5% CO_2 atmospheric environment. Plates were read on an absorbance microplate reader (Sunrise™, Tecan, Cernusco sul Naviglio (MI), Italy), complemented by universal reader control and data analysis software add-on (Magellan™, Tecan, Cernusco sul Naviglio (MI), Italy), at a wavelength of 492 nm.

2.4. Measurement of Cell Doubling Time and Cell Morphology Observation

The thawed cells were plated at a density of 2×10^4 nucleated cells/cm^2 in T25 culture flasks using the above-described culture medium. The medium was changed every 2–3 days until the adherent cell population reached 80% confluence. The adherent cAD-MSCs were passaged by digestion with Trypsin/EDTA solution (0.5%), counted with a hemocytometer, and a portion of the cells were reseeded for the subsequent passages (until P7) in T25 flasks (2×10^4 cAD-MSCs/cm^2). Cell-doubling time (DT) and cell-doubling number (CD) were calculated by hemocytometer counts and cell culture time (CT) for each passage according to the following 2 formulae:

$$CD = \ln(Nf/Ni)/\ln(2) \quad (1)$$

$$DT = CT/CD \quad (2)$$

where Nf is the final number of cells and Ni is the initial number of cells [11].

Cell-doubling time of fresh and cryopreserved cells (in medium with DMSO and 50% and 80% FBS) was compared.

Post-thaw cell morphology changes, such as cell enlargement, accumulation of vacuoles and presence of cellular debris were observed daily, at each passage, by inverted microscopy.

2.5. *Differentiation Assay*

To evaluate the stemness of established cultures, cells in P2, P4 and P6 after cryopreservation with 10% DMSO and 50% and 80% FBS were cultured in appropriate differentiation media and induced to differentiate toward adipogenic and osteogenic lineage.

2.5.1. Adipogenic Differentiation

Cells were plated at 7500 cells/cm^2 in 6-well plates and induced to differentiate in the StemMACS™AdipoDiff Media (Miltenyi Biotec, Bergisch Gladbach, Germany) differentiation medium for 21 days. Cells cultivated in basal medium (D-MEM, low glucose with 10% FBS) were used as a control. Differentiation and basal medium were changed every 48–72 h and cells were examined by inverted microscopy. The formation of lipid droplets was verified with Oil Red O staining.

2.5.2. Osteogenic Differentiation

Cells were plated at 4500 cells/cm^2 in 6-well plates and induced to differentiate in StemMACS™OsteoDiff Media (Miltenyi Biotec, Bergisch Gladbach, Germany) differentiation medium for 3 weeks. Cells cultivated in basal medium (D-MEM, low glucose with 10% FBS) were used as a control. Differentiation and basal medium were changed every 3 days and cells were examined by inverted microscopy. Calcified extracellular matrix deposits were detected by von Kossa staining.

2.6. *Reverse Transcriptase-Polymerase Chain Reaction (RT-PCR)*

Total RNAs were extracted from cells cryopreserved (in medium with DMSO and 50% and 80% FBS) using the RNAspin Mini RNA Isolation Kit (GE Healthcare, Milan, Italy) and treated with DNase I to remove contaminating DNA. cDNAs were synthesized from 1 μg total RNA using random hexamers and Superscript III reverse transcriptase (Thermo Fisher Scientific, Monza, Italy). Primers used for canine pluripotency markers (OCT4, NANOG and SOX2) were previously reported [12]. OCT4 and NANOG were amplified in 35 cycles at 94 °C for 1 min, 60 °C for 1 min, 72 °C for 1 min and followed by 72 °C for 5 min. SOX2 was amplified in 35 cycles at 94 °C for 1 min, 58 °C for 1 min, 72 °C for 1 min and followed by 72 °C for 5 min. PCR products were separated on 2% agarose gel by electrophoresis, stained with ethidium bromide and visualized under UV light.

2.7. *Microbiological Control of cAD-MSCs and Reagents*

cAD-MSCs were tested for possible contaminations during the steps of production. These quality controls included tests for bacteria and fungi, mycoplasma and viruses that were used as a part of routine and regular quality control screening procedures. To detect low level of contamination, samples from the cell cultures and reagents were inoculated onto solid media (blood agar, plate count agar and Sabouraud dextrose agar). PCR analysis were conducted for the screening of pestiviruses [13,14] and mycoplasma (MycoSensor PCR Assay Kit—Agilent Technologies, Santa Clara, CA, USA) that can contaminate cell cultures.

2.8. *Statistical Analysis*

The results are expressed as mean ± standard deviation. Data were normally distributed ($p < 0.05$, Kolmogorov–Smirnov test). Duncan's multiple post-hoc comparison test was applied.

2.9. Ethical Statement

The study did not involve any animal experiment. Specimen collection was done 7 years earlier from dogs during routine spays required and authorized by dog owners and independent of this study. We had extracted adipose tissue samples from dogs only after their owners provided written informed consent. This study was conducted as part of the IZS SI 2007 RF research project entitled "Adult mesenchymal stem cells: differentiative lineages and applications in autologous and allogenic implantation and tissue remodelling", approved by the Italian Ministry of Health.

3. Results

3.1. Culture, Expansion and Morphology of cAD-MSCs

cAD-MSCs were isolated from canine adipose tissue and grown in plastic tissue culture flasks. At P0, adherent cells grew as spindle- or star-shape cells, forming colonies 3 days after plating. Cells became semiconfluent within 5–6 days. After passage P0, the cells began to proliferate rapidly and, as soon as they reached the semiconfluency (every 3–4 days), the cAD-MSCs were split and expanded until P6. After the first passage, MSCs adopted a fibroblast-like shape. Cell size and shape persisted until P5. Cells had increased volumes from passage P6. Moreover, cells cryopreserved at each passage in 9 cryopreservation media (composed of 10% DMSO and increasing percentages of FBS from 10% to 90%) were thawed after 7 years and retrospectively compared with fresh cells isolated from the same animals. Cryopreserved cells with serum percentages less than 50% showed stunted growth in the flask. Cells of all passages, cryopreserved with serum percentages greater than 50%, maintained the ability to expand in culture and reached semi-confluence. All thawed cAD-MSCs obtained from 10 dogs presented fibroblast-like phenotype until passages P4 and P5 and demonstrated increased volume and cuboidal shape at P6 as in their fresh counterpart (Figure 1). The cryopreservation of cells in liquid nitrogen in media containing DMSO with greater than 50% FBS percentages for 7 years and their thawing did not affect the cell morphology.

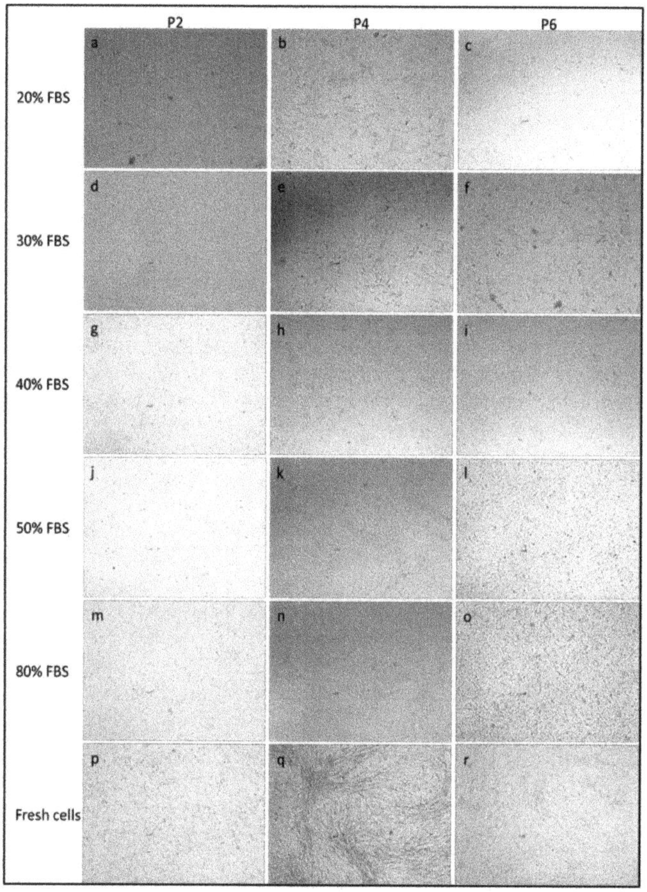

Figure 1. Morphology of cell cryopreserved with 10% DMSO—20%, 30%, 40%, 50% and 80% FBS percentages and fresh cells showed at passages P2 (**a**,**d**,**g**,**j**,**m**,**p**), P4 (**b**,**e**,**h**,**k**,**n**,**q**) and P6 (**c**,**f**,**i**,**l**,**o**,**r**). Cells frozen with FBS concentrations lower than 50%, after thawing, particularly in the passages P4–P6, showed an altered morphology, fusiform, polygonal, astroid shapes and a stunted growth in culture failing to reach semi-confluence. Cells from P2 to P6 passages, cryopreserved with serum percentages greater than 50%, maintained the ability to expand in culture and reached semiconfluence. Thawed cAD-MSCs presented fibroblast-like phenotype until passages P4 and demonstrated increased volume and cuboidal shape at P6 as in their fresh counterpart. A 100× magnification.

3.2. Cell Viability after Cryopreservation

The effects of different freezing solutions on post-thaw viability of cAD-MSCs at three passages (P2, P4 and P6) are shown in Figure 2. Cells cryopreserved with medium containing serum percentages up to 30% did not exceed 50% of cell viability after thawing at all cell passages. Frozen cells with 40% serum reached vitality values around 60%. cAD-MSCs cryopreserved at passages P2 and P4 with serum from 50% to 70% reached vitality percentages of about 80%. The viability of frozen cells with serum percentages of 80% and 90% reached 98% viability in passages P2 and P4. Cryopreserved cells at passage P6 with serum percentages from 10% to 90% showed lower viability percentages compared to passages P2 and P4.

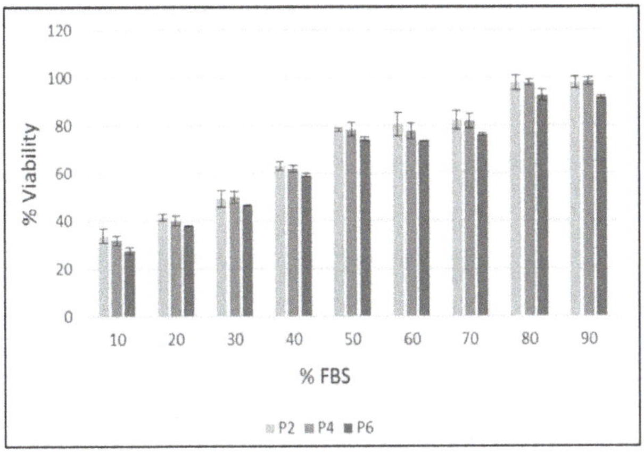

Figure 2. The effect of the different freezing solution composed of 10% DMSO and increasing percentage of FBS (from 10% to 90%) on post-thaw viability of cAD-MSCs at three passages (P2, P4 and P6).

3.3. *cAD-MSCs Proliferation*

To determine the proliferation potential, the population doubling time of cells cryopreserved with 10% DMSO and 50% and 80% FBS was calculated at passages 2, 4 and 6 and compared with that of the fresh cAD-MSCs (Figure 3). The analysis of proliferation capacity showed that doubling time is not significantly increased in passage P4 versus P2 in both fresh and cells cryopreserved with 10% DMSO and 80% FBS. However, the doubling time increased in P6. Cryopreservation with 10% DMSO and 50% FBS caused a delay of cell divisions at all passages.

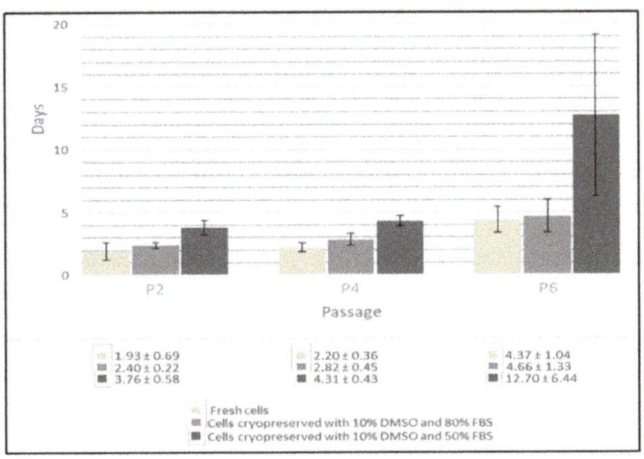

Figure 3. Analysis of population doubling time (pdt) of fresh cells and cells cryopreserved with 10% DMSO and 50% and 80% FBS (n = 10) at passages P2, P4 and P6. The pdt in fresh cells and in cells cryopreserved with 80% FBS shows the same trend in the three passages considered. Cells cryopreserved with 50% FBS show a delay in cell division at all passages compared to fresh cells.

3.4. *Differentiation Potential*

Thawed cells of all dogs differentiated up to passage 4 across the two lineages tested. Differentiation was qualitatively assessed on the basis of cell morphology and histochemical stains. The cAD-MSCs induced toward adipogenic differentiation were analyzed for presence of intracellular lipid accumulation by Oil red O stain. MSCs in passages from P2 to P4 displayed many lipid droplets when compared with those in P6 passage (Figure 4). Thawed cells up to passage P4, induced toward osteogenic differentiation, formed aggregates and shown the mineralization of extracellular matrix marked by von Kossa staining (Figure 5). No differences were found in adipogenic and osteogenic potential between fresh and frozen cells.

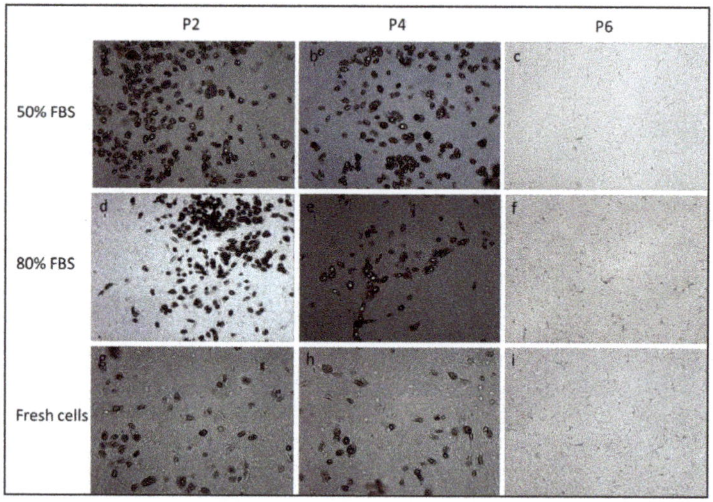

Figure 4. Fat globules appeared in cAD-MSCs, cryopreserved with 10% DMSO–50 and 80% FBS and thawed, cultured with adipogenic differentiation medium at passages P2 (**a**,**d**) and P4 (**b**,**e**). No differentiation was observed in the passage P6 (**c**,**f**). The frozen cells showed the same differentiation potential as fresh cells at each passage (**g**,**h**,**i**). A 100× magnification.

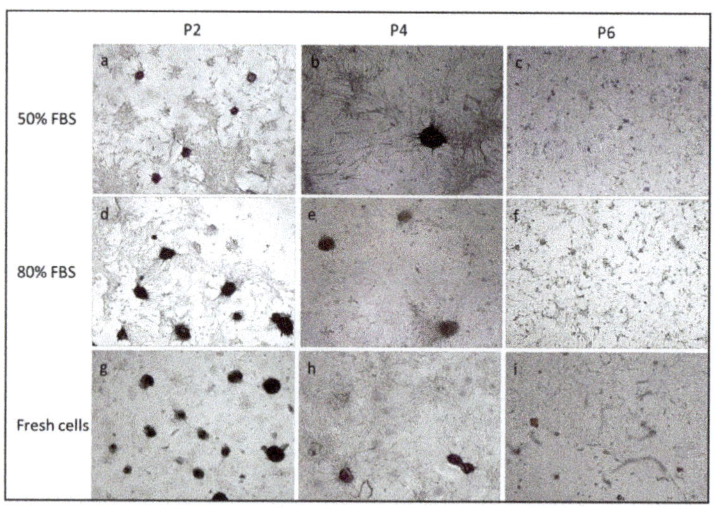

Figure 5. Deposition of calcified extracellular matrix revealed by von Kossa staining in cAD-MSCs, cryopreserved with 10% DMSO–50 and 80% FBS and thawed, cultured with osteogenic differentiation medium at passages P2 (**a**,**d**), P4 (**b**,**e**). The frozen cells showed the same differentiation potential as fresh cells at each passage (**g**,**h**,**i**). A 100× magnification.

3.5. Expression of Pluripotency Markers

A qualitative reverse-transcription PCR was used to assess the expression of the canine pluripotency associated transcription factors OCT4, SOX2 and NANOG in cell passages. These factors were not expressed in any post-thaw passage of cells cryopreserved with 10% DMSO and FBS concentration lower than 50% (data not shown). The pluripotency markers were maintained up to post-thaw passage P4 of cAD-MSCs cryopreserved with 10% DMSO and at least 50% FBS. The expression of OCT4, SOX2 and NANOG was not revealed in the P5 and P6 passages. The same results were previously observed in fresh cells (Figure 6) [12,15].

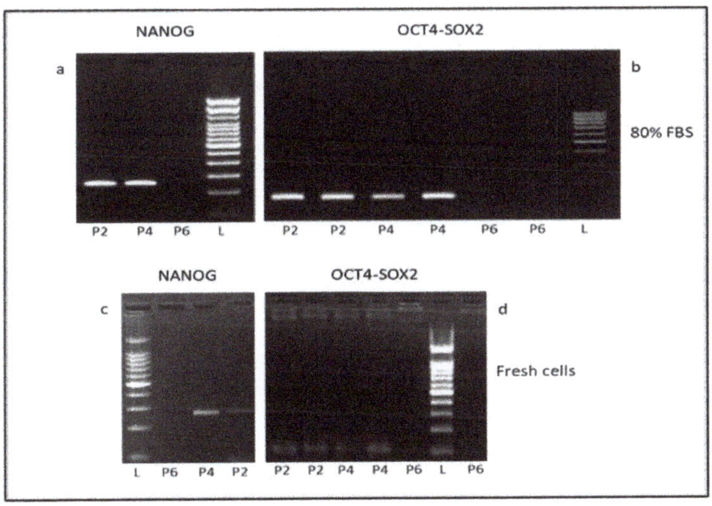

Figure 6. Representative images of expression of pluripotency-associated genes in cells cryopreserved with 10% DMSO and 80% FBS and thawed (**a**,**b**) and in fresh cells (**c**,**d**). NANOG (274 bp), OCT4 (141 bp) and SOX2 (142 bp) were expressed in passages P2 and P4 of cryopreserved and fresh cells. The gene expression was not revealed in P6.

3.6. Microbiological Control of cAD-MSCs and Reagents

All the samples from the cell cultures examined were free from bacterial, viral and fungal contaminants.

4. Discussion

Adipose tissue is easily harvested with minimal risk to patients as compared to other stem cell sources. Stem cells derived from this tissue have been increasingly used for cell therapy both in humans and animals, either as freshly isolated or as cultivated AD-MSCs [15–17]. An important advantage of adipose-derived stem cells is their abundance: from 1 g of adipose tissue an average of 0.5–2.0 $\times$ 10^6 stromal vascular fraction cells can be isolated, which gives 1–10% of stem cell yield [18]. These cells proliferate rapidly with high cellular activity and have a great potential of differentiation into multilineage cells, making them an ideal source to obtain MSCs [19]. For autologous use, the adipose tissue is collected 2 or 3 weeks before the treatment and the animal receive the cultivated cells, but long-term cultivation before therapeutic use is not recommended, since the cells may lose their progenitor characteristics [20]. Alternatively, the cells can be isolated, expanded and subjected to long-term storage, so that they can be used as needed and in a short time after thawing. Cryopreservation and biobanking allows for MSC to be prepared in large batches, under the application of accepted quality control measures to ensure their safety. To assure that cryopreservation medium does not alter the stemness characteristics and the differentiation potential of isolated MSC is also of primary importance [21]. Advances in cell therapy, stem cell research, personalized medicine and cell banking drive the need for optimize storage protocols.

To address this issue, the study was conducted to determine the effect of long-term cryopreservation on different passage of cAD-MSCs cryopreserved with ten different media composed of 10% DMSO and ten different concentrations of serum. DMSO is standard cryoprotectant used to stabilize cell protein and membrane and to prevent intracellular ice formation [22,23]. FBS stabilizes the cell membrane and adjust cell osmotic pressure [9]. Data obtained demonstrated that cells cryopreserved at the study condition with high FBS percentages show similar stem characteristics as fresh ones.

Viability is the primary indicator on cryopreservation success and its >70% value for cryopreserved cells is generally considered a post-thaw viability threshold generally accepted and the benchmark for clinical application [24]. This study demonstrated that cAD-MSCs up to passage P4 cryopreserved with 50% FBS showed >80% viability. This reached almost 100% in cells with 80–90% FBS. Cells with more than 50% FBS, cryopreserved in liquid nitrogen and thawed after 7 years showed similar morphological characteristics and proliferative ability as fresh cells. cAD-MSCs grew within a monolayer adhering to the culture flask bottom with fibroblast-like shape [25] until passages P4. Cell morphology gradually became diversified with increasing passages, accompanied by an increase in cell size changes as in their fresh counterpart [12,26]. The proliferation kinetics of cAD-MSCs from passages P1 to P7 were examined in cells cryopreserved with 10% DMSO and 50 and 80% FBS. While there were no significant differences in the population doubling time from passage P2 to P4, passage P6 displayed a longer doubling time. Moreover, doubling time increased at all passages for the cells cryopreserved with 50% FBS with respect to fresh cells.

The expression of pluripotent markers OCT-4, NANOG and SOX-2 was revealed up to passage P4 of cells cryopreserve with more than 50% FBS and not at P5 and P6 passages. These pluripotent transcription factors regulate the self-renewal and differentiation abilities of AD-MSCs [27]. OCT4 and NANOG are not only essential for the maintenance of pluripotency in embryonic stem cells but also in maintaining MSC properties. SOX2 is also important for maintaining proliferation and osteogenic differentiation potential of MSCs. OCT4 and SOX2 are usually expressed at low levels in early-passage MSCs and gradually decrease as the passage number increases [28]. The use of a quantitative RT-PCR in this study would have probably allowed to better appreciate this aspect.

The success of utilizing stem cells in tissue-engineering applications is highly dependent on maintaining a satisfactory level of differentiation potential after extensive in vitro expansion [26]. In addition, in agreement with other studies [4,29], differentiation was not affected by the cryopreservation process. Thawed cells were able to differentiate into two mesodermal lineages. Adipogenic differentiation was accompanied by cell's shape change from a fibroblast to a large rounded morphology and by accumulation of small and non-uniform lipid droplet. Osteogenic induction medium caused the formation of cell aggregates and matrix mineralization that was assessed by calcium specific von Kossa staining. Moreover, it was assessed that cell cultures were free from microbiological contamination. Specific test for the detection of bacteria, yeast, fungi, mycoplasma and viruses should be used as a part of routine and regular quality control screening procedures. Mycoplasma competes with the cells for the nutrients in the culture medium, typical signs of contamination consist in the reduction of the rate of cell proliferation, and changes in cellular physiology including gene expression, metabolism and phenotype. Biosafety assessment of cryopreserved MSCs is necessary to ensure the safe use of cells prior to clinical application. Frozen cells may have the advantage over fresh cells of being more controlled for the presence of contaminants that may come from the animal of origin or from handling. Further studies will have to be conducted to evaluate the use of cryoprotectants that allow the reduction of FBS concentrations with a view to reducing xenocontaminants.

5. Conclusions

This study demonstrated that a high percentage of FBS (at least 80%) is necessary to obtain optimal cryopreservation of cAD-MSCs with 10% DMSO. To provide the needed doses for clinical studies, cells from lower passages (P0, P1) have to be expanded in vitro and subsequently frozen for storage/banking purposes. Probably, a short recovery period post-thaw in culture may facilitate a regain of function [30]. Our data suggest that, cells thawed at from passages P2 to P4, even after seven years, could be considered in the studies on therapeutic application as well as in vitro study of cAD-MSCs, because they maintain their stem potential after cryopreservation.

Author Contributions: Conceptualization, S.D.B. and V.C.; methodology, S.D.B. and A.L.; investigation, S.D.B.; resources, A.G.; data curation, G.P.; writing—original draft preparation, S.D.B.; writing—review and editing, F.M. and F.G.; visualization, P.D.M.; supervision, G.P.; project administration, A.G.; funding acquisition, A.G. All authors have read and agreed to the published version of the manuscript.

Funding: This research received no external funding.

Institutional Review Board Statement: Not applicable.

Data Availability Statement: Data are contained within the article.

Acknowledgments: We are grateful to Laura Russotto from Istituto Zooprofilattico Sperimentale della Sicilia "A. Mirri" for her help in the laboratory activities.

Conflicts of Interest: The authors declare no conflict of interest.

References

1. Pittenger, M.F.; Discher, D.E.; Péault, B.M.; Phinney, D.G.; Hare, J.M.; Caplan, A.I. Mesenchymal stem cell perspective: Cell biology to clinical progress. *npj Regen. Med.* **2019**, *4*, 22. [CrossRef]
2. De Bakker, E.; Van Ryssen, B.; De Schauwer, C.; Meyer, E. Canine mesenchymal stem cells: State of the art, perspectives as therapy for dogs and as a model for man. *Vet. Q.* **2013**, *33*, 225–233. [CrossRef] [PubMed]
3. Vieira, N.M.; Brandalise, V.; Zucconi, E.; Secco, M.; Strauss, B.E.; Zatz, M. Isolation, characterization, and differentiation potential of canine adipose-derived stem cells. *Cell Transplant.* **2010**, *19*, 279–289. [CrossRef]
4. Martinello, T.; Bronzini, I.; Maccatrozzo, L.; Mollo, A.; Sampaolesi, M.; Mascarello, F.; Decaminada, M.; Patruno, M. Canine adipose-derived-mesenchymal stem cells do not lose stem features after a long-term cryopreservation. *Res. Vet. Sci.* **2011**, *91*, 18–24. [CrossRef]
5. Duan, W.; Lopez, M.J.; Hicok, K. Adult multipotent stromal cell cryopreservation: Pluses and pitfalls. *Vet. Surg.* **2018**, *47*, 19–29. [CrossRef] [PubMed]
6. Shahid, M.A.; Kim, W.H.; Kweon, O.K. Cryopreservation of heat-shocked canine adipose-derived mesenchymal stromal cells with 10% dimethyl sulfoxide and 40% serum results in better viability, proliferation, anti-oxidation, and in-vitro differentiation. *Cryobiology* **2020**, *92*, 92–102. [CrossRef]
7. Xiang, Y.; Zheng, Q.; Jia, B.; Huang, G.; Xie, C.; Pan, J.; Wang, J. Ex vivo expansion, adipogenesis and neurogenesis of cryopreserved human bone marrow mesenchymal stem cells. *Cell Biol. Int.* **2007**, *31*, 444–450. [CrossRef] [PubMed]
8. Carvalho, K.A.T.; Cury, C.C.; Oliveira, L.; Cattaned, R.I.I.; Malvezzi, M.; Francisco, J.C.; Pachalok, A.; Olandoski, M.; Faria-Neto, J.R.; Guarita-Souza, L.C. Evaluation of bone marrow mesenchymal stem cell standard cryopreservation procedure efficiency. *Transpl. Proc.* **2008**, *40*, 839–841. [CrossRef]
9. Bahsoun, S.; Coopman, K.; Akam, E.C. The impact of cryopreservation on bone marrow-derived mesenchymal stem cells: A systematic review. *J. Transl. Med.* **2019**, *17*, 397. [CrossRef]
10. Tripathy, S. Cryopreservation of Mesenchymal Stem Cells (MSCs): Different approaches and applications. *Int. J. Adv. Sci. Technol.* **2017**, *1*, 435–456.
11. Vidal, M.A.; Kilroy, G.E.; Johnson, J.R.; Lopez, M.J.; Moore, R.M.; Gimble, J.M. Cell growth characteristics and differentiation frequency of adherent equine bone marrow-derived mesenchymal stromal cells: Adipogenic and osteogenic capacity. *Vet. Surg.* **2006**, *35*, 601–610. [CrossRef] [PubMed]
12. Guercio, A.; Di Bella, S.; Casella, S.; Di Marco, P.; Russo, C.; Piccione, G. Canine mesenchymal stem cells (MSCs): Characterization in relation to donor age and adipose tissue-harvesting site. *Cell Biol. Int.* **2013**, *37*, 789–798. [CrossRef]
13. Studer, E.; Bertoni, G.; Candrian, U. Detection and characterization of Pestivirus contaminations in human live viral vaccines. *Biologicals* **2002**, *30*, 289–296. [CrossRef]

14. Vilcek, S.; Herring, A.J.; Herring, J.A.; Nettleton, P.F.; Lowings, J.P.; Paton, D.J. Pestiviruses isolated from pigs, cattle and sheep can be allocated into at least three genogroups using polymerase chain reaction and restriction endonuclease analysis. *Arch. Virol.* **1994**, *136*, 309–323. [CrossRef] [PubMed]
15. Guercio, A.; Di Marco, P.; Casella, S.; Cannella, V.; Russotto, L.; Purpari, G.; Di Bella, S.; Piccione, G. Production of canine mesenchymal stem cells from adipose tissue and their application in dogs with chronic osteoarthritis of the humeroradial joints. *Cell Biol. Int.* **2012**, *36*, 189–194. [CrossRef]
16. Marx, C.; Silveira, M.D.; Nardi, N.B. Adipose-derived stem cells in veterinary medicine: Characterization and therapeutic applications. *Stem Cells Dev.* **2015**, *24*, 803–813. [CrossRef]
17. Guercio, A.; Di Marco, P.; Casella, S.; Russotto, L.; Puglisi, F.; Majolino, C.; Giudice, E.; Di Bella, S.; Purpari, G.; Cannella, V.; et al. Mesenchymal Stem Cells Derived from Subcutaneous Fat and Platelet-Rich Plasma Used in Athletic Horses with Lameness of the Superficial Digital Flexor Tendon. *J. Equine Vet. Sci.* **2015**, *35*, 19–26. [CrossRef]
18. Lombardo, T.; Renzi, S.; Dotti, S.; Cinotti, S.; Ferrari, M. Isolation and Cryopreservation of Animal Mesenchymal Stromal Cells. In *Cryopreservation in Eukaryotes*; Jimenez, F.M., Akdemir, H., Eds.; IntechOpen: London, UK, 2016. [CrossRef]
19. Minonzio, G.; Corazza, M.; Mariotta, L.; Gola, M.; Zanzi, M.; Andolfi, E.; De Fazio, D.; Soldati, G. Frozen adipose derived mesenchymal stem cells maintain high capability to grow and differentiate. *Cryobiology* **2014**, *69*, 211–216. [CrossRef]
20. Geissler, S.; Textor, M.; Kühnisch, J.; Könnig, D.; Klein, O.; Ode, A.; Pfitzner, T.; Adjaye, J.; Kasper, G.; Duda, G.N. Functional comparison of chronological and in vitro aging: Differential role of the cytoskeleton and mitochondria in mesenchymal stromal cells. *PLoS ONE* **2012**, *7*. [CrossRef]
21. Martinello, T.; Bronzini, I.; Maccatrozzo, L.; Iacopetti, I.; Sampaolesi, M.; Mascarello, F.; Patruno, M. Cryopreservation does not affect the stem characteristics of multipotent cells isolated from equine peripheral blood. *Tissue Eng. Part C Methods* **2010**, *16*, 771–781. [CrossRef] [PubMed]
22. Fuller, B.J. Cryoprotectants: The essential antifreezes to protect life in the frozen state. *CryoLetters* **2004**, *25*, 375–388.
23. Ock, S.A.; Rho, G.J. Effect of dimethyl sulfoxide (DMSO) on cryopreservation of porcine mesenchymal stem cells (pMSCs). *Cell Transplant.* **2011**, *20*, 1231–1239. [CrossRef]
24. Robb, K.P.; Fitzgerald, J.C.; Barry, F.; Viswanathan, S. Mesenchymal stromal cell therapy: Progress in manufacturing and assessments of potency. *Cytotherapy* **2019**, *21*, 289–306. [CrossRef]
25. Dominici, M.; Blanc, K.L.; Mueller, I.; Slaper-Cortenbach, I.; Marini, F.C.; Krause, D.S.; Denas, R.J.; Keating, A.; Prockop, D.J.; Horwiz, E.M. Minimal criteria or defining multipotent mesenchymal stromal cells. The International Society for Cellular Therapy position statement. *Cytotherapy* **2006**, *8*, 315–317. [CrossRef] [PubMed]
26. Lee, K.S.; Kang, H.W.; Lee, H.T.; Kim, H.J.; Kim, C.L.; Song, J.Y.; Lee, K.W.; Cha, S.H. Sequential sub-passage decreases the differentiation potential of canine adipose-derived mesenchymal stem cells. *Res. Vet. Sci.* **2014**, *96*, 267–275. [CrossRef] [PubMed]
27. Reich, C.M.; Raabe, O.; Wenisch, S.; Bridger, P.S.; Kramer, M.; Arnhold, S. Isolation, culture and chondrogenic differentiation of canine adipose tissue and bone marrow-derived mesenchymal stem cells—A comparative study. *Vet. Res. Commun.* **2012**, *36*, 139–148. [CrossRef] [PubMed]
28. Han, S.M.; Han, S.H.; Coh, Y.R.; Jang, G.; Chan, R.J.; Kang, S.K.; Lee, H.W.; Youn, H.Y. Enhanced proliferation and differentiation of Oct4 and Sox2 overexpressing human adipose tissue mesenchymal stem cells. *Exp. Mol. Med.* **2014**, *46*, e101. [CrossRef]
29. Duan, W.; Lopez, M.J. Effects of Cryopreservation on Canine Multipotent Stromal Cells from Subcutaneous and Infrapatellar Adipose Tissue. *Stem Cell Rev.* **2016**, *12*, 257–268. [CrossRef]
30. Antebi, B.; Asher, A.M.; Rodriguez, L.A., 2nd; Moore, R.K.; Mohammadipoor, A.; Cancio, L.C. Cryopreserved mesenchymal stem cells regain functional potency following a 24-h acclimation period. *J. Transl. Med.* **2019**, *17*, 297. [CrossRef]

animals

Review

Chicken Mesenchymal Stem Cells and Their Applications: A Mini Review

Andrea Svoradova, Vladimir Zmrhal, Eva Venusova and Petr Slama *

Department of Animal Morphology, Physiology and Genetics, Faculty of AgriSciences, Mendel University in Brno, 613 00 Brno, Czech Republic; andrea.svoradova@mendelu.cz (A.S.); vladimir.zmrhal@mendelu.cz (V.Z.); eva.venusova@mendelu.cz (E.V.)
* Correspondence: petr.slama@mendelu.cz

Simple Summary: Mesenchymal stem cells (MSCs) are multipotent stem cells that are capable of differentiation into bone, muscle, fat, and closely related lineages and express unique and specific cell surface markers. They can be used as an avian culture model to better understand osteogenic, adipogenic, and myogenic pathways. Moreover, MSCs could also be used as a model to study various developmental and physiological processes in avian and other species. To obtain a comprehensive overview of this topic, the keywords "mesenchymal stem cells", "chicken", "disease", "chicken dermatitis", "viral infections in chicken", and "antibiotics in chicken" were searched in WOS and PUBMED databases to obtain relevant information.

Abstract: Mesenchymal stem cells (MSCs) are multipotent progenitor cells that adhere to plastic; express the specific markers CD29, CD44, CD73, CD90, and CD105; and produce cytokines and growth factors supporting and regulating hematopoiesis. MSCs have capacity for differentiating into osteocytes, chondrocytes, adipocytes, and myocytes. They are useful for research toward better understanding the pathogenic potential of the infectious bursal disease virus, mineralization during osteogenesis, and interactions between MSCs as a feeder layer to other cells. MSCs are also important for immunomodulatory cell therapy, can provide a suitable strategy model for coculture with pathogens causing dermatitis disorders in chickens, can be cultured in vitro with probiotics and prebiotics with a view to eliminate the feeding of antibiotic growth promoters, and offer cell-based meat production. Moreover, bone marrow-derived MSCs (BM-MSCs) in coculture with hematopoietic progenitor/stem cells (HPCs/HSCs) can support expansion and regulation of the hematopoiesis process using the 3D-culture system in future research in chickens. MSCs' several advantages, including ready availability, strong proliferation, and immune modulatory properties make them a suitable model in the field of stem cell research. This review summarizes current knowledge about the general characterization of MSCs and their application in chicken as a model organism.

Keywords: chicken; mesenchymal stem cells; culture; disease; probiotics; applications

Citation: Svoradova, A.; Zmrhal, V.; Venusova, E.; Slama, P. Chicken Mesenchymal Stem Cells and Their Applications: A Mini Review. *Animals* **2021**, *11*, 1883. https://doi.org/10.3390/ani11071883

Academic Editors: Eleonora Iacono and Barbara Merlo

Received: 29 May 2021
Accepted: 19 June 2021
Published: 24 June 2021

1. Introduction

As poultry meat production has increased dramatically in recent years, attention is now turning more to ensuring the high quality of that output [1]. Because genetic modification and selection for growth can cause skeletal disorders or muscle degeneration, and thus have negative impacts for the poultry industry [2], innovative methods are important for maintaining production. Mesenchymal stem cells (MSCs) seem to provide a suitable tool for examining skeletal development in poultry. MSCs are multipotent cells able to differentiate into osteocytes, chondrocytes, adipocytes, and myocytes [3]. They have been isolated from many species, including chickens, sheep, cats, dogs, rats, mice, and humans [4–11].

MSCs have the capacity to adhere to plastic surfaces under in vitro conditions and to express several surface antigens [12]. They also have the ability to produce cytokines and growth factors, commonly referred to as the MSC secretome, that support and regulate hematopoiesis [13].

MSCs can be used as a model for cell culture to better understand differentiation pathways, as well as to identify supplements that can affect these interactions, and in the areas of viral, skeletal, and immunological research. MSCs have furthermore become a subject of research due to their easy isolation, in vitro proliferation, multi-lineage differentiation, support to hematopoiesis, cytokine and growth factor production, and usefulness for immunomodulatory purposes [14]. Moreover, MSCs are able to migrate through the peripheral circulation to damaged areas, where they proliferate and differentiate, thus facilitating the healing process through the activation of several mechanisms [15]. MSCs are able to reduce cell injury by the synergistic action of small molecules, extracellular vesicles (EVs), secreted by MSCs to maintain tissue homeostasis. Studying the physiological functions of MSCs can improve their application in regenerative medicine and increase our knowledge to better understand their biological behavior [16].

This review covers what is currently known about chicken MSCs. In its first part, we summarize knowledge about their sources, culture conditions, phenotype characterization, and differentiation. In the second part, we focus on the potential for MSCs to be used in three-dimensional (3D) culture and cell-to-cell interactions, their application as a feeder layer, their usefulness for meat production *in vitro*, their cryopreservation possibilities, and their usefulness in researching diseases in chickens (Figure 1).

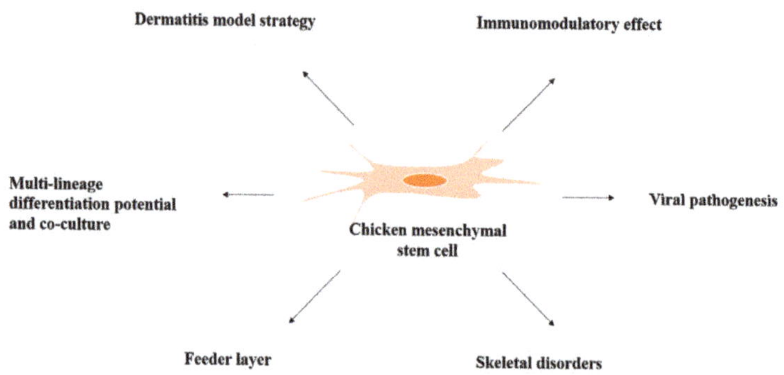

Figure 1. Multifunctional properties of chicken mesenchymal stem cells.

2. Characterization of Chicken MSCs

MSCs commonly are obtained where they were first discovered: from bone marrow as precursors for fibroblasts or stromal cells. An important part of MSC isolation is purification of samples from non-mesenchymal cell types such as hematopoietic and blood cells [17]. MSCs are usually aspirated from bone marrow (in which case they are termed bone marrow-derived MSCs, or BM-MSCs) and then isolated by sieving for plastic adherence in vitro. In addition, MSCs can be obtained from compact bones [18], Wharton's jelly (WJ-MSCs) [19], and lung (L-MSCs) [20].

The key characteristics defining MSCs have been based on their capacity for colony formation, potential for self-renewal, expression of surface markers, and multi-lineage differentiation. The availability of stem cell-specific markers in poultry has limitations; therefore, researchers have to rely on reports of cell surface markers in mammalian species. Use of markers to verify MSC identity is an important control step to eliminate experimental variability and obtain a homogenous population of MSCs [18]. In mammals, MSCs express surface markers such as CD73, CD90, and CD105 [21–23] and transcription factors that include Oct4 [24], Nanog [25], and Sox2 [26], where PouV is a chicken homolog of mammalian Oct4 [27]. MSCs isolated from chicken bone marrow exhibit features similar to those of mammalian MSCs. Furthermore, lung MSCs in chicken and other mammalian species may help in understanding the pathogenesis of infectious and non-infectious lung diseases and the mechanisms of lung injury repair [13]. All chicken MSC characteristic parameters are summarized in the Table 1.

Table 1. Comparison of chicken MSCs derived from different sources.

Sources	Type of Digestive	Separation	Morphology	Confluency	Positive Markers	Negative Markers	Transcription Factors	References
Bone marrow	-	Ficoll-Hypaque (1.090 g/mL)	spindle-shaped	14 days	CD44, CD90 CD105	CD45	PouV, Sox2, Nanog	[17]
		Percoll solution (1.073 g/mL)	spindle-shaped	2–3 days	CD44, CD29, CD71, CD73	CD31, CD34		[19]
Compact bones	0.25% collagenase		spindle-shaped	8–10 days	CD90, CD105, CD73, CD44	CD31, CD34 CD45		[18]
Lung	0.1% collagenase		spindle-shaped	5–7 days	CD29, CD73, CD90, CD105	CD34, CD45	OCT-4	[20]
	0.5 mg/mL collagenase type IV	Ficoll-Hypaque (1.090 g/mL)	spindle-shaped	14 days	CD44, CD90, CD105		PouV	[28]
Wharton's jelly	0.1% collagenase type IV		fibroblast-like shaped	5–6 days	CD29, CD44, CD71, CD73	CD31, CD34		[29]

While specific surface markers are easily evaluable, a definition of MSCs can be completed by their abilities to differentiate into classic mesodermal lineages of bone, fat, and cartilage. Because MSCs have the ability to differentiate into osteocytes, chondrocytes, myocytes, and adipocytes, they constitute a suitable and predictable source of cells for purposes of regenerative therapy [30–34]. Supplements such as ascorbic acid and dexamethasone [35] at defined concentrations are able to direct MSCs toward osteogenic differentiation. For chondrogenic differentiation, TGF- β1 often is used as an inducer [36].

The cells are stained using Alizarin Red and Von Kossa (VK) stains for determining mineralization and with alkaline phosphatase (ALP) [35] for detecting osteogenic differentiation at 7 and 14 days of treatment. Moreover, L-MSCs highly express the osteoblast-specific genes OPN and Col-1 in differentiated cells [20]. To stimulate adipogenesis, dexamethasone, indomethacin, insulin, and isobutylmethylxanthine are usually added to the cultures. After 96 h in culture in adipogenic media, Oil Red O is used to determine adipocyte induction. Adipogenic genes such as PPARγ, FABP2, c/EBPα, and c/EBPβ can be used for molecular detection of adipocyte differentiation [18,19]. Hydrocortisone and dexamethasone are well-established inducers of myogenesis. After 72 h, myogenic gene expressions of MyoD, Pax7, Myf5, and myogenin are used in combination with quantitative reverse transcription polymerase chain reaction for detecting myogenesis [18,19]. Vascular endothelial growth factor (VEGF), basic fibroblast growth factor (bFGF), and insulin-like growth factor (IGF-1) are used for endothelial induction, and endothelial markers such as CD34 and CD31 are used for its determination [19]. Moreover, neurogenic differentiation is induced by addition of glial cell-derived neurotrophic factor. Wang et al. (2018) [20] established that nestin and MAP-2, the markers of neural cells, are highly expressed in differentiated L-MSCs. Supplements such as antibiotics and growth factors can affect the phenotypic properties of MSCs and their multi-lineage potential.

MSCs are able to also influence functions of such major immune cells as dendritic cells, T cells, B cells, and natural killer (NK) cells. Although the immunomodulatory mechanism of MSCs is not yet fully understood, there are some mechanisms by which MSCs can be influenced. High levels of pro-inflammatory cytokines can activate MSCs to produce immunosuppressive cytokines, chemokines, as well as nitric oxide (NO) [37].

NO is one of the factors that can suppress T cell proliferation. Moreover, it has been noted that MSCs and macrophages also suppress T cell proliferation via NO inhibition of Stat5 phosphorylation [38,39], and the production of NO during MSC differentiation into chondrocytes has been observed. MSCs have typical in vitro modulatory functions, such as to inhibit T cell and B cell proliferation, as well as DCs differentiation [40,41], thereby resulting in inhibition of immune responses both in vitro and in vivo. Cytokines and growth factors are specific modulators by which MSCs can influence inflammatory responses. This makes MSCs a very promising tool for immunomodulatory cell therapy in immune-mediated diseases. T cells are characterized by the effect of cell proliferation and cytokine secretion. MSC inhibition of T cell proliferation is especially important for immune homeostasis and self-tolerance maintenance [42]. Like mammalian MSCs, chicken MSCs also have immunoregulatory function and inhibit in vitro the mitogenic response of T cells. In chickens, a correlation between NO production and T cell suppression in coculture with MSCs has been noted. Nevertheless, the role of NO in the MSC and T cell coculture system remains unknown.

This review now directs its focus to chickens as a model platform for application of available techniques previously used in other animals that can expand and enrich our knowledge in the area of avian research.

3. Biological Properties of Chicken MSCs

3.1. Three-Dimensional (3D) Culture and MSC Interactions

Hematopoietic progenitor/stem cells (HPCs/HSCs), MSCs, and multiple elements of the extracellular matrix are components belonging to the microenvironment. It has been noted that MSCs play a crucial role in HPC/HSC function and self-renewal [43,44]. In a study [45] involving the 3D collagen-based culture model, MSCs were used in coculture with HPCs/HSCs. In another study, a significant effect of BM-MSCs upon HPCs/HSCs in terms of self-renewal, maintenance, and differentiation was recorded [46]. BM-MSCs are used as stromal cells for coculture with HPCs/HSCs obtained from umbilical cord blood, and WJ-MSCs are applied as stromal support for HSCs. BM-MSCs in coculture with HPCs/HSCs using the 3D-culture system enable HPC/HSC expansion and regulation of the hematopoiesis process [45]. Study [45] also observed fibronectin production by BM-MSCs to support synthesis of collagen type I and production of osteopontin, which is important for osteogenic differentiation. Moreover, HPCs in the collagen gel containing MSCs revealed initial differentiation into the myeloid lineage, as proven by positivity for CD45. This differentiation was shown by comparison to stromal-cell free conditions as previously described [47].

On the whole, MSCs constitute a promising tool for cellular modulation by secretion and interaction of appropriate molecules to improve regeneration processes in many types of tissues. There are available today many scaffolds, such as micro-/nano- electrospun (EFs) fibers [48,49] or polycaprolactone (PCL) EFs, that modulate paracrine signaling to support cell attachment, proliferation, as well as maintenance of cell stemness and pluripotency [50]. Based on these findings, 3D culture can be used to study the effects of various substances in coculture with MSCs also in chickens in order to better understand the interactions and substitute for the natural environment.

3.2. Feeder Cells Layer

Primordial germ cells (PGCs) are progenitors of germ cells, and they have important roles in spermatozoa and egg formation in the adult organism [51]. The most commonly used feeder cells for PGCs culture are xeno-animal buffalo rat liver (BRL) cells, Sandoz inbred mouse-derived thioguanine-resistant and ouabain-resistant (STO) cells, or mouse embryonic fibroblast (MEF) cells, but these cells have their limitations, as contamination of various types can disturb the potential of PGCs. PGC proliferation in vitro depends upon the feeder cells having a powerful capacity to proliferate and secrete cytokines [52,53]. The authors [54] reported that feeder cells can promote proliferation of circulating blood

PGCs (cPGCs) and gonadal PGCs (gPGCs) in vitro and that they have the characteristics of an effective feeder layer. Several studies have shown that an MSC-feeder layer ensures all conditions for human-induced pluripotent stem cell (hiPSC), human embryonic stem cell (hESC), and mouse embryonic stem cell (mESC) proliferation and expansion and, moreover, while maintaining cell pluripotency [55–57]. MSCs have potential as a feeder cell layer in vitro to provide for expansion of chicken PGCs [54]. Further research is needed to analyze the potential interactions of MSCs via MSC-secreted cytokines with the different chicken cell types.

3.3. Infectious Bursal Disease Virus

Poultry are surrounded today by numerous bacterial and viral agents [58]. Infectious bursal disease virus (IBDV), also known as Gumboro disease, was first observed about 60 years ago as an immunosuppressive disorder in young chickens. This virus infects the bursa of Fabricius in young chickens at early ages, and a subclinical form of infection occurs in older birds [59]. The infection leads to morbidity, mortality, and immunosuppression [60]. In vivo presence of IBDV has been detected in several tissues, including bone marrow [61,62]. It is known that MSCs are important cells with the ability to support hematopoiesis and modulate differentiation of hematopoietic stem cells via the expression of cell adhesion molecules necessary for cell-to-cell interactions that result in cytokine and growth factor release [59]. Different types of immune cells are able to modify the host response to IBDV through release of cytokines such as interferon (IFN)- α and IFN-γ [63–67] and pro-inflammatory cytokines such as interleukin (IL)-2 [64,65], IL-18 [65], and IL-6 [65,66]. Inasmuch as the virus does not proliferate in chicken fibroblast cells [68], it therefore can be hypothesized that MSCs could also be a target for IBDV infection. In study [13] they discovered an interaction between IBDV and MSCs. Because MSCs participate in the regulation of hematopoietic precursor differentiation and proliferation, examining these interactions can contribute to better understanding of the virus's pathogenesis. Moreover, we believe that such stem cells or cell-based vaccines will provide a promising platform or strategy for controlling IBDV and other viruses (zoonoses) with unknown potential risk as a means of preventing potential pandemic diseases.

3.4. Skeletal Diseases

Vitamin D3, calcitriol (1,25-(OH)2D3), plays an important role as a nutritional factor relevant to poultry bone strength [69]. Differences between birds and mammals in the formation of the long bones relate to dietary aspects and metabolic activity of 1,25-(OH)2D3. Specifically, avian species have long bone development without secondary ossification until hatching. Ossification only occurs in the proximal and distal ends of the tibiotarsus and tarsometatarsus [70]. The calcium (Ca) level in blood increases the bone strength, and vitamin D facilitates Ca absorption. Therefore, dietary supplementation with adequate Ca and vitamin D is important. It is known that egg laying causes large Ca losses, and so it is necessary to supplement vitamin D to maintain optimal bone structure in laying hens. Tibial dyschondroplasia (TD) is a skeletal disorder in growing chickens characterized by an avascular and non-mineralized growth plate that can lead to deformed tibial bone and lameness [71]. Mineral deficiencies in the diet, and especially insufficient supply of Ca, also can cause keel bone fracture [72–77]. Stimulatory effects of 1,25-(OH)2D3 on osteogenic differentiation and mineralization have been recorded in humans [78–81], in rat osteoblasts [82,83], in mouse osteoblasts [84–87], and in chicken osteoblasts [88,89]. The amount of an administered substance is important, however, due to the possibility of negative effects. Because MSCs have the abilities for self-renewal and multi-lineage differentiation to osteogenic lineages, they can be used in the study of mineralization during osteogenesis. Vitamin D3 is related to immunoregulation, anti-oxidation, anti-cancer actions, cardiovascular benefits, and such aspects of bone development as osteogenic differentiation and mineralization [89]. Available data suggest that administration of 1,25-(OH)2D3 is important for optimizing bone health in the poultry industry and that avian

BM-MSCs constitute a useful tool for examining underlying effects [90]. As we also know, poultry and pig meats have shown the greatest consumption increases [91], and poultry meat consumption has increased in all regions of the world [92].

3.5. Probiotics and Prebiotics

Antibiotic growth promoters (AGPs) are widely used in protecting poultry against pathogens and disease and improving growth performance. Prolonged administration of AGPs, however, can lead to bacterial resistance and result in drug residues in poultry products as well as a prohibition against using antibiotics in poultry production. Therefore, it is necessary to develop an alternative pathway for improving production [93].

The chicken gastrointestinal tract is home to a population of microorganisms living in symbiotic relationship with their host, and this relationship is important for the host's nutrition, metabolism, and immunity—indeed its homeostasis. Although the intestinal microbial environment in adult chickens is highly stable, it can be influenced by feed or stress. One of the major causes of deterioration in meat quality relates to interactions between macronutrients and medications [94].

Live microbial feed supplements known as probiotics, such as *Lactobacillus*, *Bifidobacterium*, or yeasts [95], can confer health benefits to the host. On the other hand, prebiotics, which are non-digestible food ingredients, can enhance lipid metabolism and support polyunsaturated fatty acid levels in chicken meat [96] while resulting in increased levels of health-promoting bacteria within the intestinal tract. Administration of probiotics and prebiotics in feed can improve a flock's immunity by reducing harmful microbes in the intestine and thus the need for antibiotics [20,93]. Although the efficacy of probiotics and prebiotics in poultry has not yet been sufficiently studied, it is known that MSCs interact with a wide range of intestinal bacteria having significant effects on MSC function [97,98]. MSCs have the ability to home and engraft in the lamina propria of the gastrointestinal tract during intestinal inflammation and exert potent immunomodulatory functions [99]. From this point of view, therefore, it would be pertinent to analyze in vitro interactions between MSCs isolated from AGP-treated animals and applied pro- and prebiotics in order to assess the MSCs profile in terms of, for example, viability, immunomodulatory properties, and values of inflammatory cytokines. With this in mind, the studies would progress to examining the possible effects of pro- and prebiotics on MSC physiology as a possible future replacement for AGPs.

3.6. Chicken Dermatitis

Gangrenous dermatitis (GD) is a disorder that affects broiler chickens and results in economic losses in the poultry industry worldwide [100]. GD is primarily caused by the Gram-positive anaerobic bacilli *Clostridium perfringens* type A [101–104], *Clostridium septicum*, and *Staphylococcus aureus* [105]. The disease is characterized by hemorrhage, congestion, and necrosis of the skin as indicated by edema. The breast, abdomen, back, thighs, tail, and wings are the most significantly affected body areas [106]. The dermatitis results in lower meat quality [107]. Typical symptoms of GD are poor appetite, decreased muscle coordination, skin edema, leg weakness, and ultimately crepitus [108]. GD in chickens causes decreased splenocyte proliferation in response to concanavalin A (Con A) or lipopolysaccharide (LPS); greater levels of serum NO and a-1-acid glycoprotein (a-1-AGP); higher levels of T cells, B cells, and macrophages; as well as increased levels of transcripts encoding IL-8, IFN-a, TNFSF-15, and LITAF compared with GD-free chickens [94]. Therefore, MSCs can be useful in evaluating a variety of diseases, including mainly tissue-related and immune-mediated diseases, due to their ability to modulate the innate and adaptive immune systems. Several experimental models have been used to clarify the role of *C. septicum*, *C. perfringens*, and *S. aureus* in the pathogenesis of GD in vivo [109], but MSC coculture in vitro with the aforementioned pathogens can be a promising approach for assessing their interactions as well as their immunomodulatory properties.

3.7. Meat "In Vitro"

Nowadays, meat production is the major source of pollution. World meat production contributes between 15 and 24% of greenhouse gas emissions [110]. It is known that chicken meat production requires 3, 918m^3/ton of water [111]. Therefore, satisfying the demand for meat in the future will be a challenge when we intend on maximizing the use of agricultural sources and reducing the greenhouse gas production. Meat worldwide consumption was calculated by Fiala [112], who predicts a 72% increase in meat production in 2030 compared to 2000. The 1918 H1N1 Spanish Influenza pandemic [113] as well as SARS-CoV-2 are of utmost concern as they have spread to almost all countries and killed thousands of people worldwide [114]. The correlation between meat production and outbreaks of diseases cannot be overlooked.

In vitro meat culturing seems to be a suitable substitution for conventional meat production. In vitro cultured meat from stem cells in controlled culture and physiological conditions in the laboratory uses MSCs that are able to differentiate to myocyte or induced pluripotent stem cells (iPSCs) by genome reprogramming of somatic cells [115]. In vitro meat culturing has many advantages: (a) in vitro cultivation is faster than growth, (b) the impact of cultivation on the environment is lower, and (c) muscle tissue is cultivated without affecting the skeleton [116]; moreover, cultured meat is a healthier, cleaner, and disease-free animal protein source compared to commercial farming [117] (Figure 2).

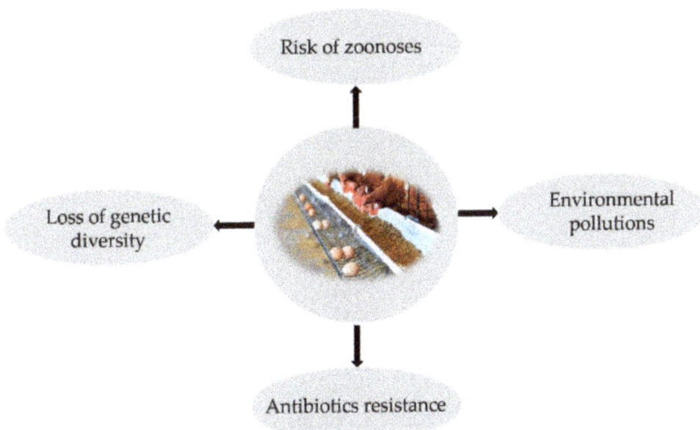

Figure 2. Disadvantages of commercial farming of livestock.

Cell culture also involves scaffold-based cell and tissue approaches relying on the isolation, culture, and differentiation into myoblasts. However, there is possibility to coculture stem cells with adipocytes, which support them to differentiate, and then form myofibers [118]. However, cell culture requires many growth factors such as IGF, bFGF, HGF, Wnt3a, and Wnt7a to promote differentiation into myotubes and myofibers [119]. Moreover, scaffolds have an important role in the cell culture-based approach for in vitro meat production. Scaffolds are commonly made of natural and edible polymeric biomaterials such as collagen that allow 3D tissue culture and subsequent complex structuring of synthetic meat [120,121].

We believe that this transformation in the farm process will be inevitable. Moreover, this technology offers an opportunity for non-meat eaters because this meat is safe and free from animal slaughtering and cruelty. Since in vitro production is controlled, it is feasible to alter high-quality meat production on a sustainable basis. From this point of view, extra embryonic and adult MSCs could be navigated toward myocyte differentiation, which ultimately forms myotubes and are, thus, suggested as a potential starting point for in vitro chicken meat production.

3.8. Cryopreservation

In recent years, several animal stem cell banks have been established worldwide. Although these banks are not used for therapeutic purposes, they are considered a promising way for the storage of animal genetic resources [108]. For this reason, we have noticed the increased demand for national gene banks to preserve either native or worldwide biodiversity. These banks preserve genetic information from many important (already endangered) livestock species. However, in addition to embryos and gametes, adult stem cells also represent a significant genetic resource that can be obtained from various biological sources.

The stem cell banks also give a source of stem cell lines that have high quality and safety standards [122]. Many animal embryonic and adult stem cell lines have been made and cryopreserved, including primordial, bone marrow mesenchymal, neural, cardiac, endothelial, adipose, and umbilical cord mesenchymal stem cells [123,124]. Biological research could use the method of cryopreservation of stem cells in other fields as well, as it has already given promising results [123]. Cryopreservation of stem cells is important to provide storage of high cell numbers, fast transport, and to preserve cells for long periods. Due to the increased level of endangered animals, it is important to preserve genetic material for future applications.

The chicken was the first farm animal with a completely sequenced genome. Because of its in ovo embryonic development rather than in utero, the chicken is a suitable model for embryology and development studies. The chicken provides a model organism for the study of cancer and viruses. The first tumor virus and oncogene (src), Rous sarcoma virus, was identified in the chicken. The immune system of chickens provides the first indication of the distinctions between T and B cells, with the B-cell based avian bursa of Fabricius [125]. Therefore, the chicken is an important model for evolution, embryology, cell biology, immunology, virology, oncology, and gene regulation studies. From this point of view, it is also important to cryopreserve MSCs for subsequent assessment of MSCs in vitro and further usage in the future (e.g., iPSCs).

4. Conclusions

This review has emphasized the importance of chicken MSCs for their self-renewal potential and multi-lineage differentiation as well as current knowledge concerning their usefulness for examining pathogenic potential of infectious bursal disease virus, studying mineralization during osteogenesis, and using MSCs as a feeder layer. MSCs also constitute a very promising tool for immunomodulatory cell therapy in immune-mediated diseases due to their inhibition effect on T cell proliferation via NO production. Moreover, MSC-based treatment could be a model strategy in studying chicken dermatitis disorders as well as for reducing the need to administer AGPs. Next, BM-MSCs in coculture with hematopoietic progenitor/stem cells can provide expansion and regulation of the hematopoiesis process using the 3D-culture system for future research in chickens. Finally, current meat production methods are associated with many problems such as animal welfare issues, risk of infectious diseases, biodiversity loss, and environmental pollution. Therefore, MSCs provide an alternative way to eliminate these problems using in vitro meat culturing. Based on these findings, we may conclude that MSCs can provide a useful model in the field of chicken stem cell research.

Author Contributions: Conceptualization, A.S. and V.Z.; formal analysis, P.S.; investigation, A.S. and P.S.; writing—original draft preparation, A.S.; writing—review and editing, V.Z., E.V. and P.S.; visualization, P.S.; supervision, P.S.; project administration, P.S.; funding acquisition, P.S. All authors have read and agreed to the published version of the manuscript.

Funding: This research was funded by the Technology Agency of the Czech Republic (TJ04000511).

Institutional Review Board Statement: Not applicable.

Data Availability Statement: No new data were created or analyzed in this study.

Conflicts of Interest: The authors declare no conflict of interest.

References

1. Mir, N.A.; Rafiq, A.; Kumar, F.; Singh, V.; Shukla, V. Determinants of broiler chicken meat quality and factors affecting them: A review. *J. Food Sci. Technol.* **2017**, *54*, 2997–3009. [CrossRef] [PubMed]
2. Fornari, M.B.; Zanella, R.; Ibelli, A.M.; Fernandes, L.T.; Cantão, M.E.; Thomaz-Soccol, V.; Ledur, M.C.; Peixoto, J.O. Unraveling the associations of osteoprotegerin gene with production traits in a paternal broiler line. *SpringerPlus* **2014**, *3*, 1–8. [CrossRef]
3. Prockop, D.J. Marrow stromal cells as stem cells for nonhematopoietic tissues. *Science* **1997**, *276*, 71–74. [CrossRef]
4. Crigler, L.; Kazhanie, A.; Yoon, T.J.; Zakhari, J.; Anders, J.; Taylor, B.; Virador, V.M. Isolation of a mesenchymal cell population from murine dermis that contains progenitors of multiple cell lineages. *FASEB J.* **2007**, *21*, 2050–2063. [CrossRef] [PubMed]
5. Dumas, A.; Le Drévo, M.A.; Moreau, M.F.; Guillet, C.; Baslé, M.F.; Chappard, D. Isolation of osteoprogenitors from murine bone marrow by selection of CD11b negative cells. *Cytotechnology* **2008**, *58*, 163. [CrossRef]
6. Kar, S.; Mitra, S.; Banerjee, E.R. Isolation and culture of embryonic stem cells, mesenchymal stem cells, and dendritic cells from humans and mice. *Methods. Mol. Biol.* **2015**, *1516*, 145–152. [CrossRef]
7. Kumar, K.; Agarwal, P.; Das, K.; Mili, B.; Madhusoodan, A.P.; Kumar, A.; Bag, S. Isolation and characterization of mesenchymal stem cells from caprine umbilical cord tissue matrix. *Tissue Cell* **2016**, *48*, 653–658. [CrossRef]
8. Li, H.; Ghazanfari, R.; Zacharaki, D.; Lim, H.C.; Scheding, S. Isolation and characterization of primary bone marrow mesenchymal stromal cells. *Ann. N. Y. Acad. Sci.* **2016**, *1370*, 109–118. [CrossRef]
9. Krešić, N.; Šimić, I.; Lojkić, I.; Bedeković, T. Canine adipose derived mesenchymal stem cells transcriptome composition alterations: A step towards standardizing therapeutic. *Stem Cells Int.* **2017**. [CrossRef]
10. Nakamura, M.; Nishida, H.; Yoshizaki, K.; Akiyoshi, H.; Hatoya, S.; Sugiura, K.; Inaba, T. Canine mesenchymal stromal cell-conditioned medium promotes survival and neurite outgrowth of neural stem cells. *J. Vet. Med. Sci.* **2020**. [CrossRef]
11. Munoz, J.L.; Greco, S.J.; Patel, S.A.; Sherman, L.S.; Bhatt, S.; Bhatt, R.S.; Shrensel, J.A.; Guan, Y.Z.; Xie, G.; Ye, J.H.; et al. Feline bone marrow-derived mesenchymal stromal cells (MSCs) show similar phenotype and functions with regards to neuronal differentiation as human MSCs. *Differentiation* **2012**, *84*, 214–222. [CrossRef] [PubMed]
12. Dominici, M.L.B.K.; Le Blanc, K.; Mueller, I.; Slaper-Cortenbach, I.; Marini, F.C.; Krause, D.S.; Horwitz, E.M. Minimal criteria for defining multipotent mesenchymal stromal cells. The International Society for Cellular Therapy position statement. *Cytotherapy* **2006**, *8*, 315–317. [CrossRef]
13. Majumdar, M.K.; Thiede, M.A.; Mosca, J.D.; Moorman, M.; Gerson, S.L. Phenotypic and functional comparison of cultures of marrow–derived mesenchymal stem cells (MSCs) and stromal cells. *J. Cell. Physiol.* **1998**, *176*, 57–66. [CrossRef]
14. Eleuteri, S.; Fierabracci, A. Insights into the secretome of mesenchymal stem cells and its potential applications. *Inter. J. Mol. Sci.* **2020**, *20*, 4597. [CrossRef]
15. Fu, X.; Liu, G.; Halim, A.; Ju, Y.; Luo, Q.; Song, G. Mesenchymal stem cell migration and tissue repair. *Cells* **2019**, *8*, 784. [CrossRef]
16. Zannetti, A.; Benga, G.; Brunetti, A.; Napolitano, F.; Avallone, L.; Pelagalli, A. Role of Aquaporins in the Physiological Functions of Mesenchymal Stem Cells. *Cells* **2020**, *9*, 2678. [CrossRef] [PubMed]
17. Khatri, M.; Sharma, J.M. Susceptibility of chicken mesenchymal stem cells to infectious bursal disease virus. *J. Virol. Methods.* **2009**, *160*, 197–199. [CrossRef] [PubMed]
18. Adhikari, R.; Chen, C.; Waters, E.; West, F.D.; Kim, W.K. Isolation and differentiation of mesenchymal stem cells from broiler chicken compact bones. *Front. Physiol.* **2019**, *9*, 1892. [CrossRef] [PubMed]
19. Bai, C.; Hou, L.; Ma, Y.; Chen, L.; Zhang, M.; Guan, W. Isolation and characterization of mesenchymal stem cells from chicken bone marrow. *Cell Tissue Bank.* **2013**, *14*, 437–451. [CrossRef]
20. Wang, X.; Wang, J.J.; Ji, H.; Guan, W.; Zhao, Y. Isolation, culture, and characterization of chicken lung-derived mesenchymal stem cells. *Can. J. Vet. Res.* **2018**, *82*, 225–235.
21. Teresa Conconi, M.; Di Liddo, R.; Tommasini, M.; Calore, C.; Paolo Parnigotto, P. Phenotype and differentiation potential of stromal populations obtained from various zones of human umbilical cord: An overview. *J. Tissue Eng. Regen. Med.* **2011**, *4*, 6–20. [CrossRef]
22. Lin, C.S.; Xin, Z.C.; Dai, J.; Lue, T.F. Commonly used mesenchymal stem cell markers and tracking labels: Limitations and challenges. *Histol. Histopathol.* **2013**, *28*, 1109. [CrossRef]
23. Cruz, F.F.; Rocco, P.R.M. The potential of mesenchymal stem cell therapy for chronic lung disease. *Expert. Rev. Respir. Med.* **2020**, *14*, 31–39. [CrossRef] [PubMed]
24. Nichols, J.; Zevnik, B.; Anastassiadis, K.; Niwa, H.; Klewe-Nebenius, D.; Chambers, I.; Schöler, H.; Smith, A. Formation of pluripotent stem cells in the mammalian embryo depends on the POU transcription factor Oct4. *Cell* **1998**, *95*, 379–391. [CrossRef]
25. Chambers, I.; Colby, D.; Robertson, M.; Nichols, J.; Lee, S.; Tweedie, S.; Smith, A. Functional expression cloning of Nanog, a pluripotency sustaining factor in embryonic stem cells. *Cell* **2003**, *113*, 643–655. [CrossRef]
26. Avilion, A.A.; Nicolis, S.K.; Pevny, L.H.; Perez, L.; Vivian, N.; Lovell-Badge, R. Multipotent cell lineages in early mouse development depend on SOX2 function. *Genes Dev.* **2003**, *17*, 126–140. [CrossRef]
27. Lavial, F.; Acloque, H.; Bertocchini, F.; MacLeod, D.J.; Boast, S.; Bachelard, E.; Montillet, G.; Thenot, S.; Sang, H.M.; Stern, C.D.; et al. The Oct4 homologue PouV and Nanog regulate pluripotency in chicken embryonic stem cells. *Development* **2007**, *134*, 3549–3563. [CrossRef] [PubMed]
28. Khatri, M.; O'Brien, T.D.; Goyal, S.M.; Sharma, J.M. Isolation and characterization of chicken lung mesenchymal stromal cells and their susceptibility to avian influenza virus. *Dev. Comp. Immunol.* **2010**, *34*, 474–479. [CrossRef]

29. Bai, C.; Li, X.; Hou, L.; Zhang, M.; Guan, W.; Ma, Y. Biological characterization of chicken mesenchymal stem/progenitor cells from umbilical cord Wharton's jelly. *Mol. Cell. Biochem.* **2013**, *376*, 95–102. [CrossRef] [PubMed]
30. Krampera, M.; Glennie, S.; Dyson, J.; Scott, D.; Laylor, R.; Simpson, E.; Dazzi, F. Bone marrow mesenchymal stem cells inhibit the response of naive and memory antigen-specific T cells to their cognate peptide. *Blood* **2003**, *101*, 3722–3729. [CrossRef]
31. Maitra, B.; Szekely, E.; Gjini, K.; Laughlin, M.J.; Dennis, J.; Haynesworth, S.E.; Koç, O.N. Human mesenchymal stem cells support unrelated donor hematopoietic stem cells and suppress T-cell activation. *Bone Marrow Transplant.* **2004**, *33*, 597–604. [CrossRef] [PubMed]
32. Beyth, S.; Borovsky, Z.; Mevorach, D.; Liebergall, M.; Gazit, Z.; Aslan, G.E.; Rachmilewitz, J. Human mesenchymal stem cells alter antigen-presenting cell maturation and induce T-cell unresponsiveness. *Blood* **2005**, *105*, 2214–2219. [CrossRef]
33. Groh, M.E.; Maitra, B.; Szekely, E.; Koç, O.N. Human mesenchymal stem cells require monocyte-mediated activation to suppress alloreactive T cells. *Ex. Hematol.* **2005**, *33*, 928–934. [CrossRef] [PubMed]
34. Meisel, R.; Zibert, A.; Laryea, M.; Göbel, U.; Däubener, W.; Dilloo, D. Human bone marrow stromal cells inhibit allogeneic T-cell responses by indoleamine 2, 3-dioxygenase–mediated tryptophan degradation. *Blood* **2004**, *103*, 4619–4621. [CrossRef] [PubMed]
35. Parhami, F.; Morrow, A.D.; Balucan, J.; Leitinger, N.; Watson, A.D.; Tintut, Y.; Berliner, J.A.; Demer, L.L. Lipid oxidation products have opposite effects on calcifying vascular cell and bone cell differentiation: A possible explanation for the paradox of arterial calcification in osteoporotic patients. *Arterioscler. Tthromb. Vasc. Biol.* **1997**, *17*, 680–687. [CrossRef]
36. Kocamaz, E.; Gok, D.; Cetinkaya, A.; Tufan, A.C. Implication of C-type natriuretic peptide-3 signaling in glycosaminoglycan synthesis and chondrocyte hypertrophy during TGF-β1 induced chondrogenic differentiation of chicken bone marrow-derived mesenchymal stem cells. *J. Mol. His.* **2012**, *43*, 497–508. [CrossRef]
37. Kyurkchiev, D.; Bochev, I.; Ivanova-Todorova, E.; Mourdjeva, M.; Oreshkova, T.; Belemezova, K.; Belemezova, K.; Kyurkchiev, S. Secretion of immunoregulatory cytokines by mesenchymal stem cells. *World J. Stem Cells* **2014**, *6*, 552–570. [CrossRef]
38. Mazzoni, A.; Bronte, V.; Visintin, A.; Spitzer, J.H.; Apolloni, E.; Serafini, P.; Zanovello, P.; Segal, D.M. Myeloid suppressor lines inhibit T cell responses by an NO-dependent mechanism. *J. Immunol.* **2002**, *168*, 689–695. [CrossRef]
39. Mais, A.; Klein, T.; Ullrich, V.; Schudt, C.; Lauer, G. Prostanoid pattern and iNOS expression during chondrogenic differentiation of human mesenchymal stem cells. *J. Cell. Biochem.* **2006**, *98*, 798–809. [CrossRef]
40. Glennie, S.; Soeiro, I.; Dyson, P.J.; Lam, E.W.F.; Dazzi, F. Bone marrow mesenchymal stem cells induce division arrest anergy of activated T cells. *Blood* **2005**, *105*, 2821–2827. [CrossRef]
41. Corcione, A.; Benvenuto, F.; Ferretti, E.; Giunti, D.; Cappiello, V.; Cazzanti, F.; Risso, M.; Gualandi, F.; Luigi, G.; Pistoia, M.V.; et al. Human mesenchymal stem cells modulate B-cell functions. *Blood* **2006**, *107*, 367–372. [CrossRef]
42. Djouad, F.; Plence, P.; Bony, C.; Tropel, P.; Apparailly, F.; Sany, J.; Jorgensen, C. Immunosuppressive effect of mesenchymal stem cells favors tumor growth in allogeneic animals. *Blood* **2003**, *102*, 3837–3844. [CrossRef] [PubMed]
43. Adams, G.B.; Scadden, D.T. The hematopoietic stem cell in its place. *Nat. Immunol.* **2006**, *7*, 333–337. [CrossRef] [PubMed]
44. Wagner, W.; Roderburg, C.; Wein, F.; Diehlmann, A.; Frankhauser, M.; Schubert, R.; Eckstein, V.; Ho, A.D. Molecular and secretory profiles of human mesenchymal stromal cells and their abilities to maintain primitive hematopoietic progenitors. *Stem Cells* **2007**, *25*, 2638–2647. [CrossRef] [PubMed]
45. Leisten, I.; Kramann, R.; Ferreira, M.S.V.; Bovi, M.; Neuss, S.; Ziegler, P.; Wagner, W.; Knüchel, R.; Schneider, R.K. 3D co-culture of hematopoietic stem and progenitor cells and mesenchymal stem cells in collagen scaffolds as a model of the hematopoietic niche. *Biomaterials* **2012**, *33*, 1736–1747. [CrossRef] [PubMed]
46. Walenda, T.; Bokermann, G.; Ferreira, M.S.V.; Piroth, D.M.; Hieronymus, T.; Neuss, S.; Zenke, M.; Ho, A.D.; Müller, A.M.; Wagner, W. Synergistic effects of growth factors and mesenchymal stromal cells for expansion of hematopoietic stem and progenitor cells. *Exp. Hematol.* **2011**, *39*, 617–628. [CrossRef] [PubMed]
47. Walenda, T.; Bork, S.; Horn, P.; Wein, F.; Saffrich, R.; Diehlmann, A.; Eckstein, V.; Ho, A.D.; Wagner, W. Co–culture with mesenchymal stromal cells increases proliferation and maintenance of haematopoietic progenitor cells. *J. Cell Mol. Med.* **2010**, *14*, 337–350. [CrossRef]
48. Rustad, K.C.; Wong, V.W.; Sorkin, M.; Glotzbach, J.P.; Major, M.R.; Rajadas, J.; Longaker, M.T.; Gurtner, G.C. Enhancement of mesenchymal stem cell angiogenic capacity and stemness by a biomimetic hydrogel scaffold. *Biomaterials* **2012**, *33*, 80–90. [CrossRef] [PubMed]
49. Li, Z.; Tian, X.; Yuan, Y.; Song, Z.; Zhang, L.; Wang, X.; Li, T. Effect of cell culture using chitosan membranes on stemness marker genes in mesenchymal stem cells. *Mol. Med. Rep.* **2013**, *7*, 1945–1949. [CrossRef]
50. Su, N.; Gao, P.L.; Wang, K.; Wang, J.Y.; Zhong, Y.; Luo, Y. Fibrous scaffolds potentiate the paracrine function of mesenchymal stem cells: A new dimension in cell-material interaction. *Biomaterials* **2017**, *141*, 74–85. [CrossRef]
51. Qian, C.; Zhou, Z.; Han, H.; Zhao, C.; Jin, X.; Zhao, H.; Zhang, Y.; Chen, W.; Yang, N.; Li, Z. Influence of microgravity on the concentration of circulating primordial germ cells in Silky chicken offspring. *J. Poult. Sci.* **2009**, *47*, 65–70. [CrossRef]
52. Naeemipour, M.; Dehghani, H.; Bassami, M.; Bahrami, A. Expression dynamics of pluripotency genes in chicken primordial germ cells before and after colonization of the genital ridges. *Mol. Reprod. Dev.* **2013**, *80*, 849–861. [CrossRef]
53. Tonus, C.; Cloquette, K.; Ectors, F.; Piret, J.; Gillet, L.; Antoine, N.; Grobet, L. Long term-cultured and cryopreserved primordial germ cells from various chicken breeds retain high proliferative potential and gonadal colonisation competency. *Reprod. Fertil. Dev.* **2016**, *28*, 628–639. [CrossRef]

54. Li, D.; Chen, Z.; Chen, S.; Ji, H.; Zhan, X.; Luo, D.; Luo, H.; Wang, B. Chicken Mesenchymal Stem Cells as Feeder Cells Facilitate the Cultivation of Primordial Germ Cells from Circulating Blood and Gonadal Ridge. *Stem Cell Discov.* **2019**, *9*, 1–14. [CrossRef]
55. Xie, H.; Sun, L.; Zhang, L.; Liu, T.; Chen, L.; Zhao, A.; Gao, F.; Zou, P.; Li, Q.; Guo, A.J.; et al. Mesenchymal stem cell-derived microvesicles support ex vivo expansion of cord blood-derived CD34+ cells. *Stem Cells* **2016**. [CrossRef]
56. Iacono, M.L.; Anzalone, R.; La Rocca, G.; Baiamonte, E.; Maggio, A.; Acuto, S. Wharton's jelly mesenchymal stromal cells as a feeder layer for the ex vivo expansion of hematopoietic stem and progenitor cells: A review. *Stem. Cell. Rev. Rep.* **2017**, *13*, 35–49. [CrossRef]
57. Chang, Y.H.; Chu, T.Y.; Ding, D.C. WNT/β-Catenin signaling pathway regulates non-tumorigenesis of human embryonic stem cells co-cultured with human umbilical cord mesenchymal stem cells. *Sci. Rep.* **2017**, *7*, 1–10. [CrossRef] [PubMed]
58. Zmrhal, V.; Slama, P. Current knowledge about interactions between avian dendritic cells and poultry pathogens. *Dev. Comp. Immunol.* **2020**, *104*, 103565. [CrossRef] [PubMed]
59. Tippenhauer, M.; Heller, D.E.; Weigend, S.; Rautenschlein, S. The host genotype influences infectious bursal disease virus pathogenesis in chickens by modulation of T cells responses and cytokine gene expression. *Dev. Comp. Immunol.* **2013**, *40*, 1–10. [CrossRef] [PubMed]
60. Dey, S.; Pathak, D.C.; Ramamurthy, N.; Maity, H.K.; Chellappa, M.M. Infectious bursal disease virus in chickens: Prevalence, impact, and management strategies. *Vet. Med. Res. Rep.* **2019**, *10*, 85. [CrossRef] [PubMed]
61. Elankumaran, S.; Heckert, R.A.; Moura, L. Pathogenesis and tissue distribution of a variant strain of infectious bursal disease virus in commercial broiler chickens. *Avian Dis.* **2002**, *46*, 169–176. [CrossRef]
62. Kabell, S.; Handberg, K.J.; Kusk, M.; Bisgaard, M. Detection of infectious bursal disease virus in various lymphoid tissues of experimentally infected specific pathogen free chickens by different reverse transcription polymerase chain reaction assays. *Avian Dis.* **2005**, *49*, 534–539. [CrossRef]
63. Kim, I.J.; You, S.K.; Kim, H.; Yeh, H.Y.; Sharma, J.M. Characteristics of bursal T lymphocytes induced by infectious bursal disease virus. *J. Virol.* **2000**, *74*, 8884–8892. [CrossRef] [PubMed]
64. Rautenschlein, S.; Yeh, H.Y.; Njenga, M.K.; Sharma, J.M. Role of intrabursal T cells in infectious bursal disease virus (IBDV) infection: T cells promote viral clearance but delay follicular recovery. *Arch. Virol.* **2002**, *147*, 285–304. [CrossRef]
65. Ruby, T.; Whittaker, C.; Withers, D.R.; Chelbi-Alix, M.K.; Morin, V.; Oudin, A.; Young, J.R.; Zoorob, R. Transcriptional profiling reveals a possible role for the timing of the inflammatory response in determining susceptibility to a viral infection. *J. Virol.* **2006**, *80*, 9207–9216. [CrossRef]
66. Eldaghayes, I.; Rothwell, L.; Williams, A.; Withers, D.; Balu, S.; Davison, F.; Kaiser, P. Infectious bursal disease virus: Strains that differ in virulence differentially modulate the innate immune response to infection in the chicken bursa. *Viral Immunol.* **2006**, *19*, 83–91. [CrossRef]
67. Liu, H.; Zhang, M.; Han, H.; Yuan, J.; Li, Z. Comparison of the expression of cytokine genes in the bursal tissues of the chickens following challenge with infectious bursal disease viruses of varying virulence. *Virol. J.* **2010**, *7*, 1–9. [CrossRef]
68. Heo, Y.T.; Lee, S.H.; Yang, J.H.; Kim, T.; Lee, H.T. Bone marrow cell-mediated production of transgenic chickens. *Lab. Investig.* **2011**, 1229–1240. [CrossRef] [PubMed]
69. Rath, N.C.; Huff, G.R.; Huff, W.E.; Balog, J.M. Factors regulating bone maturity and strength in poultry. *Poult. Sci.* **2000**, *79*, 1024–1032. [CrossRef] [PubMed]
70. Yahyaei, B.; Gilanpour, H.; Veshkini, A. Study of the ossification centers and skeletal development of pelvic limb in quail after hatching. *Adv. Environ. Biol.* **2013**, 2074–2081.
71. Iqbal, M.; Zhang, H.; Mehmood, K.; Li, A.; Jiang, X.; Wang, Y.; Zhang, J.; Iqbal, M.K.; Rehman, M.U.; Yao, W.; et al. Icariin: A potential compound for the recovery of Tibial Dyschondroplasia affected chicken via up-regulating BMP-2 expression. *Biol. Proced.* **2018**, *20*, 1–7. [CrossRef]
72. Fleming, R.H.; McCormack, H.A.; McTeir, L.; Whitehead, C.C. Incidence, pathology and prevention of keel bone deformities in the laying hen. *Brit. Poult. Sci.* **2004**, *45*, 320–330. [CrossRef]
73. Rodenburg, T.B.; Tuyttens, F.A.M.; De Reu, K.; Herman, L.; Zoons, J.; Sonck, B. Welfare assessment of laying hens in furnished cages and non-cage systems: An on-farm comparison. *Anim. Welf.* **2008**, *17*, 363–373.
74. Käppeli, S.; Gebhardt-Henrich, S.G.; Fröhlich, E.; Pfulg, A.; Stoffel, M.H. Prevalence of keel bone deformities in Swiss laying hens. Br. *Poult. Sci.* **2011**, *52*, 531–536. [CrossRef]
75. Wilkins, L.J.; McKinstry, J.L.; Avery, N.C.; Knowles, T.G.; Brown, S.N.; Tarlton, J.; Nicol, C.J. Influence of housing system and design on bone strength and keel bone fractures in laying hens. *Vet. Rec.* **2011**, *169*, 414. [CrossRef] [PubMed]
76. Petrik, M.T.; Guerin, M.T.; Widowski, T.M. On-farm comparison of keel fracture prevalence and other welfare indicators in conventional cage and floor-housed laying hens in Ontario, Canada. *Poult. Sci.* **2015**, *94*, 579–585. [CrossRef] [PubMed]
77. Toscano, M.J.; Dunn, I.C.; Christensen, J.P.; Petow, S.; Kittelsen, K.; Ulrich, R. Explanations for keel bone fractures in laying hens: Are there explanations in addition to elevated egg production? *Poult. Sci.* **2020**, *99*, 4183–4194. [CrossRef] [PubMed]
78. Chen, F.P.; Lee, N.; Wang, K.C.; Soong, Y.K.; Huang, K.E. Effect of estrogen and 1α, 25 (OH) 2-vitamin D3 on the activity and growth of human primary osteoblast-like cells in vitro. *Fertil. Steril.* **2002**, *77*, 1038–1043. [CrossRef]
79. Jørgensen, N.R.; Henriksen, Z.; Sørensen, O.H.; Civitelli, R. Dexamethasone, BMP-2, and 1, 25-dihydroxyvitamin D enhance a more differentiated osteoblast phenotype: Validation of an in vitro model for human bone marrow-derived primary osteoblasts. *Steroids* **2004**, *69*, 219–226. [CrossRef]

80. Li, X.; Liu, H.; Niu, X.; Yu, B.; Fan, Y.; Feng, Q.; Cui, F.; Watari, F. The use of carbon nanotubes to induce osteogenic differentiation of human adipose-derived MSCs in vitro and ectopic bone formation in vivo. *Biomaterials* **2012**, *33*, 4818–4827. [CrossRef]
81. Tourkova, I.L.; Liu, L.; Sutjarit, N.; Larrouture, Q.C.; Luo, J.; Robinson, L.J.; Blair, H.C. Adrenocorticotropic hormone and 1, 25-dihydroxyvitamin D 3 enhance human osteogenesis in vitro by synergistically accelerating the expression of bone-specific genes. *Lab. Investig.* **2017**, *97*, 1072–1083. [CrossRef]
82. Harrison, J.R.; Petersen, D.N.; Lichtler, A.C.; Mador, A.T.; Rowe, D.W.; Kream, B.E. 1, 25-Dihydroxyvitamin D3 inhibits transcription of type I collagen genes in the rat osteosarcoma cell line ROS 17/2.8. *Endocrinology* **1989**, *125*, 327–333. [CrossRef] [PubMed]
83. Kim, H.T.; Chen, T.L. 1, 25-Dihydroxyvitamin D3 interaction with dexamethasone and retinoic acid: Effects on procollagen messenger ribonucleic acid levels in rat osteoblast-like cells. *Mol. Endocrinol.* **1989**, *3*, 97–104. [CrossRef]
84. Van Driel, M.; Van Leeuwen, J.P. Vitamin D endocrine system and osteoblasts. *Bonekey Rep.* **2014**, *3*, 493. [CrossRef] [PubMed]
85. Chen, J.; Dosier, C.R.; Park, J.H.; De, S.; Guldberg, R.E.; Boyan, B.D.; Schwartz, Z. Mineralization of three–dimensional osteoblast cultures is enhanced by the interaction of 1α, 25–dihydroxyvitamin D3 and BMP2 via two specific vitamin D receptors. *J. Tissue. Eng. Regen. Med.* **2016**, *10*, 40–51. [CrossRef] [PubMed]
86. Kim, J.H.; Seong, S.; Kim, K.; Kim, I.; Jeong, B.C.; Kim, N. Downregulation of Runx2 by 1, 25-dihydroxyvitamin D3 induces the transdifferentiation of osteoblasts to adipocytes. *Int. J. Mol. Sci.* **2016**, *17*, 770. [CrossRef]
87. Xiong, Y.; Zhang, Y.; Xin, N.; Yuan, Y.; Zhang, Q.; Gong, P.; Wu, Y. 1α, 25-Dihydroxyvitamin D3 promotes osteogenesis by promoting Wnt signaling pathway. *J. Steroid. Biochem. Mol. Biol.* **2017**, *174*, 153–160. [CrossRef] [PubMed]
88. Broess, M.; Riva, A.; Gerstenfeld, L.C. Inhibitory effects of 1, 25 (OH) 2 vitamin D3 on collagen type I, osteopontin, and osteocalcin gene expression in chicken osteoblasts. *J. Cell. Bibiochem.* **1995**, *57*, 440–451. [CrossRef]
89. Pande, V.V.; Chousalkar, K.C.; Bhanugopan, M.S.; Quinn, J.C. Super pharmacological levels of calcitriol (1, 25-(OH) 2 D3) inhibits mineral deposition and decreases cell proliferation in a strain dependent manner in chicken mesenchymal stem cells undergoing osteogenic differentiation in vitro. *Poult. Sci.* **2015**, *94*, 2784–2796. [CrossRef]
90. Gil, A.; Plaza-Diaz, J.; Mesa, M.D. Vitamin D: Classic and novel actions. *Ann. Nutr. Metab.* **2018**, *72*, 87–95. [CrossRef]
91. Milford, A.B.; Le Mouël, C.; Bodirsky, B.L.; Rolinski, S. Drivers of meat consumption. *Appetite* **2019**, *141*, 104313. [CrossRef] [PubMed]
92. Basu, S. The transitional dynamics of caloric ecosystems: Changes in the food supply around the world. *Crit. Public Health.* **2015**, *25*, 248–264. [CrossRef]
93. Al-Khalaifa, H.; Al-Nasser, A.; Al-Surayee, T.; Al-Kandari, S.; Al-Enzi, N.; Al-Sharrah, T.; Ragheb, G.; Al-Qalaf, S.; Mohammed, A. Effect of dietary probiotics and prebiotics on the performance of broiler chickens. *Poult. Sci.* **2019**, *98*, 4465–4479. [CrossRef] [PubMed]
94. Sohail, M.U.; Hume, M.E.; Byrd, J.A.; Nisbet, D.J.; Ijaz, A.; Sohail, A.; Shabbir, M.Z.; Rehman, H. Effect of supplementation of prebiotic mannan-oligosaccharides and probiotic mixture on growth performance of broilers subjected to chronic heat stress. *Poult. Sci.* **2012**, *91*, 2235–2240. [CrossRef] [PubMed]
95. Alavi, S.A.N.; Zakeri, A.; Kamrani, B.; Pourakbari, Y. Effect of prebiotics, probiotics, acidfire, growth promoter antibiotics and synbiotic on humural immunity of broiler chickens. *Global Vet.* **2012**, *8*, 612–617.
96. Maiorano, G.; Stadnicka, K.; Tavaniello, S.; Abiuso, C.; Bogucka, J.; Bednarczyk, M. In ovo validation model to assess the efficacy of commercial prebiotics on broiler performance and oxidative stability of meat. *Poult. Sci.* **2017**, *96*, 511–518. [CrossRef]
97. Carrade, D.D.; Borjesson, D.L. Immunomodulation by mesenchymal stem cells in veterinary species. *Com. Med.* **2013**, *63*, 207–217.
98. Lotfinegad, P. Immunomodulatory nature and site specific affinity of mesenchymal stem cells: A hope in cell therapy. *Adv. Pharm. Bull.* **2014**, *4*, 5. [CrossRef]
99. Zimmermann, K.; Haas, A.; Oxenius, A. Systemic antibody responses to gut microbes in health and disease. *Gut Microbes* **2012**, *3*, 42–47. [CrossRef]
100. Li, G.; Lillehoj, H.S.; Lee, K.W.; Jang, S.I.; Marc, P.; Gay, C.G.; Ritter, G.D.; Bautista, D.A.; Phillips, K.; Neumann, A.P.; et al. An outbreak of gangrenous dermatitis in commercial broiler chickens. *Avian Path.* **2010**, *39*, 247–253. [CrossRef]
101. McDevitt, R.M.; Brooker, J.D.; Acamovic, T.; Sparks, N.H.C. Necrotic enteritis; a continuing challenge for the poultry industry. *World's Poul. Sci. J.* **2006**, *62*, 221–247. [CrossRef]
102. Mataragas, M.; Skandamis, P.N.; Drosinos, E.H. Risk profiles of pork and poultry meat and risk ratings of various pathogen/product combinations. *Int. J. Food Microbiol.* **2008**, *126*, 1–12. [CrossRef] [PubMed]
103. Cooper, K.K.; Songer, J.G. Necrotic enteritis in chickens: A paradigm of enteric infection by Clostridium perfringens type A. *Anaerobe* **2009**, *15*, 55–60. [CrossRef] [PubMed]
104. Van Immerseel, F.; Rood, J.I.; Moore, R.J.; Titball, R.W. Rethinking our understanding of the pathogenesis of necrotic enteritis in chickens. *Trends Microbiol.* **2009**, *17*, 32–36. [CrossRef]
105. Gornatti-Churria, C.D.; Crispo, M.; Shivaprasad, H.L.; Uzal, F.A. Gangrenous dermatitis in chickens and turkeys. *J. Vet. Diag. Investig.* **2018**, *30*, 188–196. [CrossRef]
106. Shivaprasad, H.L. Gangrenous dermatitis in poultry. In *Clostridial Diseases of Animals*; Wiley-Blackwell: Ames, IA, USA, 2016; pp. 255–264.
107. Dinev, I.; Denev, S.; Vashin, I.; Kanakov, D.; Rusenova, N. Pathomorphological investigations on the prevalence of contact dermatitis lesions in broiler chickens. *J. Appl. Anim. Res.* **2019**, *47*, 129–134. [CrossRef]

108. Li, G.; Lillehoj, H.S.; Lee, K.W.; Lee, S.H.; Park, M.S.; Jang, S.I.; Bauchan, G.R.; Gay, C.G.; Ritter, G.D.; Bautista, D.A.; et al. Immunopathology and cytokine responses in commercial broiler chickens with gangrenous dermatitis. *Avian Pathol.* **2010**, *39*, 255–264. [CrossRef]
109. Golchin, A.; Farahany, T.Z.; Khojasteh, A.; Soleimanifar, F.; Ardeshirylajimi, A. The clinical trials of mesenchymal stem cell therapy in skin diseases: An update and concise review. *Curr. Stem Cell Res. Ther.* **2019**, *14*, 22–33. [CrossRef]
110. Steinfeld, H.; Gerber, P.; Wassenaar, T.D.; Castel, V.; Rosales, M.; Rosales, M.; de Haan, C. *Livestock's Long Shadow: Environmental Issues and Options*; Food Agriculture Organization: Rome, Italy, 2006.
111. Hoekstra, A.Y.; Chapagain, A.K. Water footprints of nations: Water use by people as a function of their consumption pattern. In *Integrated Assessment of Water Resources and Global Change*; Springer: Dordrecht, The Netherlands, 2006; pp. 35–48.
112. Fiala, N. Meeting the demand: An estimation of potential future greenhouse gas emissions from meat production. *Ecol. Econom.* **2008**, *67*, 412–419. [CrossRef]
113. Sutton, T.C. The pandemic threat of emerging H5 and H7 avian influenza viruses. *Viruses* **2018**, *10*, 461. [CrossRef]
114. Park, S.E. Epidemiology, virology, and clinical features of severe acute respiratory syndrome-coronavirus-2 (SARS-CoV-2; Coronavirus Disease-19). *Clin. Exp. Pediatr.* **2020**, *63*, 119. [CrossRef] [PubMed]
115. Stanton, M.M.; Tzatzalos, E.; Donne, M.; Kolundzic, N.; Helgason, I.; Ilic, D. Prospects for the use of induced pluripotent stem cells in animal conservation and environmental protection. *Stem Cells Transl. Med.* **2019**, *8*, 7–13. [CrossRef]
116. Datar, I.; Betti, M. Possibilities for an in vitro meat production system. *Innov. Food. Sci. Emerg. Technol.* **2010**, *11*, 13–22. [CrossRef]
117. Arshad, M.S.; Javed, M.; Sohaib, M.; Saeed, F.; Imran, A.; Amjad, Z. Tissue engineering approaches to develop cultured meat from cells: A mini review. *Cogent Food Agric* **2017**, *3*, 1320814. [CrossRef]
118. Bhat, Z.F.; Kumar, S.; Fayaz, H. In Vitro meat production: Challenges and benefits over conventional meat production. *J. Integr. Agric.* **2015**, *14*, 241–248. [CrossRef]
119. Will, K.; Schering, L.; Albrecht, E.; Kalbe, C.; Maak, S. Differentiation of bovine satellite cell-derived myoblasts under different culture conditions. *In Vitro. Cell. Dev. Biol. Animal* **2015**, *51*, 885–889. [CrossRef] [PubMed]
120. Ostrovidov, S.; Ahadian, S.; Ramon–Azcon, J.; Hosseini, V.; Fujie, T.; Parthiban, S.P.; Khademhosseini, A. Three–dimensional co–culture of C2C12/PC12 cells improves skeletal muscle tissue formation and function. *J. Tissue Eng. Regen. Med.* **2017**, *11*, 582–595. [CrossRef] [PubMed]
121. Mehta, F.; Theunissen, R.; Post, M.J. Adipogenesis from bovine precursors. In *Myogenesis*; Humana Press: New York, NY, USA, 2019; pp. 111–125. [CrossRef]
122. Cremonesi, F.; Corradetti, B.; Consiglio, A.L. Fetal adnexa derived stem cells from domestic animal: Progress and perspectives. *Theriogenology* **2001**, *75*, 1400–1415. [CrossRef]
123. Wang, Y.; Han, Z.B.; Song, Y.P.; Han, Z.C. Safety of mesenchymal stem cells for clinical application. *Stem Cells Int.* **2012**, 652034. [CrossRef] [PubMed]
124. Bai, C.; Li, C.; Jin, D.; Guo, Y.; Guan, W.; Ma, Y.; Zhao, Q. Establishment and characterization of a fibroblast line from landrace. *Artif. Cell Blood Sub.* **2010**, *38*, 129–135. [CrossRef]
125. Na, R.S.; Zhao, Q.J.; Su, X.H.; Chen, X.W.; Guan, W.J.; Ma, Y.H. Establishment and biological characteristics of Ujumqin sheep fibroblast line. *Cytotechnology* **2010**, *62*, 43–52. [CrossRef] [PubMed]

animals

Article

Cell Source-Dependent In Vitro Chondrogenic Differentiation Potential of Mesenchymal Stem Cell Established from Bone Marrow and Synovial Fluid of *Camelus dromedarius*

Young-Bum Son [1], Yeon Ik Jeong [1], Yeon Woo Jeong [1], Mohammad Shamim Hossein [1], Per Olof Olsson [1], Alex Tinson [2], Kuhad Kuldip Singh [2], Sang-Yun Lee [3] and Woo Suk Hwang [1,*]

1 UAE Biotech Research Center, Abu Dhabi 30310, United Arab Emirates; ybs@adbrf.org (Y.-B.S.); youniks@adbrf.org (Y.I.J.); doctorj1@adbrf.org (Y.W.J.); shamim0976@gmail.com (M.S.H.); omolson@gmail.com (P.O.O.)
2 Hilli E.T. Cloning and Surgical Centre Presidential Camels and Camel Racing Affairs, Al-Ain 17292, United Arab Emirates; heffoundation@hotmail.com (A.T.); kskuhad@hotmail.com (K.K.S.)
3 Department of Theriogenology and Biotechnology, College of Veterinary Medicine and Research Institute of Life Science, Gyeongsang National University, Jinju 52828, Korea; sy_lee@gnu.ac.kr
* Correspondence: hwangws@adbrf.org

Simple Summary: This is the first study to demonstrate the establishment and subsequent analysis of attributes, including the chondrogenic capacity of mesenchymal stem cells (MSCs) from bone marrow (BM) and synovial fluid (SF) from the same donor *Camelus dromedarius*. MSCs of SF origin were notably more efficient in their chondrogenic capacity and represent a potential source for camel regenerative medicine addressing chondrocyte-related problems.

Citation: Son, Y.-B.; Jeong, Y.I.; Jeong, Y.W.; Hossein, M.S.; Olsson, P.O.; Tinson, A.; Singh, K.K.; Lee, S.-Y.; Hwang, W.S. Cell Source-Dependent In Vitro Chondrogenic Differentiation Potential of Mesenchymal Stem Cell Established from Bone Marrow and Synovial Fluid of *Camelus dromedarius*. *Animals* **2021**, *11*, 1918. https://doi.org/10.3390/ani11071918

Academic Editor: Ralf Einspanier

Received: 12 June 2021
Accepted: 21 June 2021
Published: 28 June 2021

Abstract: Mesenchymal stem cells (MSCs) are promising multipotent cells with applications for cartilage tissue regeneration in stem cell-based therapies. In cartilage regeneration, both bone marrow (BM-MSCs) and synovial fluid (SF-MSCs) are valuable sources. However, the cellular characteristics and chondrocyte differentiation potential were not reported in either of the camel stem cells. The in vitro chondrocyte differentiation competence of MSCs, from (BM and SF) sources of the same *Camelus dromedaries* (camel) donor, was determined. Both MSCs were evaluated on pluripotent markers and proliferation capacity. After passage three, both MSCs showed fibroblast-like morphology. The proliferation capacity was significantly increased in SF-MSCs compared to BM-MSCs. Furthermore, SF-MSCs showed an enhanced expression of transcription factors than BM-MSCs. SF-MSCs exhibited lower differentiation potential toward adipocytes than BM-MSCs. However, the osteoblast differentiation potential was similar in MSCs from both sources. Chondrogenic pellets obtained from SF-MSCs revealed higher levels of chondrocyte-specific markers than those from BM-MSCs. Additionally, glycosaminoglycan (GAG) content was elevated in SF-MSCs related to BM-MSCs. This is, to our knowledge, the first study to establish BM-MSCs and SF-MSCs from the same donor and to demonstrate in vitro differentiation potential into chondrocytes in camels.

Keywords: mesenchymal stem cells; *Camelus dromedarius*; bone marrow; synovial fluid; chondrocyte differentiation

1. Introduction

Cartilage damage to joint surfaces can result in osteoarthritis (OA), a condition where the bone under the articular cartilage is exposed and the synovial membrane around the joint is inflamed [1]. When cartilage is damaged due to inflammation or trauma, it cannot buffer between the bone, causing severe pain and deformation of the articular cartilage. OA is a chronic musculoskeletal disorder frequently occurring in racing horses and camels that adversely affects their future racing careers [2,3].

Since cartilage has no distribution of blood vessels and nerves, it is difficult to regenerate once damaged [4]. To treat damaged cartilage tissue, drugs (steroids and painkiller), chondroprotective agents (hyaluronic acid and glucosamine), and surgical approaches are used [5]. However, limited effects of non-specific alleviation of pain and inflammatory reactions are expected in drug treatment and chondroprotective agents only play a role in supplying nutrients to chondrocytes [1,6]. Microfracture and autologous chondrocyte implantation (ACI) have been performed as clinical surgical methods based on cell therapy [7]. Microfracture is a method of regenerating cartilage with blood clots containing mesenchymal stem cells (MSCs) by puncture or abrasion of damaged subchondral bone and has the disadvantages of regenerating into fibrocartilage rather than hyaline cartilage [8]. ACI, which is commonly used, induces the regeneration of cartilage tissue by transplanting autologous cartilage cell culture in vitro [8]. However, this method has specific problems, including the formation of cartilage defects in macroscopically normal-looking tissue donor sites. Furthermore, the phenotype can change due to de-differentiation during in vitro culture since the transplanted cells are already differentiated cells [9,10]. Therefore, therapeutic approaches have emerged for cartilage regeneration using stem cells.

MSCs are a potent resource for therapeutic application with a high stemness ability, anti-inflammatory, and immunosuppressive effects [11,12]. MSCs also possess multi-lineage differentiation capacity including chondrocytes [13]. Bone marrow stem cells (BM-MSCs), which were the first to be discovered, can mainly be collected from the iliac crest of the pelvis, and have been reported for a long time. However, several disadvantages have been reported, such as the limited amount of harvest and the complicated collection process [14]. Furthermore, compared to other sources, low proliferative capacity and donor-age-dependent characteristics have been reported in BM-MSCs [15]. Considering the shortcomings of BM-MSCs, studies have been performed on alternative MSCs sources.

Synovial fluid-derived MSCs (SF-MSCs) can be collected by the process of arthrocentesis and possess superior chondrogenic capacity [16]. SF-MSCs are the most similar source to articular cartilage, have a high expression of CD44 (hyaluronan receptors) and uridine disphosphoglucose dehydrogenase (UDPGD), and are required for hyaluronan synthesis [17]. Accordingly, several studies have reported on cartilage regeneration using SF-MSCs in various mammalians [18–20]. Human SF-MSCs derived from OA patients were transplanted with a collagen sponge into the joint of the anterior cruciate ligament transection of rats and cartilage defects were recovered [18]. SF-MSCs derived from equine sources have also been shown to have an analogous beneficial effect compared to BM-MSCs after implantation into articular cartilage defects in rat femurs [19]. An interesting study reported that after the transplantation of porcine SF-MSCs into collagen-induced arthritis mouse (CIA mouse) by intraperitoneal injection, cartilage damage in the joint was decreased, and inflammation was also reduced [20]. Nevertheless, a comparison of these two cell types is lacking in *Camelus dromedarius* (camel).

In this study, we established a distinctive population of BM-MSCs and SF-MSCs derived from the same donor. Additionally, to evaluate chondrogenic potential, both camel MSCs were differentiated into chondrocytes under specific induction conditions.

2. Materials and Methods

2.1. Chemicals and Media

Unless otherwise stated, all reagents were obtained from Sigma (St. Louis, MO, USA).

2.2. Collection and Culture of Bone Marrow and Synovial Fluid Derived Mesenchymal Stem Cells

Institutional Animal Care and Use Committee (IACUC) guidelines were followed and all experiments were approved by the Management of Scientific Centers and Presidential Camels (Accession No: PC4.1.5). Four female camels aged between 5 and 7 years were used to establish and compare the various MSCs. After placing the camels in a seated position, they were sedated with ketamine hydrochloride (Ilium, Hlendenning, Australia; 0.25 mg/kg body weight) and xylazine (Ceva, Libourne, France; 0.25 mg/kg body weight) through intravenous injection using an 18-gauge needle [13]. The 2% lidocaine (Jeil, Daegu, Korea) was infiltrated subcutaneously on the iliac crest (2 cm × 2 cm). Approximately 5 mL of BM was extracted using the 11-gauge, 15-cm-long biopsy needle (Argon Medical Devices, Frisco, TX, USA). A heparin-coated 10 mL syringe, was used for BM extraction and subsequently added to the same quantity of Dulbecco's phosphate-buffered saline (DPBS, Welgene, Gyeongsan, Korea). Centrifugation was performed at 400× *g* for 40 min, layered onto Ficoll-Paque (Amersham Biosciences, Uppsala, Sweden) without mixing. Mononuclear cells (MNCs) were harvested from the interface buffy layers and washed with DPBS.

We collected the SF using syringe aspiration as previously reported with minor modifications [21]. In brief, a hypodermic 18-gauge needle was used to gently aspirate 5 mL of SF from the femorotibial joint with a 10 mL syringe. A nylon filter with a pore size of 40 μm (Falcon, Franklin, NJ, USA) was used to separate debris before cell isolation through centrifugation at 400 g for 10 min. Cells were cultured with DMEM, high-glucose, 10% FBS, and 1% of each nonessential amino acids, antibiotic-antimycotic, and 0.1% β-mercaptoethanol (Thermo Fisher Scientific, Waltham, MA, USA). Incubator conditions were 38 °C with high humidity, 5% O_2, and 5% CO_2. Culture media were exchanged every second day until 80% cellular confluence was obtained. Subculture and cryopreservation were performed using a 0.25% trypsin EDTA solution (Gibco, Paisely, UK) and a cryopreservation solution using the same DMEM used for cell culture with 20% FBS and 10% dimethyl sulfoxide (DMSO).

2.3. Proliferation and Cell Cycle Assay

Cell cycle assay, including population doubling time (PDT) was conducted as previously reported with minor modifications [22]. Six-well plates (Nunc, Rochester, NY, USA) were used for seeding MSCs and cultured in an incubator as described above. A hematocytometer was used corresponding to passage intervals, at 72-h, to determine cell counts. PDT was established as follows: PDT = log2 × T/(logNC − logNI): T is culture time, logNC is the cell number at the time, and logNI is the initial cell count.

Cells were fixed at passage 3 using 70% ethanol for 1 h. A 10 μg/mL propidium iodide solution was used with 1×10^6 fixed cells for 15 min, including RNase A (100 μg/mL). Analysis and categorization into cell stage was performed via flow cytometry (BD FACSVerseTM, BD Biosciences, San Jose, CA, USA). All experiments were conducted in 4 independent replicates.

2.4. In Vitro Differentiation into Trilineage and Cytochemical Staining

Trilineage (osteoblast, adipocyte, and chondrocyte) differentiation was conducted using BM-MSCs and SF-MSCs as previously described [13]. For osteogenic differentiation, cells were cultured for a total of 21 days in a DMEM media supplemented with 10% FBS 10 nM dexamethasone, 50 μg/mL ascorbic acid, and 10 mM sodium beta-glycerophosphate. After differentiation, mineralization and calcium deposition were affirmed using Alizarin red S and von Kossa staining. Adipogenic differentiation of cultured MSCs was conducted in DMEM supplemented with10% FBS, 100 μM indomethacin, 10 μM insulin, and 1 μM dexamethasone. Differentiation into chondrocytes was accomplished utilizing a modified pellet culture method. Pellets were made in 15 mL centrifuge tubes with 1 mL cellular suspension of passage 3 cells (1×10^6) in STEMPRO chondrogenesis differentiation media with 10% supplement in a 15 mL tube. Centrifugation was performed at 450× *g* for 5 min and the resultant pellets were cultured for 3 weeks. Afterward, culture pellets were embedded in paraffin and sectioned on glass slides following dehydration. A 1% Alcian blue solubilized in 3% acetic acid was used for staining for a period of 10 min with proteoglycan deposition confirmed with a 1 min counterstaining using a 0.1% solution of nuclear fast.

2.5. Real-Time Quantitative Polymerase Chain Reaction (RT-qPCR) Analysis

The expressions of pluripotent markers, cluster of differentiation (CD) markers, and lineage-specific genes were analyzed. An easy-spin Total RNA Extraction Kit (Intron, Seongnam, Korea) was used for total RNA extraction. Nucleotide quantification was performed with a nanodrop 1000 spectrophotometer (Thermo Fisher Scientific, Waltham, MA, USA). Complementary DNA was synthesized with 2 µg purified RNA and reverse transcription with a HisenScript RT PreMix kit (Intron, Seongnam, Korea); synthesis was performed over 50 min at 42 °C with 10 µM OligodT primers. A Rotor-Gene Q cycler (Qiagen, Hilden, Germany) was used for RT qPCR and RealMODTM Green AP 5× qPCR mix (Intron, Seongnam, Korea) containing 200 nM primers (Table 1). Amplification was performed by denaturation at 95 °C for 60 s and subsequently 50 cycles of 95 for 10 min, 60 °C for 6 s, and 72 °C for 4 s. Expression was normalized to mRNA levels of glyceraldehyde-3-phosphate dehydrogenase (GAPDH). PCR products were analyzed by electrophoresis in 1% agarose gels and ethidium bromide staining. Amplified products were visualized under UV light. All experiments were conducted in 4 independent replicates.

Table 1. Lists of camel primers used in RT-qPCR analysis.

Gene name (Symbol)	Primers Sequence	Product Size (bp)	Anneal. Temp (°C)
POU class 5 homeobox 1 (OCT4)	F: CGAGAGGATTTTGAGGCTGC R: GAGTACAGTGTGGTGAAGTGAG	122	60
Sex determining region Y-box 2 (SOX2)	F: CTCGCAGACCTACATGAACG R: TGGGAGGAAGAGGAAACCAC	144	60
Nanog homeobox (NANOG)	F: AGCACAGAGAAGCAGGAAGA R: CCACCGCTTACATTTCATTC	213	60
Runt-related transcription factor 2 (Runx2)	F: GACAGAAGCTTGATGACTCT R: GTAATCTGACTCTGTCCTTG	166	60
Osteocalcin (ON)	F: AGTGAGATGGTGAAGAGACT R: TAGGTTGTGCCGTAGAAG	176	60
Lipoprotein lipase (LPL)	F: GAGAGTGTTACCTACACCAA R: GCCTTTACTCTGATCTTCTC	248	60
Fatty acid-binding protein 4 (FABP4)	F: GTGACCATCAGTGTGAATG R: GCACCTCCTTCTAAAGTTAC	152	60
The type X collagen gene (COL10A1)	F: TATCCAGCTATAGGCAGTC R: TCGTAGGTGTACATTACAGG	194	60
Aggrecan (ACAN)	F: TGTGGAGGGTGTTACTGAAC R: GACTGATGACCCTTCTACCC	154	60
Collagen type II alpha 1 chain (COL2A1)	F: GTGGTGACAAAGGTGAAAAA R: AGCCTTCTCATCAAATCCTC	154	60
Glyceraldehyde 3-phosphate dehydrogenase (GAPDH)	F: GCTGAGTACGTTGTGGAGTC R: TCACGCCCATCACAAACATG	133	60
Integrin beta-1 (CD 29)	F: CTTGCGTTGCTGCTGATTTG R: TTCTTGCGTGTCCCATTTGG	105	60
5′-nucleotidase (CD 73)	F: CAACCTCAGACATGCCGATG R: GTCAAAGGTGCCTCCAAAGG	154	60
Endoglin (CD 105)	F: TCCTCCAGACCTCCAACTCT R:CCCAAATTCAGTTGGCAGCT	115	60
Cluster of differentiation 34 molecule (CD 34)	F: GGTCTTGGCCAACAGAACAG R: CAGCTTCGACGGTTCATCAG	216	60
Protein tyrosine phosphatase, receptor type, C (CD 45)	F: AACTCTTGGCATTTGGCGTT R: TTCTGCCTACACTCAAGGGG	220	60

2.6. Glycosaminoglycan (GAG) Contents

The differentiated chondrocyte pellets from SF-MSCs and BM-MSCs were stored at −20 °C following washing prior to glycosaminoglycan (GAG) content assay. Cell pellets were digested with 0.5 mg/mL papain solution and proteoglycan content was determined using the dimethylmethylene blue (DMMB) spectrophotometric assay. Chondroitin-4 sulfate (Sigma-Aldrich, St. Louis, MO, USA) was used to establish a standard curve. The optical density of 525 nm was used on a microplate reader (VersaMax, Molecular Devices, Sunnyvale, CA, USA). A DNA quantification kit (Abcam, Cambridge, UK) was used to determine the DNA content. All experiments were conducted in 4 independent replicates.

2.7. Statistical Analysis

Data analysis was performed with SPSS version 23 (IBM) for independent T-tests and one-way analysis of variance (ANOVA), and intergroup differences were identified with Tukey's test. Data are represented as the mean ± standard deviation (SD), and a p value < 0.05 is considered significant.

3. Results

3.1. Establishment of MSCs Derived from Bone Marrow and Synovial Fluid

We isolated and cultured MSCs derived from BM and SF from the same donors. MSCs from both tissue sources exhibited spindle-like morphology and homogenous attachments to culture surfaces were confirmed after passage 3 (Figure 1a). The results of RT-PCR showed that BM-MSCs and SF-MSCs were positive for the mesenchymal stem cell markers (CD 29, CD73, and CD105) and were negative for the hematopoietic stem cell markers (CD34 and CD45) (Figure 1b). To confirm the proliferation capacity of cells, we analyzed PDT. SF-MSCs showed higher proliferation potential than BM-MSCs (Figure 2a). Cell cycle analysis was used as a measure of cellular viability, at the third passage. We observed a significantly increased proportion of S phase and diminished G0/G1 phase in SF- compared to BM-derived MSCs (Figure 2b). Compared with the BM-MSCs, the transcription factors, NANOG, SOX2, and OCT4, showed significantly increased expression in the SF-derived cultures (Figure 3).

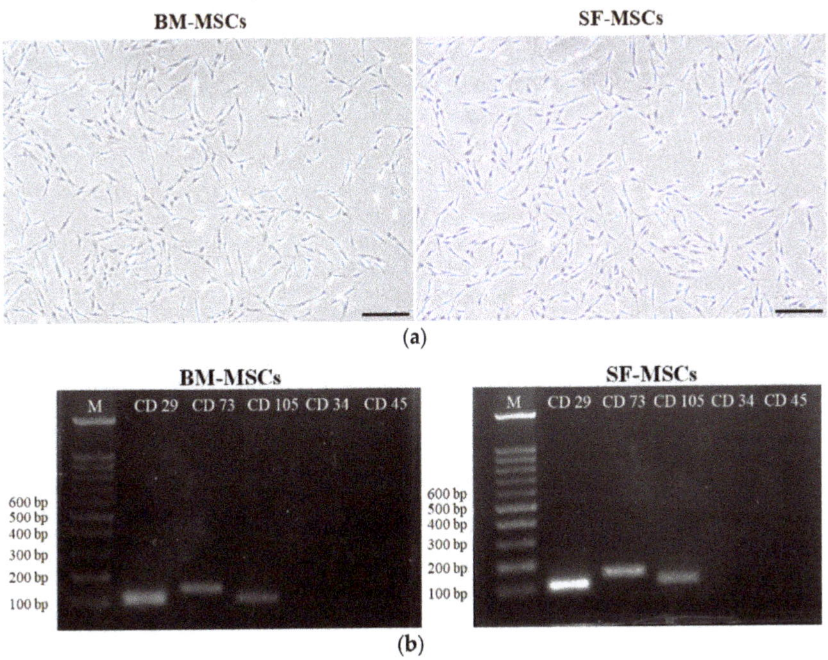

Figure 1. Establishment of BM-MSCs and SF-MSCs. (**a**) Cellular morphology of BM-MSCs and SF-MSCs from camel. Both MSCs showed a spindle-like morphology at passage 3. Scale bar = 100 μm. (**b**) Agarose gel electrophoresis of PCR products. Both BM-MSCs and SF-MSCs were positive for the mesenchymal stem cell markers (CD29, CD72, and CD105) and were negative for the hematopoietic stem cell markers (CD34 and CD45). Lane M: 100 bp DNA ladder.

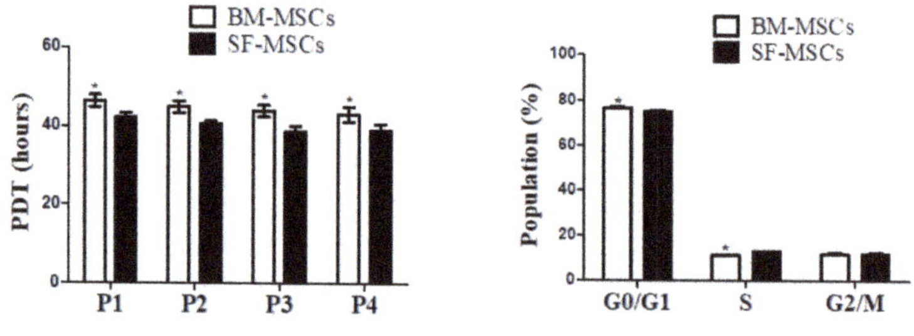

(**a**) (**b**)

Figure 2. Cell cycle and proliferation capacity of bone marrow (BM-MSCs) and synovial fluid (SF-MSCs) from the camel. (**a**) The proliferation capacity was evaluated by population doubling time (PDT) assay. SF-MSCs showed increased proliferation capacity compared to BM-MSCs during the passage. (**b**) Cell cycle analysis indicated that cell arrest and DNA replication were significantly increased in SF-MSCs compared to BM-MSCs. Bar graphs illustrate mean values +/− SD (n = 4). * denote significant ($p < 0.05$) differences (P1, P2, P3, and P4: passage 1, 2, 3, and 4; G0/G1: G0/G1 phase; S: S phase; G2/M: G2/M phase).

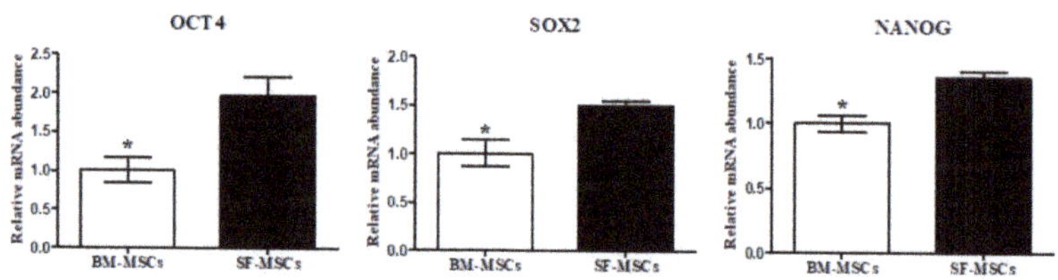

Figure 3. Stem cell transcription factor expression of BM- and SF-MSCs. SF-MSCs when related to BM-MSCs exhibited increased expression for all three transcription factors. * denote significant differences ($p < 0.05$). Bar graphs illustrate mean values +/− SD (n = 4).

3.2. *In Vitro Osteogenic and Adipogenic Lineage Differentiation Potential of MSCs*

To evaluate the influence of the source of MSCs on the differentiation capacity, BM-MSCs and SF-MSCs were differentiated into osteoblasts and adipocytes. The calcified extracellular matrix formation which is indicative of osteoblast differentiation was confirmed using Alizarin red S and von Kossa staining, and the presence of intracellular lipid droplets vacuoles, which is indicative of adipocyte differentiation, was confirmed using Oil red O staining. All differentiation processes were confirmed using both BM-MSCs and SF-MSCs (Figure 4a).

The expression of osteogenesis- and adipogenesis-related genes were analyzed pre- and post- differentiation. Gene expression associated with osteogenesis and adipogenesis was investigated pre- and post-differentiation. The expression of osteoblast-related expression, i.e., runt-related transcription factor 2 (Runx2), osteocalcin (ON), in all measured MSCs, significantly ($p < 0.05$) increased after differentiation (Figure 4b). However, there was no significant ($p < 0.05$) difference in the expression of Runx2 and ON in the differentiated osteoblasts from the two groups (Figure 4b). A significant increase was observed in adipocyte differentiation, lipoprotein lipase (LPL), and the fatty acid-binding protein 4 (FABP4) in the differentiated adipocytes derived from SF-MSCs compared to BM-MSCs (Figure 4b).

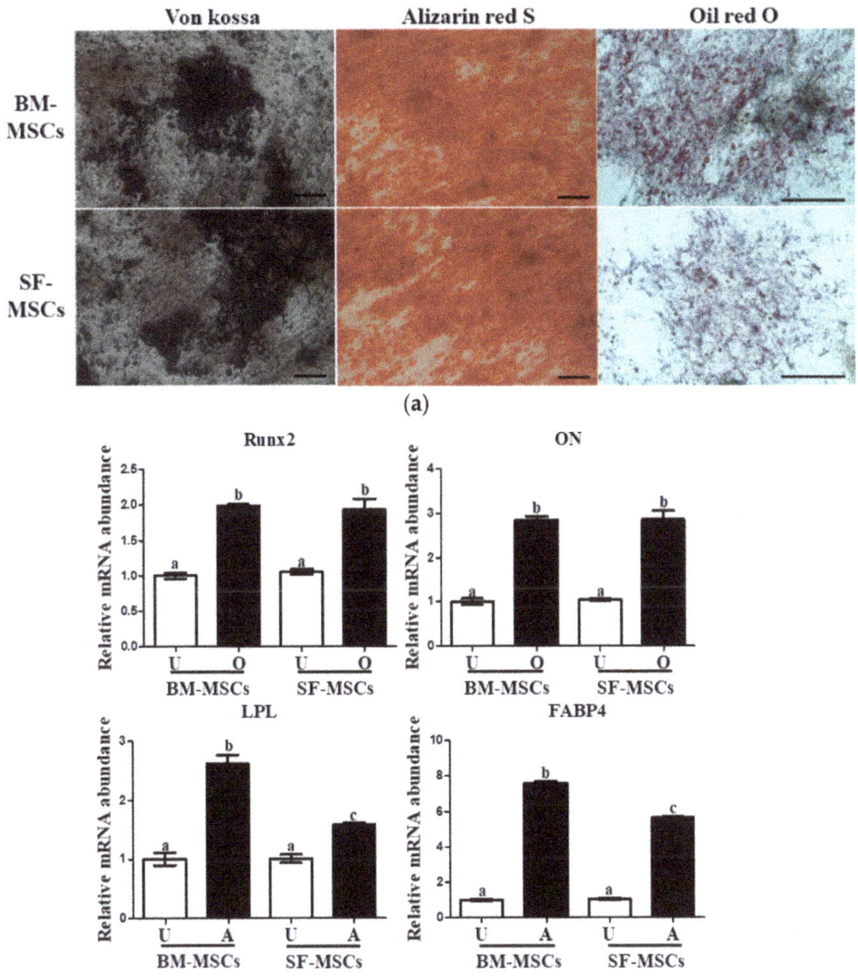

Figure 4. In vitro differentiation into osteoblast and adipocyte from BM- and SF- derived MSCs. (**a**) The osteoblast differentiation of both MSCs was confirmed by staining of calcium deposition and mineralization with von Kossa and Alizarin red S (Scale bar = 100 μm). (**b**) Osteoblast (Runx2 and ON) and adipocyte (LPL and FABP4) related gene expression in induced cells. All adipocyte-related genes were significantly increased in differentiated cells from BM-MSCs compared to those from SF-MSCs. However, there was no difference in the expression of osteoblast-related genes between the BM- and SF-MSCs. Graphs represent mean data ± SD from independent experiments (n = 4). Different superscripts (a, b, and c) denote significant differences ($p < 0.05$) among the undifferentiated control and differentiated cells using BM-MSCs and SF-MSCs (white bar: undifferentiated cells; black bar: differentiated cells; U: undifferentiated cells; O: osteogenic differentiated cells; A: adipogenic differentiated cells).

3.3. *In Vitro Chondrogenic Differentiation Capacity of MSCs*

Differentiation of BM- and SF-MSCs into chondrocytes succeeded using the pellet culture method. The accumulation of proteoglycan was observed both in pellets from BM- and SF-MSCs after 1 week by Alcian blue staining (Figure 5a). Following the differentiation process, these cells grew and the multi-layered structure was confirmed in the chondrogenic pellet. Some portion of the pellet showed hypertrophic chondrocyte formation after 2 to 3 weeks on chondrocyte differentiation using SF-MSCs (Figure 5a). The expression levels of type X collagen gene (COL10A1), aggrecan (ACAN), and the alpha 1 chain of collagen type II (COL2A1) were investigated following three weeks of the chondrogenesis procedure (Figure 5b). In BM-MSCs, the levels of chondrocyte-specific gene expressions were significantly higher in differentiated chondrocytes at 1 to 3 weeks compared to undifferentiated cells (Figure 5b). There was a more significant expression in all chondrocyte-specific markers analyzed in differentiated cells following the chondrogenesis protocol in SF- compared to BM-MSCs. The levels of ACAN gradually increased as the chondrogenesis from SF-MSCs. In the process of differentiation of BM-MSCs into chondrocytes, COL10A1, ACAN, and COL2A1 expression were not significantly different. The deposition of glycosaminoglycan (GAG) was increased in chondrogenic pellets related to undifferentiated pellets (Figure 6). Furthermore, after two weeks of differentiation into chondrocytes, SF-MSCs showed increased GAG deposition compared to BM-MSCs (Figure 6).

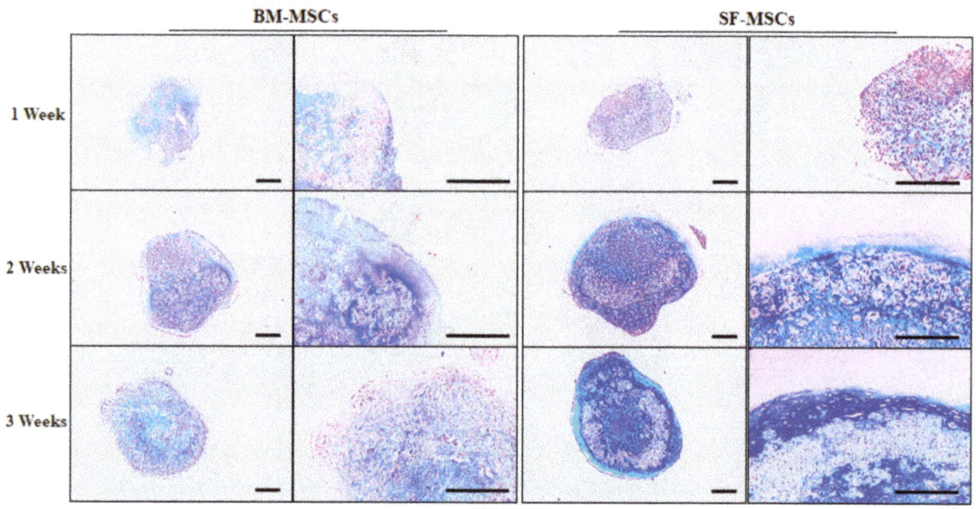

(a)

Figure 5. *Cont.*

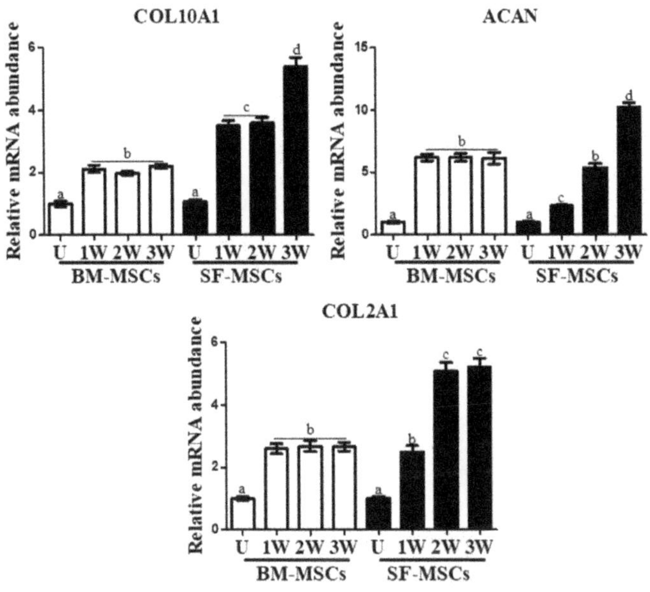

(b)

Figure 5. In vitro chondrocyte differentiation capacity by time of BM- and SF-MSCs. Chondrogenesis was evaluated by cytochemical staining and RT-qPCR analysis. (**a**) In vitro chondrocyte differentiation was demonstrated by Alcian blue staining in induced pellets one, two, and three weeks after induction. Scale bar = 500 µm. (**b**) Genes (COL10A1, ACAN, and COL2A1) expressed in chondrocytes, in induced BM- and SF-MSCs one, two, and three weeks after induction. At three weeks following chondrogenesis, all chondrocyte-specific markers were significantly increased in chondrocyte pellets in SF- versus BM-MSCs. Values displayed as mean and ± SD, independent experiments (n = 4). Different superscripts (a, b, c, and d) denote significant differences ($p < 0.05$) among undifferentiated cells, and 1, 2, and 3 weeks of differentiated cells using BM-MSCs and SF-MSCs (white bar: undifferentiated cells; black bar: differentiated cells U: undifferentiated cells; 1W: one week after chondrocyte induction; 2W: two weeks after chondrocyte induction; 3W: three weeks after chondrocyte induction).

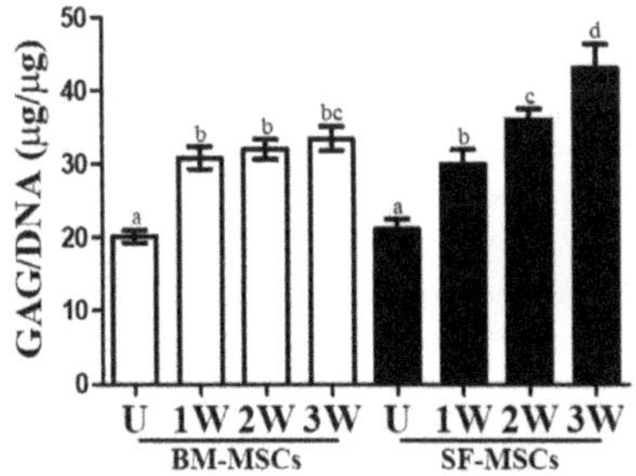

Figure 6. The values of glycosaminoglycan (GAG) content of BM- and SF-MSCs was evaluated one, two, and three weeks after induction. The GAG contents were significantly higher in groups following differentiation in SF compared to BM-MSCs after three weeks of differentiation. Values displayed are mean ± SD independent experiments (n = 4). Different superscripts (a, b, c, and d) denote significant differences ($p < 0.05$) among undifferentiated cells, and 1, 2, and 3 weeks of differentiated cells using BM-MSCs and SF-MSCs (white bar: undifferentiated cells; black bar: differentiated cells U: undifferentiated cells; 1W: one week after chondrocyte induction; 2W: two weeks after chondrocyte induction; 3W: three weeks after chondrocyte induction).

4. Discussion

Articular cartilage is fibrillar connective tissue distributed throughout the musculoskeletal system. It connects two bones and acts as a buffer against mechanical stress [4]. Chondrocytes differentiate from mesenchymal cells during development to form cartilage tissue. Unlike other tissue, cartilage consists of only one type of chondrocyte, and the formed cartilage tissue remains as a permanent cartilage tissue as with articular cartilage [4]. Cartilage damage caused by trauma has frequently been reported in racing camels that perform vigorous exercise, which develops into chronic OA in the absence of appropriate treatment. There is a limit to the natural recovering of the articular cartilage once it is damaged, and many treatments are currently applicable. Generally, these disorders are treated by autologous chondrocyte implantation (ACI) [23,24]. However, the effect of ACI is limited in broad degenerative arthritis and elderly patients. Considering the limitation of ACI, an alternative therapy based on stem cells has emerged for chondrocyte regeneration.

Ideal sources of MSCs in stem cell-based therapy for OA could have enhanced chondrogenic capacity for recovering destructed cartilage. To date, BM-MSCs and SF-MSCs therapeutic uses have predominantly been applied for the regeneration of cartilage [20,25]. No studies have been reported, on the establishment and the biological characteristics including the potential to differentiate MSCs into chondrocytes, in camels. This study aimed to evaluate biological characteristics, such as morphology; proliferation; stemness; and trilineage differentiation potency, including chondrogenesis in camel MSCs. Both BM- and SF-MSCs were successfully established from the same donors. Both types of homogenous adherent MSCs showed fibroblastic spindle-like morphology. The expression of transcription factors and the proliferation capacity were related to the self-renewal ability of MSCs [26]. OCT4 and SOX2 are crucial early transcription factors for sustaining stemness and pluripotency and are positively expressed in MSCs with NANOG. Our data showed that SF-MSCs possessed superior proliferation capacity compared to BM-MSCs as previously reported [20,27]. The levels of NANOG, SOX2, and OCT4 were significantly increased in SF compared to BM-MSCs [20].

Both osteoblast and adipocyte markers were expressed from BM- and SF-MSCs differentiation protocols. Expression of the lineage-specific genes differed between BM-MSCs and SF-MSCs. Oil red O, a cytochemical stain, along with adipocyte-associated gene expressions, indicated improvements in adipocyte differentiation capacity from SF-MSCs compared to BM-MSCs. These results are similar to previous reports in miniature pigs [20]. There was a significant increase in levels of genes; i.e., Runx2 and ON, considered to be specific to osteoblasts, followed differentiation protocols in MSCs. However, there was no difference in the capacity of osteogenesis between the two groups, which is in accordance with the previous study [28].

The chondrogenesis capacity of SF-MSCs is crucial for therapeutic recovery in OA. Chondrocytes possess variable phenotypes, such as de-differentiation that loses differentiated characteristics and re-differentiation [29,30]. When articular chondrocytes are cultured with a low-density monolayer, type II collagen expression decreases rapidly, and the expression of type I and III collagen, typically associated with fibroblasts, increases [31]. However, through three-dimensional culture, collagen type II expression was promoted, suggesting re-differentiation into chondrocytes [31]. Therefore, in this study, the chondrocyte differentiation capacity of both BM- and SF-MSCs through high cell density pellet culture, was investigated. A previous study was reported that chondrocyte-specific gene expression gradually increased during the chondrogenesis process of the SF-MSCs, whereas no gradual increase was observed in BM-MSCs [20]. The present study also showed that after three weeks of induction, chondrocyte-specific gene expression gradually increased during the chondrogenesis protocol in SF-MSCs compared to BM-MSCs. Chondrocyte-associated genes, i.e., COL10A1, ACAN, and COL2A1 were elevated in SF- compared to BM-MSCs. However, some hypertrophic chondrocytes were observed during the differentiation of SF-MSCs into chondrocytes.

Cartilage is a complex tissue comprising chondrocytes; extracellular matrix (ECM) proteins, including proteoglycans and GAGs, etc., [32,33]. Various GAGs exist, including chondroitin sulfate, dermatan sulfate, keratan sulfate, decorin, and fibromodulin in cartilage, and form proteoglycan by binding with core proteins [34]. Therefore, GAG synthesis is a crucial mark of chondrogenesis. The present study showed that the values of GAG content were significantly increased in differentiated chondrocytes compared to undifferentiated cells. Chondrocytes from SF-MSCs exhibited higher values of GAG content than those from BM-MSCs. Similarly, the intensity of Alcian blue stain was greater in SF compared to BM-MSCs. Although further studies will be necessary on the efficacy of cartilage regeneration in vivo and the suppression of cartilage hypertrophy, in which the synthesis of COL10A1 is prominent is needed but SF-MSCs appear to express superior chondrogenic capacity when compared to BM-MSCs.

The present study provides the potential stem cell-based therapy of *Camelus dromedarius* utilizing MSCs with superior chondrogenesis. The pre-establishment of MSCs can support direct treatment following cartilage-related problems such as OA caused by severe trauma. This stem cell-based therapy using MSCs promotes animal welfare by realizing injuries promptly and decreasing the risks for additional injuries.

5. Conclusions

In this paper, we showed the differences in proliferation capacity, expression of transcription factors, and the MSCs trilineage potential of both camel BM and SF. BM-MSCs were shown to have a more reliable osteogenic differentiation capacity compared to SF-MSCs. SF-MSCs had greater proliferative potential and expressed larger amounts of transcription factors than did BM-MSCs. After the completion of the in vitro chondrocyte differentiation, hypertrophic chondrocytes were observed in parts of the SF-MSCs. Nevertheless, the expression of chondrocyte-specific markers and GAG contents indicated that SF-MSCs showed enhanced chondrocyte differentiation capacity compared to BM-MSCs.

This study, to the best of and our knowledge, is the first to report the establishment and properties of camel BM- and SF-MSCs from the same donor to date. The observed chondrogenic ability alludes to the potential of SF-MSCs as a target cell source for future use in therapeutic cartilage regeneration.

Author Contributions: Conceptualization, Y.-B.S. and W.S.H.; methodology, Y.-B.S., Y.I.J., S.-Y.L. and Y.W.J.; investigation, Y.-B.S.; validation, Y.W.J.; formal analysis, Y.-B.S. and S.-Y.L.; resources, Y.I.J., A.T., P.O.O. and K.K.S.; writing—original draft preparation, Y.-B.S.; writing—review and editing, Y.-B.S., M.S.H., P.O.O. and W.S.H.; funding acquisition, W.S.H. All authors have read and agreed to the published version of the manuscript.

Funding: This project was supported by the Patronage of H.H. Sheikh Mansour bin Zayed Al Nahyan, Deputy Prime Minister of the U.A.E. and the Minister of Presidential Affairs.

Institutional Review Board Statement: All animal procedures were conducted following the animal study guidelines which were approved by the ethics committee at the Management of Scientific Centers and Presidential Camels (accession no: PC4.1.5).

Informed Consent Statement: Not applicable.

Data Availability Statement: Data access can be requested on demand from the corresponding author.

Conflicts of Interest: There are no conflict of interest to report.

Abbreviations

MSCs: Mesenchymal stem cells; BM: Bone marrow; SF: Synovial fluid; BM-MSCs: Bone marrow-derived mesenchymal stem cells; SF-MSCs: Synovial fluid-derived mesenchymal stem cells; GAG: Glycosaminoglycan; OA: Osteoarthritis; ACI: Autologous chondrocyte implantation; UDPGD: Uridine disphosphoglucose dehydrogenase; CIA mouse: Collagen-induced arthritis mouse; Camel: Camelus dromedarius; DPBS: Dulbecco's phosphate-buffered saline; MNCs: Mononuclear cells; DMSO: Dimethyl sulfoxide; PDT: Population doubling time; CD marker: Cluster of differentiation marker; DMMB: Dimethylmethylene; SD: Standard deviation; Runx2: Runt-related transcription factor 2; ON: Osteocalcin; LPL: Lipoprotein lipase; FABP4: Fatty acid-binding protein 4: COL10A1: Type X collagene gene; ACAN: Aggrecan; COL2A1: Alpha 1 chain of collagen type II; ECM: Extracellular matrix.

References

1. Martel-Pelletier, J.; Barr, A.J.; Cicuttini, F.M.; Conaghan, P.G.; Cooper, C.; Goldring, M.B.; Goldring, S.R.; Jones, G.; Teichtahl, A.J.; Pelletier, J. Osteoarthritis. *Nat. Rev. Dis. Primers* **2016**, *2*, 16072. [CrossRef]
2. Bertuglia, A.; Pagliara, E.; Grego, E.; Ricci, A.; Brkljaca-Bottegaro, N. Pro-inflammatory cytokines and structural biomarkers are effective to categorize osteoarthritis phenotype and progression in Standardbred racehorses over five years of racing career. *BMC Vet. Res.* **2016**, *12*, 246. [CrossRef] [PubMed]
3. Al-Rukibat, R.K.; Bani Ismail, Z.A.; Al-Zghoul, M.B. Cytologic analysis of synovial fluid in clinically normal tarsal joints of young camels (*Camelus dromedarius*). *Vet. Clin. Pathol.* **2006**, *35*, 326–328. [CrossRef] [PubMed]
4. Fox, A.J.S.; Bedi, A.; Rodeo, S.A. The basic science of articular cartilage: Structure, composition, and function. *Sports Health* **2009**, *1*, 461–468.
5. Ge, Z.; Hu, Y.; Heng, B.C.; Yang, Z.; Ouyang, H.; Lee, E.H.; Cao, T. Osteoarthritis and Therapy. *Arthritis Care Res.* **2006**, *55*, 493–500. [CrossRef] [PubMed]
6. Chard, J.; Dieppe, P. Glucosamine for osteoarthritis: Magic, hype, or confusion? It's probably safe-but there's no good evidence that it works. *BMJ* **2001**, *322*, 1439–1440. [CrossRef] [PubMed]
7. Bartlett, W.; Skinner, J.A.; Gooding, C.R.; Carrington, R.W.; Flanagan, A.M.; Briggs, T.W.R.; Bentley, G. Autologous chondrocyte implantation versus matrix-induced autologous chondrocyte implantation for osteochondral defects of the knee: A prospective, randomised study. *J. Bone Jt. Surg. Br.* **2005**, *87*, 640–645. [CrossRef] [PubMed]
8. Seo, S.S.; Kim, C.W.; Jung, D.W. Management of focal Chondral lesion in the knee joint. *Knee Surg. Relat. Res.* **2011**, *23*, 185–196. [CrossRef]
9. Kim, H.J.; Park, S.R.; Park, H.J.; Choi, B.H.; Min, B.H. Potential predictive markers for proliferative capacity of cultured human articular chondrocytes: PCNA and p21. *Artif. Organs.* **2005**, *29*, 393–398. [CrossRef]
10. Khang, G.; Kim, S.H.; Kim, M.S.; Rhee, J.M.; Lee, H.B. Recent and future directions of stem cells for the application of regenerative medicine. *Tissue Eng. Regen. Med.* **2007**, *4*, 441–470.
11. De Miguel, M.P.; Fuentes-Julián, S.; Blázquez-Martínez, A.; Pascual, C.Y.; Aller, M.A.; Arias, J.; Arnalich-Montiel, F. Immunosuppressive properties of mesenchymal stem cells: Advances and applications. *Curr. Mol. Med.* **2012**, *12*, 574–591. [CrossRef] [PubMed]
12. Son, Y.B.; Bharti, D.; Kim, S.B.; Bok, E.Y.; Lee, S.Y.; Ho, H.J.; Lee, S.L.; Rho, G.J. Hematological patterns and histopathological assessment of Miniature Pigs in the experiments on human mesenchymal stem cell transplantation. *Int. J. Med. Sci.* **2021**, *18*, 1259–1268. [CrossRef] [PubMed]
13. Son, Y.B.; Jeong, Y.I.; Jeong, Y.W.; Hossein, M.S.; Tinson, A.; Singh, K.K.; Hwang, W.S. Comparative study of biological characteristics, and osteoblast differentiation of mesenchymal stem cell established from camelus dromedarius skeletal muscle, dermal skin, and adipose tissues. *Animals* **2021**, *11*, 1017. [CrossRef]
14. Kern, S.; Eichler, H.; Stoeve, J.; Klüter, H.; Bieback, K. Comparative analysis of mesenchymal stem cells from bone marrow, umbilical cord blood, or adipose tissue. *Stem Cells* **2006**, *24*, 1294–1301. [CrossRef]
15. Yu, J.M.; Wu, X.; Gimble, J.M.; Guan, X.; Freitas, M.A.; Bunnell, B.A. Age-related changes in mesenchymal stem cells derived from rhesus macaque bone marrow. *Aging Cell* **2011**, *10*, 66–79. [CrossRef]
16. Chang, W.; Park, S.I.; Jun, S.; Lee, E.; Ham, H.; Bae, Y.; Kim, R.; Park, M.; Chung, Y.; Im, N.; et al. Therapeutic potential of autologous mesenchymal stem cells derived from synovial fluid in patients with degenerative arthritis. *Anim. Cells Syst.* **2013**, *17*, 315–324. [CrossRef]
17. De Sousa, E.B.; Casado, P.L.; Moura Neto, V.; Duarte, M.E.L.; Aguiar, D.P. Synovial fluid and Synovial membrane mesenchymal stem cells: Latest discoveries and therapeutic perspectives. *Stem Cell Res. Ther.* **2014**, *5*, 112. [CrossRef]
18. Neybecker, P.; Henrionnet, C.; Pape, E.; Mainard, D.; Galois, L.; Loeuille, D.; Gillet, P.; Pinzano, A. In vitro and in vivo potentialities for cartilage repair from human advanced knee osteoarthritis synovial fluid-derived mesenchymal stem cells. *Stem Cell Res. Ther.* **2018**, *9*, 329. [CrossRef]
19. Zayed, M.; Newby, S.; Misk, N.; Donnell, R.; Dhar, M. Xenogenic implantation of equine synovial fluid-derived mesenchymal stem cells leads to articular cartilage regeneration. *Stem Cells Int.* **2018**, *2018*, 1073705. [CrossRef] [PubMed]
20. Lee, W.-J.; Hah, Y.-S.; Ock, S.-A.; Lee, J.-H.; Jeon, R.-H.; Park, J.-S.; Lee, S.-I.; Rho, N.-Y.; Rho, G.-J.; Lee, S.-L. Cell source-dependent in vivo immunosuppressive properties of mesenchymal stem cells derived from the bone marrow and synovial fluid of minipigs. *Exp. Cell Res.* **2015**, *333*, 273–288. [CrossRef] [PubMed]
21. Lee, W.-J.; Park, J.-S.; Jang, S.-J.; Lee, S.-C.; Lee, H.; Lee, J.-H.; Rho, G.-J.; Lee, S.-L. Isolation and cellular phenotyping of mesenchymal stem cells derived from synovial fluid and bone marrow of minipigs. *J. Vis. Exp.* **2016**, *113*, 54077. [CrossRef]
22. Son, Y.B.; Jeong, Y.I.; Hwang, K.C.; Jeong, Y.W.; Hwang, W.S. Mitochondrial metabolism assessment of lycaon-dog fetuses in interspecies somatic cell nuclear transfer. *Theriogenology* **2021**, *165*, 18–27. [CrossRef] [PubMed]

23. Hulme, C.H.; Wilson, E.L.; Peffers, M.J.; Roberts, S.; Simpson, D.M.; Richardson, J.B.; Gallacher, P.; Wright, K.T. Autologous chondrocyte implantation-derived synovial fluids display distinct responder and non-responder proteomic profiles. *Arthritis Res. Ther.* **2017**, *19*, 150. [CrossRef]
24. Minas, T.; Gomoll, A.H.; Solhpour, S.; Rosenberger, R.; Probst, C.; Bryant, T. Autologous chondrocyte implantation for joint preservation in patients with early osteoarthritis. *Clin. Orthop. Relat. Res.* **2010**, *468*, 147–157. [CrossRef]
25. Neybecker, P.; Henrionnet, C.; Pape, E.; Grossin, L.; Mainard, D.; Galois, L.; Loeuille, D.; Gillet, P.; Pinzano, A. Respective stemness and chondrogenic potential of mesenchymal stem cells isolated from human bone marrow, synovial membrane, and synovial fluid. *Stem Cell Res. Ther.* **2020**, *11*, 316. [CrossRef] [PubMed]
26. Tsai, C.; Hung, S. Functional roles of pluripotency transcription factors in mesenchymal stem cells. *Cell Cycle* **2012**, *11*, 3711–3712. [CrossRef] [PubMed]
27. Arévalo-Turrubiarte, M.; Olmeo, C.; Accornero, P.; Baratta, M.; Martignani, E. Analysis of mesenchymal cells (MSCs) from bone marrow, synovial fluid and mesenteric, neck and tail adipose tissue sources from equines. *Stem Cell Res.* **2019**, *37*, 101442. [CrossRef]
28. Jones, E.A.; Crawford, A.; English, A.; Henshaw, K.; Mundy, J.; Corscadden, D.; Chapman, T.; Emery, P.; Hatton, P.; McGonagle, D. Synovial fluid mesenchymal stem cells in health and early osteoarthritis: Detection and functional evaluation at the single cell level. *Arthiritis Rheum.* **2008**, *58*, 1731–1740. [CrossRef] [PubMed]
29. DeLise, A.M.; Fischer, L.; Tuan, R.S. Cellular interactions and signaling in cartilage development. *Osteoarthr. Cartil.* **2000**, *8*, 309–334. [CrossRef]
30. Sandell, L.J.; Adler, P. Developmental patterns of cartilage. *Front. Biosci.* **1999**, *4*, 731–742. [CrossRef]
31. Yoon, Y.; Kim, S.; Oh, C.; Ju, J.; Song, W.K.; Yoo, Y.J.; Huh, T.; Chun, J. Maintenance of differentiated phenotype of articular chondrocytes by protein kinase C and extracellular signal-regulated protein kinase. *J. Biol. Chem.* **2002**, *277*, 8412–8420. [CrossRef] [PubMed]
32. Pomin, V.H.; Mulloy, B. Glycosaminoglycans and proteoglycans. *Pharmaceuticals* **2018**, *11*, 27. [CrossRef] [PubMed]
33. Hayes, A.J.; Melrose, J. Glycosaminoglycan and proteoglycan biotherapeutics in articular cartilage protection and repair strategies: Novel approaches to visco-supplementation in orthobiologics. *Adv. Ther.* **2019**, *2*, 1900034. [CrossRef]
34. Lozzo, R.V.; Schaefer, L. Proteoglycan form and function: A comprehensive nomenclature of proteoglycans. *Matrix Biol.* **2015**, *42*, 11–55.

animals

MDPI

Review

Current Status on Canine Foetal Fluid and Adnexa Derived Mesenchymal Stem Cells

Eleonora Iacono *, Romina Marcoccia and Barbara Merlo

Department of Veterinary Medical Sciences, University of Bologna, 40126 Bologna, Italy; romina.marcoccia2@unibo.it (R.M.); barbara.merlo@unibo.it (B.M.)
* Correspondence: eleonora.iacono2@unibo.it; Tel.: +39-051-2097-567

Simple Summary: In the last few years, dog owners have required sophisticated new treatments such as the use of MSCs for tissue engineering and regenerative medicine applications. On these topics, *Canis familiaris*, which develop many diseases with etiologies and pathogenesis similar to those that develop in humans, can be considered as a realistic preclinical model to evaluate the therapeutic potential of MSCs. The aim of the present review is to offer an update on the state of the art on canine MSCs derived from foetal adnexa and fluid, focusing on the findings in their clinical setting.

Abstract: Effective standards of care treatment guidelines have been developed for many canine diseases. However, a subpopulation of patients is partially or completely refractory to these protocols, so their owners seek novel therapies such as treatments with MSCs. Although in dogs, as with human medicine, the most studied MSCs sources have been bone marrow and adipose tissue, in recent years, many researchers have drawn attention towards alternative sources, such as foetal adnexa and fluid, since they possess many advantages over bone marrow and adipose tissue. Foetal adnexa and fluid could be considered as discarded material; therefore, sampling is non-invasive, inexpensive and free from ethical considerations. Furthermore, MSCs derived from foetal adnexa and fluid preserve some of the characteristics of the primitive embryonic layers from which they originate and seem to present immune-modulatory properties that make them a good candidate for allo- and xenotransplantation. The aim of the present review is to offer an update on the state of the art on canine MSCs derived from foetal adnexa and fluid focusing on the findings in their clinical setting.

Keywords: dog; foetal adnexa; mesenchymal stem cells; therapy

Citation: Iacono, E.; Marcoccia, R.; Merlo, B. Current Status on Canine Foetal Fluid and Adnexa Derived Mesenchymal Stem Cells. *Animals* **2021**, *11*, 2254. https://doi.org/10.3390/ani11082254

Academic Editor: Paul Dyce

Received: 5 July 2021
Accepted: 27 July 2021
Published: 30 July 2021

1. Introduction

Today, pet owners require their companion animals to be cured in veterinary hospitals, constantly monitored, using high diagnostic and medical devices. Furthermore, an increasing number of pet owners require their animals to be treated with sophisticated and new treatments. In this context, research on canine MSCs is becoming increasingly important.

On the other hand, because of the need for clinical trials in animal models for new treatments for human regenerative medicine and because humans and dogs share environmental life patterns and similar pathologies, *Canis familiaris* could be considered as a suitable model of spontaneous diseases [1–6].

For example, the most frequent and important non ischaemic cardiomyopathies, both in human and dog, showing pathological and clinical similarities, are dilated cardiomyopathy and arrhythmogenic ventricular cardiomyopathy [5]. DCM is the second most common cardiac disease, affecting a wide range of breeds, particularly Doberman Pinscher, with a prevalence of around 44% [6], while AVC has been described frequently in Boxer [7]. However, in both species, although progress in the management of symptoms has been made, the actual disease processes remain a challenge to treat, and study of the use of regenerative therapies using MSCs is very interesting.

Although the standing hindlimb angle, an aspect to consider for comparing biomechanical data, is much larger in canine than in human, dog could be an important model for osteoarthritis [8]. Indeed, compared with horse, ruminants or swine, using dog as a clinical model leads to easier post-surgical management and follow up [8]. In this context, dog can be considered the best model for SCI, cranial cruciate and meniscal injury, osteoarthritis and other skeletal muscle pathologies, both for studying mechanisms of degeneration and testing new treatments [6].

In addition, for gastro-intestinal pathologies, dog can be considered as an important clinical model, with the potential to overcome some of the major obstacles of laboratory animal modeling. Indeed, canine spontaneous lymphocytic-plasmocytic colitis, the most common form of enteropathy, included in inflammatory bowel disease syndrome, has several histopathologic and cellular-molecular features similar to human IBD [9]. This is characterized by multifactorial pathogenesis which is less compelling in rodent models [10], so dog can be essential for understanding MSCs immune modulation mechanisms, determining dose equivalence as well as biological effects of MSCs transplantation in patients affected by IBD refractory and traditional therapy [10].

Despite the importance of using the dog as model for human medicine, this raises more ethical debates than livestock animals. For this reasons, in the last decade, most studies in these animals have involved clinical cases of spontaneous pathologies observed in veterinary hospitals and clinics, highlighting their importance for clinical research [6].

To date, canine adult tissue, BM and AT have represented the most important sources of MSCs in the field of cell-based therapy [11–13] although cell harvesting is invasive, and increasing donor site morbidity and cell amount and characteristics are closely related to donor age [14–19]. Foetal fluids (amniotic fluid, umbilical cord blood), and foetal adnexa (Wharton's jelly, amniotic membrane) have been identified as ideal alternative sources of MSCs in different animal species, such as horse [20–22], bovine [23,24], goat [25,26], and others. The benefits of these cells compared to adults MSCs and embryonic SCs are determined by their origin from extraembryonic tissues, usually discarded after delivery. Moreover, due to the fact that they are at the maternal—fetal interface, they become convenient for transplantation due to their low immunogenicity and immunomodulatory properties, making them a good candidate for allo- and xenotransplantation [27].

Despite the importance of *Canis familiaris* both as a patient and disease model, to our knowledge, in the literature, there are only general reviews of MSCs derived from foetal fluid and adnexa, based on research carried out on domestic animals [28–30]. The aim of the present review is to offer an update on the state of the art of canine MSCs isolated from foetal adnexa and fluid focusing on the findings in their clinical setting, when reported.

2. Stem Cells

The term stem cell was first coined in the nineteenth century by Edmund Beecher Wilson who used this term as a synonym for mitotically quiescent primordial germ cell. In 1963, Becker et al. [31] discovered that stem cells are undifferentiated cells able to perform self-renewal, as confirmed later by Weissman [32]. Based on their differentiation ability, stem cells are classified in totipotent, pluripotent and multipotent cells. Totipotent cells are able to differentiate in all cell lines, including extra-embryonic tissues; totipotent cells include zygote and the descendants of the first three cell divisions [33]. Pluripotent stem cells are embryonic stem cells derived from blastocysts ICM. These cells are able to propagate readily [34] and are capable of forming embryoid bodies that generate a variety of specialized cells, including neural, cardiac and pancreatic cells [35]. Despite their therapeutic potential, clinical use of ESCs presents, especially in human, ethical and application problems. Different authors have observed that after ESCs' in vivo implantation, teratomas have been developed [36]. Another type of pluripotent stem cells is bioengineered IPS (Nelson et al. 2010) [37]. Utilizing IPS-based technology, all lineages of the adult body may become viable targets for replacement, avoiding immune intolerance. However, the unlimited differentiation potential of IPS is similar to ESCs, and thus the risk of dysregulated growth and teratoma formation requires stringent safeguards [37].

Many adult tissues, on the other hand, have multipotent stem cells within them, i.e., cells capable of producing a limited number of cell lines, appropriate to their location; these cells are named multipotent stem cells and the most studied among them are MSCs.

The most used mesenchymal tissues, sources of MSCs, are bone marrow and adipose tissue.

Mesenchymal stem cells are a population of multipotent stem cells which must meet the criteria established by ISCT: (i) plastic adherence when maintained in standard culture conditions; (ii) ability to differentiate in osteoblasts, adipocytes and chondroblasts when appropriately stimulated in vitro; (iii) expression of CD73, CD90 and CD105. On the contrary, MSCs lack expression of haematopoietic markers, such as CD14, CD34, CD45 and HLA-DR [38]. Due to these properties, MSCs offer a great chance for cell-based therapies and tissue-engineering applications.

3. Canine Mesenchymal Stem Cells from Foetal Fluids

3.1. Amniotic Fluid Mesenchymal Stem Cells

Amniotic fluid is characterized by a heterogeneous population of cells: cuboidal epitheloid cells, derived from foetal skin and urine, round cells, from foetal membranes and trophoblasts, and spindle-shaped fibroblastic cells, generated from mesenchymal tissues and are supposed to be the MSCs population of the AF [39]. Unlike in humans, in dogsm it is impossible to obtain AF using the amniocentesis technique. Moreover, due to its small volume, canine fluids provide a very small quantity of cells, which makes further cell expansion difficult. The use of the total recoverable volume of the AF of all foetuses improves the process of cells isolation, though there is no univocal agreement on which gestational period is the best in terms of cell yield. In 2011, Filioli-Uranio et al. [40] obtained fibroblast-like cells, with a DT of 1.12 ± 0.04 days, from AF recovered after hysterectomy in bitches between 25 and 40 days of pregnancy. These data are in contrast from those obtained by Fernandes et al. just one year later [41]. At earlier stage of gestation (25–40 days), Authors obtained fibroblast-like cells growing in adhesion but which failed to proliferate. Successful cells isolation was instead achieved from AF recovered at 50 days of gestation, as demonstrated by Choi et al. in 2013 [42]. Even if cell concentration decreased during the full-term stage of gestation [42], data recovered by Fernandes et al. [41] and Choi et al. [42] have shown the possibility of isolating AFMSCs after caesarean section in breeding bitches. As showed in Table 1, the expression of OCT4, CD44, DLA-DRA1, and DLA-79 by AFMSCs at passage P1 was observed by Filioli Uranio et al. [40], whereas from the next passage, cells expressed only CD44. On the contrary, Choi et al. [42] observed that at P5 of in vitro culture, AFMSCs express pluripotent stem cell markers OCT4, NANOG, and SOX2, as well as CD29 (β1 integrin), CD44, and CD90 (Thy1) [42]. These different results between different research groups could be determined both by the different gestational periods of samples' recoveries and by different and not fully comparable techniques employed in stem cells characterization.

Table 1. Canine foetal fluid and adnexa derived mesenchymal stem cells: potential differentiation in vitro and molecular characterization.

MSCs Source	Potential	References	Markers Positive Expression	References
AF	Osteogenic-Chondrogenic-Adipogenic	[40–42]	RT-PCR: OCT4; CD44; DLA-DRA1; DLA79	[40]
	Neurogenic	[40,41,43]	ICC: OCT4; NANOG; SOX2	[42]
	Hepatogenic	[42]	FACS: CD29; CD44; CD90	[42]

Table 1. *Cont.*

MSCs Source	Potential	References	Markers Positive Expression	References
UCB	Osteogenic	[44–47]	FACS: CD44; CD73; CD90; CD105	[44–47]
	Chondrogenic-Adipogenic	[44–47]		
	Neurogenic	[44,46]	GFAP, Tuj-1,NF160	[44]
Placenta	Osteogenic-Chondrogenic-Adipogenic	[40,48–52]	RT-PCR: OCT4; CD44; CD184; CD29	[40,48–52]
	Neurogenic	[53,54]	FACS: OCT4; SOX2; CD73; CD90; CD105; MHC1	[51,52,55]
WJ	Osteogenic	[45,48,56–69]	FACS: CD44; CD73; CD90; CD105	[45,48,56]
	Chondrogenic-Adipogenic	[45,48,56–69]		
	Neurogenic	[46,48]	PCR: CD44; CD90	[67]

The hope of cell therapy as a new clinical approach to repair tissue damage relies on the characteristics of the mesenchymal stem cells, such as their low expression of polymorphic antigens that seems to enhance transplantation tolerance, making these cells useful for allotransplant and xenotransplant [70]. Filioli-Uranio et al. [40] observed a reduced expression of MHC genes in canine AFMSCs; indeed, only DLA-DRA1 and DLA-79 were expressed at P1, as previously demonstrated in human and other animal species [20,71–73].

Despite the differences found between the different research groups in the molecular characterization of AFMSCs, there is consensus on their differentiation potential. Indeed, it was demonstrated that these cells are able to differentiate in vitro into osteogenic, chondrogenic and adipogenic lineages [40–42], but also into neuronal [40,41,43] and hepatocyte-like cells [42]. Regarding neuronal differentiation, Fernandes et al. [41] observed that undifferentiated canine AFMSCs stained positively for nestin. Nestin is expressed in neural progenitor stem cells and pluripotent stem cells, which undergo neuronal differentiation, as well as in MSCs [74,75], including human AFMSCs [76]. Filioli-Uranio et al. [40] validated these data, observing that canine AFMSCs, cultured in neuronal induced medium, stained positively for nestin and showed the presence of Nissl bodies and a neuronal-like morphology, as confirmed later by Kim et al. [43]. The expression of neural-specific genes, such as NEFL, NSE, TUBB3, and the astrocyte-specific gene, GFAP, significantly increased in AFMSCs after neural induction [43].

Liver transplantation is a last resort for patients with end-stage liver disease. In a study [42], canine AFMSCs were induced to differentiate in hepatocyte like-cells by HGF (an endocrine or paracrine factor, essential for liver development), OSM (essential for the maturation of hepatocytes), nicotinamide, and dexamethasone (essential for the development of hepatogenic morphology through the suppression of cell division). AFMSCs that underwent hepatic differentiation did not show typical hepatocyte morphological changes, but they expressed genes critical for hepatocyte differentiation [42]. Indeed, after being induced to hepatocytic differentiation, canine AFMSCs were strongly positive for TAT, α1-AT, GS, and ALB, all markers of mature hepatocytes, as well as for hepatocyte-specific markers, including ALB and TAT [42]. Taken together, the results reported by different Authors suggest that canine AFMSCs have the capacity for multilineage differentiation and have the potential to be a source for cell-based therapies in canine models of hepatic disease, as well as having the potential capacity for clinical treatment of neuronal precursor-cell transplantation.

Although the studies referenced above demonstrate the potential for canine AFMSCs for clinical uses, to our knowledge, no paper on regenerative therapies based on canine AFMSCs exists in the literature.

3.2. Umbilical Cord Blood Mesenchymal Stem Cells

The umbilical cord is the channel that connects the fetus and the placenta, considered as a physiological and inherent part of the fetus during prenatal development [77]. In 1989 [78], umbilical cord blood, flowing in the vein, was found to be a rich and readily available alternative source of primitive and unspecialized mesenchymal stem cells, probably derived from the fetal liver or bone marrow [79]. Canine UCBMSCs were isolated for the first time by Lim et al. in 2007 [80]. However, characterization and in vitro differentiation of these cells were carried out only in 2009 [44], and the results obtained were confirmed later by different Authors [45–47]. In addition to the expression of stemness markers, shown in Table 1, and in vitro differentiation in the three lineages requested by ISCT [38], canine UCBMSCs showed basically neuronal associated protein markers under the undifferentiated condition. Indeed, undifferentiated canine UCBMSCs slightly expressed GFAP, Tuj-1, and NF160 neuronal cell protein markers but they did not express Nestin and MAP2 [44]. After growing in neuronal induction medium, canine UCBMSCs exhibited morphological changes [44,46], appearing as sharp, elongated bi- or tripolar cells with primary, secondary and multi-branched processes. When inducted with neuronal differentiation media, canine UCBMSCs showed positive expression patterns for Nestin, GFAP, Tuj-1, MAP2, NF160 and NeuN [44,46], showing a percentage of cells stained with antibodies specific for NeuN higher than that of adipose tissue [46]. Unlike cells deriving from canine amniotic fluid, for which there are no references on their clinical use, as showed in Table 2, canine UCBMSCs have been tested for regenerative therapy since 2007, when Lim et al. [80] applied these cells in induced spinal cord injured dogs. Authors observed evidence of functional and sensory improvement after allogenic UCBMSCs transplantation, even though no evidence of regeneration of spinal cord tissue by magnetic resonance imaging and histology was observed [80]. However, in this first study, new neuronal formation in the injured structures of the spinal cord was observed after UCBMSCs transplantation, as well as no additional damage to the experimentally injured spinal cord such as inflammatory responses [80]. Moreover, study [80] showed that transplantation of UCBMSCs resulted in recovered nerve function in dogs after a spinal cord injury, as confirmed recently by Park et al. [81] and Ryu et al. [46]. In the research of Park et al. [81] cells transplantation was carried out 12 h, 1 week and 2 week after spinal cord injury induction. Canine UCBMSCs transplanted one week after SCI significantly improved clinical signs, evaluated using the Olby and Tarlov scales. In all groups, the scores gradually increased after 2 weeks, and decreased after 3 weeks [81]. Better results were obtained later by Ryu et al. [46] transplanting cells with Matrigel, seven days after SCI, into the parenchyma of the spinal cord, near the lesion site or directly into the injury epicenter. Matrigel maintains the microenvironment and exerts effects such as rescuing dying cells, increasing cell proliferation, blocking inflammatory and cytotoxic cytokines, promoting neuronal differentiation [82,83]. As previously observed by Park et al. [81], transplanted cell survival is increased during the subacute phase of SCI, when the lesion is not fully developed and MSCs may act as neuroprotective agent. On the contrary, when cells were transplanted 2 weeks after SCI, when fibrosis has progressed, no significant improvement in patient clinical signs was observed [81]. Moreover, the lesion epicenter may not be a favorable site for cell survival due to the presence of phagocytes, which could have been the cause of the score decreasing 3 weeks after transplantation in the previous work [81]. Regarding the influence of canine UCBMSCs on inflammation, in all studies, COX-2 protein expression was significantly decreased [46,80,81], also compared with MSCs derived from other sources [46], playing an important role in proliferation, migration and differentiation of endogenous spinal cord-derived neural progenitor cells in SCI.

Table 2. Canine foetal fluid and adnexa derived mesenchymal stem cells: clinical applications.

MSCs Source	Clinical Application	References
UCB	Bone Defect	[45]
	SCI	[46,80,81]
	AKI	[47]
Placenta	Ischemic Stroke	[84]
WJ	Bone Defect	[66]
	SCI	[46]
	TPLO	[85]
	CHF	[86]

Canine UCMSCs also showed higher osteogenic potential compared with BMMSCs and WJMSCs, as shown by greater levels of ALP activity, an early osteoblastic marker [45]. All MSCs induced substantial in vivo bone formation and significant differences in the levels of bone formation promoted in vitro by the various MSCs were not observed [45], indicating that the osteogenic potential in vitro and in vivo can be slightly different for each type of MSCs. This can be explained by several factors, such as in vivo vascularization promoted by VEGF, secreted by MSCs in different quantities. In the study of Kang et al. [45] VEGF in vitro production was determined to investigate the ability of cells to promote vascularization. Authors observed that canine BMMSCs produced higher quantities of VEGEF compared with canine UCBMSCs; however, BMMSCs had a weaker osteogenic capability in vitro but no differences were determined in bone formation after MSCs transplantation in vivo and hematopoietic tissues were observed in histological section [45]. Moreover, MSCs may affect bone formation stimulating induction and migration of endogenous cells. Previous studies by the same research group revealed that cytokines released by canine UCBMSCs 1 day after implantation can enhance bone regeneration [87,88].

The reported mortality of AKI ranges from 47% to 61% in dogs [89]. Traditional AKI treatment includes fluid administration, monitoring urine output, use of diuretics, anti-nausea agent, gastroprotectors, phosphorus absorbent, antioxidants, sodium bicarbonate for metabolic acidosis, antidote for nephrotoxin, and antibiotics for infection [90]. After nephron disruption, it is difficult for patients to overcome the disease without the aid of dialysis or renal transplantation; these therapeutic approaches in dogs have considerable limitations, including difficulty in using them in smaller animals due to their low total blood volume, immunological problems and the low availability of donor kidneys [84,91]. Recently, Lee et al. injected twice canine UCBMSCs directly into the renal corticomedullary junction of dog with induced AKI, followed by intravenous administration of gentamycin and cisplatin [47]. For evaluating renal function blood BUN and creatinine were determined. BUN levels increased, owing to elevated urea reabsorption caused by prolonged renal retention due to a decreased glomerular flow rate [92]. Creatinine, a muscular metabolic product, is a more precise indicator of renal function than BUN. Serum creatinine higher than 10 mg/dl was associated with failure to recover from AKI [93]. In the study of Lee et al. serum BUN and creatinine levels decreased in dogs treated with UCBMSCs and renal excretory function improved [47]. In transplanted dogs, these serological findings were associated with moderate renal lesions, including necrosis of tubules and glomerulus and shedding of tubular epithelium cells; on the contrary, dogs treated with PBS exhibited global cystic change of tubular tissue and massive interstitial leukocyte infiltration [47]. Up to the end of the experiment, no mortality was recorded in dogs with induced AKI and subsequently treated with UCBMSCs, suggesting that for this pathology, MSCs could be an alternative and valid treatment.

4. Canine Mesenchymal Stem Cells from Foetal Adnexa

4.1. Placenta and Foetal Membranes

The potential for the clinical application of fetal stem cells from the human amniotic membrane was first described in 2011 by Parolini and Caruso [94]. Since, interest in this tissue as a source of MSCs has also developed in dogs. As shown in Table 1, canine AMMSCs expressed embryonic and MSCs markers, such as OCT4, CD44, CD184, and CD29 and could differentiate into neurocytes, osteocytes, adipocytes and chondrocytes based on cell morphology, specific stains, and molecular analysis [40,48–52]. Regarding the immunomodulatory properties of AMMSCs, Borghesi et al. [51] found low MHC-I expression and no MHC-II expression, giving the MSCs the potential to escape recognition by CD4+T cells [55]. Moreover, at any gestational time no interleukin IL-1, IL-2, IL-6 and IL-10 labeling in AMMSCs was observed [51].

To use stem cells safely, it is necessary to know if there are risks of genetic instability, which may lead to tumorigenesis. Recently, Cardoso et al. [50] and Borghesi et al. [51] did not observe tumor formation after injecting canine AFMSCs, demonstrating that these cells are safe for in vivo application.

As previously reported for human placenta derived MSCs [95], Long et al. [53] and Amorim et al. [54], using a cytokine array demonstrated that unstimulated and stimulated DPCs secrete a number of paracrine factors. VEGF and MCP-1 are implicated in both neuroprotection and angiogenesis [56,57], IL-6 is an immunomodulatory cytokine with in vitro and in vivo demonstrated neuroprotective capabilities [58], and IL-8 is an immunomodulatory cytokine that promotes angiogenesis [59].

Long et al. [53], in 2018, after 1 week of co-culture with a neuroblastoma cell line (SH-SY5Y cells), demonstrated that DPCs could induce the formation of complex neural networks in SH-SY5Y cells, increasing the number of branching points and total segments of neurites in culture. In 2020, Amorim et al. [54] speculated that based on their potent pro-angiogenic, neuroprotective, immunomodulatory properties, MSCs could be a promising therapy for canine inflammatory brain disease. IBD syndrome comprehends idiopathic disorders subdivided based on histopathology findings, i.e., granulomatous meningoencephalomyelitis, necrotizing meningoencephalitis, and necrotizing leukoencephalitis, encompass also in the term of meningoencephalomyelitis of unknown origin, presumed to be an autoimmune disease with a genetic predisposition [60], similarly to multiple sclerosis in human [61]. In vitro experiments revealed that rat neural cells exposed to OGD conditions and co-cultured with DPCs exhibited dose-dependent improvement in cell survival and ATP production compared to the vehicle, indicating the importance of these cells and cell dose in achieving neuroprotection [62]. In vivo observation in a rat stroke model reveals that animals treated with DPCs exhibited significantly fewer behavioral deficits, with improvements in motor and neurological impairments, as confirmed by histological data. These observations highlight the possible benefits of investigation into the efficacy of autologous transplant of DPCs in dog stroke or IBD patients, considering the lack of available treatment for ischemic and IBD injury in dogs.

4.2. Wharton's Jelly or Umbilical Cord Matrix

Wharton's Jelly is a mesenchymal connective tissue developed from extraembryonic mesoderm and placed between umbilical vessels. WJ binds and encases the umbilical vessels, protecting them from twisting and compression during gestation. Since 1990, WJ has been considered an important source of MSCs in humans [63,64]. However, in canines, WJ is difficult to separate from umbilical cord matrix due to the small size of umbilical cord; in this review we will refer to cells isolated from these sources as WJMSCs. MSCs were successfully isolated from canine WJ for the first time by Seo and collaborators in 2012 [65]. These cells showed a typical mesenchymal immunophenotype and the ability to differentiate in vitro [65], as confirmed lately by other research groups [45,48,66,67] (Table 1). Recently, Souza et al. [66] demonstrated that the optimum conditions for canine WJMSCs osteogenic differentiation is co-culture with PRP and DBM, associated with optimum ALP levels and high levels of osteocalcin gene, a late marker of osteogenesis, and osteopontin gene, an important factor in bone remodeling [68,69].

Recently, the metabolic profile of canine WJMSCs was investigated and compared with that of canine ATMSCs cultured under the basal conditions [67]. In order to detect the total ATP production rates in living MSCs, serial additions of oligomycin, a specific inhibitor of the mitochondrial ATP synthase, and of rotenone plus antimycin A, inhibitors of mitochondrial complex I and III, respectively, were performed automatically and stepwise. This metabolic assay allowed the Authors to evaluate the amount of ATP produced by OXPHOS and glycolysis, which represent the two main metabolic pathways responsible for ATP production in mammalian cells. ATMSCs and WJMSCs showed a different total ATP production rate, since OXPHOS and glycolytic pathways in foetal adnexa MSCs provided a higher amount of cellular ATP than in ATMSCs. The energy map of both MSC types corroborates an aerobic energy metabolism with more active OXPHOS and glycolytic pathways in WJMSCs than ATMSCs, although the latter two showed a higher mitoATP/glycoATP ratio than WJMSCs, which highlights a prevailing oxidative phenotype. In the same paper, the key parameters of mitochondrial function, directly measured by the cell respiration profile of MSCs, were also investigated. Both basal respiration and ATP turnover showed higher values in WJMSCs than in ATMSCs. Knowledge of the mitochondrial status and especially of some bioenergetics parameters such as the spare respiratory capacity, which guarantees a metabolic flexibility, may help to select the best candidates for transplantation studies.

Different researchers have demonstrated that canine WJMSCs have a higher neurogenic potential in vitro [46,48]. Ryu et al. [46] transplanted WJMSCs with matrigel into the spinal cord parenchyma near the induced lesion site or directly into the injury epicenter 7 days after injury. The Authors focused their attention on the survival and integration of allogenic MSCs in the injured spinal cord, demonstrating that allogenic WJMSCs could successfully survive in injured spinal cords where they integrated into host tissue without using immunosuppressive agents and also improved hind-limb function following SCI [46]. In particular, after WJMSCs transplantation, reduced levels of reactive astrogliosis and macrophage infiltration into the lesion epicenter were observed [46]. However, even though many WJMSCs can survive following transplantation, very few cells can differentiate into neural-like cells in vivo [46]. Indeed, most NF160- and NeuN-positive neurons in an injured spinal cord were derived from endogenous spinal cord-derived neural progenitor cells or preserved neurons, thanks to MSCs neuroprotection and anti-inflammatory effects. This is also confirmed by the rapid recovery of treated dogs after transplantation.

Due to their anti-inflammatory potential, WJMSCs have been used in intraarticular treatment of dogs subjected to surgical tibial plateau leveling osteotomy [85]. Taroni et al., in 2017, showed that a single postoperative intraarticular injection of allogeneic WJMSCs leads to a level of postoperative lameness and pain outcome after TPLO similar to those observed in animals treated with long-term NSAIDs systemic administration. WJMSCs anti-inflammatory potential could be due to factor secreted with Extracellular Vesicles, nanoscale cellular products containing RNA, protein, and lipids [96]. WJMSC EVs present a diameter of 125 nm, low buoyant density (1.1 g/mL), and expression of EV proteins Alix and TSG101. Functionally, EVs inhibited CD4pos T cell proliferation in a dose-dependent manner and TGF-b was present on EVs as latent complexes most likely tethered to EV membrane by betaglycan. These observations demonstrate that canine WJMSC EVs utilizes TGF-b and adenosine signaling to suppress proliferation of CD4pos T cell. WJMSCs EVs could also be responsible for the results registered by Yang et al. [86] after WJMSCs intravenous injection in dogs with congestive heart failure secondary to myxomatous mitral valve disease. Indeed, decreases in blood lymphocyte, monocyte, and eosinophil counts immediately after MSC injection were observed. However, further studies are needed on these topics for clinical WJMSCs application in different pathologic conditions.

5. Conclusions

The studies presented in this review offer authoritative views on markers expression and therapeutic potential of canine MSCs from foetal tissues and fluids. As reported in human and other animal species, also in dog these sources are easily available so MSCs may have an attraction compared to other established SCs in different clinical approaches. However, more in vitro study on their metabolism and clinical applications are needed to fully understand their properties and to establish the future clinical use in the treatment of various diseases.

Author Contributions: Conceptualization, E.I. and R.M.; writing—original draft preparation, E.I.; review and editing, E.I. and B.M. All authors have read and agreed to the published version of the manuscript.

Funding: This review received no external funding.

Institutional Review Board Statement: Not applicable.

Informed Consent Statement: Not applicable.

Data Availability Statement: No new data were created or analyzed in this study. Data sharing is not applicable to this article.

Conflicts of Interest: The authors declare no conflict of interest.

Abbreviations

α1-AT	Alpha-1 Anti Trypsin
AF	Amniotic Fluid
AFMSCs	Amniotic Fluid Mesenchymal Stem Cells
AKI	Acute Kidney Injury
ALB	Albumin
ALP	Alkaline Phosphatase
AM	Amniotic Membrane
AT	Adipose Tissue
ATMSCs	Adipose Tissue Mesenchymal Stem Cells
ATP	Adenosine TriPhosphate
AVC	Arrhythmogenic Ventricular Cardiomyoapthy
BM	Bone Marrow
BMMSCs	Bone Marrow Mesenchymal Stem cells
BUN	Blood Urea Nitrogen
CD	Cluster Designation
COX	Cyclooxygenase
CHF	Congestive Heart Failure
DBM	Demineralized Bone Matrix
DCM	Dilated Cardiomyopathy
DLA	Dog Leukocyte Antigen
DPCs	Dog Placenta Cells
DT	Doubling Time
E cells	Epithelioid cells
ECSs	Embryonic Stem Cells
EV	Extracellular Vesicles
F cells	Fibroblastic cells
GFAP	Glial Fibrillar Acidic Protein
GME	Granulomatous Meningo Encephalomyelitis
GS	Glutamine synthetase
HGF	Hepatocyte Growth Factor
Hrs	Hours
IBD	Inflammatory Brain Disease
ICM	Inner Cell Mass

IL	Interleukin
IPS	Induced Pluripotent Stem Cells
ISCT	International Society For Cytotherapy
MAP	Microtube-Associated Protein
MCP	Monocyte chemoattractant protein
MHC	Major Histocompatibility Complex
MSCs	Mesenchymal Stem Cells
MS	Multiple Sclerosis
MUO	Meningo encephalomyelitis of Unknown Origin
NEFL	Neurofilament Light Chain
NLE	Necrotizing Leuko Encephalitis
NME	Necrotizing Meningo Encephalitis
NeuN	Neuronal Nuclear Antigen
NF	Neuro Filament
NSAIDs	Non-Steroidal Anti-Inflammatory Drug
NSE	Neuron Specific Enolase
OCT4	Octamer-binding Transcription factor 4
OGD	Oxygen-Glucose Deprivation
olig	Oligomycin
OSM	Oncostatin M
OXPHOS	Oxidative Phosphorylation
P	In vitro culture Passage
PBS	Phosphate Buffer Solution
PMSCs	Placenta Mesenchymal Stem Cells
PRP	Platelet Rich Plasma
Rot+AA	Rotenone plus Antimycin A
SOX2	Sex Determining Region Y-box 2 (also known as SRY)
SCI	Spinal Cord Injury
SCs	Stem Cells
R cells	Round cells
RT-PCR	Real Time Polymerase Chain reaction
TAT	Tyrosine Amino Transferase
TGF	Transforming Growth Factor
TH	Tyrosine Hydroxylase
TPLO	Tibial Plateau Leveling Osteotomy
Tuj-1	Neuronal Class III ß tubulin
TUBB3	Tubulin Beta3
UC	Umbilical Cord
UCB	Umbilical Cord Blood
UCBMSCs	Umbilical Cord Blood
VEGF	Vascular Endothelial Growth Factor
WJ	Wharton's Jelly
WJMSCs	Wharton's Jelly Mesenchymal Stem Cells

References

1. Neff, M.W.; Rine, J. A Fetching Model Organism. *Cell* **2006**, *124*, 229–231. [CrossRef]
2. Naggara, O.; Darsaut, T.E.; Salazkin, I.; Soulez, G.; Guilbert, F.; Roy, D.; Weill, A.; Gevry, G.; Raymond, J. A New Canine Carotid Artery Bifurcation Aneurysm Model for the Evaluation of Neurovascular Devices. *Am. J. Neuroradiol.* **2010**, *31*, 967–971. [CrossRef] [PubMed]
3. Schneider, M.R.; Adler, H.; Braun, J.; Kienzle, B.; Wolf, E.; Kolb, H.-J. Canine Embryo-Derived Stem Cells—Toward Clinically Relevant Animal Models for Evaluating Efficacy and Safety of Cell Therapies. *Stem Cells* **2007**, *25*, 1850–1851. [CrossRef]
4. Schneider, M.R.; Wolf, E.; Braun, J.; Kolb, H.-J.; Adler, H. Canine Embryo-Derived Stem Cells and Models for Human Diseases. *Hum. Mol. Genet.* **2008**, *17*, R42. [CrossRef]
5. Heron, M. Deaths: Leading Causes for 2014. *Natl. Vital Stat. Rep.* **2016**, *65*, 1–96.
6. Ribitsch, I.; Baptista, P.M.; Lange-Consiglio, A.; Melotti, L.; Patruno, M.; Jenner, F.; Schnabl-Feichter, E.; Dutton, L.C.; Connolly, D.J.; van Steenbeek, F.G.; et al. Large Animal Models in Regenerative Medicine and Tissue Engineering: To Do or Not to Do. *Front. Bioeng. Biotechnol.* **2020**, *8*, 972. [CrossRef] [PubMed]

7. Meurs, K.M. Arrhythmogenic Right Ventricular Cardiomyopathy in the Boxer Dog: An Update. *Vet. Clin. N. Am. Small Anim. Pract.* **2017**, *47*, 1103–1111. [CrossRef]
8. McCoy, A.M. Animal Models of Osteoarthritis: Comparisons and Key Considerations. *Vet. Pathol.* **2015**, *52*, 803–818. [CrossRef] [PubMed]
9. Hoffman, A.M.; Dow, S.W. Concise Review: Stem Cell Trials Using Companion Animal Disease Models. *Stem Cells* **2016**, *34*, 1709–1729. [CrossRef]
10. Chinnadurai, R. Challenges in Animal Modelling of Mesenchymal Stromal Cell Therapy for Inflammatory Bowel Disease. *World J. Gastroenterol.* **2015**, *21*, 4779. [CrossRef]
11. Yan, Z.; Hang, D.; Guo, C.; Chen, Z. Fate of mesenchymal stem cells transplanted to osteonecrosis of femoral head. *J. Orthop. Res.* **2009**, *27*, 442–446. [CrossRef]
12. Marx, C.; Silveira, M.D.; Selbach, I.; da Silva, A.S.; Braga, L.M.G.d.M.; Camassola, M.; Nardi, N.B. Acupoint Injection of Autologous Stromal Vascular Fraction and Allogeneic Adipose-Derived Stem Cells to Treat Hip Dysplasia in Dogs. *Stem Cells Int.* **2014**, *2014*, 1–6. [CrossRef]
13. Zubin, E.; Conti, V.; Leonardi, F.; Zanichelli, S.; Ramoni, R.; Grolli, S. Regenerative Therapy for the Management of a Large Skin Wound in a Dog. *Clin. Case Rep.* **2015**, *3*, 598–603. [CrossRef] [PubMed]
14. D'Ippolito, G.; Schiller, P.C.; Ricordi, C.; Roos, B.A.; Howard, G.A. Age-Related Osteogenic Potential of Mesenchymal Stromal Stem Cells from Human Vertebral Bone Marrow. *J. Bone Miner. Res.* **1999**, *14*, 1115–1122. [CrossRef]
15. Mueller, S.M.; Glowacki, J. Age-Related Decline in the Osteogenic Potential of Human Bone Marrow Cells Cultured in Three-Dimensional Collagen Sponges. *J. Cell. Biochem.* **2001**, *82*, 583–590. [CrossRef]
16. Stenderup, K.; Justesen, J.; Clausen, C.; Kassem, M. Aging Is Associated with Decreased Maximal Life Span and Accelerated Senescence of Bone Marrow Stromal Cells. *Bone* **2003**, *33*, 919–926. [CrossRef]
17. Choudhery, M.S.; Badowski, M.; Muise, A.; Pierce, J.; Harris, D.T. Donor Age Negatively Impacts Adipose Tissue-Derived Mesenchymal Stem Cell Expansion and Differentiation. *J. Transl. Med.* **2014**, *12*, 8. [CrossRef] [PubMed]
18. Roobrouck, V.D.; Ulloa-Montoya, F.; Verfaillie, C.M. Self-Renewal and Differentiation Capacity of Young and Aged Stem Cells. *Exp. Cell Res.* **2008**, *314*, 1937–1944. [CrossRef]
19. Alves, H.; van Ginkel, J.; Groen, N.; Hulsman, M.; Mentink, A.; Reinders, M.; van Blitterswijk, C.; de Boer, J. A Mesenchymal Stromal Cell Gene Signature for Donor Age. *PLoS ONE* **2012**, *7*, e42908. [CrossRef]
20. Iacono, E.; Brunori, L.; Pirrone, A.; Pagliaro, P.P.; Ricci, F.; Tazzari, P.L.; Merlo, B. Isolation, Characterization and Differentiation of Mesenchymal Stem Cells from Amniotic Fluid, Umbilical Cord Blood and Wharton's Jelly in the Horse. *Reproduction* **2012**, *143*, 455–468. [CrossRef] [PubMed]
21. Merlo, B.; Teti, G.; Lanci, A.; Burk, J.; Mazzotti, E.; Falconi, M.; Iacono, E. Comparison between Adult and Foetal Adnexa Derived Equine Post-Natal Mesenchymal Stem Cells. *BMC Vet. Res.* **2019**, *15*, 277. [CrossRef]
22. Iacono, E.; Pascucci, L.; Rossi, B.; Bazzucchi, C.; Lanci, A.; Ceccoli, M.; Merlo, B. Ultrastructural Characteristics and Immune Profile of Equine MSCs from Fetal Adnexa. *Reproduction* **2017**, *154*, 509–519. [CrossRef]
23. Corradetti, B.; Meucci, A.; Bizzaro, D.; Cremonesi, F.; Lange Consiglio, A. Mesenchymal Stem Cells from Amnion and Amniotic Fluid in the Bovine. *Reproduction* **2013**, *145*, 391–400. [CrossRef] [PubMed]
24. Rossi, B.; Merlo, B.; Colleoni, S.; Iacono, E.; Tazzari, P.L.; Ricci, F.; Lazzari, G.; Galli, C. Isolation and in Vitro Characterization of Bovine Amniotic Fluid Derived Stem Cells at Different Trimesters of Pregnancy. *Stem Cell Rev. Rep.* **2014**, *10*, 712–724. [CrossRef]
25. Kumar, K.; Agarwal, P.; Das, K.; Mili, B.; Madhusoodan, A.P.; Kumar, A.; Bag, S. Isolation and Characterization of Mesenchymal Stem Cells from Caprine Umbilical Cord Tissue Matrix. *Tissue Cell* **2016**, *48*, 653–658. [CrossRef]
26. Pratheesh, M.D.; Dubey, P.K.; Gade, N.E.; Nath, A.; Sivanarayanan, T.B.; Madhu, D.N.; Somal, A.; Baiju, I.; Sreekumar, T.R.; Gleeja, V.L.; et al. Comparative Study on Characterization and Wound Healing Potential of Goat (Capra Hircus) Mesenchymal Stem Cells Derived from Fetal Origin Amniotic Fluid and Adult Bone Marrow. *Res. Vet. Sci.* **2017**, *112*, 81–88. [CrossRef] [PubMed]
27. Le Blanc, K.; Tammik, C.; Rosendahl, K.; Zetterberg, E.; Ringdén, O. HLA Expression and Immunologic Properties of Differentiated and Undifferentiated Mesenchymal Stem Cells. *Exp. Hematol.* **2003**, *31*, 890–896. [CrossRef]
28. Cremonesi, F.; Corradetti, B.; Lange Consiglio, A. Fetal Adnexa Derived Stem Cells from Domestic Animal: Progress and Perspectives. *Theriogenology* **2011**, *75*, 1400–1415. [CrossRef]
29. Iacono, E.; Rossi, B.; Merlo, B. Stem Cells from Foetal Adnexa and Fluid in Domestic Animals: An Update on Their Features and Clinical Application. *Reprod. Domest. Anim.* **2015**, *50*, 353–364. [CrossRef]
30. Ambrósio, C.E.; Orlandin, J.R.; Oliveira, V.C.; Motta, L.C.B.; Pinto, P.A.F.; Pereira, V.M.; Padoveze, L.R.; Karam, R.G.; Pinheiro, A.d.O. Potential Application of Aminiotic Stem Cells in Veterinary Medicine. *Anim. Reprod.* **2019**, *16*, 24–30. [CrossRef]
31. Becker, A.J.; McCulloch, E.A.; Till, J.E. Cytological Demonstration of the Clonal Nature of Spleen Colonies Derived from Transplanted Mouse Marrow Cells. *Nature* **1963**, *197*, 452–454. [CrossRef]
32. Weissman, I.L. Stem Cells: Units of Development, Units of Regeneration, and Units in Evolution. *Cell* **2000**, *100*, 157–168. [CrossRef]
33. Alison, M.R.; Poulsom, R.; Forbes, S.; Wright, N.A. An Introduction to Stem Cells. *J. Pathol.* **2002**, *197*, 419–423. [CrossRef]
34. Thomson, J.A.; Itskovitz-Eldor, J.; Shapiro, S.S.; Waknitz, M.A.; Swiergiel, J.J.; Marshall, V.S.; Jones, J.M. Embryonic Stem Cell Lines Derived from Human Blastocysts. *Science* **1998**, *282*, 1145–1147. [CrossRef]

35. Reubinoff, B.E.; Pera, M.F.; Fong, C.Y.; Trounson, A.; Bongso, A. Embryonic Stem Cell Lines from Human Blastocysts: Somatic Differentiation in Vitro. *Nat. Biotechnol.* **2000**, *18*, 399–404. [CrossRef]
36. Erdö, F.; Bührle, C.; Blunk, J.; Hoehn, M.; Xia, Y.; Fleischmann, B.; Föcking, M.; Küstermann, E.; Kolossov, E.; Hescheler, J.; et al. Host-Dependent Tumorigenesis of Embryonic Stem Cell Transplantation in Experimental Stroke. *J. Cereb. Blood Flow Metab.* **2003**, *23*, 780–785. [CrossRef] [PubMed]
37. Nelson, T.J.; Martinez-Fernandez, A.; Yamada, S.; Ikeda, Y.; Perez-Terzic, C.; Terzic, A. Induced Pluripotent Stem Cells: Advances to Applications. *Stem Cells Cloning Adv. Appl.* **2010**, *3*, 29–37. [CrossRef]
38. Dominici, M.; Le Blanc, K.; Mueller, I.; Slaper-Cortenbach, I.; Marini, F.C.; Krause, D.S.; Deans, R.J.; Keating, A.; Prockop, D.J.; Horwitz, E.M. Minimal Criteria for Defining Multipotent Mesenchymal Stromal Cells. The International Society for Cellular Therapy Position Statement. *Cytotherapy.* **2006**, *8*, 315–317. [CrossRef]
39. Klemmt, P.A.; Vafaizadeh, V.; Groner, B. The Potential of Amniotic Fluid Stem Cells for Cellular Therapy and Tissue Engineering. *Expert Opin. Biol. Ther.* **2011**, *11*, 1297–1314. [CrossRef] [PubMed]
40. Uranio, M.F.; Valentini, L.; Lange-Consiglio, A.; Caira, M.; Guaricci, A.C.; L'Abbate, A.; Catacchio, C.R.; Ventura, M.; Cremonesi, F.; Dell'Aquila, M.E. Isolation, Proliferation, Cytogenetic, and Molecular Characterization and in Vitro Differentiation Potency of Canine Stem Cells from Foetal Adnexa: A Comparative Study of Amniotic Fluid, Amnion, and Umbilical Cord Matrix. *Mol. Reprod. Dev.* **2011**, *78*, 361–373. [CrossRef]
41. Fernandes, R.A.; Wenceslau, C.V.; Reginato, A.L.; Kerkis, I.; Miglino, M.A. Derivation and Characterization of Progenitor Stem Cells from Canine Allantois and Amniotic Fluids at the Third Trimester of Gestation. *Placenta* **2012**, *33*, 640–644. [CrossRef]
42. Choi, S.-A.; Choi, H.-S.; Kim, K.J.; Lee, D.-S.; Lee, J.H.; Park, J.Y.; Kim, E.Y.; Li, X.; Oh, H.-Y.; Lee, D.-S.; et al. Isolation of Canine Mesenchymal Stem Cells from Amniotic Fluid and Differentiation into Hepatocyte-like Cells. *In Vitro Cell. Dev. Biol. Anim.* **2013**, *49*, 42–51. [CrossRef] [PubMed]
43. Kim, E.Y.; Lee, K.-B.; Yu, J.; Lee, J.H.; Kim, K.J.; Han, K.-W.; Park, K.-S.; Lee, D.-S.; Kim, M.K. Neuronal Cell Differentiation of Mesenchymal Stem Cells Originating from Canine Amniotic Fluid. *Hum. Cell* **2014**, *27*, 51–58. [CrossRef] [PubMed]
44. Seo, M.-S.; Jeong, Y.-H.; Park, J.-R.; Park, S.-B.; Rho, K.-H.; Kim, H.-S.; Yu, K.-R.; Lee, S.-H.; Jung, J.-W.; Lee, Y.-S.; et al. Isolation and Characterization of Canine Umbilical Cord Blood-Derived Mesenchymal Stem Cells. *J. Vet. Sci.* **2009**, *10*, 181. [CrossRef] [PubMed]
45. Kang, B.-J.; Ryu, H.-H.; Park, S.S.; Koyama, Y.; Kikuchi, M.; Woo, H.-M.; Kim, W.H.; Kweon, O.-K. Comparing the Osteogenic Potential of Canine Mesenchymal Stem Cells Derived from Adipose Tissues, Bone Marrow, Umbilical Cord Blood, and Wharton's Jelly for Treating Bone Defects. *J. Vet. Sci.* **2012**, *13*, 299. [CrossRef]
46. Ryu, H.-H.; Kang, B.-J.; Park, S.-S.; Kim, Y.; Sung, G.-J.; Woo, H.-M.; Kim, W.H.; Kweon, O.-K. Comparison of Mesenchymal Stem Cells Derived from Fat, Bone Marrow, Wharton's Jelly, and Umbilical Cord Blood for Treating Spinal Cord Injuries in Dogs. *J. Vet. Med. Sci.* **2012**, *74*, 1617–1630. [CrossRef]
47. Lee, S.; Ryu, M.S.; Park, S.-B.; Ahn, J.-O.; Kang, K.-S.; Han, S.-M.; Bhang, D.-A.; Youn, H.-Y. Mesenchymal Stem Cells Contribute to Improvement of Renal Function in a Canine Kidney Injury Model. *In Vivo* **2017**, *31*, 1115–1124. [CrossRef]
48. Filioli Uranio, M.; Dell'Aquila, M.E.; Caira, M.; Guaricci, A.C.; Ventura, M.; Catacchio, C.R.; Martino, N.A.; Valentini, L. Characterization and in Vitro Differentiation Potency of Early-Passage Canine Amnion- and Umbilical Cord-Derived Mesenchymal Stem Cells as Related to Gestational Age. *Mol. Reprod. Dev.* **2014**, *81*, 539–551. [CrossRef] [PubMed]
49. Saulnier, N.; Loriau, J.; Febre, M.; Robert, C.; Rakic, R.; Bonte, T.; Buff, S.; Maddens, S. Canine Placenta: A Promising Potential Source of Highly Proliferative and Immunomodulatory Mesenchymal Stromal Cells? *Vet. Immunol. Immunopathol.* **2016**, *171*, 47–55. [CrossRef]
50. Cardoso, M.T.; Pinheiro, A.O.; Vidane, A.S.; Casals, J.B.; de Oliveira, V.C.; Gonçalves, N.; Martins, D.S.; Ambrósio, C.E. Characterization of Teratogenic Potential and Gene Expression in Canine and Feline Amniotic Membrane-Derived Stem Cells. *Reprod. Domest. Anim. Zuchthyg.* **2017**, *52* (Suppl. S2), 58–64. [CrossRef]
51. Borghesi, J.; Ferreira Lima, M.; Mario, L.C.; de Almeida da Anunciação, A.R.; Silveira Rabelo, A.C.; Giancoli Kato Cano da Silva, M.; Assunpção Fernandes, F.; Miglino, M.A.; Oliveira Carreira, A.C.; Oliveira Favaron, P. Canine Amniotic Membrane Mesenchymal Stromal/Stem Cells: Isolation, Characterization and Differentiation. *Tissue Cell.* **2019**, *58*, 99–106. [CrossRef]
52. De Oliveira Pinheiro, A.; Lara, V.M.; Souza, A.F.; Casals, J.B.; Bressan, F.F.; Fantinato Neto, P.; Oliveira, V.C.; Martins, D.S.; Ambrosio, C.E. Characterization and Immunomodulation of Canine Amniotic Membrane Stem Cells. *Stem Cells Cloning Adv. Appl.* **2020**, *13*, 43–55. [CrossRef]
53. Long, C.; Lankford, L.; Kumar, P.; Grahn, R.; Borjesson, D.L.; Farmer, D.; Wang, A. Isolation and Characterization of Canine Placenta-Derived Mesenchymal Stromal Cells for the Treatment of Neurological Disorders in Dogs. *Cytom. Part J. Int. Soc. Anal. Cytol.* **2018**, *93*, 82–92. [CrossRef] [PubMed]
54. Amorim, R.M.; Clark, K.C.; Walker, N.J.; Kumar, P.; Herout, K.; Borjesson, D.L.; Wang, A. Placenta-Derived Multipotent Mesenchymal Stromal Cells: A Promising Potential Cell-Based Therapy for Canine Inflammatory Brain Disease. *Stem Cell Res. Ther.* **2020**, *11*, 304. [CrossRef] [PubMed]
55. Götherström, C.; Ringdén, O.; Tammik, C.; Zetterberg, E.; Westgren, M.; Blanc, K.L. Immunologic Properties of Human Fetal Mesenchymal Stem Cells. *Am. J. Obstet. Gynecol.* **2004**, *190*, 239–245. [CrossRef] [PubMed]
56. Jin, K.; Zhu, Y.; Sun, Y.; Mao, X.O.; Xie, L.; Greenberg, D.A. Vascular Endothelial Growth Factor (VEGF) Stimulates Neurogenesis in Vitro and in Vivo. *Proc. Natl. Acad. Sci. USA* **2002**, *99*, 11946–11950. [CrossRef] [PubMed]

57. Madrigal, J.L.M.; Leza, J.C.; Polak, P.; Kalinin, S.; Feinstein, D.L. Astrocyte-Derived MCP-1 Mediates Neuroprotective Effects of Noradrenaline. *J. Neurosci.* **2009**, *29*, 263–267. [CrossRef]
58. Shi, X.; Liu, J.; Yang, T.; Zhang, Y.; Li, T.; Chen, J. TLR2/NFκB Signalling Regulates Endogenous IL-6 Release from Marrow-Derived Mesenchymal Stromal Cells to Suppress the Apoptosis of PC12 Cells Injured by Oxygen and Glucose Deprivation. *Mol. Med. Rep.* **2016**, *13*, 5358–5364. [CrossRef]
59. Li, A.; Dubey, S.; Varney, M.L.; Dave, B.J.; Singh, R.K. IL-8 Directly Enhanced Endothelial Cell Survival, Proliferation, and Matrix Metalloproteinases Production and Regulated Angiogenesis. *J. Immunol.* **2003**, *170*, 3369–3376. [CrossRef]
60. Coates, J.R.; Jeffery, N.D. Perspectives on Meningoencephalomyelitis of Unknown Origin. *Vet. Clin. N. Am. Small Anim. Pract.* **2014**, *44*, 1157–1185. [CrossRef]
61. Greer, K.A.; Wong, A.K.; Liu, H.; Famula, T.R.; Pedersen, N.C.; Ruhe, A.; Wallace, M.; Neff, M.W. Necrotizing Meningoencephalitis of Pug Dogs Associates with Dog Leukocyte Antigen Class II and Resembles Acute Variant Forms of Multiple Sclerosis. *Tissue Antigens* **2010**, *76*, 110–118. [CrossRef] [PubMed]
62. Yu, S.; Tajiri, N.; Franzese, N.; Franzblau, M.; Bae, E.; Platt, S.; Kaneko, Y.; Borlongan, C.V. Stem Cell-Like Dog Placenta Cells Afford Neuroprotection against Ischemic Stroke Model via Heat Shock Protein Upregulation. *PLoS ONE* **2013**, *8*, e76329. [CrossRef]
63. Takechi, K.; Kuwabara, Y.; Mizuno, M. Ultrastructural and Immunohistochemical Studies of Wharton's Jelly Umbilical Cord Cells. *Placenta* **1993**, *14*, 235–245. [CrossRef]
64. Nanaev, A.K.; Kohnen, G.; Milovanov, A.P.; Domogatsky, S.P.; Kaufmann, P. Stromal Differentiation and Architecture of the Human Umbilical Cord. *Placenta* **1997**, *18*, 53–64. [CrossRef]
65. Seo, M.-S.; Park, S.-B.; Kang, K.-S. Isolation and Characterization of Canine Wharton's Jelly-Derived Mesenchymal Stem Cells. *Cell Transplant.* **2012**, *21*, 1493–1502. [CrossRef]
66. Souza, T.F.B.; Sakamoto, S.S.; de Andrade, A.L.; Cardoso, T.C. Osteogenic Potential of Mesenchymal Cells Derived from Canine Umbilical Cord Matrix Co-Cultured with Platelet-Rich Plasma and Demineralized Bone Matrix. *J. Vet. Sci.* **2015**, *16*, 381–384. [CrossRef] [PubMed]
67. Marcoccia, R.; Nesci, S.; Merlo, B.; Ballotta, G.; Algieri, C.; Pagliarani, A.; Iacono, E. Biological Characteristics and Metabolic Profile of Canine Mesenchymal Stem Cells Isolated from Adipose Tissue and Umbilical Cord Matrix. *PLoS ONE* **2021**, *16*, e0247567. [CrossRef]
68. Chaudhary, L.R.; Hofmeister, A.M.; Hruska, K.A. Differential Growth Factor Control of Bone Formation through Osteoprogenitor Differentiation. *Bone* **2004**, *34*, 402–411. [CrossRef]
69. Kärner, E.; Unger, C.; Sloan, A.J.; Ahrlund-Richter, L.; Sugars, R.V.; Wendel, M. Bone Matrix Formation in Osteogenic Cultures Derived from Human Embryonic Stem Cells in Vitro. *Stem Cells Dev.* **2007**, *16*, 39–52. [CrossRef]
70. Weiss, M.L.; Anderson, C.; Medicetty, S.; Seshareddy, K.B.; Weiss, R.J.; VanderWerff, I.; Troyer, D.; McIntosh, K.R. Immune Properties of Human Umbilical Cord Wharton's Jelly-Derived Cells. *Stem Cells* **2008**, *26*, 2865–2874. [CrossRef] [PubMed]
71. De Coppi, P.; Bartsch, G.; Siddiqui, M.M.; Xu, T.; Santos, C.C.; Perin, L.; Mostoslavsky, G.; Serre, A.C.; Snyder, E.Y.; Yoo, J.J.; et al. Isolation of Amniotic Stem Cell Lines with Potential for Therapy. *Nat. Biotechnol.* **2007**, *25*, 100–106. [CrossRef] [PubMed]
72. Ilancheran, S.; Michalska, A.; Peh, G.; Wallace, E.M.; Pera, M.; Manuelpillai, U. Stem Cells Derived from Human Fetal Membranes Display Multilineage Differentiation Potential. *Biol. Reprod.* **2007**, *77*, 577–588. [CrossRef]
73. Iacono, E.; Lanci, A.; Merlo, B.; Ricci, F.; Pirrone, A.; Antonelli, C.; Mariella, J.; Castagnetti, C. Effects of Amniotic Fluid Mesenchymal Stem Cells in Carboxymethyl Cellulose Gel on Healing of Spontaneous Pressure Sores: Clinical Outcome in Seven Hospitalized Neonatal Foals. *Turk. J. Biol.* **2016**, *40*, 484–492. [CrossRef]
74. Vogel, W.; Grünebach, F.; Messam, C.A.; Kanz, L.; Brugger, W.; Bühring, H.-J. Heterogeneity among Human Bone Marrow-Derived Mesenchymal Stem Cells and Neural Progenitor Cells. *Haematologica* **2003**, *88*, 126–133.
75. Nagai, A.; Kim, W.K.; Lee, H.J.; Jeong, H.S.; Kim, K.S.; Hong, S.H.; Park, I.H.; Kim, S.U. Multilineage Potential of Stable Human Mesenchymal Stem Cell Line Derived from Fetal Marrow. *PLoS ONE* **2007**, *2*, e1272. [CrossRef]
76. Prusa, A.-R.; Marton, E.; Rosner, M.; Bettelheim, D.; Lubec, G.; Pollack, A.; Bernaschek, G.; Hengstschläger, M. Neurogenic Cells in Human Amniotic Fluid. *Am. J. Obstet. Gynecol.* **2004**, *191*, 309–314. [CrossRef]
77. Sultana, T.; Lee, S.; Yoon, H.-Y.; Lee, J.I. Current Status of Canine Umbilical Cord Blood-Derived Mesenchymal Stem Cells in Veterinary Medicine. *Stem Cells Int.* **2018**, *2018*, 1–14. [CrossRef] [PubMed]
78. Gluckman, E.; Broxmeyer, H.A.; Auerbach, A.D.; Friedman, H.S.; Douglas, G.W.; Devergie, A.; Esperou, H.; Thierry, D.; Socie, G.; Lehn, P. Hematopoietic Reconstitution in a Patient with Fanconi's Anemia by Means of Umbilical-Cord Blood from an HLA-Identical Sibling. *N. Engl. J. Med.* **1989**, *321*, 1174–1178. [CrossRef]
79. Zhang, X.; Hirai, M.; Cantero, S.; Ciubotariu, R.; Dobrila, L.; Hirsh, A.; Igura, K.; Satoh, H.; Yokomi, I.; Nishimura, T.; et al. Isolation and Characterization of Mesenchymal Stem Cells from Human Umbilical Cord Blood: Reevaluation of Critical Factors for Successful Isolation and High Ability to Proliferate and Differentiate to Chondrocytes as Compared to Mesenchymal Stem Cells from Bone Marrow and Adipose Tissue. *J. Cell. Biochem.* **2011**, *112*, 1206–1218. [CrossRef]
80. Lim, J.-H.; Byeon, Y.-E.; Ryu, H.-H.; Jeong, Y.-H.; Lee, Y.-W.; Kim, W.H.; Kang, K.-S.; Kweon, O.-K. Transplantation of Canine Umbilical Cord Blood-Derived Mesenchymal Stem Cells in Experimentally Induced Spinal Cord Injured Dogs. *J. Vet. Sci.* **2007**, *8*, 275–282. [CrossRef]

81. Park, S.-S.; Byeon, Y.-E.; Ryu, H.-H.; Kang, B.-J.; Kim, Y.; Kim, W.-H.; Kang, K.-S.; Han, H.-J.; Kweon, O.-K. Comparison of Canine Umbilical Cord Blood-Derived Mesenchymal Stem Cell Transplantation Times: Involvement of Astrogliosis, Inflammation, Intracellular Actin Cytoskeleton Pathways, and Neurotrophin-3. *Cell Transplant.* **2011**, *20*, 1867–1880. [CrossRef] [PubMed]
82. Mirza, B.; Krook, H.; Andersson, P.; Larsson, L.C.; Korsgren, O.; Widner, H. Intracerebral Cytokine Profiles in Adult Rats Grafted with Neural Tissue of Different Immunological Disparity. *Brain Res. Bull.* **2004**, *63*, 105–118. [CrossRef] [PubMed]
83. Uemura, M.; Refaat, M.M.; Shinoyama, M.; Hayashi, H.; Hashimoto, N.; Takahashi, J. Matrigel Supports Survival and Neuronal Differentiation of Grafted Embryonic Stem Cell-Derived Neural Precursor Cells. *J. Neurosci. Res.* **2010**, *88*, 542–551. [CrossRef]
84. Bellomo, R.; Ronco, C.; Kellum, J.A.; Mehta, R.L.; Palevsky, P.; Acute Dialysis Quality Initiative Workgroup. Acute Renal Failure—Definition, Outcome Measures, Animal Models, Fluid Therapy and Information Technology Needs: The Second International Consensus Conference of the Acute Dialysis Quality Initiative (ADQI) Group. *Crit. Care* **2004**, *8*, R204–R212. [CrossRef]
85. Taroni, M.; Cabon, Q.; Fèbre, M.; Cachon, T.; Saulnier, N.; Carozzo, C.; Maddens, S.; Labadie, F.; Robert, C.; Viguier, E. Evaluation of the Effect of a Single Intra-Articular Injection of Allogeneic Neonatal Mesenchymal Stromal Cells Compared to Oral Non-Steroidal Anti-Inflammatory Treatment on the Postoperative Musculoskeletal Status and Gait of Dogs over a 6-Month Period after Tibial Plateau Leveling Osteotomy: A Pilot Study. *Front. Vet. Sci.* **2017**, *4*, 83. [CrossRef]
86. Yang, V.K.; Meola, D.M.; Davis, A.; Barton, B.; Hoffman, A.M. Intravenous Administration of Allogeneic Wharton Jelly-Derived Mesenchymal Stem Cells for Treatment of Dogs with Congestive Heart Failure Secondary to Myxomatous Mitral Valve Disease. *Am. J. Vet. Res.* **2021**, *82*, 487–493. [CrossRef] [PubMed]
87. Jang, B.-J.; Byeon, Y.-E.; Lim, J.-H.; Ryu, H.-H.; Kim, W.H.; Koyama, Y.; Kikuchi, M.; Kang, K.-S.; Kweon, O.-K. Implantation of Canine Umbilical Cord Blood-Derived Mesenchymal Stem Cells Mixed with Beta-Tricalcium Phosphate Enhances Osteogenesis in Bone Defect Model Dogs. *J. Vet. Sci.* **2008**, *9*, 387. [CrossRef] [PubMed]
88. Byeon, Y.-E.; Ryu, H.-H.; Park, S.S.; Koyama, Y.; Kikuchi, M.; Kim, W.H.; Kang, K.-S.; Kweon, O.-K. Paracrine Effect of Canine Allogenic Umbilical Cord Blood-Derived Mesenchymal Stromal Cells Mixed with Beta-Tricalcium Phosphate on Bone Regeneration in Ectopic Implantations. *Cytotherapy* **2010**, *12*, 626–636. [CrossRef] [PubMed]
89. Nielsen, L.K.; Bracker, K.; Price, L.L. Administration of Fenoldopam in Critically Ill Small Animal Patients with Acute Kidney Injury: 28 Dogs and 34 Cats (2008–2012). *J. Vet. Emerg. Crit. Care* **2015**, *25*, 396–404. [CrossRef]
90. Ross, L. Acute Kidney Injury in Dogs and Cats. *Vet. Clin. N. Am. Small Anim. Pract.* **2011**, *41*, 1–14. [CrossRef] [PubMed]
91. Harel, Z.; Bell, C.M.; Dixon, S.N.; McArthur, E.; James, M.T.; Garg, A.X.; Harel, S.; Silver, S.; Wald, R. Predictors of Progression to Chronic Dialysis in Survivors of Severe Acute Kidney Injury: A Competing Risk Study. *BMC Nephrol.* **2014**, *15*, 114. [CrossRef]
92. Dossetor, J.B. Creatininemia versus Uremia. The Relative Significance of Blood Urea Nitrogen and Serum Creatinine Concentrations in Azotemia. *Ann. Intern. Med.* **1966**, *65*, 1287–1299. [CrossRef]
93. Vaden, S.L.; Levine, J.; Breitschwerdt, E.B. A Retrospective Case-Control of Acute Renal Failure in 99 Dogs. *J. Vet. Intern. Med.* **1997**, *11*, 58–64. [CrossRef]
94. Parolini, O.; Caruso, M. Review: Preclinical Studies on Placenta-Derived Cells and Amniotic Membrane: An Update. *Placenta* **2011**, *32*, S186–S195. [CrossRef] [PubMed]
95. Lankford, L.; Selby, T.; Becker, J.; Ryzhuk, V.; Long, C.; Farmer, D.; Wang, A. Early Gestation Chorionic Villi-Derived Stromal Cells for Fetal Tissue Engineering. *World J. Stem Cells* **2015**, *7*, 195–207. [CrossRef] [PubMed]
96. Kourembanas, S. Exosomes: Vehicles of Intercellular Signaling, Biomarkers, and Vectors of Cell Therapy. *Annu. Rev. Physiol.* **2015**, *77*, 13–27. [CrossRef] [PubMed]

animals

MDPI

Article

Platelet Lysate for Mesenchymal Stromal Cell Culture in the Canine and Equine Species: Analogous but Not the Same

Alina Hagen [1], Heidrun Holland [2], Vivian-Pascal Brandt [2], Carla U. Doll [1], Thomas C. Häußler [3], Michaela Melzer [1], Julia Moellerberndt [1], Hendrik Lehmann [4,†] and Janina Burk [1,*,†]

1 Equine Clinic (Surgery, Orthopedics), Justus-Liebig-University Giessen, 35390 Giessen, Germany; alina.hagen@vetmed.uni-giessen.de (A.H.); carla.u.doll@vetmed.uni-giessen.de (C.U.D.); michaela.melzer@vetmed.uni-giessen.de (M.M.); julia.moellerberndt@vetmed.uni-giessen.de (J.M.)

2 Saxon Incubator for Clinical Translation (SIKT), University of Leipzig, 04109 Leipzig, Germany; heidrun.holland@medizin.uni-leipzig.de (H.H.); vivian-pascal.brandt@uni-leipzig.de (V.-P.B.)

3 Small Animal Clinic (Surgery), Justus-Liebig-University Giessen, 35390 Giessen, Germany; thomas.haeussler@vetmed.uni-giessen.de

4 Department of Veterinary Clinical Sciences, Small Animal Clinic, Justus-Liebig-University Giessen, 35390 Giessen, Germany; hendrik.lehmann@vetmed.uni-giessen.de

* Correspondence: janina.burk@vetmed.uni-giessen.de

† These authors contributed equally to this work.

Simple Summary: Regenerative medicine using platelet-based blood products or adult stem cells offers the prospect of better clinical outcomes with many diseases. In veterinary medicine, most progress has been made with the development and therapeutic use of these regenerative therapeutics in horses, but the clinical need is given in dogs as well. Our aim was to transfer previous advances in the development of horse regenerative therapeutics, specifically the use of platelet lysate for feeding stem cell cultures, to the dog. Here, we describe the scalable production of canine platelet lysate, which could be used in regenerative biological therapies. We also evaluated the canine platelet lysate for its suitability in feeding canine stem cell cultures in comparison to equine platelet lysate used for equine stem cell cultures. Platelet lysate production from canine blood was successful, but the platelet lysate did not support stem cell culture in dogs in the same beneficial way observed with the equine platelet lysate and stem cells. In conclusion, canine platelet lysate can be produced in large scales as described here, but further research is needed to improve the cultivation of canine stem cells.

Abstract: Platelet lysate (PL) is an attractive platelet-based therapeutic tool and has shown promise as xeno-free replacement for fetal bovine serum (FBS) in human and equine mesenchymal stromal cell (MSC) culture. Here, we established a scalable buffy-coat-based protocol for canine PL (cPL) production (n = 12). The cPL was tested in canine adipose MSC (n = 5) culture compared to FBS. For further comparison, equine adipose MSC (n = 5) were cultured with analogous equine PL (ePL) or FBS. During canine blood processing, platelet and transforming growth factor-β1 concentrations increased ($p < 0.05$ and $p < 0.001$), while white blood cell concentrations decreased ($p < 0.05$). However, while equine MSC showed good results when cultured with 10% ePL, canine MSC cultured with 2.5% or 10% cPL changed their morphology and showed decreased metabolic activity ($p < 0.05$). Apoptosis and necrosis in canine MSC were increased with 2.5% cPL ($p < 0.05$). Surprisingly, passage 5 canine MSC showed less genetic aberrations after culture with 10% cPL than with FBS. Our data reveal that using analogous canine and equine biologicals does not entail the same results. The buffy-coat-based cPL was not adequate for canine MSC culture, but may still be useful for therapeutic applications.

Keywords: platelet lysate; canine; mesenchymal stromal cells (MSC); equine; cell fitness

Citation: Hagen, A.; Holland, H.; Brandt, V.-P.; Doll, C.U.; Häußler, T.C.; Melzer, M.; Moellerberndt, J.; Lehmann, H.; Burk, J. Platelet Lysate for Mesenchymal Stromal Cell Culture in the Canine and Equine Species: Analogous but Not the Same. *Animals* **2022**, *12*, 189. https://doi.org/10.3390/ani12020189

Academic Editors: Eleonora Iacono and Barbara Merlo

Received: 8 December 2021
Accepted: 10 January 2022
Published: 13 January 2022

1. Introduction

Regenerative medicine has gained tremendous attention in recent years in the veterinary field, including a growing number of studies and clinical applications in dogs.

In regenerative medicine, biological materials such as cells, growth factors, and matrix substances are used to regenerate defective tissues and organs, aiming to regain their full functionality. Cell-based therapy is mostly performed with multipotent mesenchymal stromal cells (MSC), alone or combined with other biologicals. Orthobiologic blood products include autologous conditioned serum, platelet-rich-plasma (PRP), platelet concentrate, or platelet lysate (PL).

MSC-based therapies are supported by a growing body of evidence in veterinary medicine. Particularly in horses, the treatment of orthopedic diseases with MSC has a well-documented history [1–7]. The use of MSC in canine medicine is also becoming more popular. Canine MSC have been used successfully in several in vivo studies, especially for the treatment of osteoarthritis [8–10], but also for various other conditions such as atopic dermatitis [11,12], diabetes mellitus [13], inflammatory bowel disease [14], or to support neuroregeneration following vertebral compression fractures [15]. At the same time, there is a need to improve and harmonize canine MSC production processes. One of the critical factors is the in vitro cultivation of MSC before administration to the patient. In this context, the cell culture medium represents a crucial element of the cell culture process and may strongly affect efficacy and quality of therapy [16]. The critically discussed fetal bovine serum (FBS) is still the gold standard for in vitro cultivation of MSC in animal species. However, the trend in cell culture points to the use of xeno-free culture supplements, for which blood products from the same species appear most promising.

Among the orthobiologic blood products, particularly platelet products are already frequently used in clinical practice. In canine medicine, PRP or platelet concentrates have been used to treat several conditions, including osteoarthritis [17–20], lumbosacral stenosis [21], wounds [22,23], corneal ulcera [24], and aural hematoma [25]. While clinical benefits were shown, a major disadvantage of platelet concentrate is its limited long-term storage, as it cannot be frozen [26]. Furthermore, there are different methods and commercially available kits for PRP or platelet concentrate production, which limits the comparability and reproducibility of treatment results. Especially, there are major differences in terms of quality regarding the platelet and growth factor concentrations and also sterility of the product. For canine platelet therapies, it has so far been shown that the platelet concentration and leukocyte removal strongly differ between products [27,28].

PL could be a feasible off-the-shelf alternative to PRP or platelet concentrate for clinical applications [26], and at the same time represents a high-quality xeno-free replacement of FBS for cell culture [29,30]. In PL, the platelet-derived growth factors have already been released and cell membranes removed, thus PL can be stored for a long time in the freezer, while most likely offering the same benefits as platelet concentrates. Therefore, while replacing the critically discussed FBS in MSC production, positive and synergistic effects of PL might not only be achieved in cell culture, but also in the subsequent therapeutic application when combining MSC and PL. While no in vivo studies have been published using PL in dogs so far, a first study reported reduced lameness in horses that had been injected with PL for treatment of coffin joint osteoarthritis [31]. However, this benefit was transient and it is obvious that also in the case of PL, the quality of the product can vary, e.g., in terms of its growth factor concentrations, depending on the manufacturing procedures. With respect to using PL as a cell culture supplement replacing FBS, substantial progress has already been made in human medicine [30,32,33], and promising results have also been obtained in the equine species [29,34–39]. In contrast, in the canine species, studies are still rare and conflicting findings were reported when using PL for MSC culture [9,40], warranting further research.

Recently, we have established the first buffy-coat-based protocol for equine PL (ePL) production in 100% plasma devoid of additive solutions [29], which we believe improved PL production in the equine species in terms of reproducible quality and scalability. We also showed that the obtained ePL supported equine MSC expansion and basic characteristics [29]. The aim of the current study was to establish a corresponding buffy-coat-based procedure for the production of canine PL (cPL) and to evaluate its effects on canine MSC in comparison with equine MSC cultured with ePL, with the prospect of utilizing the obtained cPL in therapeutic applications as well as in MSC culture.

2. Materials and Methods

2.1. Blood Collection

Whole blood for cPL preparation was collected from 12 healthy dogs (6 males and 6 females) aged 1.4–8.1 years (median: 3.4 years; interquartile range (IQR): 3.7) after approval by the local regulatory authority (i.e., regional council Giessen, Germany, A 24/2017). The health status of the donor dogs was evaluated prior to blood collection by clinical examination and blood tests with complete blood counts from ethylenediaminetetraacetic acid (EDTA) whole blood, blood chemistry from Li-heparin blood and serum, as well as microbiological tests, as described below. For these blood tests, the blood was collected from the vena jugularis, cephalica antebrachia, or vena saphena lateralis under aseptic conditions.

After the health status was assessed, whole blood for cPL preparation was then obtained aseptically from the jugular vein. Here, 450 mL whole blood was collected from each donor in commercially available blood bags loaded with 63 mL of citrate–phosphate–dextrose (CPD; Composelect, Fresenius Kabi, Bad Homburg, Germany). In this process, the filling volume of the blood bags was standardized to 450 mL by a blood donation scale (Compoguard, Fresenius Kabi, Bad Homburg, Germany). In order to cool the temperature of the whole blood down to 20 °C within a short period of time and to improve temperature uniformity, immediately after blood collection, the blood was placed upright in a box (CompoCool®, Fresenius Kabi, Bad Homburg, Germany) containing butane-1,4-diol cooling plates. The blood was left there for a minimum of 2 h and a maximum of 3 h until processing in the laboratory.

2.2. Platelet Concentrate and Lysate Preparation

The canine whole blood was separated into its individual components of plasma, buffy coat, and erythrocyte concentrate by centrifugation in a blood separation centrifuge (Hettich Rotanta 460R, Andreas Hettich GmbH and Co.KG, Tuttlingen, Germany) at 2845× *g* for 20 min (acceleration settings 7 and deceleration settings 1) at 22 °C. Using a blood-separating device (Compomat 4G, Fresenius Kabi, Bad Homburg, Germany), the buffy coat was obtained by separating the plasma and erythrocyte concentrate using a top-bottom method. The buffy coat was left to rest for 1 h and then resuspended with 110 mL of plasma, resulting in a median buffy coat volume of 187.36 mL (IQR: 5.6) prior to the next centrifugation. The resuspended buffy coat was then centrifuged again at 266× *g* for 9.3 min, acceleration settings 7 and deceleration settings 1, at 22 °C. Subsequently, the resulting supernatant, which represented the platelet (PLT) concentrate, was recovered by the same blood-separating device. This PLT concentrate was frozen at −80 °C. Lysis of PLT to release growth factors and chemokines was induced by three freeze–thaw cycles. In this process, the PLT concentrate was thawed at 37 °C for 4 h in a dry heating device designed for thawing frozen products intended for infusion under continuous agitation (Plasmatherm, Barkey GmbH and Co., KG, Leopoldshoehe, Germany) and then refrozen at −80 °C for 20 h. Following the repeated freeze–thawing, the bags were centrifuged again in the centrifuge for blood separation at 4000× *g* for 30 min, acceleration settings 9 and deceleration settings 2, at 22 °C. To remove the cell debris, the resulting supernatant, corresponding to the lysate, was filtered through a Macopharma Plas-4 filter (Lot 11290588BM, Macopharma, Langen, Germany) using gravity. The final PLs from all dogs (n = 12) were pooled under aseptic conditions to obtain the cPL used in the cell culture experiments.

2.3. Microbiological Assessment

The blood samples from each animal, the PLT concentrates, and also the final pooled cPL were tested for absence of pathogens. Bacteriological analysis with Oxoid signal blood culture system (BC0100M, Oxoid Limited, Hampshire, UK) incubated at 37 °C for 7 days was performed. The cultured blood was streaked on blood agar (Blood Agar Base,

Oxoid Limited, Hampshire, UK) containing 5% defibrinated sheep blood and on water-blue metachrome-yellow lactose agar (Water-blue Metachrome-yellow Lactose Agar acc. to Gassner, Sifin Diagnostics, Berlin, Germany) on days 1, 2, 4, and 6. These plates were incubated under aerobic conditions at 37 °C for 48 h. Brain–Heart Infusion Agar (Brain–Heart Infusion Agar, Oxoid Limited, Hampshire, UK) was additionally incubated under microaerobic conditions (10% CO_2, 37 °C) for analysis after 24 and 48 h. Furthermore, we incubated Schaedler agar (BBL™ Schaedler Agar, Becton Dickinson GmbH, Heidelberg, Germany) and Columbia agar (Columbia Agar (base), E. Merck, Darmstadt, Germany) for 72 h at 37 °C under anaerobic conditions in a jar using AnaeroGen™ gas bags (AnaeroGen™ 2.5 L, Oxoid Limited, Hampshire, UK). Kimmig agar (Agar for fungi (base) acc. to Kimmig modified, E. Merck, Darmstadt, Germany) was incubated for 72 h at 28 °C under aerobic conditions for selective culturing of fungi. In order to confirm the absence of mycoplasma, a 16S ribosomal RNA gene polymerase chain reaction analysis [41] was conducted.

2.4. Platelet and Leukocyte Counts

After the different processing steps, samples were taken to generate complete blood counts using an automated flow cytometric hematology analyzer (ADVIA 2120i, Siemens Healthcare GmbH, Erlangen, Germany) with the multispecies software MS 6.11.7. These included EDTA blood samples, citrated whole blood samples from each blood collection bag prior to further processing, PLT concentrate and lysate samples from each dog, as well as a sample from the final pooled cPL from all dogs.

2.5. Growth Factor Quantification and Chemical Analyses

Growth factor concentrations were analyzed in serum, PLT concentrate, and PL before and after the last filtration step from each dog, as well as in the final pooled cPL. Samples were stored at −80 °C until ELISA measurement.

The concentration of platelet-derived growth factor (PDGF-BB) was analyzed using a commercially available canine ELISA Kit (ELC-PDGFB, Ray Biotech, Norcross, GA, USA). Transforming growth factor beta 1 (TGF-β1) was quantified using a Quantikine ELISA kit (R&D Systems, Minneapolis, MN, USA). We followed the manufacturer's instructions, which included TGF-β1 activation with hydrochloric acid for the TGF-β1 ELISA. Finally, the ELISAs were read on an Infinite M PLEX plate reader with corresponding Magellan software (Tecan Ltd., Maennedorf, Switzerland).

Chemical quality analyses, similar to those routinely performed with FBS, were also performed on the same samples. The electrolyte content was determined using a blood gas and electrolyte analyzer (Cobas b 123 POC system, Roche Diagnostics GmbH, Mannheim, Germany). The total protein and albumin contents were determined using a clinical chemistry analyzer C400 (Pentra C400 Option I.S.E, HORIBA ABX SAS, Montpellier, France).

2.6. MSC Culture with FBS and PL Media Supplements

To evaluate the final pooled cPL in comparison with FBS as a cell culture supplement, adipose-derived MSC were obtained from five healthy dogs aged 7 months to 8 years (median: 4 years; IQR: 4.5). The dogs used for MSC production differed from the dogs used for cPL production. For MSC recovery, subcutaneous fat was collected as a waste material from routine surgeries and collagenase digestion was used to isolate the cells in accordance with the protocol used by Gittel et al. [42]. The plastic-adherent MSC were then cultured in FBS-supplemented culture medium until cryopreservation. For cryopreservation, MSC were frozen in cryomedium consisting of Dulbecco's modified Eagle´s medium (DMEM, 1 g/L glucose; Gibco®, ThermoFisher Scientific, Darmstadt, Germany) with 40% FBS and 10% dimethyl sulfoxide (DMSO, Sigma Aldrich GmbH, München, Germany) using a freezing container (Mr Frosty, Nalgene, ThermoFisher Scientific, Darmstadt, Germany) and then stored in liquid nitrogen.

The cells were thawed and seeded in DMEM supplemented with either 10% FBS (Lot: 2078409, Gibco®, ThermoFisher Scientific, Darmstadt, Germany) or 10 and 2.5% cPL, 1%

penicillin–streptomycin, and 0.1% gentamycin. Additionally, 1 U/mL heparin–natrium (B. Braun, Melsungen, Germany) was added to the culture medium when using cPL. Until the beginning of the experiments, the MSC were maintained under standard culture conditions (humified atmosphere, 37 °C, 5% CO_2) for one passage (P) in the respective culture medium so that possible adaptations to the medium could take place. The experiments were performed from P3 to P5.

The basic MSC characterization experiments, namely the assessment of cell proliferation and differentiation, were performed in direct comparison to corresponding equine MSC cultures. For this purpose, adipose-derived MSC from five horses aged 3 to 8 years (median: 5 years; IQR: 2) and ePL pooled from 19 other horses, produced as previously described [29], were used accordingly. Thereafter, canine MSC were further characterized with regard to apoptosis, necrosis, and senescence markers, as well as their genetic stability.

2.7. Cell Proliferation and Metabolic Activity

For calculation of the generation time, canine and equine MSC population doublings in P3, P4, and P5 were analyzed. For this purpose, MSC were seeded at a density of 3000 cells/cm^2 in cell-culture-treated flasks with different culture media and incubated under standard culture conditions for 5 days. After 3 days, a medium change was performed. On day 5, MSCs were trypsinized and then counted with a hemocytometer, excluding dead cells by trypan blue staining. The following formula was used to calculate the generation time:

$$Generationtime = \frac{daysinculture}{\frac{ln\left(\frac{cellcountharvest}{cellcountseeding}\right)}{ln2}} \quad (1)$$

In addition, the metabolic activity levels of canine and equine MSC were measured on days 1 and 5 of P3, P4, and P5 by performing a tetrazolium compound (MTS) assay according to the manufacturer's instructions (CellTiter 96® AQueous One Solution Cell Proliferation Assay, Promega, Mannheim, Germany). To calculate the metabolic activity, the mean absorbance on day 5 was divided by the mean absorbance on day 1.

2.8. Trilineage Differentiation

Prior to inducing differentiation, canine and equine MSC were cultured in 10% FBS, 10% cPL or ePL, and 2.5% cPL or ePL. In vitro differentiation of equine MSC was performed in P2 or P3, while differentiation of canine MSC was always performed in P3.

Adipogenic differentiation was induced using StemPro™ adipogenic differentiation medium (catalog number A1007001, Gibco®, ThermoFisher Scientific) with 0.1% gentamycin and 5% rabbit serum. The MSC were seeded at a density of 1500 cells/cm^2 in a 24 well plate and incubated for 3 days with their standard medium, until the medium was replaced by differentiation medium. Samples were fixed after 7 days of incubation with 50% ethanol for 20 min and then stained with Oil Red O and hematoxylin counterstain. The intensity of adipogenic differentiation was determined by blinded observers using a scoring system, which included the number of differentiated cells and the size and arrangement of lipid droplets in the differentiated cells [42].

For osteogenic differentiation, MSCs were seeded at a density of 1000 cells/cm^2 in a 24 well plate, incubated for 3 days in their standard medium, and then in osteogenic differentiation medium (catalog number A1007201, Gibco®, ThermoFisher Scientific) containing 0.1% gentamycin for 21 days. For analysis of extracellular mineralization, after fixation of the differentiated cells with 4% paraformaldehyde for 10 min, von Kossa staining was performed and bright-field photomicrographs were taken. On these images, the mean grayscale values were extracted using Fiji ImageJ software version 2.1.0/1.5.3c.

For chondrogenic differentiation of MSC, they were washed in PBS and chondrogenic differentiation medium (Catalog number A1007101, Gibco®, ThermoFisher Scientific) containing 0.1% gentamycin was added. MSC were then centrifuged at 280× g at 4 °C for 5 min to form a cell pellet containing 500,000 cells. The pellets were cultured in centrifuge tubes for 21 days with medium changed twice weekly and then fixed with 4% paraformaldehyde for 12 h. Paraffin sections were prepared and stained with Alcian blue and Masson's trichrome. Bright-field photomicrographs were taken and analyzed using Fiji ImageJ software [43]. The images were color-deconvolved and binarized, then the percentages of the areas stained with each staining component were analyzed and the ratios of cartilage matrix staining and counterstaining (i.e., turquoise to violet staining for Alcian blue and bluish to red staining for Masson's trichrome) were determined.

2.9. *Apoptosis, Necrosis, and Senescence Assays*

In canine MSC, the RealTime-Glo™ Annexin V Apoptosis and Necrosis Live-Cell Assay (Promega, Mannheim, Germany) was conducted to measure the exposure of phosphatidylserine (PS) on the outer leaflet of the cell membrane during the apoptotic process and to detect necrosis using a cell-impermeant and pro-fluorescent DNA dye. MSC were seeded at 5000 cells/cm^2 in a 96 well plate and the assay was performed according to the manufacturer's instructions. On day 5, the intensity of apoptosis was determined by luminescence measurement and the intensity of necrosis was determined by fluorescence measurement, using the Infinite M PLEX plate reader.

A Cellular Senescence Activity Assay (Enzo Life Sciences (ELS) AG, Lausen, Switzerland) was performed to analyze the aging process of canine MSC. MSC were seeded at a density of 3000 cells/cm^2 in a 12 well plate and cultured for 5 days. MSC were then lysed on ice using lysis buffer containing 0.5% phenylmethylsulfonyl fluoride (PMSF) and a cell scraper. The lysate was centrifuged at 14,000× g for 10 min at 4 °C and then the supernatant was frozen at −80°C until all samples were collected. After thawing, SA-β-galactosidase activity was measured according to the manufacturer's instructions using a fluorometric substrate in the Infinite M PLEX plate reader.

2.10. *Fluorescence In Situ Hybridisation (FISH) Analyses*

Cell preparation was performed on canine MSC monolayer cultures in FBS and cPL 10% medium in P5, using standard cytogenetic techniques (colcemid treatment, hypotonic treatment, and methanol–acetic acid fixation according to [44,45]), and FISH according to the manufacturer's instructions (CSL, Sapporo, Japan) for FISH analyses on interphase cells. FISH analyses applying the *Dog Chromosome XY FISH probe* (centromeric alpha satellite DNA probe; chromosome X—spectrum red; chromosome Y—spectrum green) were accomplished in all samples cultured with FBS, but only in three canine MSC cultures after cultivation in 10% cPL, as the remaining two samples showed no visible signal patterns. In total, 827 interphase cells (54–120 per sample) were analyzed. To exclude technical artefacts in the detection of gonosome aberrations, a gonosomal chromosome loss was counted only if the other gonosome was detected.

2.11. *Statistical Analyses*

Statistical analysis was performed using IBM SPSS Statistics 28 software. Possible correlations between parameters were analyzed based on the Spearman's rank correlation. Differences between groups were analyzed using non-parametric tests for paired samples, except for the comparisons between age groups, where the samples were not related and Mann–Whitney U tests were used. When more than two groups were compared, Bonferroni-corrected p-values were used for the post hoc tests. Differences were considered significant at $p \leq 0.05$.

3. Results

3.1. *Canine Platelet Lysate Production*

3.1.1. Absence of Pathogen Contamination

Microbial contaminations were detected neither in the canine whole blood and PLT concentrates from the different donors, nor in the final pooled cPL.

3.1.2. Platelet Concentration and WBC Removal

During the preparation of cPL, a median PLT concentrate volume of 129.8 mL (IQR: 14.1) was recovered from a median whole-blood volume of 519.2 mL (IQR: 13.5), while 62.3% (IQR: 9.1) of PLTs and 4.3% (IQR: 3.1) of WBC were recovered from the whole blood. In concentrate, the PLT concentration was increased 2.6-fold compared to whole blood ($p < 0.05$) and the WBC concentration was decreased 0.2-fold ($p < 0.05$). After lysis but before filtration, the PLT number in the lysate was strongly decreased ($p < 0.001$), but still low numbers of PLT could be detected (Figure 1). The number of WBCs in the lysate was negligible compared to the concentrate ($p < 0.05$) and whole blood ($p < 0.001$). The PLT concentrations in whole blood and concentrate correlated strongly ($p < 0.001$ and r = 0.820) (Figure 1).

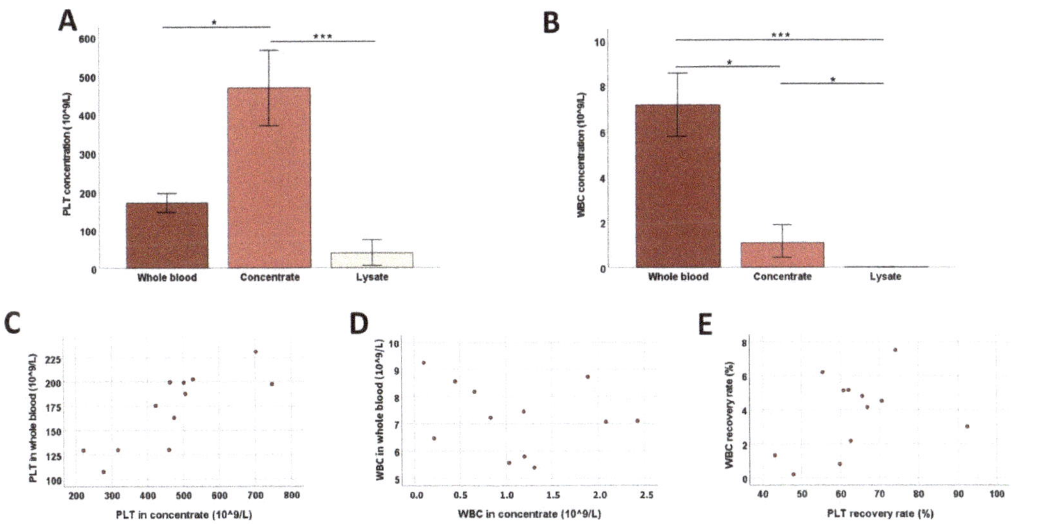

Figure 1. Platelet and white blood cell counts during blood processing. Median platelet (PLT; (**A**)) and white blood cell (WBC; (**B**)) counts at the different stages of blood processing (whole blood, concentrate, and lysate before filtration) are presented; error bars display the 95% confidence intervals. Friedman tests for group comparisons and subsequent post hoc tests were performed; the asterisks describe the significant differences between groups (* corresponds to $p < 0.05$; *** corresponds to $p < 0.001$). The dot plots show PLT (**C**) and WBC (**D**) counts in whole blood vs. concentrate and WBC vs. PLT recovery rates (**E**); the correlation shown in (**C**) was significant ($p < 0.001$ and r = 0.820, based on Spearman's rank correlation). All data were obtained from n = 12 dogs.

3.1.3. Growth Factor Concentrations and Chemical Analyses

Compared to serum, the canine PLT concentrate contained higher growth factor concentrations of PDGF-BB and TGF-β1 ($p < 0.01$ for TGF-β 1). The lysate also showed significantly higher TGF-β1 concentrations compared to serum ($p < 0.001$), but the PDGF-BB concentration decreased from the concentrate to the lysate ($p < 0.05$) (Figure 2).

PLT concentration in canine whole blood showed a strong correlation with PDGF-BB in concentrate ($p < 0.01$ and r = 0.729) and a moderate correlation with PDGF-BB and TGF-β1 in lysate ($p < 0.05$ and r = 0.641 for PDGF-BB and $p < 0.05$ and r = 0.676 for TGF-β1). Similarly, PLT concentration in concentrate showed a strong correlation with PDGF-BB concentration in concentrate ($p < 0.01$ and r = 0.804) and lysate ($p < 0.01$ and r = 0.748) and also with TGF-β1 concentration in lysate ($p < 0.05$ and r = 0.671) (Figure 2).

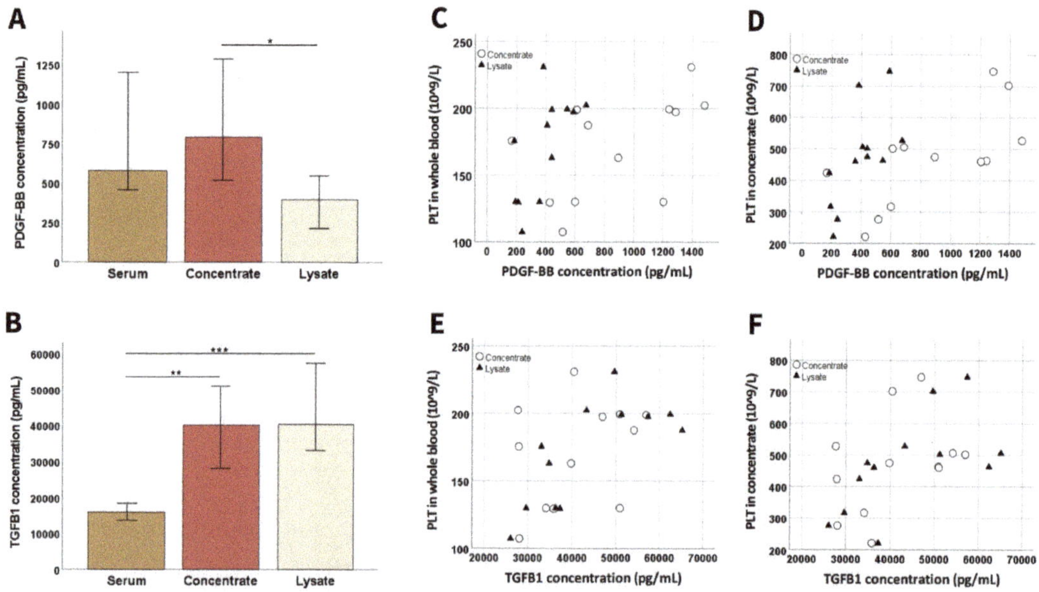

Figure 2. Growth factor concentrations during blood processing. Median platelet-derived growth factor (PDGF-BB; (**A**)) and transforming growth factor-β1 (TGFB1; (**B**)) concentrations at the different stages of blood processing (whole blood, concentrate, and lysate before filtration; samples had been stored at −80 °C before analysis) are presented; error bars display the 95% confidence intervals. Friedman tests for group comparisons and subsequent post hoc tests were performed; the asterisks indicate the significant differences between the corresponding groups (* corresponds to $p < 0.05$; ** corresponds to $p < 0.01$; *** corresponds to $p < 0.001$). The dot plots show platelet (PLT) concentrations in whole blood (**C**,**E**) or concentrate (**D**,**F**) vs. PDGF-BB (**C**,**D**) or TGFB1 (**E**,**F**) concentrations in concentrate (circles) and lysate (triangles); the correlations shown in (**C**) ($p < 0.01$ and r = 0.729 for concentrate; $p < 0.05$ and r = 0.641 for lysate), (**D**) ($p < 0.01$ and r = 0.804 for concentrate; $p < 0.01$ and r = 0.748 for lysate), (**E**) ($p < 0.05$ and r = 0.676 for lysate), and (**F**) ($p < 0.05$ and r = 0.671 for lysate) were significant based on Spearman's rank correlation. All data were obtained from n = 12 dogs.

The evaluation of the chemical composition of the samples from the different production steps showed that the pH value was not stable, as it was lower in the concentrate ($p < 0.01$) and lysate ($p < 0.01$) than in the serum. Differences were also evident in electrolyte, glucose, total protein, and albumin concentrations between serum and concentrate or lysate samples ($p < 0.01$), which is explained due to binding and/or dilution by the anticoagulant CPD, which was used in the latter (Table 1).

Table 1. Chemical analyses during blood processing. Data from n = 12 donors are presented as median (IQR) values.

Sample	pH	Na^+ (mmol/L)	K^+ (mmol/L)	Ca^{2+} (mmol/L)	Cl^-(mmol/L)	HCO_3 (mmol/L)	Glucose (mmol/L)	Lactate (mmol/L)	Total Protein (g/L)	Albumin (g/L)
Serum	7.52 (0.13)	149.15 (2.75)	4.87 (0.65)	1.33 (0.06)	110.40 (1.63)	19.60 (2.85)	3.70 (1.83)	4.95 (1.90)	64.35 (5.55)	30.10 (2.80)
Concentrate	7.19 (0.06)	156.80 (3.20)	2.86 (0.26)	0	81.20 (3.07)	16.65 (1.58)	26.35 (1.80)	2.45 (0.45)	47.25 (2.65)	24.35 (1.88)
Lysate	7.34 (0.03)	154.55 (1.50)	3.29 (0.28)	0	80.45 (2.55)	13.60 (1.85)	29.65 (0.82)	2.75 (0.53)	49.30 (3.08)	24.45 (2.33)

3.1.4. Donor-Related Parameters

There was no significant correlation between donor age and PLT concentration in whole blood or concentrate. However, the PDGF-BB concentration in lysate ($p < 0.05$ and $r = 0.676$) correlated positively with age, and the same trend was observed for TGF-β1. Splitting the canine donors by age groups (1–3 vs. 4–8 years) showed that the older donors had higher PLT concentrations in the concentrate ($p < 0.05$) and also higher growth factor concentrations in serum and lysate ($p < 0.01$ for TGF-β1 in serum, $p < 0.05$ for PDGF-BB in lysate) (Figure 3).

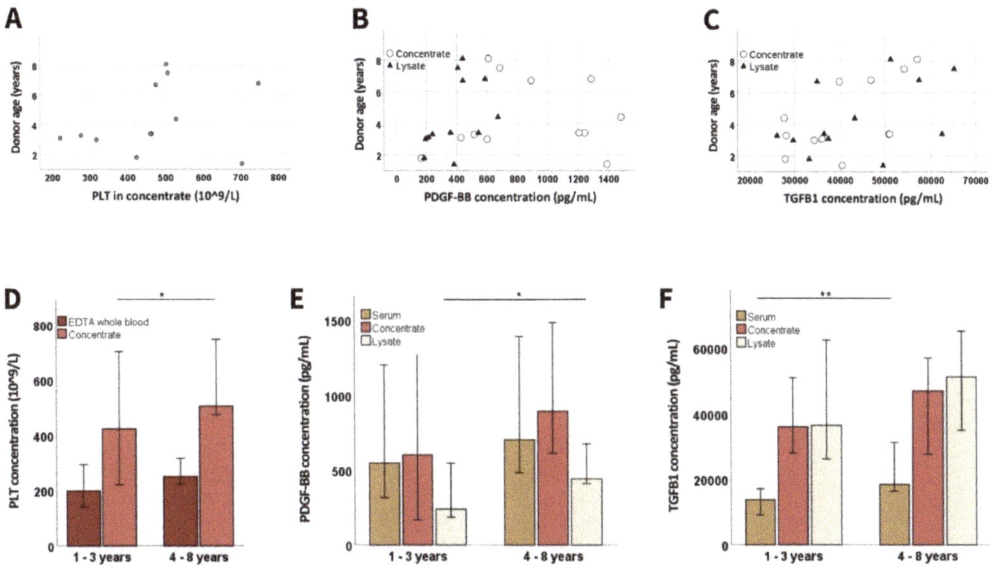

Figure 3. Age-related differences in platelet and growth factor concentrations. The dot plots show donor age vs. platelet (PLT; (**A**)) concentration in concentrate, as well as platelet-derived growth factor (PDGF-BB; (**B**)) and transforming growth factor (TGFB1; (**C**)) concentrations in concentrate (circles) and lysate (triangles); the correlation shown in (**B**) was significant for lysate ($p < 0.001$ and $r = 0.820$, based on Spearman's rank correlation). The bar plots show the median PLT (**D**), PDGF-BB (**E**), and TGFB1 (**F**) concentrations in EDTA whole blood or serum samples drawn directly from the donor dogs, as well as in samples from different processing stages, obtained from younger and older donors; data from the younger vs. the older animals were compared using Mann–Whitney U tests; the asterisks indicate the significant differences between the corresponding groups (* corresponds to $p < 0.05$; ** corresponds to $p < 0.01$). Data were obtained from n = 7 younger and n = 5 older dogs.

3.2. *Growth Factor Concentrations and Chemical Compositions in the Cell Culture Supplements*

The pooled cPL, the corresponding pooled ePL [29], and the FBS cell culture supplements showed some differences in their chemical analysis and composition. With possibly most relevance, there were differences in pH and in potassium, calcium, glucose, lactate, and growth factor concentrations. Regarding the latter, PDGF-BB was higher in ePL than in cPL or FBS, while TGF-β1 was much higher in cPL (Table 2).

Table 2. Chemical analysis of the cell culture supplements.

Parameter	cPL [1]	ePL [1]	FBS
pH	7.33	7.52	7.45
Na^+ (mmol/L)	153.70	147.80	138.60
K^+ (mmol/L)	3.25	3.68	11.21
Ca^{2+} (mmol/L)	0	<0.10	1.268
CL^- (mmol/L)	80.70	84.50	106.60
HCO_3 (mmol/L)	13.8	16.70	12.70
Glucose (mmol/L)	>30	23.10	2.20
Lactate (mmol/L)	2.70	2.90	17.70
Total protein (g/L)	49.20	54.10	36.80
Albumin (g/L)	25.50	27.90	23.00
PDGF-BB concentration (pg/mL)	307	3783	547
TGF-β1 concentration (pg/mL)	36,575	3966	3379

Note: [1] cPL and ePL data were obtained after the lysates from all donors (n = 12 for cPL and n = 19 for ePL) had been pooled.

3.3. Platelet Lysate in MSC Culture

3.3.1. Canine and Equine MSC Morphology, Proliferation, and Metabolic Activity

In FBS medium, the canine MSC showed a typical spindle shape, whereas in cPL medium, canine MSC lost the spindle shape and appeared huge, rounded, and multi-shaped. In addition, canine MSC secreted a matrix-like substance during their cultivation with 10% and 2.5% cPL medium, making detachment of MSC by trypsin difficult, as the cells formed streaks and agglomerates. In contrast to these altered characteristics of canine MSC, equine MSC showed a typical spindle shape in all three media (Figure 4A,B).

With cPL medium, due to the poor proliferation and the difficulties in MSC detachment, canine MSC counts in P4 and P5 were so low that the generation time calculation yielded negative values in several samples (3 out of 5 for cPL 10% in P4 and P5 and 3 out of 5 for cPL 2.5% in P4 and 2 out of 5 for cPL 2.5% in P5). Therefore, these data are not displayed and statistical analysis was not attempted. In the FBS group, canine MSC continued to proliferate normally, but it was evident that the generation time increased slightly over time ($p < 0.05$ from P3 to P4). In contrast, in equine MSC, the shortest generation times were observed with ePL 10% medium ($p < 0.01$ in P3 and P4 compared to ePL 2.5% in P3 and P4 and $p < 0.05$ in P5 compared to ePL 2.5% in P5), while proliferation remained consistent in FBS and ePL 10% media throughout all passages investigated (Figure 4C,E).

Confirming the findings described above, the metabolic activity of canine MSC was decreased in cPL 10% medium as compared to FBS medium ($p < 0.05$ in P5), and even more decreased in cPL 2.5% medium ($p < 0.05$ in P3, $p < 0.01$ in P4). Furthermore, the metabolic activity decreased continuously from P3 to P5 in FBS and cPL 10% media ($p < 0.05$) (Figure 4D,F).

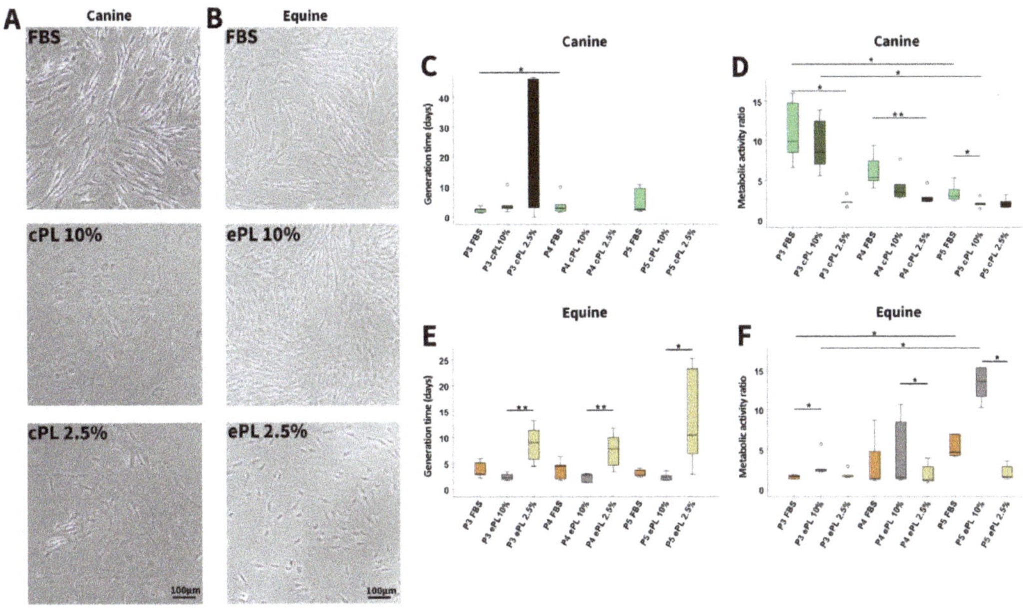

Figure 4. Mesenchymal stromal cell culture with different media supplements. Representative phase-contrast photomicrographs show canine (**A**) and equine (**B**) mesenchymal stromal cells (MSC) in passage 4 and at day 5 after seeding for population doubling assays in media supplemented with fetal bovine serum (FBS) or canine/equine platelet lysate (cPL/ePL). The boxplots display the generation times calculated for canine (**C**) and equine (**E**) MSC, as well as their metabolic activity as determined by MTS tetrazolium-based cell proliferation assay for canine (**D**) and equine (**F**) MSC in passages 3 to 5 (P3 to P5); note that missing generation time data were due to insufficient proliferation in these samples. Friedman tests for group comparisons and subsequent post hoc tests were performed; the asterisks indicate the significant differences between the corresponding groups (* corresponds to $p < 0.05$; ** corresponds to $p < 0.01$). Data were obtained using MSC from the same n = 5 donor dogs or horses in each group.

3.3.2. Canine and Equine MSC Trilineage Differentiation

No significant differences were observed regarding trilineage differentiation potential after MSC culture in the different media, neither in canine nor in equine MSC. However, it is of note that adipogenic and chondrogenic differentiation was weaker in canine than in equine MSC, irrespective of the medium supplement used (Figures 5 and 6).

3.3.3. Canine MSC Apoptosis, Necrosis, and Senescence

Despite their altered morphology, canine MSC showed less apoptosis in cPL 10% than in FBS or cPL 2.5% medium, but this was only significant when compared to the cPL 2.5% ($p < 0.05$ in P3 and P5). Additionally, significant differences over time between P3 and P5 were found, with a decrease in apoptosis with FBS but an increase with cPL 10% ($p < 0.05$).

Most necrosis was observed in cPL 2.5% MSC ($p < 0.05$ compared to FBS in P3), while no other significant differences were evident.

The highest SA-β-galactosidase activity, indicating senescence, was measured in MSC cultured with FBS and then decreased via cPL 10% to cPL 2.5%, but only in P3 ($p < 0.05$ for FBS vs. cPL 2.5%). In P5, however, this trend was reversed and the senescence value was lower with FBS medium than with cPL 10% medium. The senescence values in both P3 and P5 were lowest in MSC cultured with cPL 2.5%, but with an increase between P3 and P5 ($p < 0.05$) (Figure 7).

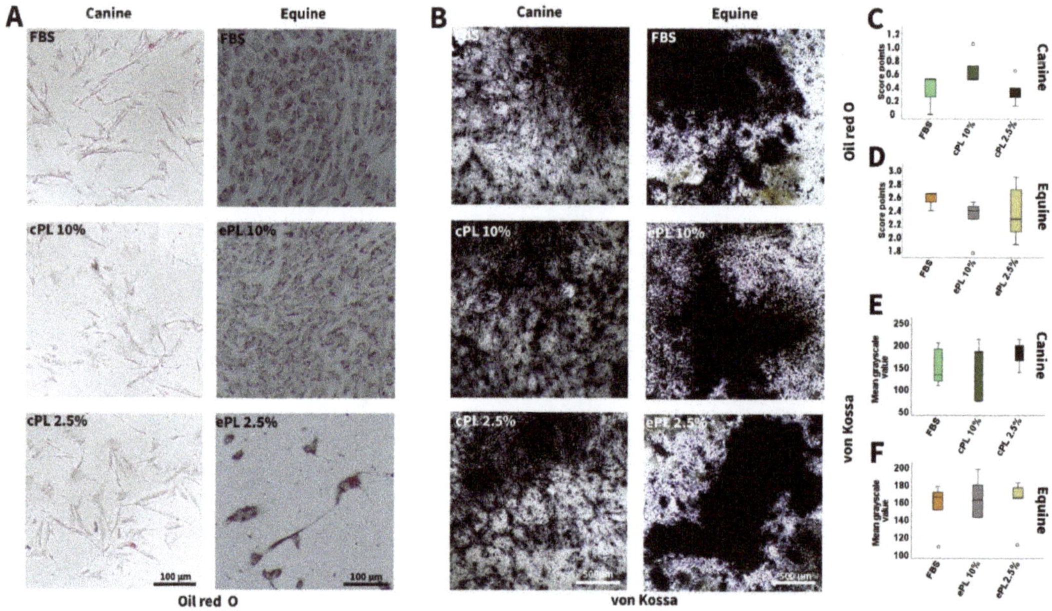

Figure 5. Adipogenic and osteogenic differentiation. Representative bright-field photomicrographs show canine (left column) and equine (right column) mesenchymal stromal cells (MSC) after adipogenic (**A**) and osteogenic (**B**) differentiation, with Oil Red O and von Kossa staining, respectively. Boxplots display the corresponding data obtained by scoring adipogenic differentiation of canine (**C**) and equine (**D**) MSC and image analysis using Fiji ImageJ after osteogenic differentiation of canine (**E**) and equine (**F**) MSC. MSC were cultured in the media indicated before differentiation was induced (FBS: fetal bovine serum; cPL: canine platelet lysate; ePL: equine platelet lysate). Data were obtained using MSC from the same n = 5 donor dogs or horses in each group.

3.3.4. Canine MSC Genetic Stability

Using molecular cytogenetic analyses (FISH), canine MSC cultured with 10% cPL showed an aberrant signal pattern in 3.5–4.2% of the interphase cells analyzed. Canine MSC cultured with 10% FBS showed an aberrant signal pattern more frequently, in 6.2–10% of the interphase cells. Nevertheless, the results did not suggest the presence of clonal numeric aberrations in the gonosomes, neither in MSC cultured with cPL nor in MSC cultured with FBS. The aberrant signal patterns observed included monosomy chromosome X, trisomy chromosome X, tetrasomy chromosome X, pentasomy chromosome X, nullisomy chromosome Y, and XYY (Figure 8).

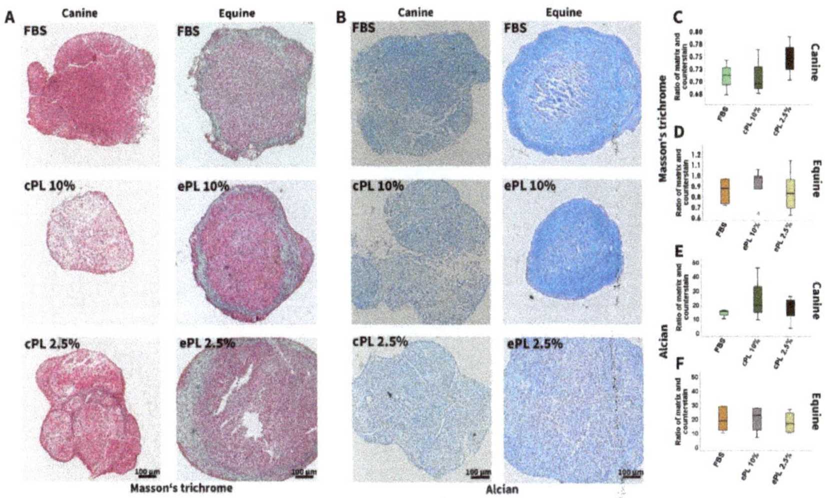

Figure 6. Chondrogenic differentiation. Representative bright-field photomicrographs show canine (left column) and equine (right column) mesenchymal stromal cells (MSC) after chondrogenic differentiation and Masson's trichrome (**A**) and Alcian (**B**) staining. Boxplots display the corresponding data obtained by image analysis using Fiji ImageJ (**C**–**F**). MSC were cultured in the media indicated before differentiation was induced (FBS: fetal bovine serum; cPL: canine platelet lysate; ePL: equine platelet lysate). Data were obtained using MSC from the same n = 5 donor dogs or horses in each group.

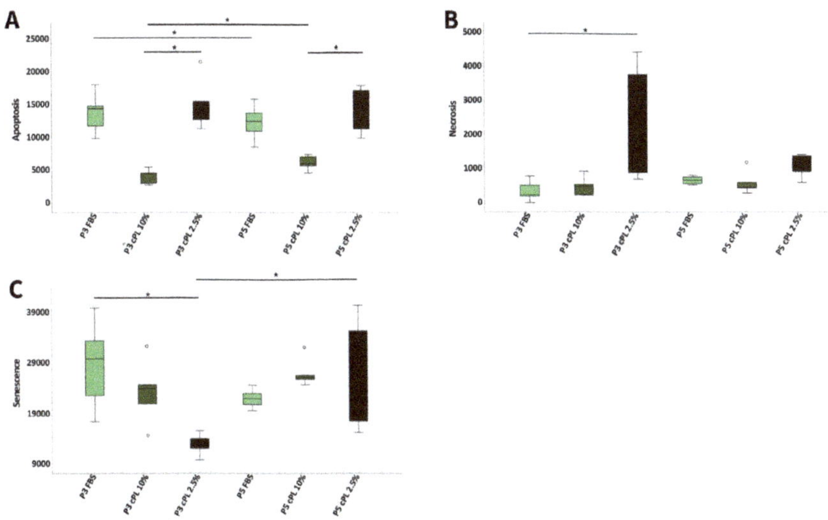

Figure 7. Apoptosis, necrosis, and senescence in canine cells. Boxplots display (**A**) apoptosis levels of canine mesenchymal stromal cells (MSC), measured by a luminescence-based Annexin V assay in passage 3 (P3) and 5 (P5) at day 5; (**B**) necrosis levels of canine MSC, measured using a cell-impermeant and pro-fluorescent DNA dye in passage 3 (P3) and 5 (P5) at day 5; (**C**) senescence levels of canine MSC in passage 3 (P3) and 5 (P5) at day 5, measured based on SA-β-galactosidase activity. Friedman tests for group comparisons and subsequent post hoc tests were performed; the asterisks indicate the significant differences between the corresponding groups (* corresponds to $p < 0.05$). Data were obtained using MSC from the same n = 5 donor dogs in each group.

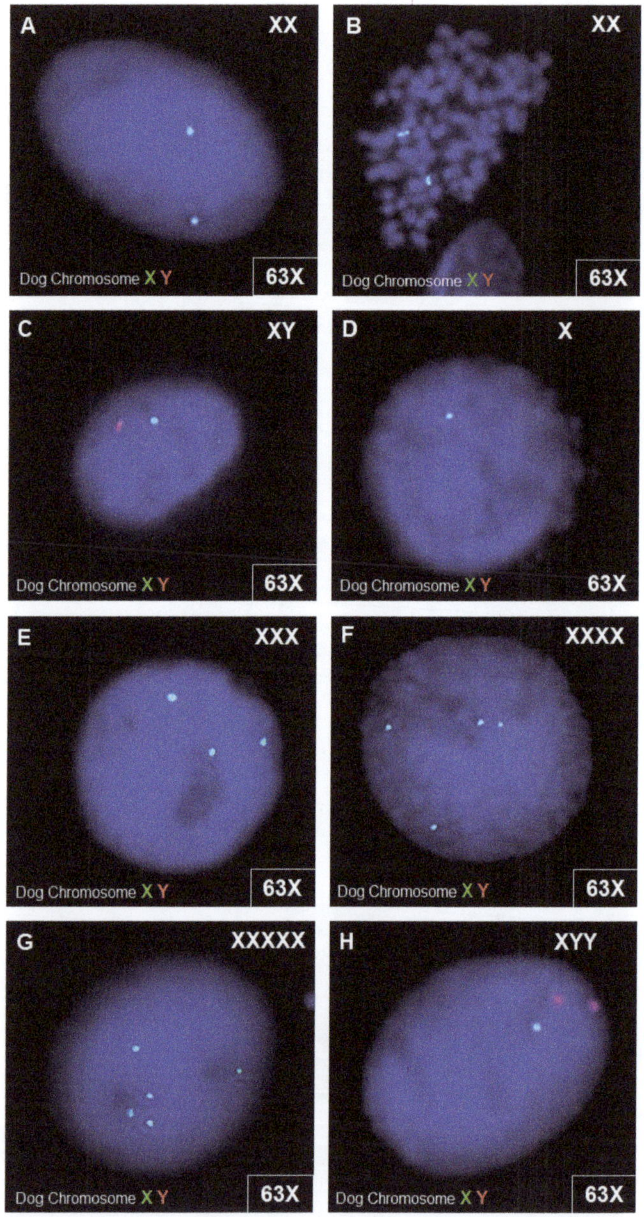

Figure 8. Cytogenetic analyses in canine cells. FISH analyses were performed with dog chromosome XY FISH probes (centromeric alpha satellite DNA probe; chromosome X—spectrum green; chromosome Y—spectrum red) on interphase cells (**A**,**C**–**F**) and one metaphase cell (**B**). Exemplary images show the detected signal constellations. In (**A**–**C**), microscopic images of interphase cells and metaphase cells with a normal gonosomal karyotype are shown. In (**D**–**H**), examples of interphase cells with different signal patterns are shown, with one signal for X (**D**), three signals for X (**E**), four signals for X (**F**), and five signals for X (**G**). For some interphase cells, a combination of signals from one X-chromosome and two Y-chromosomes were detected (**H**).

4. Discussion

Platelet lysate is not only an attractive off-the-shelf alternative to PLT concentrates for orthobiologic therapies, but has also shown great promise as a cell culture supplement replacing FBS in human and equine species. As data for the canine species is still scarce, in the current study, we first established a scalable procedure for cPL production from canine whole blood, building on our previously established protocol for the equine species [29]. Furthermore, we tested the effects of the obtained cPL on canine MSC in comparison to their equine counterparts, yet not with unambiguous results.

PL can be produced in three different ways regarding the PLT concentrate used as starting material. The latter can be obtained by a buffy-coat-based or a platelet-rich plasma method or directly by plateletpheresis. As the buffy coat method is best established in European human medicine [46,47], we had previously developed the first buffy-coat-based method to produce concentrate and ePL from equine whole blood collected in commercial blood bags [29]. This approach is scalable and led to similar or better results as compared to previous studies using the PRP method [34–37,39,48] or plateletpheresis [38,49]. In particular, while PLT were retained well, most WBC were removed, which we considered as an improvement compared to PRP-based protocols. Aiming to transfer this development to other relevant species, here we adapted the equine buffy-coat-based protocol [29] to generate canine PLT concentrates from whole blood. In contrast to the study presented here, the other studies on cPL which are currently available [9,40] have used PRP-based concentrates, either prepared in large volumes in blood bags [40,50] or in smaller volumes using centrifuge tubes [9]. In the presented study, we obtained a median PLT count of 469 G/L and a WBC count of 1.1 G/L in the canine concentrates. Others used concentrates with higher PLT counts for cPL production [9,40], but WBC counts, which are typically higher with the PRP as compared to the buffy coat method, were not given. There were also differences in the lysis procedures between the studies, as Russel et al. [40] used only one freeze-thaw cycle, while Lima et al. [9] used three cycles, comparable to our study. However, as no data are available on the resulting cPL in these previous studies, the outcomes cannot be compared.

Comparing our previously described ePL process [29] and the cPL manufacturing process described here and the resulting blood products, a few differences were evident. Regarding the processing, based on preliminary work with canine blood, harder centrifugation forces were used as compared to the equine protocol. Furthermore, while a hand press was used to produce the equine PLT concentrate after the second centrifugation, an automated blood separating device was used for this step in cPL production to further improve standardization [51]. The resulting concentrates had similar PLT counts in both species and even lower WBC counts in the canine concentrate as compared to the equine concentrate. Nevertheless, differences between the canine and equine blood products were observed regarding the concentrations of growth factors in the respective concentrates and lysates. This included much lower PDGF-BB concentrations but at the same time much higher TGF-β1 concentrations in canine concentrates and lysates as compared to the equine counterparts [29]. These species' differences in growth factor concentrations and canine growth factor levels widely correspond to the literature, although the results described in different studies are diverse [52–55]. In addition, the cPL obtained after lysis showed a significantly lower PDGF-BB concentration in comparison to the canine concentrate. The TGF-β1 concentrations of the cPL were in a similar range as in the concentrate. We suggest that PLT lysis is already evident after one freezing step, which was necessary to store the concentrates until the ELISA analyses were performed, or there could be an effect of the filter or plastic surface of the storage tubes on the growth factors concentrations. This will require further studies and it could also be considered to replace the plasma in the PLT concentrate production completely or partially by additive solutions, on the one hand to preserve plasma for the patients, and on the other hand because it has already been shown that these solutions improve the stability of the product [56]. Last but not least, an interesting difference between dogs and horses was observed regarding the correlations between donor age and PLT and growth factor concentrations, which were negative in horses [29,57] but positive in dogs. This could have practical relevance with respect to the choice of donor animals for off-the-shelf PL products.

Apart from the prospect of directly using cPL in therapies, we particularly aimed to use it for MSC culture. Given that the previously used gold standard cell culture supplement, FBS, is afflicted by several problems, a reduction or replacement of FBS is necessary from both an ethical and a scientific point of view. This was also recommended by the European Medicines Agency (EMA) [58] and the International Society for Cellular Therapy (ISCT) [16]. While FBS has already been replaced by human PL in the majority of good manufacturing practice procedures in human medicine [59], and some progress has already been made in the equine field, there were only two conflicting studies on cPL in canine cell culture so far [9,40]. However, when evaluating cPL as a cell culture supplement in canine MSC culture, the results were poorer than with ePL and equine MSC, which were analyzed for comparison.

Using cPL for canine MSC culture strongly altered the cell morphology and growth behavior. In contrast, using ePL for equine MSC culture again demonstrated that when used at the same concentration (10%), ePL is a promising alternative to FBS. The canine MSC cultured with cPL, however, showed almost no resemblance with the typical spindle shape of the MSC observed when supplemented with FBS. The canine MSC cultured in cPL medium also produced a high amount of extracellular material, so that individual cells were hardly distinguishable and passaging did not lead to reliable single-cell suspensions. While the latter partly compromised the quantification of generation times, assessment of the metabolic activity confirmed that the use of cPL compromised canine MSC viability and proliferation. These disappointing results for canine MSC are in accordance with one of the two previous studies [40], although we had hoped for improvement based on the different cPL production procedures. In disagreement with our and Russell's findings, Lima et al. [9] described canine MSC proliferation in cPL medium as improved compared to FBS medium, yet this was based on a different analytical approach with other possible influencing factors. While canine MSC expansion was only satisfactory in FBS but not in cPL medium, trilineage differentiation of canine MSC was similar after culture in FBS and cPL media. However, it should be acknowledged that canine MSC differentiation was generally rather poor, as already experienced by others [60–62]. Overall, considering the basic MSC characteristics, unlike equine MSC with ePL, canine MSC apparently suffered from cultivation with cPL.

To elucidate the observed effects of cPL on canine MSC culture expansion in more detail, the canine MSC were subjected to analyses of cell death, senescence, and genetic stability. Normal somatic cells, including MSC, proliferate for a limited number of doublings in culture and then reach a senescent state. In this senescent state, the cells are not dead but mitotically arrested while remaining metabolically active. Furthermore, they change their phenotype, increase significantly in size, and adopt a "fried egg morphology" [63–65]. If cells are too large, as observed here with cPL, they show loss of membrane and cytoskeletal integrity, increasing intracellular distances, as well as reduced surface areas for nutrient exchange and thus a reduced fitness [66]. The cell cycle of MSC can be terminated physiologically by senescence or apoptosis, which can be triggered by the same stressors [67]. The choice of pathway is strongly related to the stress level and the resulting levels of p53. Low levels of p53 promote transient cell cycle arrest and senescence. High levels of p53 and additional cooperativity of DNA-binding domains within the p53 tetramer lead to transcription of pro-apoptotic genes. In addition, at high p53 levels, pro-senescence signals are blocked, leading to apoptotic death of the cell caused by a combination of all these factors [68]. Apoptosis is the programmed cell death. In contrast, irreversible cell damage leads to passive cell death, i.e., necrosis, which in consequence leads to inflammation. In the presented study, both apoptosis and (secondary) necrosis mostly occurred in MSC cultured with 2.5% cPL. At the same time, SA-β galactosidase activity, used to indicate senescence, was lowest in these cultures. Thus, cPL used at low concentrations is likely to cause strong stress, as it initiated cell death but not senescence. However, neither increased cell death nor senescence was observed in MSC cultured with 10% cPL, and apoptosis in these cells was the lowest. This was surprising, as the canine cells showed their morphological and growth alterations not only in 2.5% but also in 10% cPL. As a final parameter for assessing cellular integrity, possible genetic aberrations in the gonosomes after cultivation with either FBS or cPL were investigated. Interestingly, canine MSC cultured with 10% cPL showed higher genetic stability than following cultivation with FBS. This is in accordance with findings in the human species [69], but was again unexpected considering the distinctly visible alterations of the MSC, which had even hampered parts of the karyotype analyses.

5. Conclusions

The buffy-coat-based protocol caused increased PLT concentrations and decreased WBC concentrations. Therefore, it delivers a concentrate suitable as starting material for cPL production, which could be used to offer off-the-shelf cPL therapies. However, when aiming to use the cPL obtained as MSC culture supplement, the results were not as convincing as observed with ePL and equine MSC. Even if part of the data suggested that using 10% cPL did not lead to cell damage, considering the strong alteration of MSC morphology and expansion characteristics, the use of cPL cannot be recommended for canine MSC culture in its current form.

Author Contributions: Conceptualization, A.H., T.C.H., H.L. and J.B.; methodology, A.H., H.H. and H.L; validation, A.H., H.H., H.L. and J.B.; formal analysis, A.H., H.H. and J.B.; investigation, A.H., H.H., V.-P.B., C.U.D., T.C.H., M.M. and J.M.; resources, H.H., T.C.H. and J.B.; data curation, A.H. and J.B.; writing—original draft preparation, A.H.; writing—review and editing, J.B.; visualization, A.H.; supervision, H.L. and J.B.; project administration, J.B.; funding acquisition, A.H., T.C.H. and J.B. All authors have read and agreed to the published version of the manuscript.

Funding: This research was supported by a scholarship from the Animal Health Academy, Germany (Akademie fuer Tiergesundheit; AfT), and by funding of the Society for the Promotion of Cynological Research (Gesellschaft zur Förderung Kynologischer Forschung e.V.(gkf)).

Institutional Review Board Statement: Sampling of canine blood was approved by the local authority (regional council Giessen, Germany, A 24/2017). Sampling of the canine and equine adipose tissues used for MSC isolation was not considered as animal experimentation by the local authority, as waste materials were used, or was approved within the framework of a previous study (provincial headquarters Leipzig, TV 34/13).

Informed Consent Statement: Not applicable.

Data Availability Statement: The data presented in this study are available on request from the corresponding author.

Conflicts of Interest: The authors declare no conflict of interest. The funders had no role in the design of the study; in the collection, analyses, or interpretation of data; in the writing of the manuscript, or in the decision to publish the results.

References

1. Pacini, S.; Spinabella, S.; Trombi, L.; Fazzi, R.; Galimberti, S.; Dini, F.; Carlucci, F.; Petrini, M. Suspension of bone marrow-derived undifferentiated mesenchymal stromal cells for repair of superficial digital flexor tendon in race horses. *Tissue Eng.* **2007**, *13*, 2949–2955. [CrossRef]
2. Godwin, E.E.; Young, N.J.; Dudhia, J.; Beamish, I.C.; Smith, R.K.W. Implantation of bone marrow-derived mesenchymal stem cells demonstrates improved outcome in horses with overstrain injury of the superficial digital flexor tendon. *Equine Vet. J.* **2012**, *44*, 25–32. [CrossRef]
3. Renzi, S.; Riccò, S.; Dotti, S.; Sesso, L.; Grolli, S.; Cornali, M.; Carlin, S.; Patruno, M.; Cinotti, S.; Ferrari, M. Autologous bone marrow mesenchymal stromal cells for regeneration of injured equine ligaments and tendons: A clinical report. *Res. Vet. Sci.* **2013**, *95*, 272–277. [CrossRef]
4. Smith, R.K.W.; Werling, N.J.; Dakin, S.G.; Alam, R.; Goodship, A.E.; Dudhia, J. Beneficial effects of autologous bone marrow-derived mesenchymal stem cells in naturally occurring tendinopathy. *PLoS ONE* **2013**, *8*, e75697. [CrossRef]

5. Broeckx, S.; Zimmerman, M.; Crocetti, S.; Suls, M.; Mariën, T.; Ferguson, S.J.; Chiers, K.; Duchateau, L.; Franco-Obregón, A.; Wuertz, K.; et al. Regenerative therapies for equine degenerative joint disease: A preliminary study. *PLoS ONE* **2014**, *9*, e85917. [CrossRef]
6. Ferris, D.J.; Frisbie, D.D.; Kisiday, J.D.; McIlwraith, C.W.; Hague, B.A.; Major, M.D.; Schneider, R.K.; Zubrod, C.J.; Kawcak, C.E.; Goodrich, L.R. Clinical outcome after intra-articular administration of bone marrow derived mesenchymal stem cells in 33 horses with stifle injury. *Vet. Surg.* **2014**, *43*, 255–265. [CrossRef]
7. Broeckx, S.Y.; Seys, B.; Suls, M.; Vandenberghe, A.; Mariën, T.; Adriaensen, E.; Declercq, J.; Van Hecke, L.; Braun, G.; Hellmann, K.; et al. Equine Allogeneic Chondrogenic Induced Mesenchymal Stem Cells Are an Effective Treatment for Degenerative Joint Disease in Horses. *Stem Cells Dev.* **2019**, *28*, 410–422. [CrossRef]
8. Shah, K.; Drury, T.; Roic, I.; Hansen, P.; Malin, M.; Boyd, R.; Sumer, H.; Ferguson, R. Outcome of Allogeneic Adult Stem Cell Therapy in Dogs Suffering from Osteoarthritis and Other Joint Defects. *Stem Cells Int.* **2018**, *2018*, 7309201. [CrossRef]
9. Lima, V.P.; Tobin, G.C.; Jesus Pereira MR de Silveira, M.D.; Witz, M.I.; Nardi, N.B. Chondrogenic effect of liquid and gelled platelet lysate on canine adipose-derived mesenchymal stromal cells. *Res. Vet. Sci.* **2019**, *124*, 393–398. [CrossRef]
10. Maki, C.B.; Beck, A.; Wallis, C.-B.C.C.; Choo, J.; Ramos, T.; Tong, R.; Borjesson, D.L.; Izadyar, F. Intra-articular Administration of Allogeneic Adipose Derived MSCs Reduces Pain and Lameness in Dogs With Hip Osteoarthritis: A Double Blinded, Randomized, Placebo Controlled Pilot Study. *Front. Vet. Sci.* **2020**, *7*, 570. [CrossRef] [PubMed]
11. Klinger, C. Mesenchymal stem cells: A potential therapy for canine atopic dermatitis? *Vet. Rec.* **2018**, *183*, 651. [CrossRef] [PubMed]
12. Villatoro, A.J.; Hermida-Prieto, M.; Fernández, V.; Fariñas, F.; Alcoholado, C.; Rodríguez-García, M.I.; Mariñas-Pardo, L.; Becerra, J. Allogeneic adipose-derived mesenchymal stem cell therapy in dogs with refractory atopic dermatitis: Clinical efficacy and safety. *Vet. Rec.* **2018**, *183*, 654. [CrossRef]
13. Rhew, S.-Y.; Park, S.-M.; Li, Q.; An, J.-H.; Chae, H.-K.; Lee, J.-H.; Ahn, J.-O.; Song, W.-J.; Youn, H.-Y. Efficacy and safety of allogenic canine adipose tissue-derived mesenchymal stem cell therapy for insulin-dependent diabetes mellitus in four dogs: A pilot study. *J. Vet. Med. Sci.* **2021**, *83*, 592–600. [CrossRef]
14. Pérez-Merino, E.M.; Usón-Casaús, J.M.; Zaragoza-Bayle, C.; Duque-Carrasco, J.; Mariñas-Pardo, L.; Hermida-Prieto, M.; Barrera-Chacón, R.; Gualtieri, M. Safety and efficacy of allogeneic adipose tissue-derived mesenchymal stem cells for treatment of dogs with inflammatory bowel disease: Clinical and laboratory outcomes. *Vet. J.* **2015**, *206*, 385–390. [CrossRef]
15. Sharun, K.; Rawat, T.; Kumar, R.; Chandra, V.; Saxena, A.C.; Pawde, A.M.; Kinjavdekar, P.; Amarpal; Sharma, G.T. Clinical evaluation following the percutaneous transplantation of allogenic bone marrow-derived mesenchymal stem cells (aBM-MSC) in dogs affected by vertebral compression fracture. *Vet. Anim. Sci.* **2020**, *10*, 100152. [CrossRef]
16. Karnieli, O.; Friedner, O.M.; Allickson, J.G.; Zhang, N.; Jung, S.; Fiorentini, D.; Abraham, E.; Eaker, S.S.; Yong, T.K.; Chan, A.; et al. A consensus introduction to serum replacements and serum-free media for cellular therapies. *Cytotherapy* **2017**, *19*, 155–169. [CrossRef] [PubMed]
17. Fahie, M.A.; Ortolano, G.A.; Guercio, V.; Schaffer, J.A.; Johnston, G.; Au, J.; Hettlich, B.A.; Phillips, T.; Allen, M.J.; Bertone, A.L. A randomized controlled trial of the efficacy of autologous platelet therapy for the treatment of osteoarthritis in dogs. *J. Am. Vet. Med. Assoc.* **2013**, *243*, 1291–1297. [CrossRef]
18. Catarino, J.; Carvalho, P.; Santos, S.; Martins, Â.; Requicha, J. Treatment of canine osteoarthritis with allogeneic platelet-rich plasma: Review of five cases. *Open Vet. J.* **2020**, *10*, 226–231. [CrossRef]
19. Venator, K.P.; Frye, C.W.; Gamble, L.-J.; Wakshlag, J.J. Assessment of a Single Intra-Articular Stifle Injection of Pure Platelet Rich Plasma on Symmetry Indices in Dogs with Unilateral or Bilateral Stifle Osteoarthritis from Long-Term Medically Managed Cranial Cruciate Ligament Disease. *Vet. Med.* **2020**, *11*, 31–38. [CrossRef]
20. Sanghani-Kerai, A.; Black, C.; Cheng, S.O.; Collins, L.; Schneider, N.; Blunn, G.; Watson, F.; Fitzpatrick, N. Clinical outcomes following intra-articular injection of autologous adipose-derived mesenchymal stem cells for the treatment of osteoarthritis in dogs characterized by weight-bearing asymmetry. *Bone Jt. Res.* **2021**, *10*, 650–658. [CrossRef]
21. Hernández-Guerra, Á.M.; Carrillo, J.M.; Sopena, J.J.; Vilar, J.M.; Peláez, P.; Cuervo, B.; Santana, A.; Rubio, M. Platelet-Rich Plasma for the Treatment of Degenerative Lumbosacral Stenosis: A Study with Retired Working Dogs. *Animals* **2021**, *11*, 2965. [CrossRef]
22. Iacopetti, I.; Patruno, M.; Melotti, L.; Martinello, T.; Bedin, S.; Badon, T.; Righetto, E.M.; Perazzi, A. Autologous Platelet-Rich Plasma Enhances the Healing of Large Cutaneous Wounds in Dogs. *Front. Vet. Sci.* **2020**, *7*, 575449. [CrossRef]
23. Berni, P.; Leonardi, F.; Conti, V.; Ramoni, R.; Grolli, S.; Mattioli, G. Case Report: A Novel Ventilated Thermoplastic Mesh Bandage for Post-operative Management of Large Soft Tissue Defects: A Case Series of Three Dogs Treated With Autologous Platelet Concentrates. *Front. Vet. Sci.* **2021**, *8*, 704567. [CrossRef]
24. Farghali, H.A.; AbdElKader, N.A.; AbuBakr, H.O.; Ramadan, E.S.; Khattab, M.S.; Salem, N.Y.; Emam, I.A. Corneal Ulcer in Dogs and Cats: Novel Clinical Application of Regenerative Therapy Using Subconjunctival Injection of Autologous Platelet-Rich Plasma. *Front. Vet. Sci.* **2021**, *8*, 641265. [CrossRef]
25. Perego, R.; Spada, E.; Moneta, E.; Baggiani, L.; Proverbio, D. Use of Autologous Leucocyte- and Platelet-Rich Plasma (L-PRP) in the Treatment of Aural Hematoma in Dogs. *Vet. Sci.* **2021**, *8*, 172. [CrossRef]
26. Zamani, M.; Yaghoubi, Y.; Movassaghpour, A.; Shakouri, K.; Mehdizadeh, A.; Pishgahi, A.; Yousefi, M. Novel therapeutic approaches in utilizing platelet lysate in regenerative medicine: Are we ready for clinical use? *J. Cell Physiol.* **2019**, *234*, 17172–17186. [CrossRef]

27. Carr, B.J.; Canapp, S.O.; Mason, D.R.; Cox, C.; Hess, T. Canine Platelet-Rich Plasma Systems: A Prospective Analysis. *Front. Vet. Sci.* **2015**, *2*, 73. [CrossRef]
28. Franklin, S.P.; Birdwhistell, K.E.; Strelchik, A.; Garner, B.C.; Brainard, B.M. Influence of Cellular Composition and Exogenous Activation on Growth Factor and Cytokine Concentrations in Canine Platelet-Rich Plasmas. *Front. Vet. Sci.* **2017**, *4*, 40. [CrossRef]
29. Hagen, A.; Lehmann, H.; Aurich, S.; Bauer, N.; Melzer, M.; Moellerberndt, J.; Patané, V.; Schnabel, C.L.; Burk, J. Scalable Production of Equine Platelet Lysate for Multipotent Mesenchymal Stromal Cell Culture. *Front. Bioeng. Biotechnol.* **2020**, *8*, 613621. [CrossRef]
30. Schallmoser, K.; Henschler, R.; Gabriel, C.; Koh, M.B.C.; Burnouf, T. Production and Quality Requirements of Human Platelet Lysate: A Position Statement from the Working Party on Cellular Therapies of the International Society of Blood Transfusion. *Trends Biotechnol.* **2020**, *38*, 13–23. [CrossRef]
31. Tyrnenopoulou, P.; Diakakis, N.; Karayannopoulou, M.; Savvas, I.; Koliakos, G. Evaluation of intra-articular injection of autologous platelet lysate (PL) in horses with osteoarthritis of the distal interphalangeal joint. *Vet. Q.* **2016**, *36*, 56–62. [CrossRef]
32. Mojica-Henshaw, M.P.; Jacobson, P.; Morris, J.; Kelley, L.; Pierce, J.; Boyer, M.; Reems, J.-A. Serum-converted platelet lysate can substitute for fetal bovine serum in human mesenchymal stromal cell cultures. *Cytotherapy* **2013**, *15*, 1458–1468. [CrossRef]
33. Becherucci, V.; Piccini, L.; Casamassima, S.; Bisin, S.; Gori, V.; Gentile, F.; Ceccantini, R.; De Rienzo, E.; Bindi, B.; Pavan, P.; et al. Human platelet lysate in mesenchymal stromal cell expansion according to a GMP grade protocol: A cell factory experience. *Stem Cell Res. Ther.* **2018**, *9*, 124. [CrossRef]
34. Del Bue, M.; Riccò, S.; Conti, V.; Merli, E.; Ramoni, R.; Grolli, S. Platelet lysate promotes in vitro proliferation of equine mesenchymal stem cells and tenocytes. *Vet. Res. Commun.* **2007**, *31* (Suppl. S1), 289–292. [CrossRef]
35. Seo, J.; Tsuzuki, N.; Haneda, S.; Yamada, K.; Furuoka, H.; Tabata, Y.; Sasaki, N. Comparison of allogeneic platelet lysate and fetal bovine serum for in vitro expansion of equine bone marrow-derived mesenchymal stem cells. *Res. Vet. Sci.* **2013**, *95*, 693–698. [CrossRef]
36. Russell, K.A.; Koch, T.G. Equine platelet lysate as an alternative to fetal bovine serum in equine mesenchymal stromal cell culture—Too much of a good thing? *Equine Vet. J.* **2016**, *48*, 261–264. [CrossRef] [PubMed]
37. Gilbertie, J.M.; Long, J.M.; Schubert, A.G.; Berglund, A.K.; Schaer, T.P.; Schnabel, L.V. Pooled Platelet-Rich Plasma Lysate Therapy Increases Synoviocyte Proliferation and Hyaluronic Acid Production While Protecting Chondrocytes From Synoviocyte-Derived Inflammatory Mediators. *Front. Vet. Sci.* **2018**, *5*, 150. [CrossRef]
38. Naskou, M.C.; Sumner, S.M.; Chocallo, A.; Kemelmakher, H.; Thoresen, M.; Copland, I.; Galipeau, J.; Peroni, J.F. Platelet lysate as a novel serum-free media supplement for the culture of equine bone marrow-derived mesenchymal stem cells. *Stem Cell Res. Ther.* **2018**, *9*, 75. [CrossRef]
39. Yaneselli, K.; Barrachina, L.; Remacha, A.R.; Algorta, A.; Vitoria, A.; Cequier, A.; Romero, A.; Vázquez, F.J.; Maisonnave, J.; Rodellar, C. Effect of allogeneic platelet lysate on equine bone marrow derived mesenchymal stem cell characteristics, including immunogenic and immunomodulatory gene expression profile. *Vet. Immunol. Immunopathol.* **2019**, *217*, 109944. [CrossRef]
40. Russell, K.A.; Gibson, T.W.G.; Chong, A.; Co, C.; Koch, T.G. Canine Platelet Lysate Is Inferior to Fetal Bovine Serum for the Isolation and Propagation of Canine Adipose Tissue- and Bone Marrow-Derived Mesenchymal Stromal Cells. *PLoS ONE* **2015**, *10*, e0136621. [CrossRef]
41. Van Kuppeveld, F.J.; van der Logt, J.T.; Angulo, A.F.; van Zoest, M.J.; Quint, W.G.; Niesters, H.G.; Galama, J.M.; Melchers, W.J. Genus- and species-specific identification of mycoplasmas by 16S rRNA amplification. *Appl. Environ. Microbiol.* **1992**, *58*, 2606–2615. [CrossRef]
42. Gittel, C.; Brehm, W.; Burk, J.; Juelke, H.; Staszyk, C.; Ribitsch, I. Isolation of equine multipotent mesenchymal stromal cells by enzymatic tissue digestion or explant technique: Comparison of cellular properties. *BMC Vet. Res.* **2013**, *9*, 221. [CrossRef] [PubMed]
43. Ruifrok, A.C.; Johnston, D.A. Quantification of Histochemical Staining by Color Deconvolution. *Anal. Quant. Cytol. Histol.* **2001**, *23*, 291–299. [PubMed]
44. Seabright, M. The use of proteolytic enzymes for the mapping of structural rearrangements in the chromosomes of man. *Chromosoma* **1972**, *36*, 204–210. [CrossRef] [PubMed]
45. Seabright, M. Human chromosome banding. *Lancet* **1972**, *299*, 967. [CrossRef]
46. Gulliksson, H. Platelets from platelet-rich-plasma versus buffy-coat-derived platelets: What is the difference? *Rev. Bras. Hematol. Hemoter.* **2012**, *34*, 76–77. [CrossRef]
47. Burnouf, T.; Strunk, D.; Koh, M.B.C.; Schallmoser, K. Human platelet lysate: Replacing fetal bovine serum as a gold standard for human cell propagation? *Biomaterials* **2016**, *76*, 371–387. [CrossRef]
48. Bozorgmanesh, R.; Magdesian, K.G.; Sutton-Burges, J.W.; Owens, S.D.; Tablin, F. Equine platelet concentrate preparation and validation. *J. Vet. Intern. Med.* **2019**, *33*, 1500–1506. [CrossRef]
49. Sumner, S.M.; Naskou, M.C.; Thoresen, M.; Copland, I.; Peroni, J.F. Platelet lysate obtained via plateletpheresis performed in standing and awake equine donors. *Transfusion* **2017**, *57*, 1755–1762. [CrossRef]
50. Abrams-Ogg, A.C.; Kruth, S.A.; Carter, R.F.; Valli, V.E.; Kamel-Reid, S.; Dubé, I.D. Preparation and transfusion of canine platelet concentrates. *Am. J. Vet. Res.* **1993**, *54*, 635–642.
51. Van der Meer, P.; Pietersz, R.; Hinloopen, B.; Dekker, W.; Reesink, H. Automated separation of whole blood in top and bottom bags into components using the Compomat G4. *Vox Sanguinis* **1999**, *76*, 90–99. [CrossRef]

52. Bowen-Pope, D.F.; Hart, C.E.; Seifert, R.A. Sera and conditioned media contain different isoforms of platelet-derived growth factor (PDGF) which bind to different classes of PDGF receptor. *J. Biol. Chem.* **1989**, *264*, 2502–2508. [CrossRef]
53. Stief, M.; Gottschalk, J.; Ionita, J.-C.; Einspanier, A.; Oechtering, G.; Böttcher, P. Concentration of platelets and growth factors in canine autologous conditioned plasma. *Vet. Comp. Orthop. Traumatol.* **2011**, *24*, 122–125. [CrossRef] [PubMed]
54. Hatakeyama, I.; Marukawa, E.; Takahashi, Y.; Omura, K. Effects of platelet-poor plasma, platelet-rich plasma, and platelet-rich fibrin on healing of extraction sockets with buccal dehiscence in dogs. *Tissue Eng. Part A* **2014**, *20*, 874–882. [CrossRef] [PubMed]
55. Perego, R.; Spada, E.; Baggiani, L.; Martino, P.A.; Proverbio, D. Efficacy of a Semi Automated Commercial Closed System for Autologous Leukocyte- and Platelet-Rich Plasma (l-prp) Production in Dogs: A Preliminary Study. *Animals* **2020**, *10*, 1342. [CrossRef] [PubMed]
56. Haines, J.M.; Hwang, J.K.; Wardrop, K.J. The effects of additive solutions on the development of storage lesions in stored canine platelet concentrates. *J. Vet. Emerg. Crit. Care* **2021**, *31*, 247–255. [CrossRef]
57. Giraldo, C.E.; López, C.; Álvarez, M.E.; Samudio, I.J.; Prades, M.; Carmona, J.U. Effects of the breed, sex and age on cellular content and growth factor release from equine pure-platelet rich plasma and pure-platelet rich gel. *BMC Vet. Res.* **2013**, *9*, 29. [CrossRef]
58. European Medicines Agency, London Committee for Medicinal Products for Human Use (CHMP) (Ed.) Guideline on the Use of Bovine Serum in the Manufacture of Human Biological Medicinal Products. 2013. Available online: https://www.ema.europa.eu/en/documents/scientific-guideline/guideline-use-bovine-serum-manufacture-human-biological-medicinal-products_en.pdf (accessed on 12 January 2022).
59. Trento, C.; Bernardo, M.E.; Nagler, A.; Kuçi, S.; Bornhäuser, M.; Köhl, U.; Strunk, D.; Galleu, A.; Sanchez-Guijo, F.; Gaipa, G.; et al. Manufacturing Mesenchymal Stromal Cells for the Treatment of Graft-versus-Host Disease: A Survey among Centers Affiliated with the European Society for Blood and Marrow Transplantation. *Biol. Blood Marrow Transplant.* **2018**, *24*, 2365–2370. [CrossRef] [PubMed]
60. Neupane, M.; Chang, C.-C.; Kiupel, M.; Yuzbasiyan-Gurkan, V. Isolation and characterization of canine adipose-derived mesenchymal stem cells. *Tissue Eng. Part A* **2008**, *14*, 1007–1015. [CrossRef]
61. Volk, S.W.; Wang, Y.; Hankenson, K.D. Effects of donor characteristics and ex vivo expansion on canine mesenchymal stem cell properties: Implications for MSC-based therapies. *Cell Transpl.* **2012**, *21*, 2189–2200. [CrossRef]
62. Bearden, R.N.; Huggins, S.S.; Cummings, K.J.; Smith, R.; Gregory, C.A.; Saunders, W.B. In-vitro characterization of canine multipotent stromal cells isolated from synovium, bone marrow, and adipose tissue: A donor-matched comparative study. *Stem Cell Res. Ther.* **2017**, *8*, 218. [CrossRef]
63. Lee, B.Y.; Han, J.A.; Im, J.S.; Morrone, A.; Johung, K.; Goodwin, E.C.; Kleijer, W.J.; DiMaio, D.; Hwang, E.S. Senescence-associated β-galactosidase is lysosomal β-galactosidase. *Aging Cell* **2006**, *5*, 187–195. [CrossRef]
64. Wagner, W.; Bork, S.; Lepperdinger, G.; Joussen, S.; Ma, N.; Strunk, D.; Koch, C. How to track cellular aging of mesenchymal stromal cells? *Aging* **2010**, *2*, 224–230. [CrossRef]
65. Bertolo, A.; Guerrero, J.; Stoyanov, J. Autofluorescence-based sorting removes senescent cells from mesenchymal stromal cell cultures. *Sci. Rep.* **2020**, *10*, 19084. [CrossRef]
66. Miettinen, T.P.; Caldez, M.J.; Kaldis, P.; Björklund, M. Cell size control—A mechanism for maintaining fitness and function. *Bioessays* **2017**, *39*, 1700058. [CrossRef]
67. Herranz, N.; Gil, J. Mechanisms and functions of cellular senescence. *J. Clin. Investig.* **2018**, *128*, 1238–1246. [CrossRef] [PubMed]
68. Childs, B.G.; Baker, D.J.; Kirkland, J.L.; Campisi, J.; van Deursen, J.M. Senescence and apoptosis: Dueling or complementary cell fates? *EMBO Rep.* **2014**, *15*, 1139–1153. [CrossRef] [PubMed]
69. Trojahn Kølle, S.-F.; Oliveri, R.S.; Glovinski, P.V.; Kirchhoff, M.; Mathiasen, A.B.; Elberg, J.J.; Andresen, P.S.; Drzewiecki, K.T.; Fischer-Nielsen, A. Pooled human platelet lysate versus fetal bovine serum-investigating the proliferation rate, chromosome stability and angiogenic potential of human adipose tissue-derived stem cells intended for clinical use. *Cytotherapy* **2013**, *15*, 1086–1097. [CrossRef] [PubMed]

animals

MDPI

Article

Peptide Mediated Adhesion to Beta-Lactam Ring of Equine Mesenchymal Stem Cells: A Pilot Study

Barbara Merlo [1,2,*], Vito Antonio Baldassarro [1,3], Alessandra Flagelli [2], Romina Marcoccia [2], Valentina Giraldi [2,4], Maria Letizia Focarete [2,4], Daria Giacomini [2,4] and Eleonora Iacono [1,2]

1 Department of Veterinary Medical Sciences, University of Bologna, Via Tolara di Sopra, 50, 40064 Ozzano Emilia, BO, Italy; vito.baldassarro2@unibo.it (V.A.B.); eleonora.iacono2@unibo.it (E.I.)

2 Interdepartmental Center for Industrial Research in Health Sciences and Technologies, University of Bologna, Via Tolara di Sopra, 41/E, 40064 Ozzano Emilia, BO, Italy; alessandra.flagelli2@unibo.it (A.F.); romina.marcoccia.rm@gmail.com (R.M.); valentina.giraldi2@unibo.it (V.G.); marialetizia.focarete@unibo.it (M.L.F.); daria.giacomini@unibo.it (D.G.)

3 IRET Foundation, Via Tolara di Sopra, 41/E, 40064 Ozzano Emilia, BO, Italy

4 Department of Chemistry "Giacomo Ciamician" and INSTM UdR of Bologna, University of Bologna, Via Selmi 2, 40126 Bologna, BO, Italy

* Correspondence: barbara.merlo@unibo.it

Simple Summary: In recent years, stem cell therapy has emerged as a promising potential treatment for chronic wounds in both human and veterinary medicine. Particularly, mesenchymal stem cells (MSCs) may be an attractive therapeutic tool for regenerative medicine and tissue engineering because these cells play a critical role in wound repair and tissue regeneration due to their immunosuppressive properties and multipotency. The use of biomaterials with integrin agonists could promote cell adhesion increasing tissue repair processes. This pilot study focuses on the adhesion ability of equine adult (adipose tissue) and fetal adnexa (Wharton's jelly) derived MSCs mediated by GM18, an $\alpha 4\beta 1$ integrin agonist, alone and combined with a biodegradable polymeric scaffold. Results show that a 24 h exposition to soluble GM18 affects equine MSCs adhesion ability with a donor-related variability and might suggest that WJ-MSCs more easily adhere to poly L-lactic acid (PLLA) nanofibers combined with GM18. These preliminary results need to be confirmed by further studies on the interactions between the different types of equine MSCs and GM18 incorporated PLLA scaffolds before drawing definitive conclusions on which cells and scaffolds could be successfully used for the treatment of decubitus ulcers.

Abstract: Regenerative medicine applied to skin lesions is a field in constant improvement. The use of biomaterials with integrin agonists could promote cell adhesion increasing tissue repair processes. The aim of this pilot study was to analyze the effect of an $\alpha 4\beta 1$ integrin agonist on cell adhesion of equine adipose tissue (AT) and Wharton's jelly (WJ) derived MSCs and to investigate their adhesion ability to GM18 incorporated poly L-lactic acid (PLLA) scaffolds. Adhesion assays were performed after culturing AT- and WJ-MSCs with GM18 coating or soluble GM18. Cell adhesion on GM18 containing PLLA scaffolds after 20 min co-incubation was assessed by HCS. Soluble GM18 affects the adhesion of equine AT- and WJ-MSCs, even if its effect is variable between donors. Adhesion to PLLA scaffolds containing GM18 is not significantly influenced by GM18 for AT-MSCs after 20 min or 24 h of culture and for WJ-MSCs after 20 min, but increased cell adhesion by 15% GM18 after 24 h. In conclusion, the $\alpha 4\beta 1$ integrin agonist GM18 affects equine AT- and WJ-MSCs adhesion ability with a donor-related variability. These preliminary results represent a first step in the study of equine MSCs adhesion to PLLA scaffolds containing GM18, suggesting that WJ-MSCs might be more suitable than AT-MSCs. However, the results need to be confirmed by increasing the number of samples before drawing definite conclusions.

Keywords: equine; mesenchymal stem cell; adhesion; $\alpha 4\beta 1$ integrin; β-lactam agonist; poly L-lactic acid (PLLA) scaffold

Citation: Merlo, B.; Baldassarro, V.A.; Flagelli, A.; Marcoccia, R.; Giraldi, V.; Focarete, M.L.; Giacomini, D.; Iacono, E. Peptide Mediated Adhesion to Beta-Lactam Ring of Equine Mesenchymal Stem Cells: A Pilot Study. *Animals* **2022**, *12*, 734. https://doi.org/10.3390/ani12060734

Academic Editor: Milka Vrecl

Received: 11 January 2022
Accepted: 14 March 2022
Published: 15 March 2022

1. Introduction

Wound healing is a very complex physiological response to the disruption in the normal architecture of the skin and is influenced by many factors [1]. This is a stepwise process that includes a proliferative phase, in which damaged tissue is removed and granulation tissue forms in the wound, and a remodeling phase, with the formation of scar tissue indicating the completion of the wound healing process [2]. The conventional approaches applied for the instant healing of skin wounds include the use of different drugs and natural products with anti-inflammatory, anti-microbial, and antioxidant properties [3].

In recent years, wound healing has become a logical target for innovative therapies, such as regenerative medicine strategies, which have the potential to restore tissue, perhaps equaling or exceeding pre-damage levels, resulting in improved outcomes and quality of life [4]. Stem cell therapy has emerged as a promising potential treatment for chronic wounds [5]. Particularly, mesenchymal stem cells (MSCs) may be an attractive therapeutic tool for regenerative medicine and tissue engineering because they play a critical role in wound repair and tissue regeneration due to their immunosuppressive properties and multipotency [6].

Regenerative medicine applied to skin lesions has been a field of constant improvement for both human and veterinary medicine [7–10]. The objective of regenerative medicine is to stimulate the self-repair of tissues and organs with stem cells alone or in conjunction with biomaterials [11]. Biomaterials have a very important role in tissue engineering but, unlike natural polymers, synthetic polymeric biomaterials used in tissue engineering applications lack biological activity and typically do not promote excellent cell adhesion and growth [12,13]. Cell–cell and cell–extracellular matrix (ECM) interactions in the niche are mediated by different cell adhesion molecules, and integrins are one of the main players [14].

Integrins are transmembrane receptors comprised of two subunits, alpha (α) and beta (β). The molecular family includes 18 α and 8 β subunits, leading to the formation of 24 different heterodimeric transmembrane receptors [15,16]. They mediate cellular interactions with the ECM and surrounding cells by binding specific ligands [14,16] regulating crucial aspects of cellular functions, including adhesion, differentiation, growth, gene expression, and survival [17]. The ability of integrins to bind and associate with various soluble ligands largely depends on the structural conformations of the α and β subunits.

To modulate integrins' action, a novel series of β-lactam targeting RGD (arginine–glycine–aspartic acid) fragment and leukocyte integrins were designed [18]. β-lactams have two specific structural features that are of interest with regard to biological activity: a constrained four-membered cyclic amide, which could easily undergo ring-opening reactions by nucleophilic residues in the active sites of enzymes, and a rigid core structure that, by reducing the conformational degrees of freedom, could favor and actually enhance directional noncovalent bonding for ligand–receptor recognition [19]. On their feature, Baiula et al. [18] obtained selective and potent agonists that could induce cell adhesion and promote cell signaling mediated by $\alpha v\beta 3$, $\alpha v\beta 5$, $\alpha 5\beta 1$, or $\alpha 4\beta 1$ integrin, and antagonists for the integrins $\alpha v\beta 3$ and $\alpha 5\beta 1$, as well as $\alpha 4\beta 1$ and $\alpha L\beta 2$, preventing the effects elicited by the respective endogenous agonists [18].

Electrospinning is a highly versatile method to process polymer solutions or melts into continuous fibers in the form of non-woven porous mats that find application in the biomedical field where they are used as tissue engineering scaffolds [20,21]. The use of biomaterials, such as electrospun scaffolds, conjugated with integrin agonists could promote cell adhesion by influencing tissue repair processes and therapeutic progress. The modulation of the activity of integrins can represent an excellent strategy applicable to cell therapies that aim to stimulate tissue repair using MSCs.

New scaffolds based on electrospun poly L-lactic acid (PLLA) and agonist ligands of monocyclic β-lactam compounds of specific integrins have recently been tested [22]. Incorporation of GM18 β-lactam into PLLA scaffolds has been shown to support increased cell proliferation of human MSCs from bone marrow [22]. Studies on the use of scaffolds with integrin agonists based on β-lactams in equines have never been reported. Therefore, the aim of this pilot study was to analyze the effect of an α4β1 integrin agonist on cell adhesion of equine adult (adipose tissue; AT) and fetal membrane (Wharton's jelly; WJ) derived MSCs and to preliminarily investigate the adhesion of these cells to GM18-incorporated PLLA scaffolds, to evaluate their potential use for the treatment of decubitus ulcers.

2. Materials and Methods

2.1. Materials

Poly L-Lactic Acid (PLLA) (Lacea H.100-E Mw 8.4 × 104 g/mol) was purchased from Mitsui Fine Chemicals (Duesseldorf, Germany). Dichloromethane (DCM) and dimethylformamide (DMF) were purchased from Sigma–Aldrich (Milan, Italy) and used without further purification. β-Lactam GM18 was prepared accordingly to a previously reported multistep synthesis [18]. Other chemicals were purchased from Sigma–Aldrich, culture media from Life Technologies (Monza, Italy), and laboratory plastic was from Sarstedt Inc. (Verona, Italy) unless otherwise stated.

2.2. Fabrication of GM18-Incorporated PLLA Scaffolds

Scaffolds were fabricated using a homemade electrospinning apparatus, consisting of a high-voltage power supply (Spellman SL 50 P 10/CE/230), a syringe pump (KD Scientific 200 series), a glass syringe containing the polymer solution and connected to a stainless steel blunt-ended needle (inner diameter = 0.51 mm) through a PTFE tube. Electrospinning was performed at room temperature (RT) and with a relative humidity of 50−60%. Blends of the β-lactam compound and the polymer were prepared by dissolving the two components in a mixed solvent of DCM:DMF = 65:35 *v*/*v* at a polymer concentration of 13% *w*/*v* and a concentration of β-lactam of 0, 5, 10, and 15 wt% with respect to the polymer. The polymeric solutions were electrospun by applying the following processing conditions: applied voltage = 22 kV, feed rate = 1 mL/h, needle-to-collector distance = 15 cm. The scaffolds were dried on P_2O_5 under vacuum for three days to remove any solvent residue, sterilized by irradiation with γ rays, and finally stored at 4 °C.

2.3. Animals

Intra-abdominal AT was collected from horses during colic surgery (n = 3) and umbilical cord (UC) samples (n = 3) were collected after physiological birth. All owners spontaneously referred the animals to the Department of Veterinary Medical Sciences (University of Bologna) and gave written consent to allow for the use of removed tissue for research purposes. Experimental procedures were approved by the Ethics Committee on animal use of the University of Bologna (Prot. 55948-X/10).

2.4. Sample Collection and Cell Isolation

AT and UC samples were stored in DPBS (Dulbecco's Phosphate Buffered Saline) supplemented with antibiotics (100 IU/mL penicillin, 100 μg/mL streptomycin) immediately after removal, and kept at 4 °C until the transfer to the lab. Under a laminar flow hood, the richest portion of WJ was immediately isolated from the cord tissue. For both tissues, MSCs were isolated as previously described [23]. Briefly, samples were washed by repeated immersion in DPBS, weighed, and cut into 0.5 cm pieces using sterile scissors. Minced tissue was transferred into a 50 mL polypropylene tube and processed by enzymatic digestion using a 0.1% collagenase type I solution in DPBS (1 mL solution/1 g tissue). The suspension was vigorously mixed every 10 min while kept in a 37 °C water bath for an overall 30 min period. Then, collagenase was inactivated by diluting the suspension 1:1 with DPBS supplemented with 10% FBS. The resulting solution was filtered through a stainless-steel mesh to discard the undigested tissue and centrifuged at 470× *g* for 10 min at 25 °C. The cell pellet was suspended in a culture medium consisting of DMEM/MEM 1:1, plus 10% FBS and antibiotics (100 U/mL penicillin, 100 μg/mL streptomycin). Cells were plated as "Passage 0" (P0) in a 25 cm^2 flask containing 5 mL of culture medium and cultured in humidified air with 5% CO_2 at 38.5 °C. After 48 h of in vitro culture, non-adherent cells were removed by completely re-placing the culture medium. Then cell culture medium was changed twice a week until cell growth reached 80 to 90% confluence.

2.5. Cell Freezing and Thawing

In order to perform all tests at the same time for all samples, reducing experimental/technical effects, P0 cells were cryopreserved and stored in liquid nitrogen. AT- and WJ-MSCs were deep-frozen as previously described [24]. Briefly, when cells reached 80 to 90% confluence, they were trypsinized (0.25% trypsin) for 10 min. Then DPBS plus 10% FBS was added 2:1 to inactivate trypsin and the cell suspension was centrifuged at 470× g for 10 min at 25 °C. The pellet was suspended in 0.5 mL of FBS and transferred into a 1.5 mL cryogenic tube kept at 5 to 8 °C for 10 min. Refrigerated cells were diluted 1:1 with FBS plus 16% DMSO (dimethyl sulfoxide) reaching a final concentration of 8% DMSO. The suspension was further kept at 5 to 8 °C for 10 min, then the cryogenic tube was put in Mr Frosty (Nalgene) at −80 °C for 24 h and finally stored in liquid nitrogen. For thawing, AT- and WJ-MSCs vials were immersed in a water bath at 37 °C, the suspension was transferred into a 50 mL tube and dropwise diluted with 20 mL of culture medium, then centrifuged at 470× g for 10 min at 25 °C. The pellet was suspended in 1 mL of culture medium and cell concentration was evaluated by using a hemocytometer. Cells were plated in a 25 cm^2 flask (5000 cells/cm^2) as "Passage 1" (P1). Cells were allowed to proliferate until 80 to 90% confluence before trypsinization and successive passage.

2.6. Growth Curve

At Passage 3 (P3) of in vitro culture, the effect of GM18 on the growth capacity of AT- and WJ-MSCs was evaluated with a growth curve. After trypsinization, nucleated cells were centrifuged at 470× g for 10 min at 25 °C and the pellet was suspended in 1 mL of culture medium. Cell concentration was evaluated by using a hemocytometer and 5000 cells/cm^2 were plated in 35 mm Petri dishes and cultured for 6 days in presence of different concentrations (0 µg/mL, 5 µg/mL, 10 µg/mL, 20 µg/mL) of GM18 solubilized with DMSO. Cells were trypsinized every 24 h and the concentration was evaluated by using a hemocytometer. The same procedure was repeated up to 120 h. Cell doubling number (CD) was calculated according to the following formula:

$$CD = \ln (Nf/Ni)/\ln (2)$$

where Nf and Ni are the final and the initial number of cells, respectively.

The experiment was done for all donors using three replicate wells for each treatment at each time point.

2.7. Adhesion Assay with GM18 Coating

GM18 was solubilized with DMSO and then different concentrations (0 µg/mL, 5 µg/mL, 10 µg/mL, 20 µg/mL) of GM18 were prepared with PBS. GM18 dilutions were placed into a 24-multiwell plate and incubated under a laminar flow hood to form the coating. After removing the supernatant, AT- and WJ-MSCs were added to the wells (10,000 cells/well) and incubated at 37 °C for 20 min, 2 h, 4 h, and 6 h. At the end of this time, the AT- and WJ-MSCs were fixed with 4% paraformaldehyde (PAF 4%) and allowed to incubate for 30 min at RT. MSCs were then stained with a nuclear dye (Hoechst 33,258) and the 24-multiwell plates were analyzed using a cell-based high-content screening (HCS). The experiment was done for all donors using three replicate wells for each treatment at each time point.

2.8. Adhesion Assay with Soluble GM18 in Pre-Treatment

GM18 was solubilized and diluted as described above. In order to investigate the influence of soluble GM18 on the subsequent cell adhesion ability, a 24 h pre-treatment was considered suitable to stimulate a cell response. AT and WJ-MSCs were initially seeded in 35 mm Petri dishes (5000 cells/cm^2) and cultured for 24 h to allow for standard cell adhesion on plastic. Then, different concentrations of GM18 (0 μg/mL, 5 μg/mL, 10 μg/mL, and 20 μg/mL) were added to the culture medium. After 24 h of in vitro culture in presence of soluble GM18, MSCs were detached and passed into a 24-multiwell plate (10,000 cells/well) and incubated at 38.5 °C for 20 min, 2 h, 4 h, and 6 h. At the end of the incubation period, the AT- and WJ-MSCs were fixed with PAF 4% at RT for 30 min. MSCs were stained with a nuclear dye (Hoechst 33,258) and the multiwell plates were analyzed using a HCS. The experiment was done for all donors using three replicate wells for each treatment at each time point.

2.9. *Adhesion Assay on Scaffolds*

Four different scaffolds were used for this assay: PLLA without GM18; PLLA + GM18 5%; PLLA + GM18 10%; PLLA + GM18 15%. The sterile scaffolds were cut and included inside rings that allowed them to be locked inside a 24-multiwell plate. Circular cover glasses were used as control supports for cell detection analysis. Scaffolds were pre-wetted by adding 0.5 mL of 30% ethanol in DPBS for 2 s, then washed twice with 0.5 mL of DPBS for 5 min. AT- and WJ-MSCs were plated on the mounted scaffolds and incubated in humidified air with 5% CO_2 at 38.5 °C for 20 min and 24 h. At the end of the incubation, the MSCs were fixed with PAF 4% for 30 min at RT and stained with a nuclear dye (Hoechst 33,258) and the 24-multiwell plates were analyzed using HCS. The experiment was done using three replicate wells for each treatment at each time point. Only samples that showed downregulation of integrin genes after 24 h exposure to GM18 (CV 6-20 and CV 6-15) were chosen for these experiments.

2.10. *Cell-Based High Content Screening Analysis*

For all the adhesion assays, the HCS technology (Cell Insight NXT, Thermo Scientific, Waltham, MA, USA) was used to detect and count all the cells present in each analyzed well. Analysis was performed by selecting the general intensity measurement assay (Compartmental analysis) from the software algorithm list (HCS Studio v. 6.6.0, Thermo Scientific). Using nuclear staining, cells were detected and counted in each well of the whole cultures.

2.11. *Cell Viability Assay*

A viability test was performed to test the effect of DMSO, used to solubilize GM18, on equine MSCs. Cells were cultured for 24 h in standard medium, then DMSO was added at the 3 concentrations used for GM18 dilutions. After 24 h of culture, cell viability was evaluated by staining cells with 0.05% Eosin.

2.12. *Molecular Characterization*

To analyze the effects of GM18 on gene expression, MSCs were cultured in vitro for 24 h with a culture medium supplemented with GM18 (0 μg/mL, 5 μg/mL, 10 μg/mL, 20 μg/mL) and then lysed.

RNA Extraction, Reverse Transcription, and qPCR

The RNeasy Micro Kit (QIAGEN, Milan, Italy) was used for the total RNA extraction, then quantified with Nanodrop 2000 spectrophotometer (Thermo Scientific). First-strand cDNA was produced using the iScript gDNA Clear cDNA Synthesis Kit (BioRad, Hercules, CA, USA). An RNA sample with no reverse transcriptase enzymes in the reaction mix was processed as a no-reverse transcription control sample. Semi-quantitative real-time PCR reactions were performed in a final volume of 20 μL (1× SYBR Green qPCR master mix—BioRad—and 0.5 μM forward and reverse primers), using the CFX96 real-time PCR system (BioRad). To test possible genomic DNA contaminations, the no-reverse transcriptase sample (NoRT) was processed in parallel with the others.

Details of the primer sequences are included in Table 1.

Table 1. Primer sequences for real-time PCR.

Gene	Accession Number	Primer Sequence
GAPDH	NM_001163856	F: 5′-GAT GCC CCA ATG TTT GTG A-3′ R: 5′-AAG CAG GCA TGA TGT TCT GG-3′
ITGA4	XM_023622141	F: 5′-CAG ATG CCG GAT CGG AAA GA-3′ R: 5′-GCC CAC AAG TCA CGA TGG AT-3′
ITGB1	XM_023631884	F: 5′-CCA AAT GGG ACA CGC AAG AA-3′ R: 5′-GCA CAG CGA GTG CTC ATT TT-3′

PCR reactions were performed using the following steps and thermal profile: denaturation step (98 °C for 3 min) and amplification (95 °C for 10 s and 60 °C for 60 s; 40 cycles), followed by the melting curve of the amplified products (55 °C to 95 °C, ΔT = 0.5 °C/s).

Primer efficiency values for all primers were 95–102%; therefore the $2^{(-\Delta\Delta Ct)}$ method was used to perform the analysis.

2.13. Statistical Analysis

Data are reported as mean ± SEM. Prism software (v.9; GraphPad Software, San Diego, CA, USA) was used for statistical analyses and graph generation. Data were collected from at least three independent experiments. Two-Way ANOVA was used for the population doubling, while One-Way ANOVA was used for all the other analyses. The results were considered significant when the probability of their occurrence as a result of chance alone was <5% ($p < 0.05$).

3. Results

3.1. GM18 Treatment Affects the Adhesion of Adipose Tissue-Derived Mesenchymal Stem Cells

We analyzed the effect of the integrin α4β1 ligand GM18 on cells derived by adult equine AT, isolated from three different animals (CV6-20, CV4-20, and CV24-19). We used the adhesion assay, counting by HCS the cells present in three replicate wells at specific times after the cell seeding, from two different conditions: using GM18 as a coating (Figure 1A,C,E), or pre-treating the cells for 24 h with GM18, and after detaching and seeding back on 24 multiwell plates (Figure 1B,D,F).

The GM18 coating only showed a significant increase in adhesion in CV6-20 derived cells cultures after 2 h at all the analyzed doses (One Way ANOVA, $F(3,8) = 23.34$; $p = 0.0003$; followed by Dunnett's post-test; 5 μg/mL, $p = 0.0018$; 10 μg/mL, $p = 0.0060$; 20 μg/mL, $p = 0.0001$) (Figure 1A).

An increase in adhesion for all the three cultures preparation at different time points was detected after GM18 was in solution 24 h and cells were detached and analyzed for their adhesion capacity.

In particular, cells derived from CV6-20 showed an increased adhesion at 2 h (One Way ANOVA, $F(3,8) = 47$; $p < 0.0001$), 4 h (One Way ANOVA, $F(3,8) = 21.40$; $p = 0.0004$), and 6 h (One Way ANOVA, $F(3,8) = 9.245$; $p < 0.0056$) from seeding, when treated with 10 μg/mL (Dunnett's post-test; 2 h, $p < 0.0001$; 4 h, $p = 0.0005$; 6 h, $p = 0.0046$). At 6 h, a 20 μg/mL dose was also effective in increasing the adhesion ($p = 0.0329$) (Figure 1B).

The cultures isolated from CV4-20 respond at the 24 h treatment with an increased adhesion in a short time (20 min; One-Way ANOVA, $F(3,8) = 15.18$; $p = 0.0012$; Dunnett's post-test, 5 μg/mL, $p = 0.0004$; 20 μg/mL, $p = 0.0137$) (Figure 1D).

In addition, the cultures produced from CV24-19 showed an increase in adhesion at 20 min (One-Way ANOVA, $F(3,8) = 7.934$, $p = 0.0088$, 10 μg/mL $p = 0.0250$) but also at the longer time points, at 2 h (One-Way ANOVA, $F(3,8) = 10.14$, $p = 0.0042$) and 4 h (One-Way ANOVA, $F(3,8) = 5.704$, $p = 0.0219$), at the highest dose (Dunnett's post-test; 20 μg/mL; 2 h, $p = 0.0140$; 4 h, $p = 0.0272$) (Figure 1F).

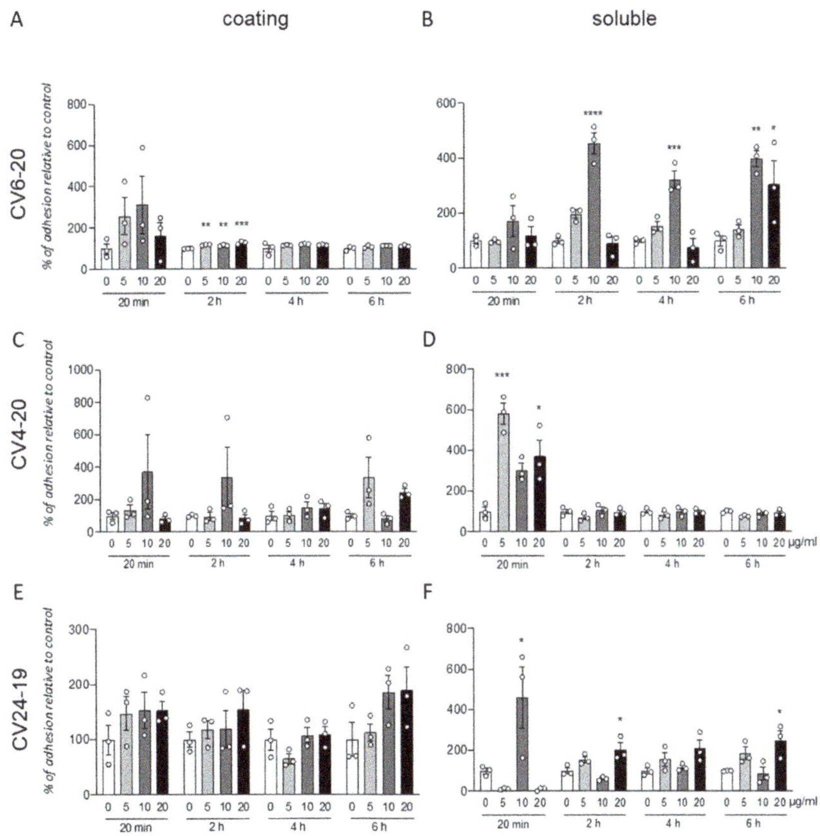

Figure 1. Cell adhesion assay for cells derived from adipose tissues. Graphs show the percentage of cell adhesion relative to control after 20 min, 2, 4, or 6 h from seeding. Cells were seeded on a surface coated with GM18 or vehicle (**A**,**C**,**E**) or exposed for 24 h to the molecule and afterward detached and seeded again (**B**,**D**,**F**). Results are expressed as a percentage of adhesion compared to the control (0 µg/mL; 100%) for each condition and each time point. Each column represents the mean value ± SEM. Statistical analysis. One-Way ANOVA within each condition and each time point, followed by Dunnett's post-test. Asterisks represent the differences compared to the control group (* $p < 0.05$; ** $p < 0.01$; *** $p < 0.001$; **** $p < 0.0001$).

All data were corrected by the effect of the vehicle analyzed by the viability test (data not shown).

Representative panels of representative images from statistically significant results are included in Supplementary Figures S1 and S2.

3.2. *GM18 Treatment Impacts on the Adhesion of Wharton's Jelly Derived Mesenchymal Stem Cells*

The same protocol was used for the analysis of the GM18 effect on cells isolated from WJ (Figure 2). Further, in this case, the coating was not effective in influencing the cell adhesion, even if a single significant increase was described for cells isolated from CV6-15, at 20 min (One-Way ANOVA, $F(3,8) = 5.461$, $p = 0.0245$) and the lowest dose (Dunnett's post-test, 5 µg/mL, $p = 0.0104$) (Figure 2C).

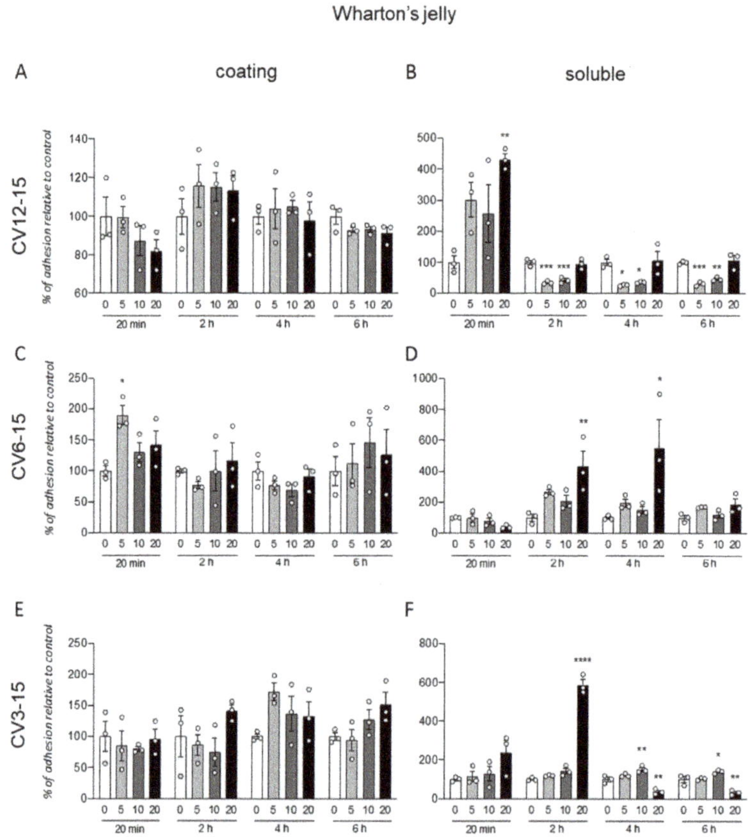

Figure 2. Cell adhesion assay for cells derived from Wharton's jelly. Graphs show the percentage of cell adhesion relative to control after 20 min, 2, 4, or 6 h from seeding. Cells were seeded on a surface coated with GM18 or vehicle (**A**,**C**,**E**) or exposed for 24 h to the molecule and afterward detached and seeded again (**B**,**D**,**F**). Results are expressed as a percentage of adhesion compared to the control (0 µg/mL; 100%) for each condition and each time point. Each column represents the mean value ± SEM is shown. Statistical analysis. One-Way ANOVA within each condition and each time point, followed by Dunnett's post-test. Asterisks represent the differences compared to the control group (* $p < 0.05$; ** $p < 0.01$; *** $p < 0.001$; **** $p < 0.0001$).

For these cells, the effect of the pre-treatment seems to induce both an increase or decrease in adhesion, depending on the origin of cells and the analyzed time.

In fact, for CV12-15 (Figure 2B), with at early time point (20 min, One-Way ANOVA, $F(3,8) = 0.0185$, $p = 6.075$) at highest dose (20 µg/mL, Dunnett's post-test, $p = 0.0074$), the GM18 exposure generated an increase in cell adhesion, while it was decreased at all the other analyzed times (One-Way ANOVA, 2 h, $F(3,8) = 0.0001$; $p = 29.09$; 4 h, $F(3,8) = 7.939$, $p = 0.0088$; 6 h, $F(3,8) = 12.95$, $p = 0.0003$) at 5 µg/mL (Dunnett's post-test, 2 h, $p = 0.0003$; 4 h, $p = 0.0205$; 6 h, $p = 0.0008$) and 10 µg/mL (2 h, $p = 0.0005$; 4 h, $p = 0.0381$; 6 h, $p = 0.0046$).

CV6-15 cells, instead, showed only an increase in adhesion at 2 and 4 h (One-Way ANOVA, 2 h, $F(3,8) = 6.279$, $p = 0.0169$; 4 h, $F(3,8) = 4.745$, $p = 0.0348$) at the highest concentration (Dunnett's post-test, 20 µg/mL, 2 h, $p = 0.0073$; 4 h, $p = 0.0237$) (Figure 2D). For the CV3-15 samples (Figure 2F), the effect favored the adhesion at 2 h with the highest dose (One-Way ANOVA, $F(3,8) = 209.5$, $p < 0.0001$; Dunnett's post-test, 20 µg/mL $p < 0.0001$), and at 4 h (One-Way ANOVA, $F(3,8) = 29.46$, $p = 0.0001$) and 6 h ($F(3,8) = 23.22$, $p = 0.0003$)

at 10 μg/mL (Dunnett's post-test, 4 h, $p = 0.0085$; 6 h, $p = 0.0348$). However, the use of the highest concentration in these cells reduced the adhesion (20 μg/mL, 4 h, $p = 0.0028$; 6 h, $p = 0.0025$).

Representative panels of representative images from statistically significant results are included in Supplementary Figures S3–S5.

3.3. *GM18 Effect Is Highly Variable between Cell Isolation from Different Donors*

Following the analysis of the individual data produced by treatment of cells isolated from different animals, we pooled the data from different isolations. Both for AT (Figure 3A,B) and WT (Figure 3C,D) derived cell, and for both protocols (coating or soluble treatment), the cells exposed to the molecule resulted in a highly variable response.

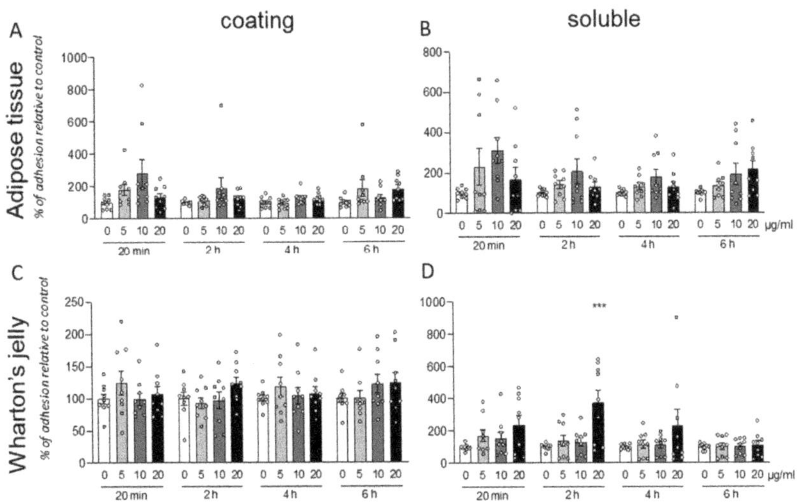

Figure 3. Cumulative analysis for cell adhesion assay. Graphs show the percentage of cell adhesion relative to control after 20 min, 2, 4, or 6 h from seeding. Cells were seeded on a surface coated with GM18 or vehicle (**A**,**C**) or exposed for 24 h to the molecule and afterward detached and seeded again (**B**,**D**). Results are expressed as a percentage of adhesion compared to the control (0 μg/mL; 100%) for each condition and each time point. Each column represents the mean value of all the tested wells for each group, pooling the results obtained from cells isolated from different animals ± SEM is shown. Statistical analysis. One-Way ANOVA within each condition and each time point, followed by Dunnett's post-test. Asterisks represent the differences compared to the control group (*** $p < 0.001$).

Due to this high variability, it is not possible to describe any statistically significant effect of the GM18 exposure. However, the results were significant (One-Way ANOVA, $F(3,32) = 7.721$, $p = 0.0005$) at 20 μg/mL (Dunnett's post-test, $p = 0.0005$) only for soluble treatment in WJ derived cells, analyzed after 2 h from the seeding.

3.4. *GM18 Exposure Deregulates the Gene Expression of the Target Integrins*

Of the three samples analyzed for both AT and WJ, for two of them, exposure to GM18 for 24 h did not result in gene expression regulation of the target integrins (Supplementary Figure S6). However, for cells isolated from CV6-20 (AT) and CV6-15 (WJ), the exposure to the molecule resulted in the deregulation of the target genes. For these two samples, all the concentrations were tested (Figure 4).

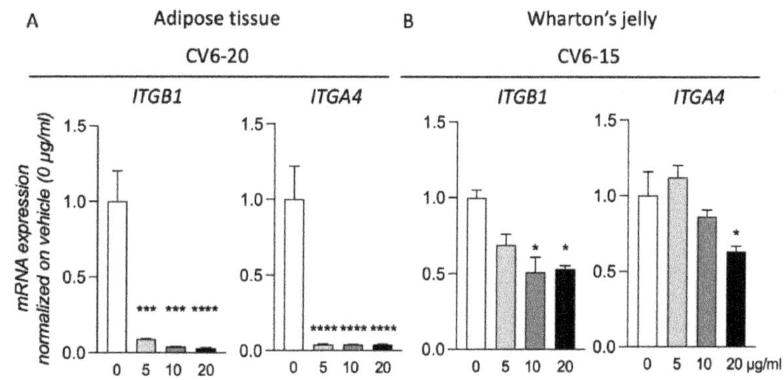

Figure 4. Gene expression regulation of the two target integrins. Graphs show the gene expression regulation of *ITGB1* and *ITGA4* genes in cells derived from CV6-20 adipose tissue (**A**) and CV6-15 Wharton's jelly (**B**), treated for 24 h with different concentrations of GM18 (0, 5, 10, and 20 µg/mL). Columns represent the mean value + SEM. Statistical analysis. One-Way ANOVA followed by Dunnett's post-test. Asterisks represent the differences compared to the control group (* $p < 0.05$; *** $p < 0.001$; **** $p < 0.0001$).

MSCs of the CV6-20 showed a reduction in the expression of *ITGB1* (One-Way ANOVA, $F(3,8) = 39.24$, $p < 0.0001$) and *ITGA4* (One-Way ANOVA, $F(3,8) = 206.7$, $p < 0.0001$) genes at all the concentrations tested (Dunnett's post-test, Itgb1, 5 µg/mL, $p = 0.0006$; 10 µg/mL, $p = 0.0002$; 20 µg/mL, $p < 0.0001$; Itga4, 5 µg/mL, $p < 0.0001$; 10 µg/mL, $p < 0.0001$; 20 µg/mL, $p < 0.0001$).

For the cells isolated from WJ of CV6-15, the treatment generated a reduction in the expression of both integrin subunits (One-Way ANOVA, *ITGB1*, $F(3,8) = 6.017$, $p = 0.0237$; *ITGA4*, $F(3,8) = 12.11$, $p = 0.0037$). However not all doses were effective. For *ITGB1* expression, the reduction was caused by the treatment with a concentration of 10 µg/mL (Dunnett's post-test, $p = 0.0155$) and 20 µg/mL ($p = 0.0202$), while for *ITGA4*, only the highest dose was effective ($p = 0.0106$).

3.5. GM18 Exposure Affects the Population Doubling

To investigate if the effect on the cell adhesion and gene expression may affect the cell division, we analyzed the population doubling of each cell preparation (Figure 5). For the AT, cells derived from CV6-20 and CV24-19 showed an effect related to drug concentration (Two-Way ANOVA, CV6-20, $F(3,8) = 5.042$, $p = 0.0299$; CV24-19, $F(3,8) = 5.747$, $p = 0.0214$). For WJ derived cells, all samples were significantly affected by the presence of the GM18 in the culture medium (Two-Way ANOVA, CV6-15, $F(3,8) = 4.068$, $p = 0.0500$; CV3-15, GM18 concentration, $F(3,8) = 17.53$, $p = 0.0007$; CV12-15, $F(3,8) = 12.42$, $p = 0.0022$).

Moreover, an interaction effect between time and concentration was reported for the same cell preparations which resulted more affected by GM18: one sample of the AT (CV6-20; $F(15,14) = 2.314$, $p = 0.0175$) and one sample of the WJ (CV6-15; $F(15,14) = 1.939$, $p = 0.0481$).

The Dunnett's post-test revealed differences at single time points for AT in the sample CV24-19 (day 4, 5 µg/mL, $p = 0.0203$; day 5, 5 µg/mL, $p = 0.0145$) and for WJ in the sample CV3-15 (day 2, 5 µg/mL, $p = 0.0086$; 10 µg/mL, $p = 0.0298$).

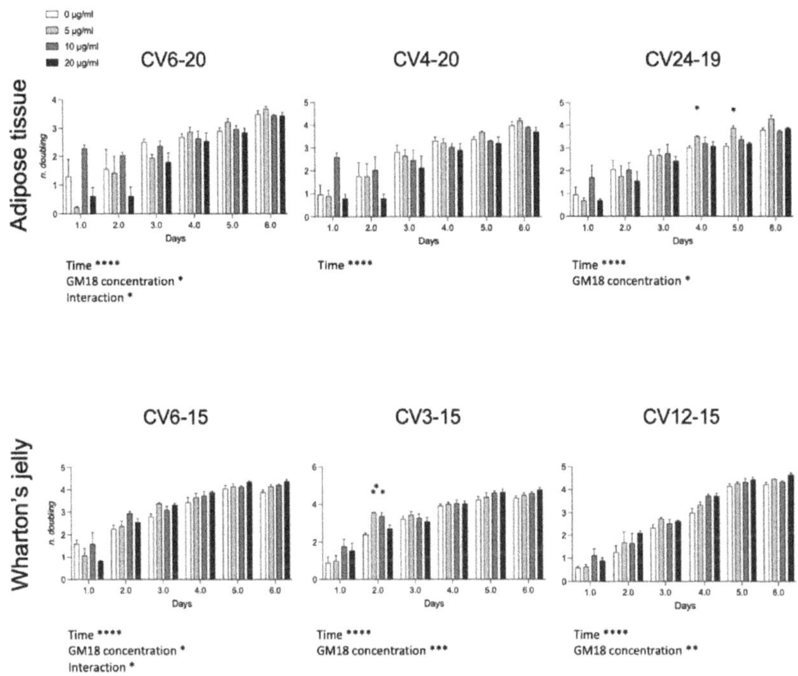

Figure 5. Effect of GM18 exposure on population doubling. Graphs show the number of doublings measured each day for 6 consecutive days, for cells isolated from adipose tissue and Wharton's jelly exposed to GM18 at different concentrations (0, 5, 10, and 20 µg/mL). Statistical analysis. Two-Way ANOVA followed by Dunnett's post-test. Asterisks represent the differences in the ANOVA parameters as indicated (* $p < 0.05$; ** $p < 0.01$; *** $p < 0.001$; **** $p < 0.0001$).

Analyzing data from AT and WJ pooled samples (Figure 6), the statistical analysis revealed a strong effect of time, GM18 concentration, and the interaction between the two variables in AT-MSCs (Two-way ANOVA, time, $F(2.480,79.36) = 182.3$, $p < 0.0001$; GM18 concentration, $F(3,32) = 6.523$, $p = 0.0014$; interaction, $F(15,160) = 5.278$, $p < 0.0001$), while only the effect of time and concentration for WJ-MSCs (Two-way ANOVA, time, $F(2.884,92.29) = 346.2$, $p < 0.0001$; GM18 concentration, $F(3,32) = 3.418$, $p = 0.0289$). For AT derived MSCs there are time point differences at day 1 (Dunnett's post-test, 10 µg/mL, $p = 0.0082$) and day 5 (5 µg/mL, $p = 0.0112$), while for WJ derived MSCs, there were differences only at longer time points, namely at day 5 (20 µg/mL, $p = 0.0158$) and day 6 (20 µg/mL, $p = 0.0043$).

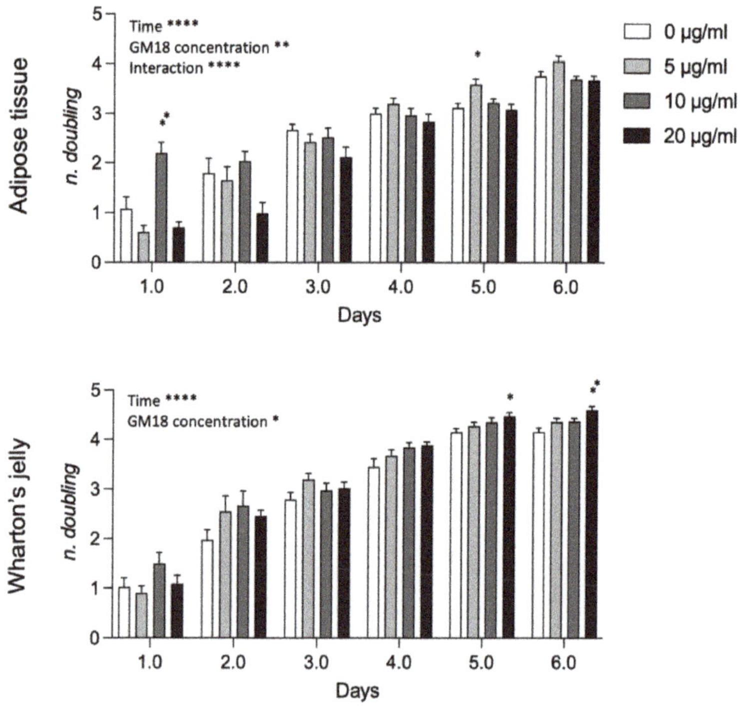

Figure 6. Effect of GM18 exposure on population doubling. Graphs show the number of doublings measured each day for six consecutive days, for cells isolated from adipose tissue and Wharton's jelly exposed to GM18 at different concentrations (0, 5, 10, and 20 µg/mL). Statistical analysis. Two-Way ANOVA followed by Dunnett's post-test. Asterisks represent the differences in the ANOVA parameters as indicated (* $p < 0.05$; ** $p < 0.01$; **** $p < 0.0001$).

3.6. PLLA Scaffolds Containing GM18

The two types of cells, AT cells and WJ derived cells, which were more sensitive to GM18 treatments, both as adhesion assay and gene expression regulation (CV6-20 for AT and CV6-15 for WJ), were used to investigate the effect of a PLLA scaffold containing different concentrations of GM18 on the cell adhesion (Figure 7). Cultures seeded on scaffolds containing GM18 were compared to cells cultured on PLLA control scaffolds. The method was also tested on standard cultures seeded on glass (Supplementary Figure S7).

For cells derived from AT (CV6-20), the presence of GM18 did not affect the adhesion capacity, mainly due to a high variability detected at a concentration of 10% (Figure 7A). Moreover, analyzing the differences in the cell number between the two analyzed time points, the cultures seeded on the PLLA control scaffold were not able to increase the cell number after 24 h, reflecting an impairment in the early proliferation (Student's t-test = 0.0641). However, the presence of GM18 in all the analyzed concentrations was able to induce an increase in cell number after one DIV (Student's t-test, 5%, $p = 0.0170$; 10%, $p = 0.0144$; 15%, $p = 0.0395$) (Figure 7A).

For the cultures obtained from WJ (CV6-15), the cell adhesion was affected after 24 h (One-way ANOVA, $F(3,8) = 4.671$; $p = 0.0427$), with a drastic increase due to the highest concentration of GM18 (Dunnett's posttest, $p = 0.0368$). The same concentration was also the only condition producing an increase in cell number after one DIV (Student's t-test, $p = 0.0021$) (Figure 7B).

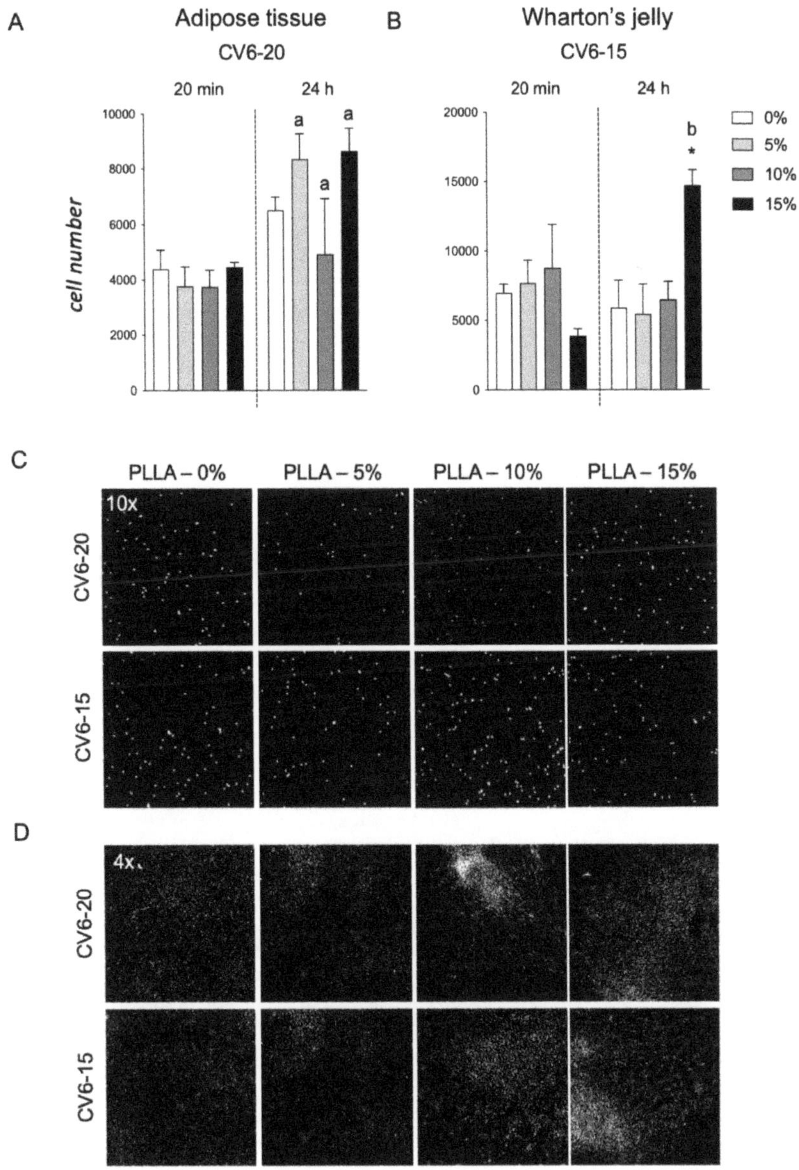

Figure 7. Cell adhesion assay for cells seeded on PLLA-GM18 scaffolds. Graphs show the number of counted cells per well in 20 min and 24 h after seeding the cells isolated from CV6-20 adipose tissue (**A**) and CV6-15 Wharton's jelly (**B**) on PLLA scaffolds containing different concentrations of GM18 (0, 5, 10, and 15%). Representative images of Hoechst staining are included in the figure, showing the cultures at 20 min acquired with a 10× objective (**C**) and 24 h acquired a with a 4× objective (**D**) from the seeding. Statistical analysis. One-Way ANOVA followed by Dunnett's post-test within the same time point; asterisk represents the differences compared to the control group seeded on PLLA 0% GM18 (* $p < 0.05$). Student's t-test within the same experimental group between the two time points, letters represent the statistically significant differences (a = $p < 0.05$; b = $p < 0.01$).

Representative images of the cultures are included in Figure 7C,D, showing the morphology of the cultures with different magnifications (10× or 4×).

To validate the analysis, control cells cultured on glass coverslips at the same time points were included (Supplementary Figure S7A,B), also proving the capacity of the HCS software to visualize and detect the nuclei, at low magnification objective, with a correct segmentation algorithm identifying the single nuclei (Supplementary Figure S7C).

4. Discussion

The number of investigations into the interaction of MSCs with integrin α4β1 ligands is still limited. The agonist used in this pilot study, GM18, was designed with an amine, a carboxylate side chain, and the β-lactam ring as a site of conformational restriction, to assure the necessary integrin affinity and selectivity by improving the alignment on the receptor [18]. Among different β-lactams, GM18 was the best at enhancing the adhesion of human bone marrow-derived MSCs to PLLA electrospun scaffolds [22]. In another research, a high-affinity integrin α4β1 ligand was successfully tested on human chorionic villus-derived MSCs [25]. The adherence of cells to polymeric electrospun scaffolds increased when the surface was treated with the ligand [25].

Here we present the first preliminary results about the effects of an integrin α4β1 ligand on equine MSCs. GM18's influence on the adhesion ability of both adult (AT) and fetal membrane (WJ) derived MSCs was studied when used as a coating or as a pre-treatment before standard seeding. Furthermore, the potential use of GM18 to increase cell adhesion to a PLLA scaffold was investigated from the perspective of its application for the treatment of decubitus ulcers. Significant results were observed, despite the limitations of a pilot study and the related critical issues, emerging from the donor variability in a relatively low number of samples. Nonetheless, these findings represent a fundamental phase that seems to validate the feasibility of similar studies at a larger scale.

In the horse, a few studies about α4β1 integrin are available [26–32], and are related to its presence on leukocytes [26,29] and endometrial epithelial cells [27,28] or equine herpesvirus 1 infection mechanism [30–32]. Integrins represent one of the most important families of cell adhesion receptors that mediate cell–cell and cell–ECM interactions. α4β1 integrin (also known as CD49d/CD29 or very late antigen-4, VLA-4) plays a crucial role in inflammatory diseases, cancer development, metastasis, and stem cell mobilization or retention [33]. It is involved in the regulation of hematopoietic stem cell homing and retention within the bone marrow niche [34] and its activation may be a promising strategy to improve cell retention and engraftment in stem cell-based therapies [35].

There is a wide range of MSCs, which can be isolated from different tissues. So far, donor-to-donor variations can be present even when MSCs are derived from the same tissue [36,37]. This is confirmed in the present study, where the effect of GM18 on both equine AT- and WJ-derived MSCs was highly variable in relation to the different donors. Even if only three donors for each tissue were used in this pilot study, the number was sufficient to preliminarily corroborate the hypothesis of donor variability and to highlight the critical issues of testing molecules and scaffolds with biological materials, such as MSCs. When observing the pooled data from AT-MSCs samples, it was not possible to detect a significant effect of GM18 on the cell adhesion ability. On the other hand, when considering results from each donor, it was evident that the GM18 coating increased adhesion only in one sample (CV6-20) after 2 h, while soluble GM18 was able to improve adhesion in the other samples. For WJ-MSCs, a positive effect of soluble GM18 at 20 μg/mL after 2 h was also observed in pooled data. As for AT-MSCs, the GM18 coating increased adhesion only in one WJ-MSCs sample (CV6-15), while soluble GM18 increased cell adhesion for all donors. However, differently from AT-MSCs, a decrease in cell adhesion was also observed in the other two samples at different times and GM18 concentrations. Overall, these preliminary results suggest that GM18 affects MSCs' adhesion ability also in the horse and that a donor-based and a tissue-based response should be considered when selecting samples for cell-based therapies.

Gene expression analysis for target integrins confirmed what was observed by the adhesion assays. In fact, only in the samples that were more influenced in vitro by the presence of GM18 was it possible to observe a reduction in the expression of both integrin subunits. While AT-MSCs were sensible to all agonist concentrations, WJ-MSCs appeared to require a higher GM18 dosage to downregulate α4β1 integrin genes.

The population doublings of cells were also affected by the presence of GM18 over time. On the whole, a higher number of doublings was observed for AT-MSCs on day 1, with 10 μg/mL of GM18, and on day 5, with 5 μg/mL of GM18, while for WJ-MSCs, the number of doublings was higher on day 5 and 6 with the highest concentration of agonist (20 μg/mL of GM18), confirming adhesion and gene expression findings. Integrins have been identified as regulators of mitotic events [38]. It has been demonstrated that the β1 integrin is involved in the orientation of the mitotic spindle and adhesion in basal epidermal cells [39] and HeLa cells [40]. Downregulation of the β1 integrin resulted in severe misorientation of spindles [40]. In nonpolarized adherent cells, the β1 integrin-dependent mechanism for spindle orientation may ensure the attachment of both daughter cells to the substratum after cell division and prevent their mitosis-associated detachment [40]. These findings might explain the behavior of equine MSCs growth curves in presence of an integrin α4β1 ligand, especially for those cells where the β1 subunit gene expression was more deregulated.

PLLA scaffolds containing different concentrations of GM18 were seeded with the same two samples that were more sensitive to GM18 treatments and cell adhesion was evaluated. It was evident that equine AT-MSCs' adhesion ability to PLLA scaffold was not influenced by the presence of GM18. However, the presence of GM18 increased cell adhesion after 24 h of culture. WJ-MSCs seem more suitable for PLLA scaffold adhesion, and the presence of 15% GM18 increased cell adhesion at 24 h. As the results are derived from only one suitable sample for each tissue, these preliminary results need to be confirmed before drawing definite conclusions. Research on equine MSCs and PLLA scaffolds is aimed at osteogenic differentiation. It has been observed that the osteogenic differentiation capacity of equine adipose tissue-derived MSCs on a nano-bioactive glass-coated PLLA nanofibers scaffold was higher than on the uncoated PLLA scaffold [41]. In another study using equine MSCs from adipose tissue and bone marrow, the addition of minerals to polymer scaffolds enhanced equine MSC osteogenesis over polymer alone [42]. Similarly, coating PLLA scaffolds with zinc silicate mineral nanoparticles improved in vitro osteogenic differentiation of equine adipose tissue-derived MSCs compared to uncoated PLLA scaffold [43]. The present work is the first report of equine WJ-MSCs cultivation on PLLA scaffolds. Human WJ-derived MSCs were seeded into polyglycolic acid (PGA) and PLLA scaffolds to investigate the potential ability of chondrogenesis in vitro [44]. The adhesion rate of human WJ-MSCs on PLLA scaffolds was 58 ± 6% and 75 ± 4% 3 and 6 h after seeding, respectively [44]. The same GM18 coated PLLA scaffolds used in the present study were tested with human bone marrow-derived MSCs and GM18 increased the rate of cell adhesion after 2 h of co-incubation [22]. It might be speculated that for equine AT-MSCs a longer co-incubation is needed than for WJ-MSCs in order to achieve higher adhesion rates. Nevertheless, further studies are necessary to better assess equine AT- and WJ-MSCs' adhesion ability on PLLA scaffolds over time and to test the concurrent effect of the presence of GM18.

5. Conclusions

The α4β1 integrin agonist GM18 affects equine AT- and WJ-derived MSCs' adhesion capability. Soluble GM18 pre-treatment was more effective in enhancing cell adhesion than GM18 coating. However, there was a high variability related to different donors and it should be carefully considered when selecting samples for cell-based therapies. Samples more responsive to GM18 showed a downregulation of the target integrins' genes and an effect on cell divisions. Preliminary results on equine MSCs adhesion to PLLA scaffolds containing GM18 might suggest that WJ-MSCs are more suitable than AT-MSCs for the development of cell-based scaffolds to be used for wound healing, but additional research is required to examine in depth the various cell characteristics after seeding and prolonged culture on these scaffolds.

Supplementary Materials: The following supporting information can be downloaded at: https://www.mdpi.com/article/10.3390/ani12060734/s1, Figure S1: panel for Figure 1A,B, Figure S2: panel for Figure 1C–F, Figure S3: panel for Figure 2B, Figure S4: panel for Figure 2D, Figure S5: panel for Figure 2F, Figure S6: gene expression analysis.

Author Contributions: Conceptualization, B.M., V.A.B. and E.I.; methodology, B.M., V.A.B. and E.I.; formal analysis, V.A.B. and B.M.; investigation, B.M., V.A.B., A.F. and R.M.; resources, V.G., M.L.F., D.G. and E.I.; data curation, B.M.,V.A.B. and A.F.; writing—original draft preparation, B.M., V.A.B., R.M., V.G., M.L.F. and D.G.; writing—review and editing, A.F., V.G., M.L.F., D.G. and E.I.; project administration, B.M. and E.I.; funding acquisition, E.I. All authors have read and agreed to the published version of the manuscript.

Funding: This research was funded by Emilia Romagna POR-FESR 2014-20 and FSC, project "Mat2Rep" and by the Fondazione Cassa di Risparmio di Imola, project "Cellule staminali mesenchimali per la terapia rigenerativa cutanea: l'importanza della medicina veterinaria".

Institutional Review Board Statement: The study was approved by the Ethics Committee on animal use of the University of Bologna (Prot. 55948-X/10).

Informed Consent Statement: Not applicable.

Data Availability Statement: The data presented in this study are available within the article and supplementary materials.

Acknowledgments: The authors wish to thank (1) Laura Calzà and Luciana Giardino, IRET Foundation, responsible for the "Mat2Rep" project and funding; (2) Carolina Castagnetti and the staff of the Equine Clinical Service, DIMEVET, University of Bologna, for the equine tissue samples recovery.

Conflicts of Interest: The authors declare no conflict of interest.

References

1. Kashpur, O.; Smith, A.; Gerami-Naini, B.; Maione, A.G.; Calabrese, R.; Tellechea, A.; Theocharidis, G.; Liang, L.; Pastar, I.; Tomic-Canic, M.; et al. Differentiation of Diabetic Foot Ulcer–Derived Induced Pluripotent Stem Cells Reveals Distinct Cellular and Tissue Phenotypes. *FASEB J.* **2019**, *33*, 1262–1277. [CrossRef] [PubMed]
2. Lawrence, W.T. Physiology of the Acute Wound. *Clin. Plast. Surg.* **1998**, *25*, 321–340. [CrossRef]
3. Mogoşanu, G.D.; Grumezescu, A.M. Natural and Synthetic Polymers for Wounds and Burns Dressing. *Int. J. Pharm.* **2014**, *463*, 127–136. [CrossRef] [PubMed]
4. Gurtner, G.C.; Chapman, M.A. Regenerative Medicine: Charting a New Course in Wound Healing. *Adv. Wound Care* **2016**, *5*, 314–328. [CrossRef] [PubMed]
5. Gorecka, J.; Kostiuk, V.; Fereydooni, A.; Gonzalez, L.; Luo, J.; Dash, B.; Isaji, T.; Ono, S.; Liu, S.; Lee, S.R.; et al. The Potential and Limitations of Induced Pluripotent Stem Cells to Achieve Wound Healing. *Stem Cell Res. Ther.* **2019**, *10*, 87. [CrossRef] [PubMed]
6. Fu, X.; Li, H. Mesenchymal Stem Cells and Skin Wound Repair and Regeneration: Possibilities and Questions. *Cell Tissue Res.* **2008**, *335*, 317. [CrossRef]
7. Lopes, B.; Sousa, P.; Alvites, R.; Branquinho, M.; Sousa, A.; Mendonça, C.; Atayde, L.M.; Maurício, A.C. The Application of Mesenchymal Stem Cells on Wound Repair and Regeneration. *Appl. Sci.* **2021**, *11*, 3000. [CrossRef]
8. Iacono, E.; Merlo, B.; Pirrone, A.; Antonelli, C.; Brunori, L.; Romagnoli, N.; Castagnetti, C. Effects of Mesenchymal Stem Cells Isolated from Amniotic Fluid and Platelet-Rich Plasma Gel on Severe Decubitus Ulcers in a Septic Neonatal Foal. *Res. Vet. Sci.* **2012**, *93*, 1439–1440. [CrossRef] [PubMed]
9. Volk, S.W.; Theoret, C. Translating Stem Cell Therapies: The Role of Companion Animals in Regenerative Medicine. *Wound Repair Regen.* **2013**, *21*, 382–394. [CrossRef]
10. Lanci, A.; Merlo, B.; Mariella, J.; Castagnetti, C.; Iacono, E. Heterologous Wharton's Jelly Derived Mesenchymal Stem Cells Application on a Large Chronic Skin Wound in a 6-Month-Old Filly. *Front. Vet. Sci.* **2019**, *6*, 9. [CrossRef]
11. Ji, X.; Yuan, X.; Ma, L.; Bi, B.; Zhu, H.; Lei, Z.; Liu, W.; Pu, H.; Jiang, J.; Jiang, X.; et al. Mesenchymal Stem Cell-Loaded Thermosensitive Hydroxypropyl Chitin Hydrogel Combined with a Three-Dimensional-Printed Poly(ε-Caprolactone)/Nano-Hydroxyapatite Scaffold to Repair Bone Defects via Osteogenesis, Angiogenesis and Immunomodulation. *Theranostics* **2020**, *10*, 725–740. [CrossRef] [PubMed]
12. Tayalia, P.; Mooney, D.J. Controlled Growth Factor Delivery for Tissue Engineering. *Adv. Mater.* **2009**, *21*, 3269–3285. [CrossRef] [PubMed]
13. Marklein, R.A.; Burdick, J.A. Controlling Stem Cell Fate with Material Design. *Adv. Mater.* **2010**, *22*, 175–189. [CrossRef] [PubMed]

14. Isomursu, A.; Lerche, M.; Taskinen, M.E.; Ivaska, J.; Peuhu, E. Integrin Signaling and Mechanotransduction in Regulation of Somatic Stem Cells. *Exp. Cell Res.* **2019**, *378*, 217–225. [CrossRef] [PubMed]
15. Campbell, I.D.; Humphries, M.J. Integrin Structure, Activation, and Interactions. *Cold Spring Harb. Perspect. Biol.* **2011**, *3*, a004994. [CrossRef] [PubMed]
16. Prowse, A.B.J.; Chong, F.; Gray, P.P.; Munro, T.P. Stem Cell Integrins: Implications for Ex-Vivo Culture and Cellular Therapies. *Stem Cell Res.* **2011**, *6*, 1–12. [CrossRef] [PubMed]
17. Hynes, R.O. Integrins: Bidirectional, Allosteric Signaling Machines. *Cell* **2002**, *110*, 673–687. [CrossRef]
18. Baiula, M.; Galletti, P.; Martelli, G.; Soldati, R.; Belvisi, L.; Civera, M.; Dattoli, S.D.; Spampinato, S.M.; Giacomini, D. New β-Lactam Derivatives Modulate Cell Adhesion and Signaling Mediated by RGD-Binding and Leukocyte Integrins. *J. Med. Chem.* **2016**, *59*, 9721–9742. [CrossRef] [PubMed]
19. Alcaide, B.; Almendros, P.; Aragoncillo, C. Highly Reactive 4-Membered Ring Nitrogen-Containing Heterocycles: Synthesis and Properties. *Curr. Opin. Drug Discov. Devel.* **2010**, *13*, 685–697. [PubMed]
20. Greiner, A.; Wendorff, J.H. Electrospinning: A Fascinating Method for the Preparation of Ultrathin Fibers. *Angew. Chem. Int. Ed.* **2007**, *46*, 5670–5703. [CrossRef]
21. Wang, C.; Wang, J.; Zeng, L.; Qiao, Z.; Liu, X.; Liu, H.; Zhang, J.; Ding, J. Fabrication of Electrospun Polymer Nanofibers with Diverse Morphologies. *Molecules* **2019**, *24*, 834. [CrossRef] [PubMed]
22. Martelli, G.; Bloise, N.; Merlettini, A.; Bruni, G.; Visai, L.; Focarete, M.L.; Giacomini, D. Combining Biologically Active β-Lactams Integrin Agonists with Poly(l-Lactic Acid) Nanofibers: Enhancement of Human Mesenchymal Stem Cell Adhesion. *Biomacromolecules* **2020**, *21*, 1157–1170. [CrossRef] [PubMed]
23. Iacono, E.; Pascucci, L.; Bazzucchi, C.; Cunto, M.; Ricci, F.; Rossi, B.; Merlo, B. Could Hypoxia Influence Basic Biological Properties and Ultrastructural Features of Adult Canine Mesenchymal Stem /Stromal Cells? *Vet. Res. Commun.* **2018**, *42*, 297–308. [CrossRef] [PubMed]
24. Merlo, B.; Pirondi, S.; Iacono, E.; Rossi, B.; Ricci, F.; Mari, G. Viability, in vitro differentiation and molecular characterization of equine adipose tissue-derived mesenchymal stem cells cryopreserved in serum and serum-free medium. *Cryo Lett.* **2016**, *37*, 243–252.
25. Hao, D.; Ma, B.; He, C.; Liu, R.; Farmer, D.L.; Lam, K.S.; Wang, A. Surface Modification of Polymeric Electrospun Scaffolds via a Potent and High-Affinity Integrin A4β1 Ligand Improved the Adhesion, Spreading and Survival of Human Chorionic Villus-Derived Mesenchymal Stem Cells: A New Insight for Fetal Tissue Engineering. *J. Mater. Chem. B* **2020**, *8*, 1649–1659. [CrossRef]
26. Foster, A.P.; McCabe, P.J.; Sanjar, S.; Cunningham, F.M. Agonist-Induced Adherence of Equine Eosinophils to Fibronectin. *Vet. Immunol. Immunopathol.* **1997**, *56*, 205–220. [CrossRef]
27. Al-Ramadan, S.Y.; Brinsko, P.; Rigby, S.L.; Jaegera, L.A.; Burghardt, R.C. Analysis of MUC-1 and Integrin Subunit Expression in Equine Uterine Epithelium and Trophectoderm. *Theriogenology* **2002**, *58*, 829–832. [CrossRef]
28. Grant, D.M.; Macedo, A.; Toms, D.; Klein, C. Fibrinogen in Equine Pregnancy as a Mediator of Cell Adhesion, an Epigenetic and Functional Investigation. *Biol. Reprod.* **2020**, *102*, 170–184. [CrossRef]
29. Treonze, K.M.; Alves, K.; Fischer, P.; Hagmann, W.K.; Hora, D.; Kulick, A.; Vakerich, K.; Smith, N.D.; Lingham, R.B.; Maniar, S.; et al. Characterization of A4β1 (CD49d/CD29) on Equine Leukocytes: Potential Utility of a Potent A4β1 (CD49d/CD29) Receptor Antagonist in the Treatment of Equine Heaves (Recurrent Airway Obstruction). *Vet. Immunol. Immunopathol.* **2009**, *130*, 79–87. [CrossRef]
30. Claessen, C.; Favoreel, H.; Ma, G.; Osterrieder, N.; De Schauwer, C.; Piepers, S.; Van de Walle, G.R. Equid Herpesvirus 1 (EHV1) Infection of Equine Mesenchymal Stem Cells Induces a PUL56-Dependent Downregulation of Select Cell Surface Markers. *Vet. Microbiol.* **2015**, *176*, 32–39. [CrossRef]
31. Laval, K.; Favoreel, H.W.; Poelaert, K.C.K.; Van Cleemput, J.; Nauwynck, H.J. Equine Herpesvirus Type 1 Enhances Viral Replication in CD172a+ Monocytic Cells upon Adhesion to Endothelial Cells. *J. Virol.* **2015**, *89*, 10912–10923. [CrossRef] [PubMed]
32. Kolyvushko, O.; Kelch, M.A.; Osterrieder, N.; Azab, W. Equine Alphaherpesviruses Require Activation of the Small GTPases Rac1 and Cdc42 for Intracellular Transport. *Microorganisms* **2020**, *8*, 1013. [CrossRef] [PubMed]
33. Baiula, M.; Spampinato, S.; Gentilucci, L.; Tolomelli, A. Novel Ligands Targeting A4β1 Integrin: Therapeutic Applications and Perspectives. *Front. Chem.* **2019**, *7*, 489. [CrossRef] [PubMed]
34. Grassinger, J.; Haylock, D.N.; Storan, M.J.; Haines, G.O.; Williams, B.; Whitty, G.A.; Vinson, A.R.; Be, C.L.; Li, S.; Sørensen, E.S.; et al. Thrombin-Cleaved Osteopontin Regulates Hemopoietic Stem and Progenitor Cell Functions through Interactions with A9β1 and A4β1 Integrins. *Blood* **2009**, *114*, 49–59. [CrossRef] [PubMed]
35. Vanderslice, P.; Biediger, R.J.; Woodside, D.G.; Brown, W.S.; Khounlo, S.; Warier, N.D.; Gundlach, C.W.; Caivano, A.R.; Bornmann, W.G.; Maxwell, D.S.; et al. Small Molecule Agonist of Very Late Antigen-4 (VLA-4) Integrin Induces Progenitor Cell Adhesion*. *J. Biol. Chem.* **2013**, *288*, 19414–19428. [CrossRef] [PubMed]
36. Colleoni, S.; Bottani, E.; Tessaro, I.; Mari, G.; Merlo, B.; Romagnoli, N.; Spadari, A.; Galli, C.; Lazzari, G. Isolation, Growth and Differentiation of Equine Mesenchymal Stem Cells: Effect of Donor, Source, Amount of Tissue and Supplementation with Basic Fibroblast Growth Factor. *Vet. Res. Commun.* **2009**, *33*, 811. [CrossRef]
37. Bagge, J.; MacLeod, J.N.; Berg, L.C. Cellular Proliferation of Equine Bone Marrow- and Adipose Tissue-Derived Mesenchymal Stem Cells Decline With Increasing Donor Age. *Front. Vet. Sci.* **2020**, *7*, 933. [CrossRef] [PubMed]

38. LaFlamme, S.E.; Nieves, B.; Colello, D.; Reverte, C.G. Integrins as Regulators of the Mitotic Machinery. *Curr. Opin. Cell Biol.* **2008**, *20*, 576–582. [CrossRef]
39. Lechler, T.; Fuchs, E. Asymmetric Cell Divisions Promote Stratification and Differentiation of Mammalian Skin. *Nature* **2005**, *437*, 275–280. [CrossRef]
40. Toyoshima, F.; Nishida, E. Integrin-Mediated Adhesion Orients the Spindle Parallel to the Substratum in an EB1- and Myosin X-Dependent Manner. *EMBO J.* **2007**, *26*, 1487–1498. [CrossRef]
41. Mahdavi, F.S.; Salehi, A.; Seyedjafari, E.; Mohammadi-Sangcheshmeh, A.; Ardeshirylajimi, A. Bioactive Glass Ceramic Nanoparticles-Coated Poly(l-Lactic Acid) Scaffold Improved Osteogenic Differentiation of Adipose Stem Cells in Equine. *Tissue Cell* **2017**, *49*, 565–572. [CrossRef] [PubMed]
42. Duan, W.; Chen, C.; Haque, M.; Hayes, D.; Lopez, M.J. Polymer-Mineral Scaffold Augments in Vivo Equine Multipotent Stromal Cell Osteogenesis. *Stem Cell Res. Ther.* **2018**, *9*, 60. [CrossRef] [PubMed]
43. Bageshlooyafshar, B.; Vakilian, S.; Kehtari, M.; Eslami-Arshaghi, T.; Rafeie, F.; Ramezanifard, R.; Rahchamani, R.; Mohammadi-Sangcheshmeh, A.; Mostafaloo, Y.; Seyedjafari, E. Zinc Silicate Mineral-Coated Scaffold Improved in Vitro Osteogenic Differentiation of Equine Adipose-Derived Mesenchymal Stem Cells. *Res. Vet. Sci.* **2019**, *124*, 444–451. [CrossRef] [PubMed]
44. Zhao, L.; Detamore, M.S. Chondrogenic Differentiation of Stem Cells in Human Umbilical Cord Stroma with PGA and PLLA Scaffolds. *J. Biomed. Sci. Eng.* **2010**, *3*, 1041–1049. [CrossRef] [PubMed]

animals

MDPI

Systematic Review

Efficacy of Stem Cell Therapy in Large Animal Models of Ischemic Cardiomyopathies: A Systematic Review and Meta-Analysis

Debora La Mantia [1], Chiara Bernardini [1,*], Augusta Zannoni [1,2], Roberta Salaroli [1], Changzhen Wang [1], Silvia Bencivenni [1] and Monica Forni [1,2]

[1] Department of Veterinary Medical Sciences, University of Bologna, Ozzano dell'Emilia, 40064 Bologna, Italy; debora.lamantia2@unibo.it (D.L.M.); augusta.zannoni@unibo.it (A.Z.); roberta.salaroli@unibo.it (R.S.); changzhen.wang@studio.unibo.it (C.W.); silvia.bencivenni2@unibo.it (S.B.); monica.forni@unibo.it (M.F.)

[2] Health Sciences and Technologies—Interdepartmental Center for Industrial Research (CIRI-SDV), Alma Mater Studiorum—University of Bologna, 40126 Bologna, Italy

* Correspondence: chiara.bernardini5@unibo.it; Tel.: +39-051-2097914

Simple Summary: The present work focuses on stem-cell assessment as a therapeutic approach on cardiovascular diseases, both in terms of safety and efficacy. In particular, this is a systematic review of the relevant literature about the use of stem-cell treatment against acute or chronic ischemic cardiomyopathies in large animal models and a meta-analysis on collected data with regard to the left ventricular ejection fraction (LVEF) as functional parameter. This approach is compliant with the "3Rs" (replacement, reduction and refinement) principle about the use of animal experimentation in preclinical trials to predict evidences and perform the future translational researches.

Abstract: Stem-cell therapy provides a promising strategy for patients with ischemic heart disease. In recent years, numerous studies related to this therapeutic approach were performed; however, the results were often heterogeneous and contradictory. For this reason, we conducted a systematic review and meta-analysis of trials, reporting the use of stem-cell treatment against acute or chronic ischemic cardiomyopathies in large animal models with regard to Left Ventricular Ejection Fraction (LVEF). The defined research strategy was applied to the PubMed database to identify relevant studies published from January 2011 to July 2021. A random-effect meta-analysis was performed on LVEF mean data at follow-up between control and stem-cell-treated animals. In order to improve the definition of the effect measure and to analyze the factors that could influence the outcomes, a subgroup comparison was conducted. Sixty-six studies (n = 1183 animals) satisfied our inclusion criteria. Ischemia/reperfusion infarction was performed in 37 studies, and chronic occlusion in 29 studies; moreover, 58 studies were on a pig animal model. The meta-analysis showed that cell therapy increased LVEF by 7.41% (95% Confidence Interval 6.23–8.59%; $p < 0.001$) at follow-up, with significative heterogeneity and high inconsistency ($I^2 = 82\%$, $p < 0.001$). By subgroup comparison, the follow-up after 31–60 days ($p = 0.025$), the late cell injection (>7 days, $p = 0.005$) and the route of cellular delivery by surgical treatment ($p < 0.001$) were significant predictors of LVEF improvement. This meta-analysis showed that stem-cell therapy may improve heart function in large animal models and that the swine specie is confirmed as a relevant animal model in the cardiovascular field. Due to the significative heterogeneity and high inconsistency, future translational studies should be designed to take into account the evidenced predictors to allow for the reduction of the number of animals used.

Keywords: stem cells; cell therapy; large animal models; ischemic cardiomyopathies; myocardial infarction

Citation: La Mantia, D.; Bernardini, C.; Zannoni, A.; Salaroli, R.; Wang, C.; Bencivenni, S.; Forni, M. Efficacy of Stem Cell Therapy in Large Animal Models of Ischemic Cardiomyopathies: A Systematic Review and Meta-Analysis. *Animals* **2022**, *12*, 749. https://doi.org/10.3390/ani12060749

Academic Editor: Paul Dyce

Received: 27 January 2022
Accepted: 14 March 2022
Published: 16 March 2022

1. Introduction

Myocardial infarction is a leading cause of mortality and morbidity worldwide [1]. This pathology leads to death by necrosis of myocardial cells, due to prolonged ischemia, usually following coronary atherosclerosis [2]. In particular, coronary occlusion causes a loss of myocardial perfusion with consequent morphological, biochemical and functional alterations of the affected area, thus establishing ischemia, which, based on its extent and duration, can cause cell necrosis. Cell death leads to dramatic consequences because it triggers an acute inflammatory reaction. Subsequently, the damaged area is replaced by intensely vascularized granulation tissue, which then evolves into a process of fibrosis and, consequently, scar formation. Hyperplastic scar tissue is not functional, but the surviving patient's heart must still find a way to function while maintaining adequate cardiac output. To do this, it undergoes a series of structural and dynamic changes which are referred to as "ventricular remodeling". In fact, both the necrotic area and the non-infarcted segment of the ventricle progressively change in size, thickness and shape. All of this can then lead to heart failure [3]. Effective treatment strategies for myocardial infarction are designed to limit adverse ventricular remodeling to attenuate myocardial scar expansion and promote improvement of cardiac function and myocardial regeneration [4,5]. Among many therapies proposed, stem cells represent a promising option to repair the injured heart. Several cell types, including embryonic stem cells, skeletal myoblasts, mesenchymal stem cells (MSCs), cardiac stem cells (CSCs) and induced pluripotent stem cells (iPSCs), have been employed to re-functionalize the injured heart [5]. It has been shown that CSCs can differentiate into endothelial cells (ECs), vascular smooth-muscle cells (VSMCs) and cardiomyocytes (CMs) [6,7]. MSCs can differentiate into cardiomyocytes and induce angiogenesis [8,9]. Nevertheless, in vivo studies show that the percentage of inoculated stem cells that are stably implanted in the infarcted region and the related rate of cardiomyogenesis and angiogenesis are very slow to support myocardial regeneration [10]. However, studies from the past 20 years have clearly shown that it has been demonstrated that transplanted stem cells are able to release soluble factors that act in a paracrine way, contributing to the repair and regeneration of the infarcted myocardium [11]. These factors include a variety of growth factors, cytokines and extracellular matrix proteins [12]. Moreover, paracrine effects also include the recruitment activation and proliferation of resident endothelial progenitor cells (EPCs), cardiac progenitor cells (CPCs) and/or resident CSCs [12,13]. Furthermore, paracrine factors influence the contractile abilities of CM [6], promote cytoprotection (inhibition of apoptosis and necrosis) and formation of new blood vessels [7,12], prevent degradation of extracellular matrix (ECM), inhibit fibrosis and release of granulation factors [7]. Currently the most promising results have been obtained through the paracrine action rather than the direct action of cell differentiation [10,14–16].

In the last 20 years, in addition to the numerous experiments performed with in vitro models, numerous studies have been performed with large animal models with ischemic cardiomyopathies. These preclinical studies evaluated the risk of this new cell therapy, considering safety, feasibility and efficacy. In addition, they tried to answer the unsolved problems in clinical cell therapy (cell-type selection, number of cells, method of administration, time of administration and follow-up after cell transplantation); however, the results obtained were often heterogeneous and contradictory [17]. Studies based on large animal models often suffer from extremely limited sample sizes, due to ethical reasons, costs and management difficulties. Systematic reviews and meta-analysis substantially increase the statistical power, and the different experimental settings in the studies make it possible to obtain an estimate with a much higher external validity of the model. The present systematic review is an update of the previous work published by van der Spoel and colleagues [17] and aims to summarize trials reported in the literature about the use of stem-cell treatment against acute or chronic ischemic cardiomyopathies in large animal models and perform a meta-analysis on collected data from 2011 to 2021, with regard to Left Ventricular Ejection Fraction (LVEF) as a functional parameter.

2. Materials and Methods

2.1. Search Strategy and Selection Criteria

The international principles of preferred reporting items for systematic reviews and meta-analyses (PRISMA) 2020 guidelines were followed throughout the study [18]. Research was conducted in PubMed [19] to identify all the relevant publication from the period January 2011 to July 2021 by using the following search terms: "(pig OR porcine OR swine OR canine OR dog OR sheep OR ovine) AND (stem cells OR progenitor cells OR bone marrow) AND (myocardial infarction OR heart failure OR coronary artery disease OR cardiac repair OR myocardial regeneration)". Only articles published in English were included. The collected studies were carefully examined, and duplicates were removed.

2.2. Eligibility Criteria

The primary literature used to conduct the systematic review was compliant with the following inclusion and exclusion criteria. Studies that used large animal models with acute myocardial infarction (MI) or chronic ischemic cardiomyopathies, randomized controlled trials (RCTs) or no RCT studies were included to investigate the effect of stem-cell therapy on cardiac function as determined by left ventricular ejection fraction (LVEF). In addition, a placebo or sham-operated control group was included in the study. Studies using reporter genes (for stem-cell-imaging purposes only) were also included. In vitro studies, studies using genetically engineered or transfected stem cells with altered cellular behavior and studies using only conditioned media were excluded. Reviews, editorials, comments, letters and reports were excluded.

2.3. Data Extraction

Two reviewers (D.L.M. and C.W.) independently selected the studies by reading titles, abstracts and full manuscripts and applying the criteria mentioned above, and the resulting list of studies was approved by a third reviewer (M.F.). Then the following information was extracted from the full text of the selected studies: basal characteristics of the studies and left ventricular ejection fraction (LVEF) outcomes. If necessary, LVEF data were recalculated as follows: $(EDV - ESV)/EDV \times 100\%$ (EDV, end-diastolic volume, ESV, end-systolic volume). Accordingly, the standards deviations (SD) were determined or recalculated from the standard errors of mean (SEM).

2.4. Statistical Analysis

Primary analysis consisted of calculating the LVEF mean difference (reported in %) at follow-up between the control and stem-cell-treated groups when exposed to acute myocardial infarction or chronic ischemic cardiomyopathies. Continuous variables were reported as weighted mean differences with the 95% confidence intervals (CIs) between the treated and control groups. In the presence of multiple experimental groups alongside the control group within a study, the control group was used as control for each experimental group. A random-effect model (DerSimonian–Laird) was applied for the meta-analysis. Heterogeneity was assessed by using the I^2 statistics. Values for 25%, 50% and 75% for I^2 represented low, moderate and high heterogeneity, respectively [20]. In addition, the following subgroup analyses were performed: type of study (RCT or cohort); MI model (ischemia/reperfusion (I/R) or chronic occlusion); location of infarct-related artery (left anterior descending artery (LAD) or left circumflex artery (LCX)); autologous cell therapy (yes or no); cell type (adipose-derived stem cells (ASCs), bone-marrow mononuclear cells (BMMNC), bone-marrow-derived mononuclear cells (BMDMNCs), bone-marrow stem cells (BMSCs), cardiosphere-derived cells (CDCs), cardiac-derived progenitor cells (CPCs), cardiac stem cells (CSCs), multipotent adult progenitor cells (MAPCs), mesenchymal precursor cells (MPCs), induced pluripotent stem cells (iPSCs), mesenchymal stem cells (MSCs) or other types of stem cells (SC)); number of cells injected ($<10^7$, 10^7–10^8, $\geq 10^8$); timing of cell therapy after heart attack (<1 day, 1–7 days, >7 days); follow-up after cell therapy ($\leq$30 days, 31–60 days, >60 days); type of animal (pig, dog or sheep); and route of delivery (intramyocardial (IM), intracoronary (IC), trans-endocardial (TE), surgical or other routes)). Welch's *t*-test or ANOVA test was applied to compare subgroups. A funnel plot for LVEF was drawn to explore publication bias. All analyses were performed by using JASP software (JASP Team, 2022; version 0.14.1; Amsterdam, The Netherlands).

3. Results

3.1. Study Selection

The database search yielded 435 publications. After removing articles not in English and reviews, 401 publications were identified and assessed for eligibility. Based on the defined criteria, 289 studies were excluded and 112 studies were reviewed in detail. Only 66 studies met our inclusion criteria. The study search and selection processes are described in detail in Figure 1.

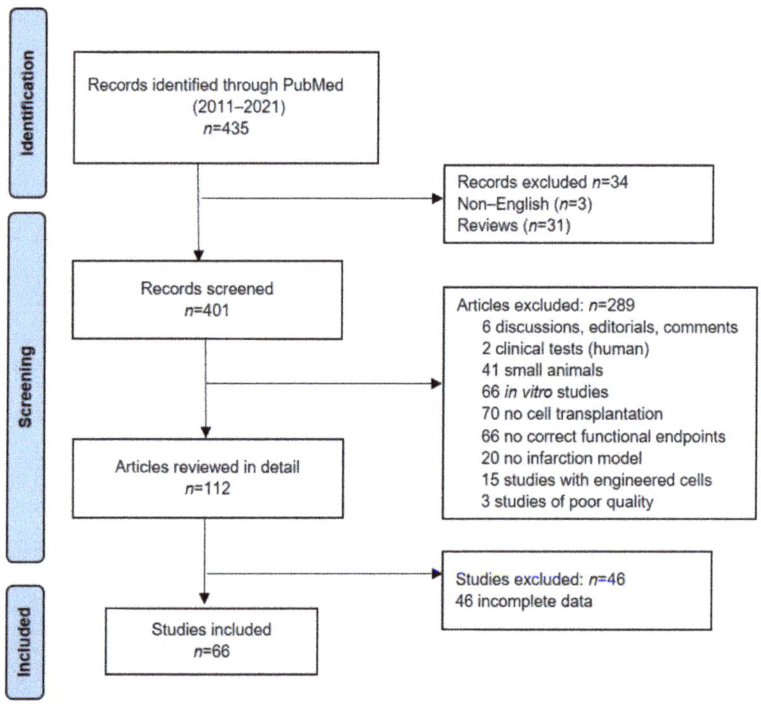

Figure 1. PRISMA workflow of the study selection process, records screened and studies included.

3.2. Included Studies Characteristics

In total, 1183 animals met the inclusion criteria, and the data derived from them were analyzed. Table 1 provides the characteristics of the included studies. Most studies used the porcine model (58 studies). In 37 studies, ischemia/reperfusion was used as an MI model. MI was mainly induced in the LAD (60 studies), but the site of ligation/constriction of the vessel (proximal, mid or distal) varied. Ten different types of cells were studied (25 studies used MSCs), but the number of stem cells administered varied (from 10^6 to 10^9); 26 studies used autologous cells. The main routes of delivery were IC infusion, IM, TE injection and surgical. Cell therapy was performed at different times after MI: <1 day (21 studies), 1–7 days (12 studies) and >7 days (33 studies). Follow-ups after cell therapy varied from 1 day to 180 days. The median and interquartile range of follow-up time was 51 days (28–60 days).

Table 1. Characteristics of the included studies.

Author	*n*	Type of Animal	Type of Study	Type of Infarction	MI Model	Cell Type	Number of Cells	Autologous Cells (Yes or No)	Route of Delivery	Timing of Cell Therapy after MI [a]	Follow-Up (Days)
Alestalo et al. 2015 [21]	24	Pig	RCT	LCX	I/R	BMMNC	6.2×10^7–1.43×10^8	Yes	Surgical	1.5 h	21
Bobi et al. 2017 [22]	14	Pig	RCT	LAD	I/R	ATMSC	1×10^7	No	IC	1 h	60
Bolli et al. 2013 [23]	21	Pig	No RCT	LAD	I/R	CSC	5×10^5	Yes	IC	90 d	30
Cai et al. 2016 [24]	20	Pig	RCT	LAD	No I/R	BMMSC	3×10^7	Yes	IM	0.5 h	28
Chang et al. 2015 [25]	12	Dog	RCT	LAD	No I/R	BMSC	2×10^7	Yes	IC	4 h	28
Chen et al. 2014 [26]	50	Pig	RCT	LAD	No I/R	BMSC	1×10^7	Yes	IC	3 h or 1 d or 3 d or 7 d or 14 d	28
Cheng et al. 2013 [27]	39	Sheep	RCT	LAD	I/R	MPC	2.5×10^7, 7.5×10^7, 2.25×10^8	No	TE	28 d	56
Crisostomo et al. 2019 [28]	25	Pig	RCT	LAD	I/R	CPC	2.5×10^7, 5.0×10^7	No	IC	7 d	70
Crisostomo et al. 2015 [29]	17	Pig	No RCT	LAD	I/R	CSC	2.5×10^7	No	IC	2 h or 7 d	70
Dariolli et al. 2017 [30]	25	Pig	RCT	LCX	No I/R	pASC	1×10^6, 2×10^6, 4×10^6	No	Surgical	30 d	30
Emmert et al. 2017 [31]	18	Pig	RCT	LAD	I/R	cardiopoietic stem cells	5×10^7	No	IM	30 d	30
Fanton et al. 2015 [32]	18	Pig	RCT	LAD	I/R	CASC	$8.3 \times 10^7 \pm 1.26 \times 10^8$	Yes	IM	2 h	60
Gahremanpour et al. 2013 [33]	30	Pig	RCT	LAD	I/R	USSC	$3.02 \times 10^8 \pm 2.3 \times 10^7$	No	TE	10 d	28–56
Haenel et al. 2019 [34]	17	Pig	RCT	LAD	I/R	ADRC	1.8×10^7	Yes	RCV	28 d	42
Hao et al. 2015 [35]	12	Dog	RCT	LAD	I/R	MSC	1×10^7	No	IC	2–3 h	70
Houtgraaf et al. 2013 [36]	34	Sheep	RCT	LAD	I/R	MPC	1.25×10^7–3.75×10^7	No	IC	1.5 h	56
Ishigami et al. 2018 [37]	10	Pig	RCT	LAD	No I/R	hiPSC	1×10^7	No	Surgical	14 d	28
Jansen of Lorkeers et al. 2015 [38]	16	Pig	RCT	LAD	I/R	hCMPC	1×10^7	No	IC	28 d	28
Jun Hong et al. 2015 [39]	21	Pig	RCT	LAD	No I/R	ASC	1.5×10^8, $5 \times 10^7 \times 3$	No	IV	1 h	2–28
Kanazawa et al. 2015 [40]	14	Pig	RCT	LAD	I/R	CDC	5×10^6, 7.5×10^6, 1×10^7, 8.7×10^6	No	IC	0.5 h	2
Karantalis et al. 2015 [41]	20	Pig	RCT	LAD	I/R	MSC/MSC+CSC	$2 \times 10^8/2 \times 10^8+1 \times 10^6$	Yes	TE	90 d	90
Kawamura et al. 2015 [42]	12	Pig	RCT	LAD	No I/R	BMMSC	1×10^8	No	Surgical	28 d	28–56
Kawamura et al. 2017 [43]	11	Pig	RCT	LAD	No I/R	hiPS-CM	3.5×10^7	No	Surgical	28 d	30–60–90
Kawamura et al. 2012 [44]	12	Pig	RCT	LAD	No I/R	hiPS-CM	3.2×10^7	No	Surgical	28 d	28–56
Kim et al. 2017 [45]	18	Pig	No RCT	LAD	No I/R	ATMSC	1×10^7	No	percutaneous	7 d	21
Ko et al. 2011 [46]	12	Pig	RCT	LAD	No I/R	BMDMNC	3×10^7	Yes	Surgical	0.25 h	3–90
Lee et al. 2015 [47]	28	Pig	No RCT	LAD	I/R	ADSC	2×10^6	Yes	IC	0.5 h	28
Lee et al. 2011 [48]	21	Pig	RCT	LAD	I/R	CDC	1×10^7	Yes	Surgical	28 d	56
Leu et al. 2011 [49]	12	Pig	No RCT	LAD	No I/R	BMDMNC	3×10^7	Yes	Surgical	Immediately	90
Li et al. 2013 [50]	24	Pig	RCT	LAD	No I/R	iPS	2×10^7	No	IM	7 d	7–42
Liao et al. 2019 [51]	24	Pig	RCT	LCX	No I/R	CM/MSC	$2 \times 10^8/2 \times 10^8$	No	IM	56 d	56
Lin et al. 2015 [52]	10	Pig	No RCT	LAD	No I/R	MNC	1×10^8	Yes	IM	Immediately	90
Liu et al. 2016 [53]	12	Pig	RCT	LCX	No I/R	UC-MSC	$3 \times 10^7 + 3 \times 10^7$	No	IC+IV	28 d + 35–42 d	28
Liu et al. 2015 [54]	12	Pig	No RCT	LAD	No I/R	PDMC	1×10^7	No	Surgical	Immediately	56

Table 1. *Cont.*

Author	n	Type of Animal	Type of Study	Type of Infarction	MI Model	Cell Type	Number of Cells	Autologous Cells (Yes or No)	Route of Delivery	Timing of Cell Therapy after MI [a]	Follow-Up (Days)
Locatelli et al. 2015 [55]	16	Sheep	RCT	LAD	No I/R	MSC	2×10^7	No	Intramyocardial transepicardial	7 d	30
Lu et al. 2012 [56]	24	Pig	RCT	LAD	I/R	MSC	3×10^7	Yes	IC	7 d	3–42
Malliaras et al. 2013 [57]	10	Pig	RCT	LAD	I/R	CDC	1.25×10^7	No	IC	14–21 d	60
Mao et al. 2014 [58]	16	Pig	RCT	LAD	I/R	MSC	1.5×10^7	No	IM	7 d	28
Mazo et al. 2012 [59]	16	Pig	RCT	LAD	I/R	ADSC	$2.1 \times 10^8 \pm 4.2 \times 10^7$	Yes	Percutaneous myocardial	9 d	90
Medicetty et al. 2012 [60]	19	Pig	No RCT	LAD	I/R	MAPC	$2 \times 10^7, 2 \times 10^8$	No	Percutaneous adventitial	2 d	2–30–90
Mori et al. 2018 [61]	12	Pig	RCT	LAD	No I/R	ADSC	1×10^8	No	Cell spray	28 d	28
Natsumeda et al. 2017 [62]	25	Pig	RCT	LAD	I/R	MSC/CSC/ MSC+CSC	$2 \times 10^8/1 \times 10^6/2 \times 10^8 + 1 \times 10^6$	No	TE	90 d	90
Ozawa et al. 1 2016 [63]	10	Juvenile pig	No RCT	LAD	I/R	SSC	$4.5 \times 10^7 - 6 \times 10^7$	Yes	Surgical	28 d	28–56
Ozawa et al. 2 2016 [63]	10	Adult Pig	No RCT	LAD	I/R	SSC	1.5×10^8	Yes	Surgical	28 d	28–56
Peng et al. 2013 [64]	10	Pig	RCT	LAD	I/R	MSC	$1 \times 10^8 - 2.3 \times 10^8$	Yes	IC	7–14 d	7–56
Prifti et al. 2016 [65]	25	Pig	No RCT	LAD	I/R	Mouse skeletal C2C12 myoblasts	NA	No	Venous coronary sinus retrograde infusion	30 d	30
Rabbani et al. 2017 [66]	18	Sheep	No RCT	LAD	No I/R	MSC/EC	2.7×10^7	Yes	Surgical	Immediately	60
Rigol et al. 2014 [67]	24	Pig	No RCT	LAD	I/R	ATMSC	1×10^7	No	IC	0.25 h or 7 d	21
Romagnuolo et al. 2019 [68]	10	Pig	No RCT	LAD	I/R	hESC-CM	1×10^9	No	Transepicardial	21 d	28
Schuleri et al. 2011 [69]	22	Pig	RCT	LAD	I/R	MSC	2×10^8	No	IM	84 d	84
Sheu et al. 2015 [70]	12	Pig	No RCT	LAD	No I/R	BMMSC	3×10^7	Yes	IM	1 h	4–60
Shudo et al. 2013 [71]	12	Pig	RCT	LAD	I/R	SMB	4.5×10^8	Yes	Cell sheets transepicardial	28 d	28–56
Song et al. 2013 [72]	14	Pig	RCT	LAD	I/R	BMMSC	3×10^7	Yes	IM	2 h	28
Sun et al. 2016 [73]	14	Dog	RCT	LAD	No I/R	MSC	1×10^7	Yes	RCV	7 d	40
Sun et al. 2020 [74]	16	Pig	RCT	LCX	No I/R	hiPSC-MSC	2×10^8	No	IM	56 d	56
Suzuki et al. 2016 [75]	11	Pig	RCT	LAD	No I/R	CDC	2×10^7	No	IC	60 d	28
Tseliou et al. 1 2016 [76]	15	Pig	RCT	LAD	No I/R	CDC	1.25×10^7	No	Single-vessel intracoronary (stop-flow or continuous-flow)	21 d	28

Table 1. *Cont.*

Author	*n*	Type of Animal	Type of Study	Type of Infarction	MI Model	Cell Type	Number of Cells	Autologous Cells (Yes or No)	Route of Delivery	Timing of Cell Therapy after MI [a]	Follow-Up (Days)
Tseliou et al. 2 2016 [76]	15	Pig	RCT	LAD	No I/R	CDC	1.25×10^7	No	Multi-vessel intracoronary (stop-flow or continuous-flow)	21 d	28
van der Spoel et al. 2015 [77]	17	Pig	No RCT	LCX	I/R	MSC/ BMMNC+MSC	$1 \times 10^7/1 \times 10^7 + 1 \times 10^7$	Yes	TE	28 d/28 d + 56 d	28–56
Wang et al. 2021 [3]	30	Pig	No RCT	LAD	I/R	MSC	1×10^8	No	IM	60 d	90
Wang et al. 2015 [78]	8	Dog	RCT	LAD	No I/R	MSC	1×10^8	No	RCV	7 d	28
Williams et al. 2013 [79]	20	Pig	No RCT	LAD	I/R	MSC/CSC/ MSC+CSC	$2 \times 10^8/1 \times 10^6/2 \times 10^8 + 1 \times 10^6$	No	IM	14 d	14–28
Winkler et al. 2020 [80]	13	Pig	RCT	LAD	I/R	CDC	1×10^7	No	IC	0.25 h	30
Yang et al. 2011 [81]	25	Pig	RCT	LAD	I/R	MSC	$9 \times 10^7 – 1.8 \times 10^8$	No	IC	14 d	42
Yin et al. 2014 [82]	10	Pig	RCT	LAD	I/R	ASC	4×10^7	Yes	IC	7 d	56
Zhang et al. 2015 [83]	12	Pig	RCT	LAD	No I/R	BMSC	2×10^7	Yes	IM	NA	42
Zhang et al. 2011 [84]	12	Pig	RCT	LAD	No I/R	BMSC	2×10^7	Yes	IM	NA	42
Zhao et al. 2014 [85]	20	Pig	RCT	LAD	No I/R	BMSC	1×10^7	NA	IM	Immediately	180

ADRC, adipose-derived regenerative cells; ADSC, adipose tissue-derived stem cells; AMSC, amniotic-membrane-derived mesenchymal stromal cell; ASC, adipose-derived stem cells; ATMSC, adipose tissue–derived mesenchymal stem cells; BMDMNC, bone-marrow-derived mononuclear cell; BMSC, bone-marrow stem cells; BMMSC, bone-marrow mesenchymal stem cells; BM-MNC, bone-marrow mononuclear cells; CASC, cardiac atrial appendage stem cells; CB-MNC, human cord blood mononuclear cells; CBSC, cortical-bone stem cells; CDC, cardiosphere-derived cells; CPC, cardiac-derived progenitor cells; CSC, cardiac stem cells; EC, endothelial cells; EPC, endothelial progenitor cells; ESC, embryonic stem cells; hCMPC, human cardiomyocyte progenitor cells; hESC-CM, human embryonic stem-cell-derived cardiomyocytes; hiPSC-MSC, human-induced pluripotent stem cell-derived mesenchymal stem cells; hiPSC, human-induced pluripotent stem cell; hiPS-CM, human-induced pluripotent stem-cell-derived cardiomyocytes; HPC, hematopoietic progenitor cells; iPS, induced pluripotent stem cells; IC, intracoronary infusion; IM, intramyocardial injection; I/R, ischemia/reperfusion; IV, peripheral intravenous; LAD, left anterior descending artery; LCX, left circumflex artery; MAPC, multipotent adult progenitor cells; MI, myocardial infarction; MNC, mononuclear cells; MPC, mesenchymal precursor cells; MSC, mesenchymal stem cells; *n*, number of animals (treated group and control group); NA, not applicable; pASC, porcine-adipose-tissue-derived mesenchymal stem cells; PDMC, placenta-derived multipotent cells; RCT, randomized controlled trail; RCV, retrograde coronary transvenous injection; SMB, skeletal myoblast; SSC, skeletal stem cells; TE, trans-endocardial injection; UC-MSC, umbilical-cord-derived mesenchymal stromal cells; USSC, unrestricted somatic stem cells. [a] Timing in hours (h) or days (d).

3.3. Meta-Analysis

Meta-analysis showed a LVEF difference of 7.41% at follow-up after stem-cell therapy vs. control (95% CI, 6.23–8.59%; $p < 0.001$), with significative heterogeneity ($p < 0.001$) and high inconsistency (I^2: 82%) (Figure 2). At follow-up, the mean LVEF after stem-cell treatment and control was 48% and 40.7%, respectively.

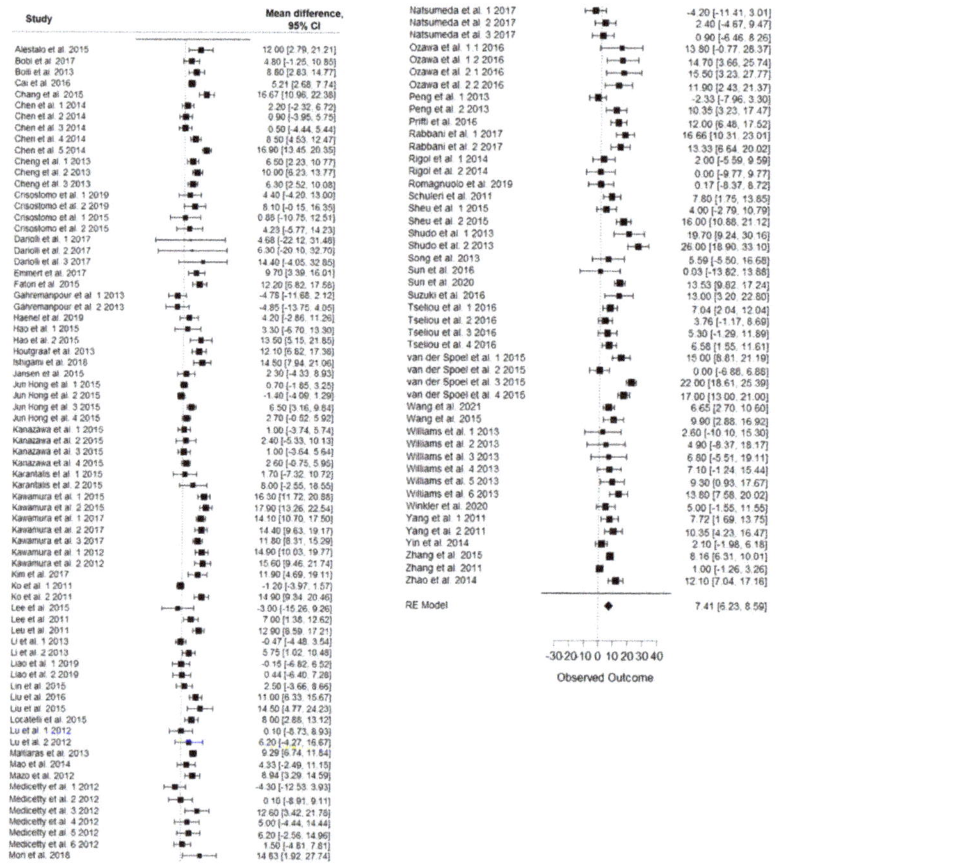

Figure 2. Forest plot showing the effect of stem-cell therapy on LVEF improvement compared with controls. Note: 95% CI, 95% confidence interval.

3.4. Subgroup Analysis

LVEF mean difference values were compared for subgroup analysis, and a Welch's *t*-test or ANOVA test was applied. The analysis showed that follow-up after cell therapy ($p = 0.025$), time between infarction and cell injection ($p = 0.005$) and the route of delivery ($p < 0.001$) are independent significant predictors of LVEF improvement. Figure 3d,f,i shows a trend toward greater improvements after cell therapy in the following aspects: follow-up at 31–60 days, since, after that period, the effect of cell therapy appeared to decline over time; and the late cell injection after MI (>7 days) and the surgical treatment. In addition, less benefit was observed in the ischemia/reperfusion MI model compared to the chronic MI models ($p = 0.063$), and there was an improvement with autologous cell treatment ($p = 0.079$) (Figure 3b,c). No significant differences in LVEF were observed in the following cases: animal model ($p = 0.355$), type of infarction ($p = 0.257$), type of study ($p = 0.345$), cell number (p = 0.39) and cell type (p = 0.361) (Figure 3a,e,g,h,j). An additional subgroup analysis was performed to analyze the three predictors at the significant levels (Follow up 31–60 days, timing of cell therapy after MI > 7 days and surgical as route of delivery) to understand whether the effect on LVEF improvement was related to a specific cell type. No significative differences were detected (Figure 4).

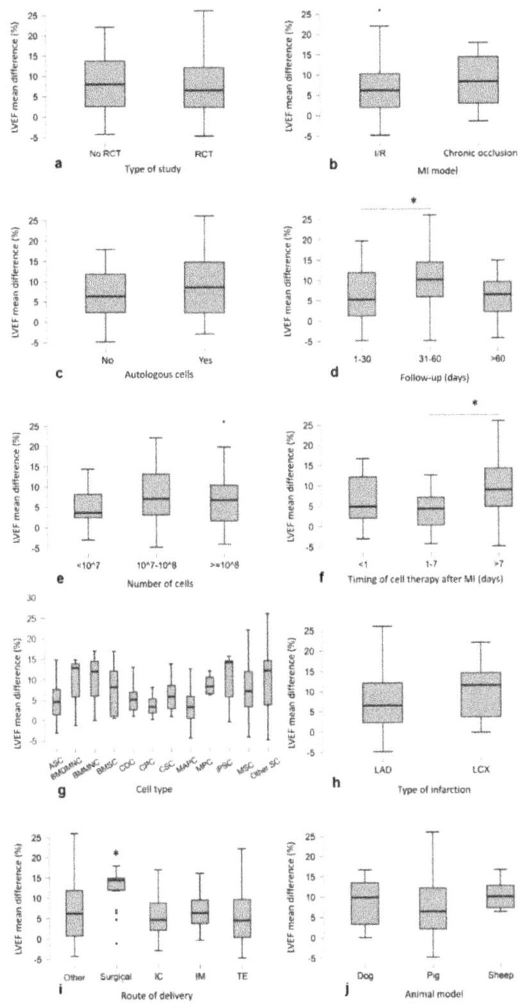

Figure 3. Subgroup analysis showing the LVEF trends toward more improvements after cell therapy compared with control: (**d**) follow-up at 31–60 days (p = 0.025), (**f**) the late cell injection (>7 days, p = 0.005), (**i**) surgical administration (p < 0.001), (**b**) chronic occlusion model (p = 0.063) and (**c**) autologous cells (p = 0.079). No significant differences were observed in (**j**) animal model (p = 0.355), (**h**) type of infarction (p = 0.257), (**a**) type of study (p = 0.345), (**e**) number of cells (p = 0.39) and (**g**) cell type (n $\geq$ 3 studies) (p = 0.361). Graphs are represented as Boxplots; the two segments that delimit the rectangle represent the 25th and 75th percentiles; the central segment is the median; the bars represent the minimum and maximum values, respectively; and the external points are the outliers. Note: (* p <0.05) represents statistical significance resulting from one-way ANOVA, followed by post hoc Tukey comparison test.

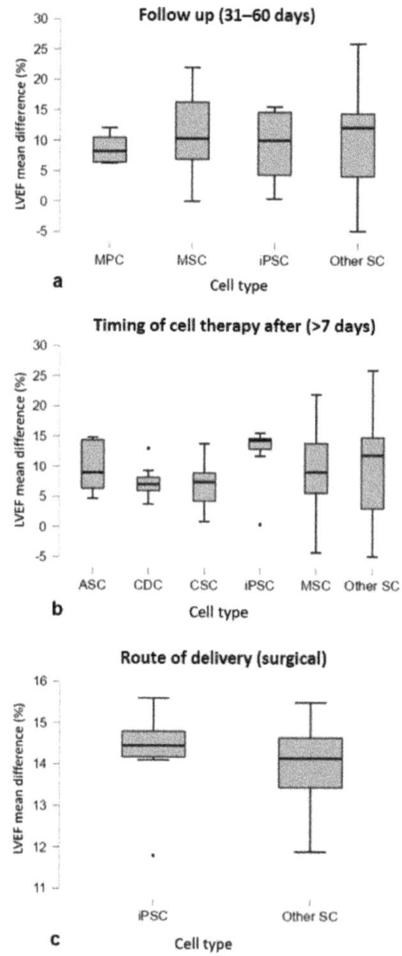

Figure 4. Subgroup analysis showing LVEF trends of cell type ($n > 3$ studies) by considering studies of the three significant predictors, in particular (**a**) follow-up (30–60 days) ($p = 0.904$), (**b**) timing of cell therapy after MI (>7 days) ($p = 0.690$) and (**c**) route of delivery (surgical) ($p = 0.729$). Graphs are represented as Boxplots; the two segments that delimit the rectangle represent the 25th and 75th percentiles; the central segment is the median; the bars represent the minimum and maximum values, respectively; and the external points the outliers.

The funnel plot for LVEF mean difference shows that there is no publication bias (Figure 5), as values are evenly distributed around the effect estimate, as evidenced by the regression test for funnel plot asymmetry (Egger's test) ($p = 0.657$).

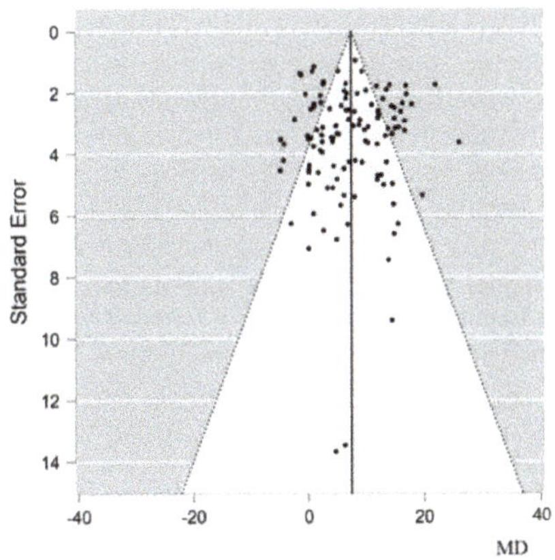

Figure 5. Funnel plot for LVEF improvement showing the absence of publication biases. The vertical solid line represents the estimated overall mean difference; black dots are the standard error of each study. MD, mean difference.

4. Discussion

In the present systematic review and meta-analysis, we assessed the effect of stem-cell therapy against ischemic cardiomyopathies in large animal models; this is an update of a previous work published by van der Spoel et al. [17] that reviewed the same topic in studies performed from 1980 to 2010. The analysis includes data from 66 published pre-clinical studies (2011–2021) that used large animal models treated with stem cells in order to study the effects of cell therapy of ischemic cardiomyopathies by reporting outcomes derived from left ventricular ejection fraction (LVEF) as functional parameter. The meta-analysis showed a significant improvement of LVEF by 7.41% (95% CI 6.23–8.59%) after stem-cell therapy against control group confirming the positive effect reported in the previous meta-analysis [17], in which LVEF effect size was 7.51% (95% CI 6.15–8.87%). Given the number of the studies included, a random-effect model was applied by resulting a significative heterogeneity with high inconsistency (I^2: 82%). For this reason, a comparison between subgroups was investigated in order to analyze clinically relevant parameters. The sub-analysis revealed that time of follow-up, time between infarction and cell injection, and the route of cellular delivery are independent significant predictors of LVEF improvement. In detail, in large animals, the effect of cell therapy achieved better results at 31–60 days, after which it fades over time; this phenomenon is in accordance with the previous analysis [17]. This finding could suggest the use of new applications and therapeutic strategies to increase cell survival over time, such as the use of slow-release molecules by cell pre-conditioning [86,87], the application of biomaterials [88,89] or the genetic stem-cell modifications [90]. Late cell injection assumed better benefit if applied 7 days after MI; our findings are comparable with the previous meta-analyses both in large animals [17] and human [91]. Optimal stem-cell therapy depends not only on engraftment and survival of the transplanted cells but also on successful delivery. By comparing different types of cellular delivery, our results demonstrated that surgical treatment is the route that significantly improves the heart functionality. In general, stem cells can be delivered by intravenous or intracoronary routes after coronary revascularization in the setting of acute MI to avoid the risk of invasive procedure; however, both IV and IC routes seems to be not applicable for patients with chronic myocardial ischemia not amenable to coronary revascularization, so direct intramyocardial injection via either surgical epicardial or transcatheter endocardial approaches may be necessary, as they allow for the direct visualization of the site of injection [92]. In addition, our findings showed that less benefit in LVEF improvement was observed in ischemia/reperfusion MI model compared to the chronic occlusion models, but without reaching significance. This is compliant with the findings obtained by van der Spoel et al. [17] both in large and small animal models [93]. Autologous cell therapy resulted in better results on LVEF improvement, but not significative; a similar effect was shown in a meta-analysis performed in large animal models about autologous and allogeneic cell therapy for ischemic heart disease by using BM-MNCs, MSCs and cardiac stem-cell types [38]. Furthermore, the use of autologous BM- or MSC-derived cells is confounded by the functional impairment of those stem cells associated with aging and because of the restricted immediate availability; the use of allogenic cell products with limited immunogenicity, such as MSC derived from different tissues, or standardized non-cellular products, such as conditional medium, may overcome these problems in terms of efficacy and safety [38,92,94].

No significative differences in LVEF were observed in animal species, infarct type, type of study, number of cells and cell type. Regarding the cell type, the result obtained contrasts with that of van der Spoel et al. [17]. We could not exclude that this result is due to the high difference in the studies' number between the group analyzed. The same consideration regarding the lack of significative results deserves to be made for the number of cells ($<10^7$ $n = 6$, 10^7–10^8 $n = 41$, and $\geq 10^8$ $n = 19$) and the animal species (porcine $n = 58$ others $n = 8$).

Although no difference was observed between species, we would sustain the widely accepted porcine animal model as the one to be recommended to evaluate the effect of the cell therapy, confirming the swine as a relevant animal model in the cardiovascular field and in translational research in a broader sense [95]. This is because most of the published studies (89% of those included in our SR) are based on this model, and the data are, therefore, available for comparison as a fundamental tool in future experimental designs, in particular, in relation to the Reduction aspects. Furthermore, we wanted to investigate what could contribute to the statistical significance of the improving predictors of LVEF. In particular, we analyzed whether the effect of LVEF improvement was attributed to a specific cell type. No significative difference was observed in cell type in large animal studies with a significative improvement in follow-up (31–60 days), timing of cell therapy after MI (>7 days) and surgical treatment (Figure 4).

Limitations

The limitations of meta-analysis are well-known [96]. Meta-analyses and systematic reviews are statistical and scientific techniques that can highlight areas where evidence is lacking, but they cannot overcome these deficiencies [97]. Publication bias and search bias are potential problems in all meta-analyses [97]; this arises from the fact that unpublished studies may contradict the results due to the tendency not to publish negative studies, thus leading to the over-representation of "positive" ones [98]. In this meta-analysis, the funnel plot (Figure 4) for the LVEF mean difference showed that there is no publication bias. Thus, based on the results we obtained, we can affirm that, in the future, stem-cell-transplant studies in large animal models with ischemic cardiomyopathies should therefore focus on late (>7 days) surgical treatments and 31–60-day follow-up. The analysis of the subgroups shows that the greater heterogeneity of the included studies could be mainly due to the different amounts of data in the different comparison groups, such as in the case of cell number, cell type and animal species.

5. Conclusions

In conclusion, in the present systematic review and meta-analysis, we evaluated the effect of stem-cell transplantation in large animal models with ischemic cardiomyopathies, showing that stem-cell therapy could improve LVEF. The SR is therefore confirmed as a reliable method for obtaining a complete contextual framework from which to start for further experimentation, and future translational studies should be designed by considering the evidenced predictors to allow for the reduction of the number of animals used in preclinical trials. Large animal models, especially the swine, are a translational step necessary to predict outcomes of clinical trials in the cardiovascular field.

Author Contributions: Conceptualization, D.L.M., C.B. and M.F.; methodology, D.L.M. and M.F.; software, D.L.M. and C.W.; validation, M.F.; writing—original draft preparation D.L.M. and C.B.; writing—review and editing, A.Z., R.S. and S.B.; supervision, C.B. and M.F.; funding acquisition, M.F. All authors have read and agreed to the published version of the manuscript.

Funding: This research received no external funding. The APC was funded by RFO 2020–University of Bologna to M.F.

Institutional Review Board Statement: Not applicable.

Informed Consent Statement: Not applicable.

Data Availability Statement: Not applicable.

Conflicts of Interest: The authors declare no conflict of interest.

References

1. Cardiovascular Diseases. Available online: https://www.who.int/westernpacific/health-topics/cardiovascular-diseases (accessed on 19 January 2022).
2. Thygesen, K.; Alpert, J.S.; Jaffe, A.S.; Simoons, M.L.; Chaitman, B.R.; White, H.D. Writing Group on behalf of the Joint For the Universal Definition of Myocardial Infarction Third Universal Definition of Myocardial Infarction. *Glob. Heart* **2012**, *7*, 275. [CrossRef] [PubMed]
3. Wang, Q.; He, X.; Wang, B.; Pan, J.; Shi, C.; Li, J.; Wang, L.; Zhao, Y.; Dai, J.; Wang, D. Injectable Collagen Scaffold Promotes Swine Myocardial Infarction Recovery by Long-Term Local Retention of Transplanted Human Umbilical Cord Mesenchymal Stem Cells. *Sci. China Life Sci.* **2021**, *64*, 269–281. [CrossRef] [PubMed]
4. Perea-Gil, I.; Prat-Vidal, C.; Bayes-Genis, A. In Vivo Experience with Natural Scaffolds for Myocardial Infarction: The Times They Are a-Changin'. *Stem Cell Res. Ther.* **2015**, *6*, 248. [CrossRef] [PubMed]
5. Dixit, P.; Katare, R. Challenges in Identifying the Best Source of Stem Cells for Cardiac Regeneration Therapy. *Stem Cell Res. Ther.* **2015**, *6*, 26. [CrossRef] [PubMed]
6. Leri, A.; Kajstura, J.; Anversa, P. Cardiac Stem Cells and Mechanisms of Myocardial Regeneration. *Physiol. Rev.* **2005**, *85*, 1373–1416. [CrossRef]
7. Gnecchi, M.; Zhang, Z.; Ni, A.; Dzau, V.J. Paracrine Mechanisms in Adult Stem Cell Signaling and Therapy. *Circ. Res.* **2008**, *103*, 1204–1219. [CrossRef]
8. Tomita, S.; Li, R.-K.; Weisel, R.D.; Mickle, D.A.G.; Kim, E.-J.; Sakai, T.; Jia, Z.-Q. Autologous Transplantation of Bone Marrow Cells Improves Damaged Heart Function. *Circulation* **1999**, *100*, 247–256. [CrossRef]
9. Pittenger, M.F.; Martin, B.J. Mesenchymal Stem Cells and Their Potential as Cardiac Therapeutics. *Circ. Res.* **2004**, *95*, 9–20. [CrossRef]
10. Yang, D.; Wang, W.; Li, L.; Peng, Y.; Chen, P.; Huang, H.; Guo, Y.; Xia, X.; Wang, Y.; Wang, H.; et al. The Relative Contribution of Paracine Effect versus Direct Differentiation on Adipose-Derived Stem Cell Transplantation Mediated Cardiac Repair. *PLoS ONE* **2013**, *8*, e59020. [CrossRef]
11. Sid-Otmane, C.; Perrault, L.P.; Ly, H.Q. Mesenchymal stem cell mediates cardiac repair through autocrine, paracrine and endocrine axes. *J. Transl. Med.* **2020**, *18*, 336. [CrossRef]
12. Burchfield, J.S.; Dimmeler, S. Role of Paracrine Factors in Stem and Progenitor Cell Mediated Cardiac Repair and Tissue Fibrosis. *Fibrogenesis Tissue Repair* **2008**, *1*, 4. [CrossRef] [PubMed]
13. Segers, V.F.M.; Tokunou, T.; Higgins, L.J.; MacGillivray, C.; Gannon, J.; Lee, R.T. Local Delivery of Protease-Resistant Stromal Cell Derived Factor-1 for Stem Cell Recruitment After Myocardial Infarction. *Circulation* **2007**, *116*, 1683–1692. [CrossRef] [PubMed]
14. Chimenti, I.; Smith, R.R.; Li, T.-S.; Gerstenblith, G.; Messina, E.; Giacomello, A.; Marbán, E. Relative Roles of Direct Regeneration Versus Paracrine Effects of Human Cardiosphere-Derived Cells Transplanted Into Infarcted Mice. *Circ. Res.* **2010**, *106*, 971–980. [CrossRef] [PubMed]
15. Sepantafar, M.; Maheronnaghsh, R.; Mohammadi, H.; Rajabi-Zeleti, S.; Annabi, N.; Aghdami, N.; Baharvand, H. Stem Cells and Injectable Hydrogels: Synergistic Therapeutics in Myocardial Repair. *Biotechnol. Adv.* **2016**, *34*, 362–379. [CrossRef] [PubMed]
16. Chien, K.R.; Frisén, J.; Fritsche-Danielson, R.; Melton, D.A.; Murry, C.E.; Weissman, I.L. Regenerating the Field of Cardiovascular Cell Therapy. *Nat. Biotechnol.* **2019**, *37*, 232–237. [CrossRef]
17. Van der Spoel, T.I.G.; Jansen of Lorkeers, S.J.; Agostoni, P.; van Belle, E.; Gyöngyösi, M.; Sluijter, J.P.G.; Cramer, M.J.; Doevendans, P.A.; Chamuleau, S.A.J. Human Relevance of Pre-Clinical Studies in Stem Cell Therapy: Systematic Review and Meta-Analysis of Large Animal Models of Ischaemic Heart Disease. *Cardiovasc. Res.* **2011**, *91*, 649–658. [CrossRef]
18. Page, M.J.; Moher, D.; Bossuyt, P.M.; Boutron, I.; Hoffmann, T.C.; Mulrow, C.D.; Shamseer, L.; Tetzlaff, J.M.; Akl, E.A.; Brennan, S.E.; et al. PRISMA 2020 Explanation and Elaboration: Updated Guidance and Exemplars for Reporting Systematic Reviews. *BMJ* **2021**, *372*, n160. [CrossRef]
19. PubMed. Available online: https://pubmed.ncbi.nlm.nih.gov/ (accessed on 24 January 2022).

20. Higgins, J.P.T.; Thompson, S.G.; Deeks, J.J.; Altman, D.G. Measuring Inconsistency in Meta-Analyses. *BMJ* **2003**, *327*, 557–560. [CrossRef]
21. Alestalo, K.; Korpi, R.; Mäkelä, J.; Lehtonen, S.; Mäkelä, T.; Yannopoulos, F.; Ylitalo, K.; Haapea, M.; Juvonen, T.; Anttila, V.; et al. High Number of Transplanted Stem Cells Improves Myocardial Recovery after AMI in a Porcine Model. *Scand. Cardiovasc. J.* **2015**, *49*, 82–94. [CrossRef]
22. Bobi, J.; Solanes, N.; Fernández-Jiménez, R.; Galán-Arriola, C.; Dantas, A.P.; Fernández-Friera, L.; Gálvez-Montón, C.; Rigol-Monzó, E.; Agüero, J.; Ramírez, J.; et al. Intracoronary Administration of Allogeneic Adipose Tissue-Derived Mesenchymal Stem Cells Improves Myocardial Perfusion But Not Left Ventricle Function, in a Translational Model of Acute Myocardial Infarction. *J. Am. Heart Assoc.* **2017**, *6*, e005771. [CrossRef]
23. Bolli, R.; Tang, X.-L.; Sanganalmath, S.K.; Rimoldi, O.; Mosna, F.; Abdel-Latif, A.; Jneid, H.; Rota, M.; Leri, A.; Kajstura, J. Intracoronary Delivery of Autologous Cardiac Stem Cells Improves Cardiac Function in a Porcine Model of Chronic Ischemic Cardiomyopathy. *Circulation* **2013**, *128*, 122–131. [CrossRef] [PubMed]
24. Cai, M.; Shen, R.; Song, L.; Lu, M.; Wang, J.; Zhao, S.; Tang, Y.; Meng, X.; Li, Z.; He, Z.-X. Erratum: Bone Marrow Mesenchymal Stem Cells (BM-MSCs) Improve Heart Function in Swine Myocardial Infarction Model through Paracrine Effects. *Sci. Rep.* **2016**, *6*, 31528. [CrossRef] [PubMed]
25. Chang, X.; Liu, J.; Liao, X.; Liu, G. Ultrasound-Mediated Microbubble Destruction Enhances the Therapeutic Effect of Intracoronary Transplantation of Bone Marrow Stem Cells on Myocardial Infarction. *Int. J. Clin. Exp. Pathol.* **2015**, *8*, 2221–2234. [PubMed]
26. Chen, Y.; Teng, X.; Chen, W.; Yang, J.; Yang, Z.; Yu, Y.; Shen, Z. Timing of Transplantation of Autologous Bone Marrow Derived Mesenchymal Stem Cells for Treating Myocardial Infarction. *Sci. China Life Sci.* **2014**, *57*, 195–200. [CrossRef]
27. Cheng, Y.; Yi, G.; Conditt, G.B.; Sheehy, A.; Kolodgie, F.D.; Tellez, A.; Polyakov, I.; Gu, A.; Aboodi, M.S.; Wallace-Bradley, D.; et al. Catheter-Based Endomyocardial Delivery of Mesenchymal Precursor Cells Using 3D Echo Guidance Improves Cardiac Function in a Chronic Myocardial Injury Ovine Model. *Cell Transplant.* **2013**, *22*, 2299–2309. [CrossRef]
28. Crisostomo, V.; Baez, C.; Abad, J.L.; Sanchez, B.; Alvarez, V.; Rosado, R.; Gómez-Mauricio, G.; Gheysens, O.; Blanco-Blazquez, V.; Blazquez, R.; et al. Dose-Dependent Improvement of Cardiac Function in a Swine Model of Acute Myocardial Infarction after Intracoronary Administration of Allogeneic Heart-Derived Cells. *Stem Cell Res. Ther.* **2019**, *10*, 152. [CrossRef]
29. Crisostomo, V.; Baez-Diaz, C.; Maestre, J.; Garcia-Lindo, M.; Sun, F.; Casado, J.G.; Blazquez, R.; Abad, J.L.; Palacios, I.; Rodriguez-Borlado, L.; et al. Delayed Administration of Allogeneic Cardiac Stem Cell Therapy for Acute Myocardial Infarction Could Ameliorate Adverse Remodeling: Experimental Study in Swine. *J. Transl. Med.* **2015**, *13*, 156. [CrossRef]
30. Dariolli, R.; Naghetini, M.V.; Marques, E.F.; Takimura, C.K.; Jensen, L.S.; Kiers, B.; Tsutsui, J.M.; Mathias, W.; Lemos Neto, P.A.; Krieger, J.E. Allogeneic PASC Transplantation in Humanized Pigs Attenuates Cardiac Remodeling Post-Myocardial Infarction. *PLoS ONE* **2017**, *12*, e0176412. [CrossRef]
31. Emmert, M.Y.; Wolint, P.; Jakab, A.; Sheehy, S.P.; Pasqualini, F.S.; Nguyen, T.D.L.; Hilbe, M.; Seifert, B.; Weber, B.; Brokopp, C.E.; et al. Safety and Efficacy of Cardiopoietic Stem Cells in the Treatment of Post-Infarction Left-Ventricular Dysfunction—From Cardioprotection to Functional Repair in a Translational Pig Infarction Model. *Biomaterials* **2017**, *122*, 48–62. [CrossRef]
32. Fanton, Y.; Robic, B.; Rummens, J.-L.; Daniëls, A.; Windmolders, S.; Willems, L.; Jamaer, L.; Dubois, J.; Bijnens, E.; Heuts, N.; et al. Cardiac Atrial Appendage Stem Cells Engraft and Differentiate into Cardiomyocytes in Vivo: A New Tool for Cardiac Repair after MI. *Int. J. Cardiol.* **2015**, *201*, 10–19. [CrossRef]
33. Gahremanpour, A.; Vela, D.; Zheng, Y.; Silva, G.V.; Fodor, W.; Cardoso, C.O.; Baimbridge, F.; Fernandes, M.R.; Buja, L.M.; Perin, E.C. Xenotransplantation of Human Unrestricted Somatic Stem Cells in a Pig Model of Acute Myocardial Infarction. *Xenotransplantation* **2013**, *20*, 110–122. [CrossRef] [PubMed]
34. Haenel, A.; Ghosn, M.; Karimi, T.; Vykoukal, J.; Shah, D.; Valderrabano, M.; Schulz, D.G.; Raizner, A.; Schmitz, C.; Alt, E.U. Unmodified Autologous Stem Cells at Point of Care for Chronic Myocardial Infarction. *World J. Stem Cells* **2019**, *11*, 831–858. [CrossRef] [PubMed]
35. Hao, L.; Hao, J.; Fang, W.; Han, C.; Zhang, K.; Wang, X. Dual Isotope Simultaneous Imaging to Evaluate the Effects of Intracoronary Bone Marrow-Derived Mesenchymal Stem Cells on Perfusion and Metabolism in Canines with Acute Myocardial Infarction. *Biomed. Rep.* **2015**, *3*, 447–452. [CrossRef] [PubMed]
36. Houtgraaf, J.H.; de Jong, R.; Kazemi, K.; de Groot, D.; van der Spoel, T.I.G.; Arslan, F.; Hoefer, I.; Pasterkamp, G.; Itescu, S.; Zijlstra, F.; et al. Intracoronary Infusion of Allogeneic Mesenchymal Precursor Cells Directly after Experimental Acute Myocardial Infarction Reduces Infarct Size, Abrogates Adverse Remodeling, and Improves Cardiac Function. *Circ. Res.* **2013**, *113*, 153–166. [CrossRef]
37. Ishigami, M.; Masumoto, H.; Ikuno, T.; Aoki, T.; Kawatou, M.; Minakata, K.; Ikeda, T.; Sakata, R.; Yamashita, J.K.; Minatoya, K. Human IPS Cell-Derived Cardiac Tissue Sheets for Functional Restoration of Infarcted Porcine Hearts. *PLoS ONE* **2018**, *13*, e0201650. [CrossRef]
38. Jansen of Lorkeers, S.J.; Gho, J.M.I.H.; Koudstaal, S.; van Hout, G.P.J.; Zwetsloot, P.P.M.; van Oorschot, J.W.M.; van Eeuwijk, E.C.M.; Leiner, T.; Hoefer, I.E.; Goumans, M.-J.; et al. Xenotransplantation of Human Cardiomyocyte Progenitor Cells Does Not Improve Cardiac Function in a Porcine Model of Chronic Ischemic Heart Failure. Results from a Randomized, Blinded, Placebo Controlled Trial. *PLoS ONE* **2015**, *10*, e0143953. [CrossRef] [PubMed]

39. Jun Hong, S.; Rogers, P.I.; Kihlken, J.; Warfel, J.; Bull, C.; Deuter-Reinhard, M.; Feng, D.; Xie, J.; Kyle, A.; Merfeld-Clauss, S.; et al. Intravenous Xenogeneic Transplantation of Human Adipose-Derived Stem Cells Improves Left Ventricular Function and Microvascular Integrity in Swine Myocardial Infarction Model. *Catheter. Cardiovasc. Interv. Off. J. Soc. Card. Angiogr. Interv.* **2015**, *86*, E38–E48. [CrossRef]
40. Kanazawa, H.; Tseliou, E.; Malliaras, K.; Yee, K.; Dawkins, J.F.; De Couto, G.; Smith, R.R.; Kreke, M.; Seinfeld, J.; Middleton, R.C.; et al. Cellular Postconditioning: Allogeneic Cardiosphere-Derived Cells Reduce Infarct Size and Attenuate Microvascular Obstruction When Administered after Reperfusion in Pigs with Acute Myocardial Infarction. *Circ. Heart Fail.* **2015**, *8*, 322–332. [CrossRef] [PubMed]
41. Karantalis, V.; Suncion-Loescher, V.Y.; Bagno, L.; Golpanian, S.; Wolf, A.; Sanina, C.; Premer, C.; Kanelidis, A.J.; McCall, F.; Wang, B.; et al. Synergistic Effects of Combined Cell Therapy for Chronic Ischemic Cardiomyopathy. *J. Am. Coll. Cardiol.* **2015**, *66*, 1990–1999. [CrossRef] [PubMed]
42. Kawamura, M.; Miyagawa, S.; Fukushima, S.; Saito, A.; Toda, K.; Daimon, T.; Shimizu, T.; Okano, T.; Sawa, Y. Xenotransplantation of Bone Marrow-Derived Human Mesenchymal Stem Cell Sheets Attenuates Left Ventricular Remodeling in a Porcine Ischemic Cardiomyopathy Model. *Tissue Eng. Part A* **2015**, *21*, 2272–2280. [CrossRef]
43. Kawamura, M.; Miyagawa, S.; Fukushima, S.; Saito, A.; Miki, K.; Funakoshi, S.; Yoshida, Y.; Yamanaka, S.; Shimizu, T.; Okano, T.; et al. Enhanced Therapeutic Effects of Human IPS Cell Derived-Cardiomyocyte by Combined Cell-Sheets with Omental Flap Technique in Porcine Ischemic Cardiomyopathy Model. *Sci. Rep.* **2017**, *7*, 8824. [CrossRef] [PubMed]
44. Kawamura, M.; Miyagawa, S.; Miki, K.; Saito, A.; Fukushima, S.; Higuchi, T.; Kawamura, T.; Kuratani, T.; Daimon, T.; Shimizu, T.; et al. Feasibility, Safety, and Therapeutic Efficacy of Human Induced Pluripotent Stem Cell-Derived Cardiomyocyte Sheets in a Porcine Ischemic Cardiomyopathy Model. *Circulation* **2012**, *126*, S29–S37. [CrossRef] [PubMed]
45. Kim, M.C.; Kim, Y.S.; Kang, W.S.; Lee, K.H.; Cho, M.; Hong, M.H.; Lim, K.S.; Jeong, M.H.; Ahn, Y. Intramyocardial Injection of Stem Cells in Pig Myocardial Infarction Model: The First Trial in Korea. *J. Korean Med. Sci.* **2017**, *32*, 1708–1712. [CrossRef] [PubMed]
46. Ko, S.-F.; Yip, H.-K.; Lee, C.-C.; Sheu, J.-J.; Sun, C.-K.; Ng, S.-H.; Huang, C.-C.; Lin, Y.-C.; Chang, L.-T.; Chen, M.-C. Immediate Intramyocardial Bone Marrow-Derived Mononuclear Cells Implantation in Minipig Myocardium after Permanent Coronary Artery Ligation: Magnetic Resonance Imaging with Histopathologic and Immunochemical Correlation. *Investig. Radiol.* **2011**, *46*, 495–503. [CrossRef]
47. Lee, H.W.; Lee, H.C.; Park, J.H.; Kim, B.W.; Ahn, J.; Kim, J.H.; Park, J.S.; Oh, J.-H.; Choi, J.H.; Cha, K.S.; et al. Effects of Intracoronary Administration of Autologous Adipose Tissue-Derived Stem Cells on Acute Myocardial Infarction in a Porcine Model. *Yonsei Med. J.* **2015**, *56*, 1522–1529. [CrossRef] [PubMed]
48. Lee, S.-T.; White, A.J.; Matsushita, S.; Malliaras, K.; Steenbergen, C.; Zhang, Y.; Li, T.-S.; Terrovitis, J.; Yee, K.; Simsir, S.; et al. Intramyocardial Injection of Autologous Cardiospheres or Cardiosphere-Derived Cells Preserves Function and Minimizes Adverse Ventricular Remodeling in Pigs with Heart Failure Post-Myocardial Infarction. *J. Am. Coll. Cardiol.* **2011**, *57*, 455–465. [CrossRef] [PubMed]
49. Leu, S.; Sun, C.-K.; Sheu, J.-J.; Chang, L.-T.; Yuen, C.-M.; Yen, C.-H.; Chiang, C.-H.; Ko, S.-F.; Pei, S.-N.; Chua, S.; et al. Autologous Bone Marrow Cell Implantation Attenuates Left Ventricular Remodeling and Improves Heart Function in Porcine Myocardial Infarction: An Echocardiographic, Six-Month Angiographic, and Molecular-Cellular Study. *Int. J. Cardiol.* **2011**, *150*, 156–168. [CrossRef] [PubMed]
50. Li, X.; Zhang, F.; Song, G.; Gu, W.; Chen, M.; Yang, B.; Li, D.; Wang, D.; Cao, K. Intramyocardial Injection of Pig Pluripotent Stem Cells Improves Left Ventricular Function and Perfusion: A Study in a Porcine Model of Acute Myocardial Infarction. *PLoS ONE* **2013**, *8*, e66688. [CrossRef]
51. Liao, S.; Zhang, Y.; Ting, S.; Zhen, Z.; Luo, F.; Zhu, Z.; Jiang, Y.; Sun, S.; Lai, W.-H.; Lian, Q.; et al. Potent Immunomodulation and Angiogenic Effects of Mesenchymal Stem Cells versus Cardiomyocytes Derived from Pluripotent Stem Cells for Treatment of Heart Failure. *Stem Cell Res. Ther.* **2019**, *10*, 78. [CrossRef]
52. Lin, Y.-D.; Chang, M.-Y.; Cheng, B.; Liu, Y.-W.; Lin, L.-C.; Chen, J.-H.; Hsieh, P.C.H. Injection of Peptide Nanogels Preserves Postinfarct Diastolic Function and Prolongs Efficacy of Cell Therapy in Pigs. *Tissue Eng. Part A* **2015**, *21*, 1662–1671. [CrossRef]
53. Liu, C.-B.; Huang, H.; Sun, P.; Ma, S.-Z.; Liu, A.-H.; Xue, J.; Fu, J.-H.; Liang, Y.-Q.; Liu, B.; Wu, D.-Y.; et al. Human Umbilical Cord-Derived Mesenchymal Stromal Cells Improve Left Ventricular Function, Perfusion, and Remodeling in a Porcine Model of Chronic Myocardial Ischemia. *Stem Cells Transl. Med.* **2016**, *5*, 1004–1013. [CrossRef] [PubMed]
54. Liu, Y.-H.; Peng, K.-Y.; Chiu, Y.-W.; Ho, Y.-L.; Wang, Y.-H.; Shun, C.-T.; Huang, S.-Y.; Lin, Y.-S.; de Vries, A.A.F.; Pijnappels, D.A.; et al. Human Placenta-Derived Multipotent Cells (HPDMCs) Modulate Cardiac Injury: From Bench to Small and Large Animal Myocardial Ischemia Studies. *Cell Transplant.* **2015**, *24*, 2463–2478. [CrossRef] [PubMed]
55. Locatelli, P.; Olea, F.D.; Hnatiuk, A.; De Lorenzi, A.; Cerdá, M.; Giménez, C.S.; Sepúlveda, D.; Laguens, R.; Crottogini, A. Mesenchymal Stromal Cells Overexpressing Vascular Endothelial Growth Factor in Ovine Myocardial Infarction. *Gene Ther.* **2015**, *22*, 449–457. [CrossRef] [PubMed]
56. Lu, M.; Zhao, S.; Liu, Q.; Jiang, S.; Song, P.; Qian, H.; Zhang, Y.; Ling, J.; Yan, C.; Cheng, H.; et al. Transplantation with Autologous Mesenchymal Stem Cells after Acute Myocardial Infarction Evaluated by Magnetic Resonance Imaging: An Experimental Study. *J. Thorac. Imaging* **2012**, *27*, 125–135. [CrossRef] [PubMed]

57. Malliaras, K.; Smith, R.R.; Kanazawa, H.; Yee, K.; Seinfeld, J.; Tseliou, E.; Dawkins, J.F.; Kreke, M.; Cheng, K.; Luthringer, D.; et al. Validation of Contrast-Enhanced Magnetic Resonance Imaging to Monitor Regenerative Efficacy after Cell Therapy in a Porcine Model of Convalescent Myocardial Infarction. *Circulation* **2013**, *128*, 2764–2775. [CrossRef] [PubMed]
58. Mao, Q.; Lin, C.; Gao, J.; Liang, X.; Gao, W.; Shen, L.; Kang, L.; Xu, B. Mesenchymal Stem Cells Overexpressing Integrin-Linked Kinase Attenuate Left Ventricular Remodeling and Improve Cardiac Function after Myocardial Infarction. *Mol. Cell. Biochem.* **2014**, *397*, 203–214. [CrossRef] [PubMed]
59. Mazo, M.; Hernández, S.; Gavira, J.J.; Abizanda, G.; Araña, M.; López-Martínez, T.; Moreno, C.; Merino, J.; Martino-Rodríguez, A.; Uixeira, A.; et al. Treatment of Reperfused Ischemia with Adipose-Derived Stem Cells in a Preclinical Swine Model of Myocardial Infarction. *Cell Transplant.* **2012**, *21*, 2723–2733. [CrossRef]
60. Medicetty, S.; Wiktor, D.; Lehman, N.; Raber, A.; Popovic, Z.B.; Deans, R.; Ting, A.E.; Penn, M.S. Percutaneous Adventitial Delivery of Allogeneic Bone Marrow-Derived Stem Cells via Infarct-Related Artery Improves Long-Term Ventricular Function in Acute Myocardial Infarction. *Cell Transplant.* **2012**, *21*, 1109–1120. [CrossRef]
61. Mori, D.; Miyagawa, S.; Yajima, S.; Saito, S.; Fukushima, S.; Ueno, T.; Toda, K.; Kawai, K.; Kurata, H.; Nishida, H.; et al. Cell Spray Transplantation of Adipose-Derived Mesenchymal Stem Cell Recovers Ischemic Cardiomyopathy in a Porcine Model. *Transplantation* **2018**, *102*, 2012–2024. [CrossRef]
62. Natsumeda, M.; Florea, V.; Rieger, A.C.; Tompkins, B.A.; Banerjee, M.N.; Golpanian, S.; Fritsch, J.; Landin, A.M.; Kashikar, N.D.; Karantalis, V.; et al. A Combination of Allogeneic Stem Cells Promotes Cardiac Regeneration. *J. Am. Coll. Cardiol.* **2017**, *70*, 2504–2515. [CrossRef]
63. Ozawa, H.; Miyagawa, S.; Fukushima, S.; Itoh, E.; Harada, A.; Saito, A.; Ueno, T.; Toda, K.; Kuratani, T.; Sawa, Y. Sirtuin1 Regulates the Stem Cell Therapeutic Effects on Regenerative Capability for Treating Severe Heart Failure in a Juvenile Animal Model. *Ann. Thorac. Surg.* **2016**, *102*, 803–812. [CrossRef] [PubMed]
64. Peng, C.; Yang, K.; Xiang, P.; Zhang, C.; Zou, L.; Wu, X.; Gao, Y.; Kang, Z.; He, K.; Liu, J.; et al. Effect of Transplantation with Autologous Bone Marrow Stem Cells on Acute Myocardial Infarction. *Int. J. Cardiol.* **2013**, *162*, 158–165. [CrossRef] [PubMed]
65. Prifti, E.; Di Lascio, G.; Harmelin, G.; Bani, D.; Briganti, V.; Veshti, A.; Bonacchi, M. Cellular Cardiomyoplasty into Infracted Swine's Hearts by Retrograde Infusion through the Venous Coronary Sinus: An Experimental Study. *Cardiovasc. Revasc. Med. Mol. Interv.* **2016**, *17*, 262–271. [CrossRef] [PubMed]
66. Rabbani, S.; Soleimani, M.; Sahebjam, M.; Imani, M.; Nassiri, S.M.; Atashi, A.; Daliri Joupari, M.; Ghiaseddin, A.; Latifpour, M.; Ahmadi Tafti, S.H. Effects of Endothelial and Mesenchymal Stem Cells on Improving Myocardial Function in a Sheep Animal Model. *J. Tehran Heart Cent.* **2017**, *12*, 65–71. [PubMed]
67. Rigol, M.; Solanes, N.; Roura, S.; Roqué, M.; Novensà, L.; Dantas, A.P.; Martorell, J.; Sitges, M.; Ramírez, J.; Bayés-Genís, A.; et al. Allogeneic Adipose Stem Cell Therapy in Acute Myocardial Infarction. *Eur. J. Clin. Investig.* **2014**, *44*, 83–92. [CrossRef] [PubMed]
68. Romagnuolo, R.; Masoudpour, H.; Porta-Sánchez, A.; Qiang, B.; Barry, J.; Laskary, A.; Qi, X.; Massé, S.; Magtibay, K.; Kawajiri, H.; et al. Human Embryonic Stem Cell-Derived Cardiomyocytes Regenerate the Infarcted Pig Heart but Induce Ventricular Tachyarrhythmias. *Stem Cell Rep.* **2019**, *12*, 967–981. [CrossRef]
69. Schuleri, K.H.; Centola, M.; Choi, S.H.; Evers, K.S.; Dawoud, F.; George, R.T.; Lima, J.A.C.; Lardo, A.C. CT for Evaluation of Myocardial Cell Therapy in Heart Failure: A Comparison with CMR Imaging. *JACC Cardiovasc. Imaging* **2011**, *4*, 1284–1293. [CrossRef]
70. Sheu, J.-J.; Lee, F.-Y.; Yuen, C.-M.; Chen, Y.-L.; Huang, T.-H.; Chua, S.; Chen, Y.-L.; Chen, C.-H.; Chai, H.-T.; Sung, P.-H.; et al. Combined Therapy with Shock Wave and Autologous Bone Marrow-Derived Mesenchymal Stem Cells Alleviates Left Ventricular Dysfunction and Remodeling through Inhibiting Inflammatory Stimuli, Oxidative Stress & Enhancing Angiogenesis in a Swine Myocardial Infarction Model. *Int. J. Cardiol.* **2015**, *193*, 69–83. [CrossRef]
71. Shudo, Y.; Miyagawa, S.; Nakatani, S.; Fukushima, S.; Sakaguchi, T.; Saito, A.; Asanuma, T.; Kawaguchi, N.; Matsuura, N.; Shimizu, T.; et al. Myocardial Layer-Specific Effect of Myoblast Cell-Sheet Implantation Evaluated by Tissue Strain Imaging. *Circ. J. Off. J. Jpn. Circ. Soc.* **2013**, *77*, 1063–1072. [CrossRef]
72. Song, L.; Yang, Y.-J.; Dong, Q.-T.; Qian, H.-Y.; Gao, R.-L.; Qiao, S.-B.; Shen, R.; He, Z.-X.; Lu, M.-J.; Zhao, S.-H.; et al. Atorvastatin Enhance Efficacy of Mesenchymal Stem Cells Treatment for Swine Myocardial Infarction via Activation of Nitric Oxide Synthase. *PLoS ONE* **2013**, *8*, e65702. [CrossRef]
73. Sun, Q.-W.; Zhen, L.; Wang, Q.; Sun, Y.; Yang, J.; Li, Y.-J.; Li, R.-J.; Ma, N.; Li, Z.-A.; Wang, L.-Y.; et al. Assessment of Retrograde Coronary Venous Infusion of Mesenchymal Stem Cells Combined with Basic Fibroblast Growth Factor in Canine Myocardial Infarction Using Strain Values Derived from Speckle-Tracking Echocardiography. *Ultrasound Med. Biol.* **2016**, *42*, 272–281. [CrossRef] [PubMed]
74. Sun, S.; Jiang, Y.; Zhen, Z.; Lai, W.-H.; Liao, S.; Tse, H.-F. Establishing a Swine Model of Post-Myocardial Infarction Heart Failure for Stem Cell Treatment. *J. Vis. Exp. JoVE* **2020**, *159*, e60392. [CrossRef] [PubMed]
75. Suzuki, G.; Young, R.F.; Leiker, M.M.; Suzuki, T. Heart-Derived Stem Cells in Miniature Swine with Coronary Microembolization: Novel Ischemic Cardiomyopathy Model to Assess the Efficacy of Cell-Based Therapy. *Stem Cells Int.* **2016**, *2016*, 6940195. [CrossRef] [PubMed]
76. Tseliou, E.; Kanazawa, H.; Dawkins, J.; Gallet, R.; Kreke, M.; Smith, R.; Middleton, R.; Valle, J.; Marbán, L.; Kar, S.; et al. Widespread Myocardial Delivery of Heart-Derived Stem Cells by Nonocclusive Triple-Vessel Intracoronary Infusion in Porcine

Ischemic Cardiomyopathy: Superior Attenuation of Adverse Remodeling Documented by Magnetic Resonance Imaging and Histology. *PLoS ONE* **2016**, *11*, e0144523. [CrossRef]
77. Van der Spoel, T.I.G.; Gathier, W.A.; Koudstaal, S.; van Slochteren, F.; Of Lorkeers, S.J.; Sluijter, J.P.G.; Hoefer, I.E.; Steendijk, P.; Cramer, M.J.M.; Doevendans, P.A.; et al. Autologous Mesenchymal Stem Cells Show More Benefit on Systolic Function Compared to Bone Marrow Mononuclear Cells in a Porcine Model of Chronic Myocardial Infarction. *J Cardiovasc. Transl. Res.* **2015**, *8*, 393–403. [CrossRef]
78. Wang, X.; Zhen, L.; Miao, H.; Sun, Q.; Yang, Y.; Que, B.; Lopes Lao, E.P.; Wu, X.; Ren, H.; Shi, S.; et al. Concomitant Retrograde Coronary Venous Infusion of Basic Fibroblast Growth Factor Enhances Engraftment and Differentiation of Bone Marrow Mesenchymal Stem Cells for Cardiac Repair after Myocardial Infarction. *Theranostics* **2015**, *5*, 995–1006. [CrossRef]
79. Williams, A.R.; Hatzistergos, K.E.; Addicott, B.; McCall, F.; Carvalho, D.; Suncion, V.; Morales, A.R.; Da Silva, J.; Sussman, M.A.; Heldman, A.W.; et al. Enhanced Effect of Combining Human Cardiac Stem Cells and Bone Marrow Mesenchymal Stem Cells to Reduce Infarct Size and to Restore Cardiac Function after Myocardial Infarction. *Circulation* **2013**, *127*, 213–223. [CrossRef]
80. Winkler, J.; Lukovic, D.; Mester-Tonczar, J.; Zlabinger, K.; Gugerell, A.; Pavo, N.; Jakab, A.; Szankai, Z.; Traxler, D.; Müller, C.; et al. Quantitative Hybrid Cardiac [18F]FDG-PET-MRI Images for Assessment of Cardiac Repair by Preconditioned Cardiosphere-Derived Cells. *Mol. Ther. Methods Clin. Dev.* **2020**, *18*, 354–366. [CrossRef]
81. Yang, K.; Xiang, P.; Zhang, C.; Zou, L.; Wu, X.; Gao, Y.; Kang, Z.; He, K.; Liu, J.; Peng, C. Magnetic Resonance Evaluation of Transplanted Mesenchymal Stem Cells after Myocardial Infarction in Swine. *Can. J. Cardiol.* **2011**, *27*, 818–825. [CrossRef]
82. Yin, Q.; Pei, Z.; Wang, H.; Zhao, Y. Cyclosporine A-Nanoparticles Enhance the Therapeutic Benefit of Adipose Tissue-Derived Stem Cell Transplantation in a Swine Myocardial Infarction Model. *Int. J. Nanomed.* **2014**, *9*, 17–26. [CrossRef]
83. Zhang, G.-W.; Gu, T.-X.; Guan, X.-Y.; Sun, X.-J.; Jiang, D.-Q.; Tang, R.; Qi, X.; Li, X.-Y. Delayed Enrichment for C-Kit and Inducing Cardiac Differentiation Attenuated Protective Effects of BMSCs' Transplantation in Pig Model of Acute Myocardial Ischemia. *Cardiovasc. Ther.* **2015**, *33*, 184–192. [CrossRef] [PubMed]
84. Zhang, G.-W.; Liu, X.-C.; Li-Ling, J.; Luan, Y.; Ying, Y.-N.; Wu, X.-S.; Zhao, C.-H.; Liu, T.-J.; Lü, F. Mechanisms of the Protective Effects of BMSCs Promoted by TMDR with Heparinized BFGF-Incorporated Stent in Pig Model of Acute Myocardial Ischemia. *J. Cell. Mol. Med.* **2011**, *15*, 1075–1086. [CrossRef] [PubMed]
85. Zhao, J.-J.; Liu, X.-C.; Kong, F.; Qi, T.-G.; Cheng, G.-H.; Wang, J.; Sun, C.; Luan, Y. Bone Marrow Mesenchymal Stem Cells Improve Myocardial Function in a Swine Model of Acute Myocardial Infarction. *Mol. Med. Rep.* **2014**, *10*, 1448–1454. [CrossRef] [PubMed]
86. Noiseux, N.; Borie, M.; Desnoyers, A.; Menaouar, A.; Stevens, L.M.; Mansour, S.; Danalache, B.A.; Roy, D.-C.; Jankowski, M.; Gutkowska, J. Preconditioning of Stem Cells by Oxytocin to Improve Their Therapeutic Potential. *Endocrinology* **2012**, *153*, 5361–5372. [CrossRef]
87. Tan, S.C.; Gomes, R.S.; Yeoh, K.K.; Perbellini, F.; Malandraki-Miller, S.; Ambrose, L.; Heather, L.C.; Faggian, G.; Schofield, C.J.; Davies, K.E.; et al. Preconditioning of Cardiosphere-Derived Cells With Hypoxia or Prolyl-4-Hydroxylase Inhibitors Increases Stemness and Decreases Reliance on Oxidative Metabolism. *Cell Transpl.* **2016**, *25*, 35–53. [CrossRef]
88. Araña, M.; Gavira, J.J.; Peña, E.; González, A.; Abizanda, G.; Cilla, M.; Pérez, M.M.; Albiasu, E.; Aguado, N.; Casado, M.; et al. Epicardial Delivery of Collagen Patches with Adipose-Derived Stem Cells in Rat and Minipig Models of Chronic Myocardial Infarction. *Biomaterials* **2014**, *35*, 143–151. [CrossRef]
89. Rashedi, I.; Talele, N.; Wang, X.-H.; Hinz, B.; Radisic, M.; Keating, A. Collagen Scaffold Enhances the Regenerative Properties of Mesenchymal Stromal Cells. *PLoS ONE* **2017**, *12*, e0187348. [CrossRef]
90. Müller, P.; Lemcke, H.; David, R. Stem Cell Therapy in Heart Diseases—Cell Types, Mechanisms and Improvement Strategies. *Cell. Physiol. Biochem.* **2018**, *48*, 2607–2655. [CrossRef]
91. Xu, J.; Liu, D.; Zhong, Y.; Huang, R. Effects of Timing on Intracoronary Autologous Bone Marrow-Derived Cell Transplantation in Acute Myocardial Infarction: A Meta-Analysis of Randomized Controlled Trials. *Stem Cell Res. Ther.* **2017**, *8*, 231. [CrossRef]
92. Siu, C.-W.; Liao, S.-Y.; Liu, Y.; Lian, Q.; Tse, H.-F. Stem Cells for Myocardial Repair. *Thromb. Haemost.* **2010**, *104*, 6–12. [CrossRef]
93. Zwetsloot, P.P.; Végh, A.M.D.; Jansen of Lorkeers, S.J.; van Hout, G.P.J.; Currie, G.L.; Sena, E.S.; Gremmels, H.; Buikema, J.W.; Goumans, M.-J.; Macleod, M.R.; et al. Cardiac Stem Cell Treatment in Myocardial Infarction. *Circ. Res.* **2016**, *118*, 1223–1232. [CrossRef] [PubMed]
94. Korf-Klingebiel, M.; Kempf, T.; Sauer, T.; Brinkmann, E.; Fischer, P.; Meyer, G.P.; Ganser, A.; Drexler, H.; Wollert, K.C. Bone Marrow Cells Are a Rich Source of Growth Factors and Cytokines: Implications for Cell Therapy Trials after Myocardial Infarction. *Eur. Heart J.* **2008**, *29*, 2851–2858. [CrossRef] [PubMed]
95. Spannbauer, A.; Mester-Tonczar, J.; Traxler, D.; Kastner, N.; Zlabinger, K.; Hašimbegović, E.; Riesenhuber, M.; Pavo, N.; Goliasch, G.; Gyöngyösi, M. Large Animal Models of Cell-Free Cardiac Regeneration. *Biomolecules* **2020**, *10*, 1392. [CrossRef] [PubMed]
96. Petitti, D.B. *Meta-Analysis, Decision Analysis, and Cost-Effectiveness Analysis: Methods for Quantitative Synthesis in Medicine*; Oxford University Press: New York, NY, USA, 2000; ISBN 978-0-19-513364-6.
97. Gurevitch, J.; Koricheva, J.; Nakagawa, S.; Stewart, G. Meta-Analysis and the Science of Research Synthesis. *Nature* **2018**, *555*, 175–182. [CrossRef] [PubMed]
98. Spector, T.D.; Thompson, S.G. The Potential and Limitations of Meta-Analysis. *J. Epidemiol. Community Health* **1991**, *45*, 89–92. [CrossRef] [PubMed]

animals

MDPI

Review

An Outstanding Role of Adipose Tissue in Canine Stem Cell Therapy

Marina Prišlin [1], Dunja Vlahović [2], Petar Kostešić [2], Ivana Ljolje [3], Dragan Brnić [1], Nenad Turk [2], Ivana Lojkić [1], Valentina Kunić [1], Tugomir Karadjole [2] and Nina Krešić [1,*]

1 Croatian Veterinary Institute, Savska Cesta 143, 10000 Zagreb, Croatia; prislin@veinst.hr (M.P.); brnic@veinst.hr (D.B.); ilojkic@veinst.hr (I.L.); kunic@veinst.hr (V.K.)
2 Faculty of Veterinary Medicine, University of Zagreb, Heinzelova 55, 10000 Zagreb, Croatia; dvlahovic@vef.unizg.hr (D.V.); kostesic@vef.hr (P.K.); turk@vef.unizg.hr (N.T.); ktugomir@vef.unizg.hr (T.K.)
3 Veterinary Clinic for Small Animals Buba, Dore Pfanove 11, 10000 Zagreb, Croatia; ivana.ljolje@gmail.com
* Correspondence: lemo@veinst.hr

Simple Summary: Throughout history, the role of adipose tissue has changed for humans, and regarding canines: the role has changed from connective tissue to restoration of physiological functions. The adipose tissue cells have extraordinary mechanisms of healing tissue function, and the most outstanding component of adipose tissue discovered are mesenchymal stem cells. It has been almost fifteen years since their discovery in canine adipose tissue. Since then, numerous studies have investigated the possibilities of adipose-derived mesenchymal stem cells in treating various canine diseases. This review summarised the progress of confirming the therapeutic role of adipose tissue components, focusing on stem cells as the most researched and with the highest potential in enabling a better quality of life for canines.

Abstract: Adipose tissue, previously known as connective tissue with a role in energy storage, is currently changing the course of treatments in veterinary medicine. Recent studies have revealed one particularly impressive function among all the newly discovered functions of adipose tissue. The interactive cells hosted by adipose tissue, the stromal vascular fraction (SVF), and their role in treating numerous diseases have provided a prospective course of research with positive outcomes in regenerative veterinary medicine (RVM). This review describes the main features of adipose tissue, emphasizing an eclectic combination of cells within the SVF and its thus far researched therapeutic possibilities in canine RVM. An afterwards focus is on a highly researched component of the SVF, adipose-derived mesenchymal stem cells (ASCs), which were shown to have an extraordinary impact relying on several proposed mechanisms of action on mitigating pathologies in canines. Furthermore, ASC therapy showed the most significant results in the orthopaedics field and in neurology, dermatology, ophthalmology, gastroenterology, and hepatology, which elevates the possibilities of ASC therapy to a whole new level. Therefore, this review article aims to raise awareness of the importance of research on cellular components, within abundant and easily accessible adipose tissue, in the direction of regenerative therapy in canines, considering the positive outcomes so far. Although the focus is on the positive aspects of cellular therapy in canines, the researchers should not forget the importance of identifying the potential negative aspects within published and upcoming research. Safe and standardized treatment represents a fundamental prerequisite for positively impacting the lives of canine patients.

Keywords: canine; adipose-derived mesenchymal stem cells; stem cell therapy; regenerative veterinary medicine; stromal vascular fraction; adipose tissue

Citation: Prišlin, M.; Vlahović, D.; Kostešić, P.; Ljolje, I.; Brnić, D.; Turk, N.; Lojkić, I.; Kunić, V.; Karadjole, T.; Krešić, N. An Outstanding Role of Adipose Tissue in Canine Stem Cell Therapy. *Animals* **2022**, *12*, 1088. https://doi.org/10.3390/ani12091088

Academic Editors: Eleonora Iacono and Barbara Merlo

Received: 18 March 2022
Accepted: 20 April 2022
Published: 22 April 2022

1. Introduction

Adipose tissue (AT) was once considered to be the only connective tissue involved in energy storage. Currently, recognition of AT function is beyond simple fat storage, and is well known as a metabolic and endocrine organ [1–3]. Consequently, this review aims to raise awareness on the importance of research on cellular components, within abundant and easily accessible adipose tissue, in the direction of regenerative therapy in canines, considering positive outcomes so far and failures of current treatment options.

Today it is accepted that adipose tissue is of mesodermal origin. However, there is evidence that craniofacial adipose deposits may originate from the neural crest [4]. Consequently, the origins of adipose tissue are complex and have to be fully explored [5]. Adipose tissue is crucial in maintaining lipid and glucose homeostasis [6]. Endocrine role turnover appears with the possibility of producing oestrogen, resistin, and leptin [1,7] and regulates food intake, body mass, reproductive functioning, foetal growth, pro-inflammatory immune responses, angiogenesis, and lipolysis [8]. Furthermore, it was discovered that AT secretes pro-inflammatory chemokines and cytokines such as interleukins (IL) 1, IL 6, IL 8, tumour necrosis factor-alpha (TNF-α), as well as proteins with a role in lipid metabolism, in vascular haemostasis or the complement system. The mechanism of action of those proteins may be autocrine, paracrine, or distant from AT [9]. To date, several AT types are identified, i.e., white (WAT), brown (BAT), and beige (BGAT) distributed in various anatomical parts throughout the organism [6]. Adult canines contain AT located mainly in subcutaneous and visceral depots [10]. In regenerative veterinary medicine (RVM), AT from the periovarian area, ligament falciform, and subcutaneous area is generally used. It can also be easily obtained during elective surgeries such as ovariotomy and gastropexy where AT is collected as medical waste (Figure 1).

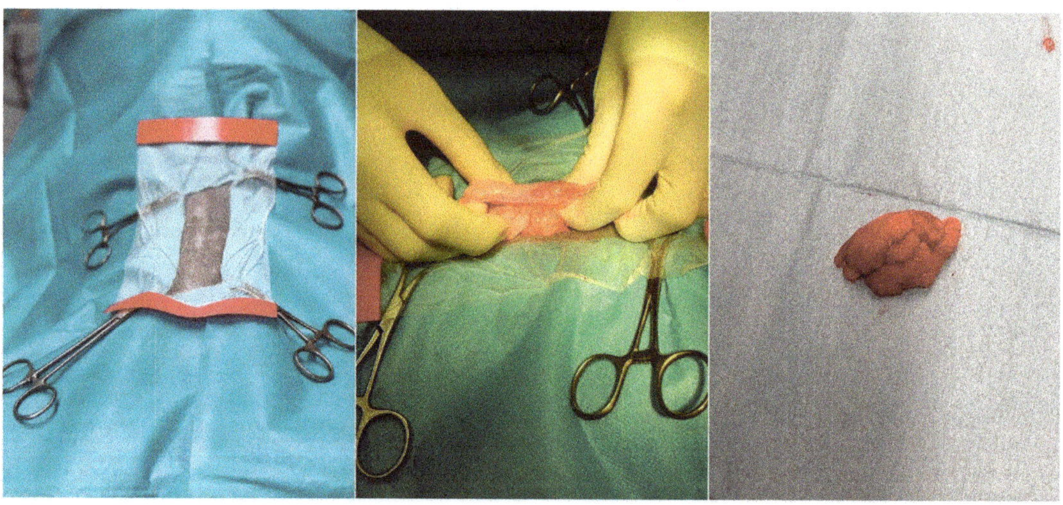

Figure 1. Adipose tissue collection during canine ovariotomy. The routine procedure commonly performed in young females presents an excellent opportunity to collect adipose tissue and store cells for future use in regenerative therapy.

2. Adipocytes—The Main Compound of AT

Adipocytes are the main compounds of AT and can exist in almost every organism structure. They occur individually or in small groups scattered throughout the connective tissue. Loose connective tissue contains adipocytes or clusters of multiple cells, but the tissue is referred to as AT when the fat cells outnumber other cell types [11].

WAT cells are formed soon after birth, and their main purpose is to store triglycerides. The formation of adipocytes starts with mesenchymal stem cells turning into adipoblasts which further differentiate into pre-adipocytes. After pre-adipocytes reach growth arrest, they change their appearance, accumulate triglycerides, and become mature adipocytes with lost ability of division [12]. BAT cells develop before birth and specialize in defending mammals against hypothermia [13]. The morphogenetic protein (BMP)-7 is responsible for the differentiation process of brown pre-adipocytes into BAT [14].

BAT is equipped with the metabolic machinery comprising the numerous mitochondria and the appropriate enzymes that allow fatty acids to oxidize at enhanced rates than that of WAT. In addition, the mitochondria of brown adipose tissue cells primarily generate heat rather than adenosine triphosphate (ATP) and can sustain body heat during prolonged periods of cold [15].

The last discovered and current highly researched type of adipocytes are beige orbrite. These have the characteristics of WAT and BAT cells [16]. The synthesis of BGAT is a highly investigated topic in diabetes and metabolism research [17,18].

As mentioned, AT was once viewed as a passive triglyceride depot, but AT is now known as a complex tissue giving residence to various interacting cells, also known as the stromal vascular fraction (SVF) [1,13].

3. Stromal Vascular Fraction—Interacting Cells Hosted by AT

SVF is an eclectic combination of cells, including adipose-derived mesenchymal stem cells (ASCs), blood cells, endothelial precursors, endothelial and smooth muscle cells, pre-adipocytes, pericytes, macrophages and adipocytes [3,19,20] (Figure 2). Although adipocytes account for >90% of AT volume, SVF predominates in overall cell number [13]. In humans, SVF cells isolated from WAT possess more hematopoietic cells, macrophages, hematopoietic progenitors, and immature cells that, together, contribute to a higher degree of plasticity than SVF cells isolated from BAT [3,21].

Isolation of SVF from AT can be obtained by mechanical disruption and enzymatic digestion (Figure 3). The AT disassociation and SVF extraction most often involve a combination of the mechanical disruption of connective tissue, followed by enzymatic digestion with collagenase [22,23]. Both procedures aim to preserve the stem cells, the vascular compartment (stromal cell niche) viability and the therapeutic benefits of SVF products [24]. However, there are differences in outcomes between these two methods. For example, enzymatic digestion provides the phenotype of individual cells, while the mechanical extraction itself preserves interaction between cells and matrix [25]. Nevertheless, enzymatic digestion is considered the "gold standard" since it provides significantly greater cell viability [26].

The tissue harvesting site also presents a challenge since it can impair SVF and ASCs viability, cellular yield and immunophenotype. Recently, Hendawy et al. (2021) found that the peri-ovarian region is the most favourable site for harvesting ASCs in dogs since it contains the highest number of viable cells per gram of AT compared to subcutaneous and falciform ligament sites and also the highest number of CD90+ cells [22]. In 2013, Astor et al. reported similar results; AT collected at the falciform location had significantly fewer viable cells per gram (VCPG) than tissue collected at the thoracic wall and inguinal sites [27]. The same authors also reported the influence of age in SVF cell viability with significantly higher VCPG in dogs up to 4.5 years old; higher VCPG was also noted in non-spayed dogs compared to spayed ones. In addition, other authors noted the significantly higher population doubling and differentiation potentials in young donors [28]. As observed, consideration of many specific factors is needed to provide the best SVF therapy solution.

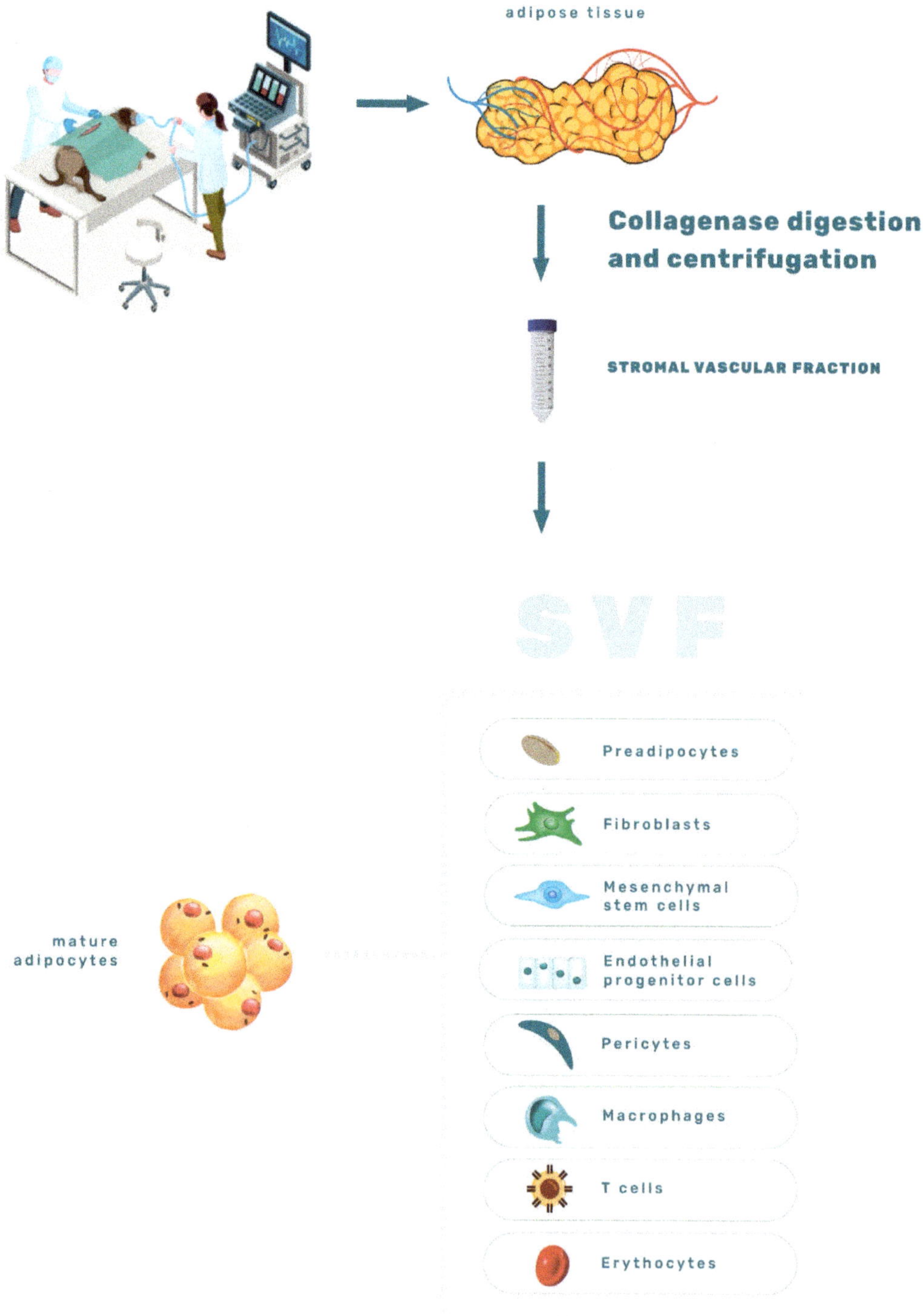

Figure 2. Graphical representation of the stromal vascular fraction components.

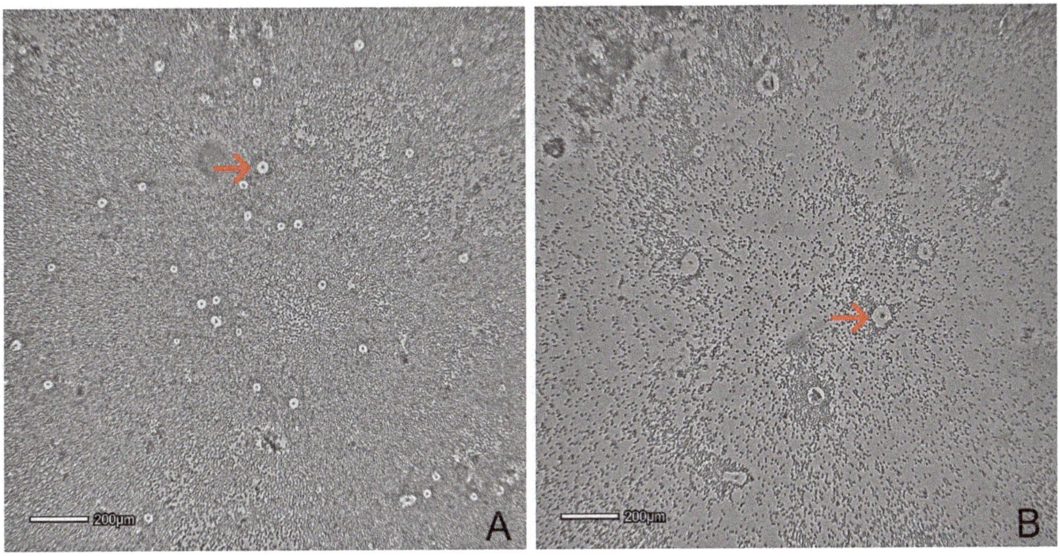

Figure 3. (**A**,**B**) represent stromal vascular fraction 24 h post isolation from peri-ovarian and subcutaneous adipose tissue, seeded in T25 flask after mechanical and enzymatic disruption. Cells pointed with a red arrow are plastic adherent cells in expansion; the surrounding cells are nonadherent. Pictures were obtained with Cytosmart Lux2 (CytoSMART Technologies B.V., The Netherlands).

3.1. Mechanism of Action

The existing literature suggests that SVF achieves regeneration and healing through pro-angiogenic and immunomodulatory mechanisms, including differentiation and extracellular matrix secretion [19]. The first study reports the effectiveness of SVF therapy in dogs in 2007 [29]. The effectiveness may be due to the presence of ASCs, the vascular niche cells, and, finally, the interactions between all cells present in SVF [24]. Senesi et al. (2019) retain that the anti-inflammatory and immunoregulatory effect of SVF for osteoarthrosis is more likely than cells' ability to differentiate in the specific cell lineage [26]. Hendawy et al. (2021) attribute the crucial effects of SVF to the presence of a sufficient number of ASCs, with preserved differentiation capacity. Because of the complex interactions between SVF and specific organs, the function of SVF in the treatment of various pathologies needs further clarification [22,26]. The mentioned mechanisms of action are elaborated in detail in the following sections of this review.

3.2. SVF Clinical Application for Various Conditions

Adipose SVF injection proved helpful in the orthopaedic field because it is a favourable, minimally invasive, non-surgical alternative for treating musculoskeletal disorders [26]. Osteoarthritis of the hip joint was significantly improved 24-weeks following treatment with simultaneous intraarticular (IA) and intravenous (IV) injection of autologous adipose-derived SVF and platelet-rich plasma (PRP) [30]. Lameness and range of motion significantly improved, as well as the overall quality of life in a double-blind study of canine hip joint osteoarthritis after 30, 60 and 90 days; although the cells in this study are named ASCs, the study indicates the application of a heterogeneous population of cells, including ASCs [29]. The same research group tested dogs suffering from elbow joint osteoarthritis. The placebo control group was not included in this study, but based on their previous analysis of hip joints, the significant improvement was attributed to the IA injected AT-derived heterogeneous cell population [31]. In four canine patients diagnosed with hip dysplasia, autologous SVF acupoint injection showed marked improvement, compared with baseline results after the first week of treatment [32].

The use of allogenic SVF in degenerative joint disease of the spine in dogs revealed an increased serum level of the vascular endothelial growth factor of affected animals in the second week of treatment. In the eighth week, the levels were decreased [33]. The same study published that decreased pain and reduced lameness were noticed a few days following therapy, overall concluding the improvement of joint regeneration capacity. The lack of research in the veterinary field indicates a significant need for further investigation of SVF benefits.

4. Adipose-Derived Mesenchymal Stem Cells—An Outstanding Component of the SVF

It is well known that stem cells provide tissues and organs with a fresh cellular compartment that can replace cells that have expired naturally and provide physiological balance in the organism. In addition, the expiration of cells due to natural processes or damage enables regeneration of the tissues [34]. The significant discovery of a stem cell system within AT occurred twenty years ago [35,36]. This finding raised considerable interest in the veterinary scientific community. The results were first documented in 2008 when scientists successfully isolated and fully described ASCs in canines [37] which laid the foundation for RVM.

The ASCs, a subpopulation within SVF, are non-hemopoietic stem cells originating from the mesoderm [38]. What makes them intriguing for cell research and therapy, among MSC properties such as self-renewal, in vitro proliferation, non-specialization, and ability to differentiate in another type of cell, is their easy accessibility. To address AT isolated cells as ASCs, the International Society for Cellular Therapy (ISCT) and The International Federation for Adipose Therapeutics (IFATS) have provided guidelines and recommendations for the minimal essential characterization of human ASCs. The established criteria were: capacity to proliferate as adherent cells in cultures, the ability of minimal three lineages in vitro differentiation (osteogenic, chondrogenic and adipogenic) (Figure 4), phenotypical positivity for CD90, CD73, CD105 and negativity for CD14, CD34, CD45, CD11b, CD19 or CD79α [24,39]. Although scientists apply those rules for canine ASCs research, the exact criteria are still not wholly established for this species. Though, numerous studies are contributing to ASCs characterization. In this context, the investigation of these changes in surface marker expression (CD73, CD90, CD29, CD44, CD271, CD45 and CD14) has been performed through six passages, providing a timeframe the ASCs cultivated in vitro possess optimal surface marker expression for use in therapy [23].

From the moment of their discovery, ASCs features were exploited in vitro to generate sufficient cell numbers to reach therapeutic doses depending on the disease for which the ASCs are being tested; meanwhile, their properties, gene expression and surface marker expression can be heavily influenced by such manipulation. Inevitably, prolonged cultivation in vitro carries side effects in terms of affection of the characteristic ASCs membrane markers responsible for their positive impact [23]. Therefore, basic research on ASCs properties in vitro is needed to further reveal their molecular signatures.

4.1. Mechanism of Action

As already well documented for MSC in general, the healing properties are probably a result of the secretion of many factors influencing the immune system, with anti-apoptotic, anti-inflammatory, chemotactic and pro-angiogenic functions [10,40–42].

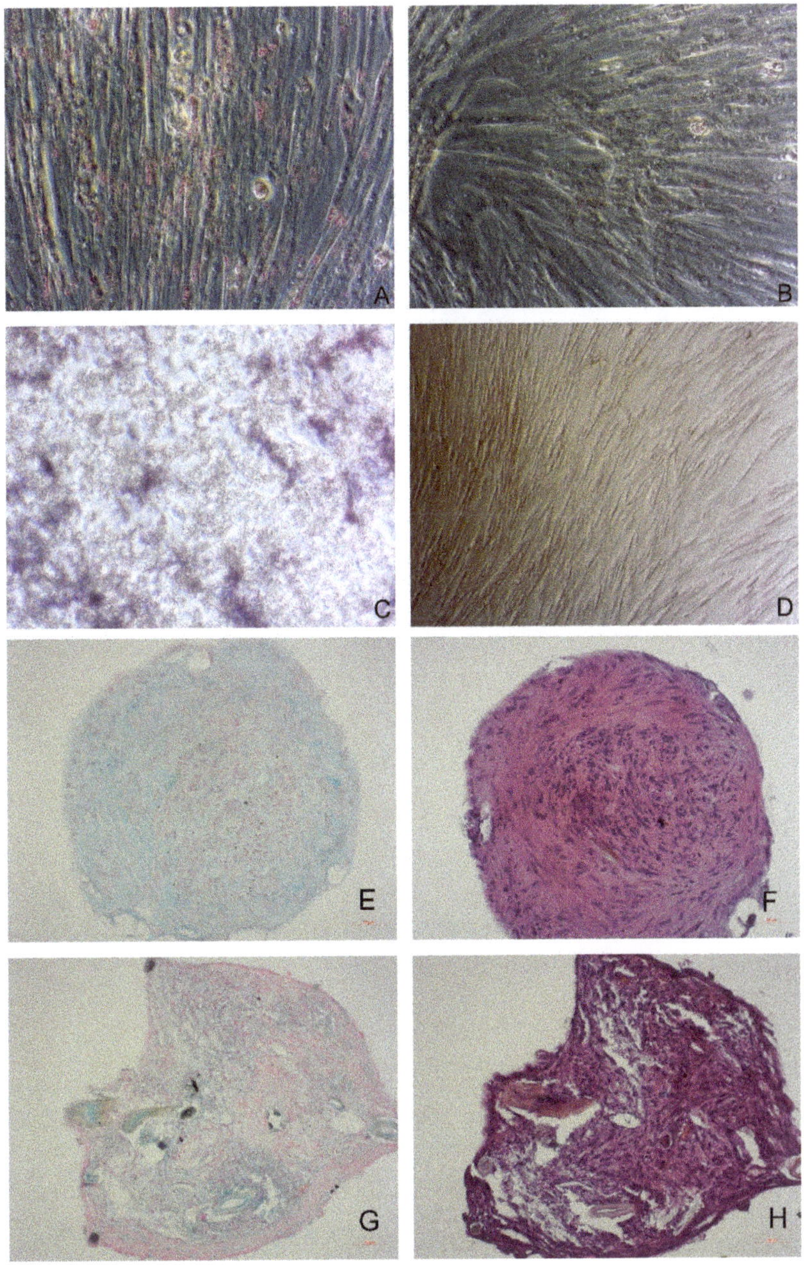

Figure 4. (**A**,**B**) Successful differentiation of canine adipose-derived mesenchymal stem cells (ASCs) in adipocyte differentiation media. When stained with Oil O Red, accumulated lipid droplets show high-intensity red staining within the cell (**A**) regarding control cultivated in basal medium (**B**). (**C**,**D**) Canine ASCs after successful osteodifferentiation; cells were stained with substrate to detect alkaline phosphatase activity. Purple strains of canine ASCs showed activity of alkaline phosphatase (**C**), while cells cultivated in basal medium (negative control) (**D**), showing low-intensity staining.

4E-H Images of histological sections of paraffin-embedded spheroids (20×, Zeiss, Germany) of canine ASCs three-dimensional culture after successful chondrodifferentiation. ASCs spheroids were stained with Alcian blue to detect the presence of aggrecan (**E**,**G**) and with H&E (**F**,**H**). Microscopic images (20×) (**A**–**H**) were taken with Zeiss Axiovert, Carl Zeiss AG, Jena, Germany.

The mechanism of action (Figure 5) operates by sending and receiving autocrine, paracrine, endocrine and intracellular signals [40]. However, the primary therapeutic effect of MSCs is paracrine signalling inducing functional changes in monocytes/macrophages, dendritic cells, T-cells, B-cells, and natural killer cells [42]. Furthermore, MSCs can transfer various molecules through the extracellular vesicles (ECV): exosomes, microvesicles, and apoptotic bodies. ECVs are vesicles produced from the plasma membrane, and carry mRNA, proteins, miRNA, and mitochondria and travel within the body [42]. Except for the two MSC mechanisms of action mentioned above, apoptosis-mediated immunomodulation and mitochondrial transfer are other possible mechanisms of MSC action [42].

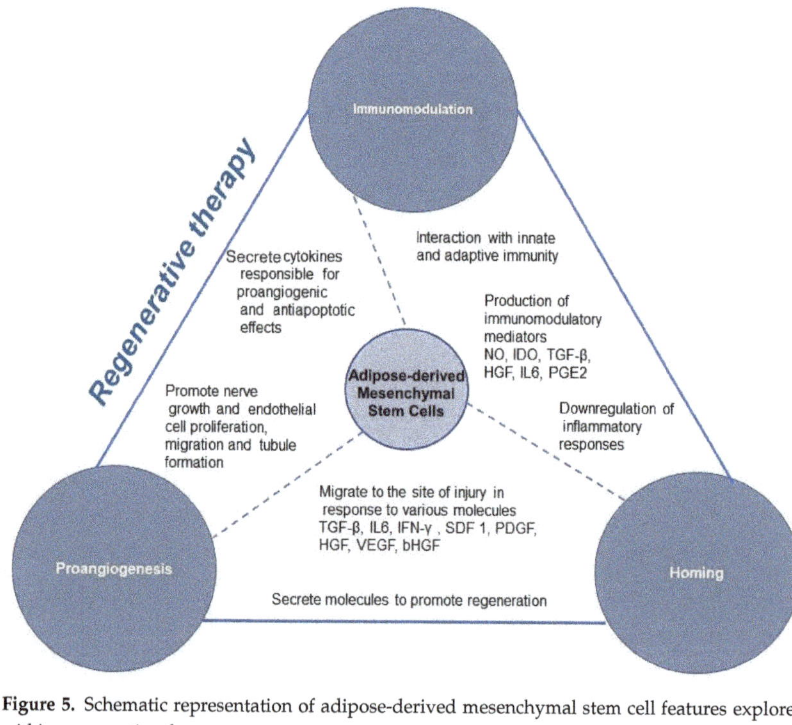

Figure 5. Schematic representation of adipose-derived mesenchymal stem cell features explored within regenerative therapy.

4.1.1. Immunomodulation

The interaction of MSCs with the innate and adaptive immune systems usually results in the downregulation of ongoing inflammatory responses, though the immune response can also be upregulated. The MSC immunomodulation is influenced by many factors such as activation, tissue of origin, dose and time of application, and interaction with immune cells [43]. MSC immunomodulation remains yet to be elucidated; however, paracrine signalling via immunomodulatory mediators such as nitric oxide (NO), indoleamine 2,3-dioxygenase (IDO), transforming growth factor-β (TGF-β), hepatocyte growth factor (HGF), hemoxygenase (HO), IL-6 and prostaglandin E2 (PGE2) is believed to be the first stage. In addition, this may also occur through direct contact between cells [43–45]. Chow et al. (2017) reported that canine MSC suppressed T cell activation by TGF-b signalling pathways and adenosine signalling [46]. This finding further indicates that canine MSC, unlike human and rodent MSC, relies primarily on cyclooxygenase and TGF-b pathways for T cell suppression rather than on NO or IDO-mediated pathways. Besides suppressing T cells, MSCs suppress B cell activation and proliferation, dendritic cells maturation, inhibit NK cell proliferation and cytotoxicity, and promote regulatory T cell generation via soluble factors or cell-cell contact [44]. T cell necrosis by canine MSC is an additional mechanism of immune modulation [46]. Canine ASCs can suppress lipopolysaccharide mediated activation/maturation of canine dendritic cells (DC). The impact in vivo of such squelched DC activation would undoubtedly result in an attenuated ability to appropriately prime T cell responses. This effect would be exacerbated if the ASCs were first activated with IFNg, suggesting that the suppressive effect would be optimal in an inflammatory environment typical of autoimmune or pro-inflammatory conditions [47].

Another "immune-privileged" MSC property is their low immunogenicity attributed to low expression of MHC I, absence of co-stimulating CD80, CD86 and CD40, MHC II deficiency and whole paracrine spectrum of biomolecules and growth factors through which they establish their action [48–50]. Each of the mentioned pathways reflects the possibilities these cells offer to treat various disorders and organ systems. However, all aforementioned mechanisms also imply the differences between species and offer space for new acknowledgements.

4.1.2. Homing

The MSCs have a remarkable ability to locate damaged tissues [3,42]. In response to chemotactic signals, MSCs reach the circulation and migrate to the site of injury, where they secrete molecules to promote regeneration. However, it is unclear which chemotactic signals guide MSCs to appropriate microenvironments [51]. The homing of MSCs is currently inefficient, and after they are systemically administered a small percentage of cells reach the target tissue [52]. The process of migration from the bloodstream to tissue involves steps for lymphocyte migration: (1) tethering and rolling, (2) activation, (3) firm adhesion, (4) transmigration or diapedesis, resulting in migration into tissue due to chemotaxis as described by Sackstein [53]. The migration of MSCs occur in response to various chemokines and growth factors, including TNF-α (tumour necrosis factor-α), IL-6, IL-8 [54]. Unlike comprehensive knowledge on blood cell homing, MSC homing remains poorly understood as tethering or rolling and transmigration.

The therapy research of MSCs bears one of the most significant aims, i.e., improving their homing efficiency. MSC homing can be categorized into (1) targeted administration—administration of ASCs at or near the target tissue, (2) magnetic guidance—cells labelled with magnetic particles are directed to the organ of interest using an external magnetic field, (3) genetic modification—permanent overexpression of homing factors via viral transduction, (4) cell surface engineering—temporarily chemical engineering by enzymes or ligands, (5) in vitro priming—altering culture conditions to affect gene expression, (6) and modification of the target tissue by direct injection of homing factors, genetic modification of target tissue, scaffold implantation, or using radiotherapeutic and ultrasound techniques[52].

4.1.3. Pro-Angiogenic and Anti-Apoptotic Mechanism of Action

MSCs secrete various cytokines responsible for pro-angiogenic and anti-apoptotic effects and in doing so, MSCs enable tissue regeneration and revascularization. Soluble angiogenic factors secreted by MSCs include fibroblast growth factors, hepatocyte growth factor and the vascular endothelial growth factor.

The lack of canine-specific antibodies has hampered identification of growth factors in the secretome of canine MSCs. Likely, secretomes of other species are similar to secretomes secreted by canine MSCs thus there is an idea on the secretome composition of canine MSCs based on this information [55]. Canine MSCs promote nerve growth and endothelial cell proliferation, migration and tubule formation by secretion of neurotrophic and angiogenic factors. Delfi et al., 2016 demonstrated MSC paracrine activity on nerves and blood vessels in the vicinity of the wound site. It was shown that MSC transplants promote increased

Table 1. *Cont.*

	Clinical Condition	Number of Canines Included	Type of Application	Route of Administration	Number of Cells ($\times 10^6$)	Reference
	Atopic dermatitis	26	Allogeneic ASCs	Intravenous	1.5	Villatoro et al. (2018) [78]
	Acute and chronic skin wound	24	Allogeneic ASCs	Intradermal	30	Enciso et al. (2020) [79]
	Atopic dermatitis	15	Allogeneic ASCs	Subcutaneous	Low dose = 0.5/kg body weight High dose = 5/kg body weight	Kaur et al. (2022) [80]
Ophthalmology	Keratoconjunctivitis sicca	12	Allogeneic ASCs	Around the lacrimal glands	5	Villatoro et al. (2015) [81]
	Keratoconjunctivitis sicca	15 (24 eyes)	Allogeneic ASCs	Intralacrimal	1	Bittencourt et al. (2016) [82]
	Keratoconjunctivitis sicca	22	Allogeneic ASCs	Topic in the conjunctival sac	1	Sgrignoli et al. (2019) [83]
	Corneal wound	26	Allogeneic ASCs	Sub-conjunctival	3	Falcao et al. (2020) [84]
Gastroenterology	Inflammatory bowel disease	11	Allogeneic ASCs	Intravenous	2/kg body weight	Perez-Merino et al. (2015) [85]
Hepatology	Acute liver injury	9	Allogeneic ASCs	Peripheral vein/splenic vein	2	Teshima et al. (2017) [86]
	Degenerative hepatopathy	10	Autologous ASCs	Portal vein	0.5/kg body weight	Gardin et al. (2018) [87]
	Acute liver injury	6	Allogeneic ASCs	Intravenous	10	Yan et al. (2019) [88]

4.2.1. Orthopaedics

Therapeutic effects of autologous and allogeneic ASCs applications in orthopaedics have improved pain and lameness in dogs with osteoarthritis [57,58,60,61,64,65,67–70]. In 2018, a group of scientists described the allogeneic ASCs application on 203 dogs and concluded that IA treatment gave better results when compared with the IV treatment in the polyarthritis condition. The age proved influential as most dogs under the age of five receiving IA treatment showed good improvement [67]. In addition, the canines' overall health and vitality are significant factors in response to the ASCs therapy. Positive therapeutic outcomes were further observed in chronic osteoarthritis (OA) of the hip and elbow joint [59,63,66]. In addition, ASCs treatment significantly improved the symptoms of hip dysplasia in 60% of treated dogs after one week [32]. This study compared the effects of SVF and ASCs therapy administered to acupuncture points and reported better results for SVF than ASCs therapy. However, it was concluded that SVF or allogeneic ASCs could be safely used as an acupoint injection for treating hip dysplasia in dogs [31]. A follow-up study highlighted the importance of cell administration before the injury becomes severe [59]. Furthermore, significant improvements following ASCs therapy of semitendinosus myopathy are documented [71,72].

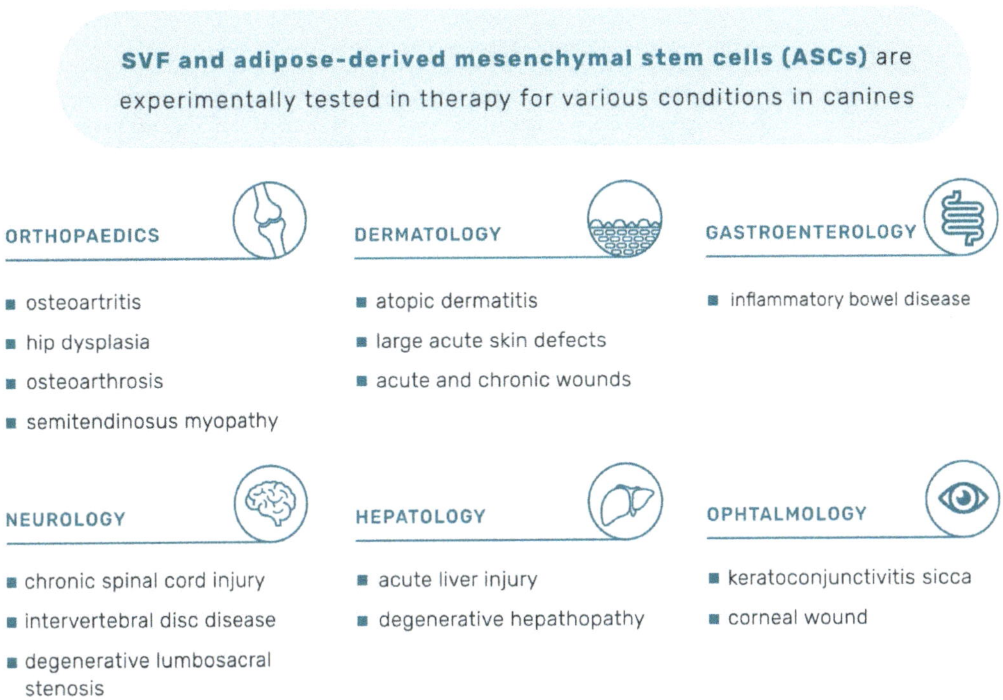

Figure 7. Pathological conditions in canines for which adipose-derived mesenchymal stem cell therapy was applied.

4.2.2. Neurology

In dogs with chronic spinal cord injury/intervertebral disc disease, percutaneous intraspinal transplantation of allogeneic ASCs had no adverse effects or complications (infection, neuropathic pain, or worsening neurological function) during the 16-week follow-up period. In addition, three animals improved locomotion, and one animal walked without support. However, no changes in deep pain perception were observed [73]. In the most recent research on lumbosacral spinal cord injury, transplantation of allogeneic ASCs with surgery in four dogs showed significant neurological improvements with normal ambulatory ability (4/4) and urinary control (3/4) three months after the surgery and the first ASCs transplantation [76]. While in the case of acute paraplegia, epidural canine ASCs transplantation with surgical decompression contributed to faster locomotor recovery and reduced the length of post-surgery hospitalization [74]. Another successful case reported the use of cultured autologous ASCs injected bilaterally at the level of L7-S1 in the external aperture of the intervertebral foramen of degenerative lumbosacral stenosis in a canine patient [75].

4.2.3. Dermatology

The stem cell treatment also gained popularity in treating skin pathologies; systemic administration of ASCs had a positive outcome for atopic dermatitis refractory to conventional medications for six months and with no side effects [78]. The prospective role in dermatology was also shown in treating large acute skin defects when corrective surgery offers no solution, as Zubin et al. (2015) [77] reported. In addition, the healing of acute and chronic wounds in 24 dogs of different ages and breeds significantly improved in a manner of contraction and re-epithelialization in the treated group. Furthermore, histopathological findings revealed an inflammatory infiltrate decrease and the presence of multiple hair follicles on day seven after treatment with ASCs [79]. Most recently, Kaur et al. (2022) performed the first double-blinded, placebo-controlled evaluation of the efficacy of allogeneic canine ASCs to treat canine atopic dermatitis. No severe side effects were observed in any patient in this study. Furthermore, the high dose ASCs treatment proved to be efficacious in alleviating the clinical signs of atopic dermatitis until 30 days after the last subcutaneous administration of MSCs[80].

4.2.4. Ophthalmology

Reviewing ophthalmological benefits, an immune-mediated condition common in humans and canines, keratoconjunctivitis sicca (KCS), was studied. Results in canines with KCS revealed that a single infusion of ACSs into lacrimal glands of 15 dogs resulted in no side effects during 12-months follow up.

Furthermore, a significant clinical improvement was observed in all patients, single administration was effective, and daily use of corticosteroids was not required [82]. In 2019, topical application into the conjunctival sac resulted in decreased expression of pro-inflammatory markers, which implies ASCs as an adjuvant therapy in treating KCS in dogs and humans. [83]. Another successful study of ASCs for KCS was reported with significant outcomes in canines where allogeneic ASCs were applied [81]. Falcao et al., in 2020, evaluated the use of sub-conjunctival applied ASCs in dogs diagnosed with deep corneal ulcers. Allogeneic ASCs therapy in 22 out of 26 dogs presented complete ulcer wound healing within 14 days, totalling 84.6%, indicating that this therapy is a simple solution to substitute surgery with satisfying results [84].

4.2.5. Gastroenterology

The ASCs therapy was also tested for currently uncurable inflammatory bowel disease (IBD); administration of a single IV ASCs infusion showed no acute reaction or side effects during the follow-up of 11 dogs. Furthermore, 9 out of 11 dogs were in clinical remission. As the primary goal of treatment is to reduce symptoms, achieve and maintain remission, and prevent complications, ASCs were well tolerated and appeared to produce clinical benefits in dogs with severe IBD [85].

4.2.6. Hepatology

Liver diseases share clinical and pathological features in humans and canines, thus, dogs may be a representative model for humans. Autologous ASCs transplantation in dogs with liver diseases significantly ameliorated liver function; decreased liver biomarkers and observed effects seem to be related to stem cells' immunomodulatory mechanism of action [87,88]. The effects of allogenic ASCs on acute liver injury by carbon tetrachloride in dogs were investigated by Teshima et al. (2017). It was observed that serum liver enzymes decreased significantly. In the liver, the mRNA expression levels of pro-inflammatory cytokines, such as IL-1, IL-6, IL-8, and IFNγ decreased significantly, but anti-inflammatory cytokines such as IL-4 and IL-10, HGF and VEGFA, were significantly increased after the first ASCs injection. The authors suggest that allogenic ASCs ameliorate acute hepatic injury in dogs [86].

5. The Importance of the Regulatory Considerations and Safety Aspects Using of Animal Cell-Based Products in Regenerative Therapy

As demonstrated by these positive examples, the use of ASCs therapy for numerous conditions holds excellent promise and encourages more research to provide safe, effective, and quality treatment. The European Medicines Agency (EMA), 2015, published the first draft problem statement agreed by Ad Hoc Expert Group on Veterinary Novel Therapies (ADVENT), which raised questions concerning the sterility of animal-cell-based products. The conclusion, published in 2019, was that sterility assurance of the finished stem-cell product is critical in light of the fact that the product may be administered prior to final sterility result being obtained [89]. The Novel Therapies and Technologies Working Party

(NTWP) of the EMA Committee for Medicinal Products for Veterinary Use is currently preparing scientific guidance on the requirements for authorization of novel therapy veterinary medicines, which involves guidelines on veterinary cell-based therapy products taking into consideration the mechanism of action, potency and clinical effects [90]. In order to implement safe new treatments for animals, the FDA published guidelines for the application and handling of animal-cell-based products, and all cell-based products require premarket review and FDA approval to be legally marketed. Precautionary steps in therapeutic use include the control of transmitting infectious agents, tumorigenicity or unintended tissue formation, immunogenicity, long-term safety, cell survival and biodistribution [91].

6. Conclusions

To conclude, while AT was once considered an energy depot, today it is well known that, among others, AT hosts components with extraordinary potential in relieving pain and treating numerous diseases. The canine SVF and ASCs treatments provide many benefits, starting with the degenerative orthopaedic pathologies, and also regenerative possibilities within other organs such as skin, bowel, and eyes. In addition, although this review focused on positive aspects of therapy in canines, the possible side effects it can carry should not be overlooked, such as the transmission of infectious agents, tumorigenicity, immunogenicity, donor selection, long-term safety, cell survival, biodistribution or ectopic tissue formation [91]. The (NTWP) of the EMA Committee for Medicinal Products for Veterinary Use is currently preparing scientific guidance on the requirements for authorization of novel therapy veterinary medicines, which involves guidelines on veterinary cell-based therapy products taking into consideration the mechanism of action, potency and clinical effects [90].

Although therapy will be available in the near future, there remains much laboratory and clinical work to undertake to better understand the complexity behind the healing mechanisms of canine ASCs. In this context, the development of regenerative veterinary medicine is essential not only for pets and their health but also for humans since canines represent an important model for human conditions. Nevertheless, it is evident that the course of research in this field is expanding, which welcomes further high quality basic, translational, and clinical research in stem cell regenerative therapy. Furthermore, in order to positively impact the lives of canine patients, adequate research for safe and standardized treatment is a fundamental prerequisite.

Author Contributions: Conceptualization: M.P. and N.K.; writing—original draft preparation: M.P. and N.K.; writing—review and editing: M.P., D.V., P.K., I.L. (Ivana Ljolje), T.K., D.B., V.K., I.L. (Ivana Lojkić), N.T. and N.K.; visualization M.P. and N.K.; supervision: N.K. and N.T.; project administration: N.K.; funding: N.K. All authors have read and agreed to the published version of the manuscript.

Funding: This review article was conducted within the Installation Research Project (UIP-2019-04-2178) "Revealing the adipose-derived mesenchymal stem cell transcriptome and secretome"—SECRET funded by Croatian Science Foundation.

Institutional Review Board Statement: Not applicable.

Informed Consent Statement: Not applicable.

Data Availability Statement: Not applicable.

Acknowledgments: Graphical illustrations were produced by graphic designer Mirna Mrcelja.

Conflicts of Interest: The authors declare no conflict of interest.

References

1. Kershaw, E.E.; Flier, J.S. Adipose tissue as an endocrine organ. *J. Clin. Endocrinol. Metab.* **2004**, *89*, 2548–2556. [CrossRef] [PubMed]
2. Esteve Ràfols, M. Adipose tissue: Cell heterogeneity and functional diversity. *Endocrinol. y Nutr. (English Ed.)* **2014**, *61*, 100–112. [CrossRef] [PubMed]
3. Ramakrishnan, V.M.; Boyd, N.L. The Adipose Stromal Vascular Fraction as a Complex Cellular Source for Tissue Engineering Applications. *Tissue Eng.—Part B Rev.* **2018**, *24*, 289–299. [CrossRef] [PubMed]

4. Billon, N.; Iannarelli, P.; Monteiro, M.C.; Glaviuex-Pardanaud, C.; Richardson, W.D.; Kessaris, N.; Dani, C.; Dupin, E. The generation of adipocytes by the neural crest. *Development* **2007**, *134*, 2283–2292. [CrossRef]
5. Trevor, L.V.; Riches-Suman, K.; Mahajan, A.L.; Thornton, M.J. Adipose tissue: A source of stem cells with potential for regenerative therapies for wound healing. *J. Clin. Med.* **2020**, *9*, 2161. [CrossRef]
6. Chait, A.; den Hartigh, L.J. Adipose Tissue Distribution, Inflammation and Its Metabolic Consequences, Including Diabetes and Cardiovascular Disease. *Front. Cardiovasc. Med.* **2020**, *7*, 22. [CrossRef]
7. Simpson, E.R. Sources of estrogen and their importance. *J. Steroid Biochem. Mol. Biol.* **2003**, *86*, 225–230. [CrossRef]
8. Obradovic, M.; Sudar-Milovanovic, E.; Soskic, S.; Essack, M.; Arya, S.; Stewart, A.J.; Gojobori, T.; Isenovic, E.R. Leptin and Obesity: Role and Clinical Implication. *Front. Endocrinol. (Lausanne)* **2021**, *12*, 585887. [CrossRef]
9. Trayhurn, P.; Beattie, J.H. Physiological role of adipose tissue: White adipose tissue as an endocrine and secretory organ. *Proc. Nutr. Soc.* **2001**, *60*, 329–339. [CrossRef]
10. Bahamondes, F.; Flores, E.; Cattaneo, G.; Bruna, F.; Conget, P. Omental adipose tissue is a more suitable source of canine Mesenchymal stem cells. *BMC Vet. Res.* **2017**, *13*, 166. [CrossRef]
11. Eurell, J.A.; Van Sickle, D.C. Connective and Supportive Tissues. In *Dellmann's Textbook of Veterinary Histology*; Eurell, J.A., Freappier, B.L., Eds.; Blackwell Publishing: Ames, IA, USA, 2006; pp. 1–405. ISBN 9780781741484.
12. Otto, T.C.; Lane, M.D. Adipose development: From stem cell to adipocyte. *Crit. Rev. Biochem. Mol. Biol.* **2005**, *40*, 229–242. [CrossRef] [PubMed]
13. Cohen, P.; Spiegelman, B.M. Cell biology of fat storage. *Mol. Biol. Cell* **2016**, *27*, 2523–2527. [CrossRef] [PubMed]
14. Tseng, Y.H.; Kokkotou, E.; Schulz, T.J.; Huang, T.L.; Winnay, J.N.; Taniguchi, C.M.; Tran, T.T.; Suzuki, R.; Espinoza, D.O.; Yamamoto, Y.; et al. New role of bone morphogenetic protein 7 in brown adipogenesis and energy expenditure. *Nature* **2008**, *454*, 1000–1004. [CrossRef]
15. Lee, J.H.; Park, A.; Oh, K.J.; Lee, S.C.; Kim, W.K.; Bae, K.H. The role of adipose tissue mitochondria: Regulation of mitochondrial function for the treatment of metabolic diseases. *Int. J. Mol. Sci.* **2019**, *20*, 4924. [CrossRef] [PubMed]
16. Wu, J.; Boström, P.; Sparks, L.M.; Ye, L.; Choi, J.H.; Giang, A.H.; Khandekar, M.; Virtanen, K.A.; Nuutila, P.; Schaart, G.; et al. Beige adipocytes are a distinct type of thermogenic fat cell in mouse and human. *Cell* **2012**, *150*, 366–376. [CrossRef]
17. Bargut, T.C.L.; Souza-Mello, V.; Aguila, M.B.; Mandarim-De-Lacerda, C.A. Browning of white adipose tissue: Lessons from experimental models. *Horm. Mol. Biol. Clin. Investig.* **2017**, *31*, 1–13. [CrossRef]
18. Kaisanlahti, A.; Glumoff, T. Browning of white fat: Agents and implications for beige adipose tissue to type 2 diabetes. *J. Physiol. Biochem.* **2019**, *75*, 1–10. [CrossRef]
19. Guo, J.; Nguyen, A.; Banyard, D.A.; Fadavi, D.; Toranto, J.D.; Wirth, G.A.; Paydar, K.Z.; Evans, G.R.D.; Widgerow, A.D. Stromal vascular fraction: A regenerative reality? Part 2: Mechanisms of regenerative action. *J. Plast. Reconstr. Aesthetic Surg.* **2016**, *69*, 180–188. [CrossRef]
20. Bora, P.; Majumdar, A.S. Adipose tissue-derived stromal vascular fraction in regenerative medicine: A brief review on biology and translation. *Stem Cell Res. Ther.* **2017**, *8*, 145. [CrossRef]
21. Prunet-Marcassus, B.; Cousin, B.; Caton, D.; André, M.; Pénicaud, L.; Casteilla, L. From heterogeneity to plasticity in adipose tissues: Site-specific differences. *Exp. Cell Res.* **2006**, *312*, 727–736. [CrossRef]
22. Hendawy, H.; Uemura, A.; Ma, D.; Namiki, R.; Samir, H.; Ahmed, M.F.; Elfadadny, A.; El-husseiny, H.M.; Chieh-jen, C.; Tanaka, R. Tissue harvesting site effect on the canine adipose stromal vascular fraction quantity and quality. *Animals* **2021**, *11*, 460. [CrossRef] [PubMed]
23. Krešić, N.; Prišlin, M.; Vlahović, D.; Kostešić, P.; Ljolje, I.; Brnić, D.; Turk, N.; Musulin, A.; Habrun, B. The expression pattern of surface markers in canine adipose-derived mesenchymal stem cells. *Int. J. Mol. Sci.* **2021**, *22*, 7476. [CrossRef] [PubMed]
24. Andia, I.; Maffulli, N.; Burgos-Alonso, N. Stromal vascular fraction technologies and clinical applications. *Expert Opin. Biol. Ther.* **2019**, *19*, 1289–1305. [CrossRef] [PubMed]
25. van Dongen, J.A.; Stevens, H.P.; Parvizi, M.; van der Lei, B.; Harmsen, M.C. The fractionation of adipose tissue procedure to obtain stromal vascular fractions for regenerative purposes. *Wound Repair Regen.* **2016**, *24*, 994–1003. [CrossRef] [PubMed]
26. Senesi, L.; De Francesco, F.; Farinelli, L.; Manzotti, S.; Gagliardi, G.; Papalia, G.F.; Riccio, M.; Gigante, A. Mechanical and enzymatic procedures to isolate the stromal vascular fraction from adipose tissue: Preliminary results. *Front. Cell Dev. Biol.* **2019**, *7*, 88. [CrossRef]
27. Astor, D.E.; Hoelzler, M.G.; Harman, R.; Bastian, R.P. Patient factors influencing the concentration of stromal vascular fraction (SVF) for adipose-derived stromal cell (ASC) therapy in dogs. *Can. J. Vet. Res.* **2013**, *77*, 177–182.
28. Lee, J.; Lee, K.S.; Kim, C.L.; Byeon, J.S.; Gu, N.Y.; Cho, I.S.; Cha, S.H. Effect of donor age on the proliferation and multipotency of canine adipose-derived mesenchymal stem cells. *J. Vet. Sci.* **2017**, *18*, 141–148. [CrossRef]
29. Black, L.L.; Gaynor, J.; Gahring, D.; Adams, C.; Aron, D.; Harman, S.; Gingerich, D.A.; Harman, R. Effect of adipose-derived mesenchymal stem and regenerative cells on lameness in dogs with chronic osteoarthritis of the coxofemoral joints: A randomized, double-blinded, multicenter, controlled trial. *Vet. Ther.* **2007**, *8*, 272–284.
30. Upchurch, D.A.; Renberg, W.C.; Roush, J.K.; Milliken, G.A.; Weiss, M.L. Effects of administration of adipose-derived stromal vascular fraction and platelet-rich plasma to dogs with osteoarthritis of the hip joints. *Am. J. Vet. Res.* **2016**, *77*, 940–951. [CrossRef]

31. Black, L.L.; Gaynor, J.; Adams, C.; Dhupa, S.; Sams, A.E.; Taylor, R.; Harman, S.; Gingerich, D.A.; Harman, R. Effect of intraarticular injection of autologous adipose-derived mesenchymal stem and regenerative cells on clinical signs of chronic osteoarthritis of the elbow joint in dogs. *Vet. Ther.* **2008**, *9*, 192–200.
32. Marx, C.; Silveira, M.D.; Selbach, I.; Da Silva, A.S.; Braga, L.M.G.D.M.; Camassola, M.; Nardi, N.B. Acupoint injection of autologous stromal vascular fraction and allogeneic adipose-derived stem cells to treat hip dysplasia in dogs. *Stem Cells Int.* **2014**, *2014*, 6. [CrossRef] [PubMed]
33. Kemilew, J.; Sobczyńska-Rak, A.; Zylińska, B.; Szponder, T.; Nowicka, B.; Urban, B. The use of allogenic stromal vascular fraction (SVF) cells in degenerative joint disease of the spine in dogs. *In Vivo (Brooklyn)* **2019**, *33*, 1109–1117. [CrossRef] [PubMed]
34. Caplan, A.I.; Dennis, J.E. Mesenchymal stem cells as trophic mediators. *J. Cell. Biochem.* **2006**, *98*, 1076–1084. [CrossRef]
35. Zuk, P.A.; Zhu, M.; Mizuno, H.; Huang, J.; Futrell, J.W.; Katz, A.J.; Benhaim, P.; Lorenz, H.P.; Hedrick, M.H. Multilineage cells from human adipose tissue: Implications for cell-based therapies. *Tissue Eng.* **2001**, *7*, 211–228. [CrossRef] [PubMed]
36. Zuk, P.A.; Min, Z.; Ashjian, P.; De Ugarte, D.A.; Huang, J.I.; Mizuno, H.; Alfonso, Z.C.; Fraser, J.K.; Benhaim, P.; Hedrick, M.H. Human Adipose Tissue Is a Source of Multipotent Stem Cells. *Mol. Biol. Cell* **2002**, *13*, 4279–4295. [CrossRef]
37. Neupane, M.; Chang, C.-C.; Kiupel, M.; Yuzbasiyan-Gurkan, V. Isolation and Characterization of Canine Adipose–Derived Mesenchymal Stem Cells. *Tissue Eng. Part A* **2008**, *14*, 1007–1015. [CrossRef]
38. Marcoccia, R.; Nesci, S.; Merlo, B.; Ballotta, G.; Algieri, C.; Pagliarani, A.; Iacono, E. Biological characteristics and metabolic profile of canine mesenchymal stem cells isolated from adipose tissue and umbilical cord matrix. *PLoS ONE* **2021**, *16*, e0247567. [CrossRef]
39. Dominici, M.; Le Blanc, K.; Mueller, I.; Slaper-Cortenbach, I.; Marini, F.C.; Krause, D.S.; Deans, R.J.; Keating, A.; Prockop, D.J.; Horwitz, E.M. Minimal criteria for defining multipotent mesenchymal stromal cells. The International Society for Cellular Therapy position statement. *Cytotherapy* **2006**, *8*, 315–317. [CrossRef]
40. Kulus, M.; Kulus, J.; Jankowski, M.; Borowiec, B.; Jeseta, M.; Bukowska, D.; Brüssow, K.P.; Kempisty, B.; Antosik, P. The use of mesenchymal stem cells in veterinary medicine. *Med. J. Cell Biol.* **2018**, *6*, 101–107. [CrossRef]
41. Gugjoo, M.B.; Amarpal, A.; Sharma, G.T. Mesenchymal stem cell basic research and applications in dog medicine. *J. Cell. Physiol.* **2019**, *234*, 16779–16811. [CrossRef]
42. Voga, M.; Adamic, N.; Vengust, M.; Majdic, G. Stem Cells in Veterinary Medicine—Current State and Treatment Options. *Front. Vet. Sci.* **2020**, *7*, 278. [CrossRef] [PubMed]
43. Dias, I.E.; Pinto, P.O.; Barros, L.C.; Viegas, C.A.; Dias, I.R.; Carvalho, P.P. Mesenchymal stem cells therapy in companion animals: Useful for immune-mediated diseases? *BMC Vet. Res.* **2019**, *15*, 358. [CrossRef] [PubMed]
44. Gao, F.; Chiu, S.M.; Motan, D.A.L.; Zhang, Z.; Chen, L.; Ji, H.L.; Tse, H.F.; Fu, Q.L.; Lian, Q. Mesenchymal stem cells and immunomodulation: Current status and future prospects. *Cell Death Dis.* **2016**, *7*, e2062. [CrossRef] [PubMed]
45. Ayala-Cuellar, A.P.; Kang, J.H.; Jeung, E.B.; Choi, K.C. Roles of mesenchymal stem cells in tissue regeneration and immunomodulation. *Biomol. Ther.* **2019**, *27*, 25–33. [CrossRef]
46. Chow, L.; Johnson, V.; Coy, J.; Regan, D.; Dow, S. Mechanisms of Immune Suppression Utilized by Canine Adipose and Bone Marrow-Derived Mesenchymal Stem Cells. *Stem Cells Dev.* **2017**, *26*, 374–389. [CrossRef]
47. Wheat, W.H.; Chow, L.; Kurihara, J.N.; Regan, D.P.; Coy, J.W.; Webb, T.L.; Dow, S.W. Suppression of Canine Dendritic Cell Activation/Maturation and Inflammatory Cytokine Release by Mesenchymal Stem Cells Occurs Through Multiple Distinct Biochemical Pathways. *Stem Cells Dev.* **2017**, *26*, 249–262. [CrossRef]
48. Herrero, C.; Pérez-Simón, J.A. Immunomodulatory effect of mesenchymal stem cells. *Brazilian J. Med. Biol. Res.* **2010**, *43*, 425–430. [CrossRef]
49. Marx, C.; Silveira, M.D.; Nardi, N.B. Adipose-Derived Stem Cells in Veterinary Medicine: Characterization and Therapeutic Applications. *Stem Cells Dev.* **2015**, *24*, 803–813. [CrossRef]
50. Krešić, N.; Šimić, I.; Lojkić, I.; Bedeković, T. Mezenhimske matične stanice u veterinarskoj medicini. *Vet. Stanica* **2019**, *50*, 149–154.
51. Ponte, A.L.; Marais, E.; Gallay, N.; Langonné, A.; Delorme, B.; Hérault, O.; Charbord, P.; Domenech, J. The In Vitro Migration Capacity of Human Bone Marrow Mesenchymal Stem Cells: Comparison of Chemokine and Growth Factor Chemotactic Activities. *Stem Cells* **2007**, *25*, 1737–1745. [CrossRef]
52. Ullah, M.; Liu, D.D.; Thakor, A.S. Mesenchymal Stromal Cell Homing: Mechanisms and Strategies for Improvement. *iScience* **2019**, *15*, 421–438. [CrossRef] [PubMed]
53. Sackstein, R. The lymphocyte homing receptors: Gatekeepers of the multistep paradigm. *Curr. Opin. Hematol.* **2005**, *12*, 444–450. [CrossRef] [PubMed]
54. Kwon, Y.W.; Heo, S.C.; Jeong, G.O.; Yoon, J.W.; Mo, W.M.; Lee, M.J.; Jang, I.H.; Kwon, S.M.; Lee, J.S.; Kim, J.H. Tumor necrosis factor-α-activated mesenchymal stem cells promote endothelial progenitor cell homing and angiogenesis. *Biochim. Biophys. Acta—Mol. Basis Dis.* **2013**, *1832*, 2136–2144. [CrossRef] [PubMed]
55. Al Delfi, I.R.; Sheard, J.J.; Wood, C.R.; Vernallis, A.; Innes, J.F.; Myint, P.; Johnson, W.E.B. Canine mesenchymal stem cells are neurotrophic and angiogenic: An in vitro assessment of their paracrine activity. *Vet. J.* **2016**, *217*, 10–17. [CrossRef] [PubMed]
56. Yao, C.; Zhou, Y.; Wang, H.; Deng, F.; Chen, Y.; Zhu, X.; Kong, Y.; Pan, L.; Xue, L.; Zhou, X.; et al. Adipose-derived stem cells alleviate radiation-induced dermatitis by suppressing apoptosis and downregulating cathepsin F expression. *Stem Cell Res. Ther.* **2021**, *12*, 447. [CrossRef]

57. Brondeel, C.; Pauwelyn, G.; de Bakker, E.; Saunders, J.; Samoy, Y.; Spaas, J.H. Review: Mesenchymal Stem Cell Therapy in Canine Osteoarthritis Research: "Experientia Docet" (Experience Will Teach Us). *Front. Vet. Sci.* **2021**, *8*, 668881. [CrossRef]
58. Yoon, H.-Y.; Lee, J.; Jeong, S. Long-term follow-up after implantation of autologous adipose tissue derived mesenchymal stem cells to treat a dog with stifle joint osteoarthrosis. *J. Vet. Clin.* **2012**, *29*, 82–86.
59. Guercio, A.; Di Marco, P.; Casella, S.; Cannella, V.; Russotto, L.; Purpari, G.; Di Bella, S.; Piccione, G. Production of canine mesenchymal stem cells from adipose tissue and their application in dogs with chronic osteoarthritis of the humeroradial joints. *Cell Biol. Int.* **2012**, *36*, 189–194. [CrossRef]
60. Vilar, J.M.; Morales, M.; Santana, A.; Spinella, G.; Rubio, M.; Cuervo, B.; Cugat, R.; Carrillo, J.M. Controlled, blinded force platform analysis of the effect of intraarticular injection of autologous adipose-derived mesenchymal stem cells associated to PRGF-Endoret in osteoarthritic dogs. *BMC Vet. Res.* **2013**, *9*, 131. [CrossRef]
61. Cuervo, B.; Rubio, M.; Sopena, J.; Dominguez, J.M.; Vilar, J.; Morales, M.; Cugat, R.; Carrillo, J.M. Hip osteoarthritis in dogs: A randomized study using mesenchymal stem cells from adipose tissue and plasma rich in growth factors. *Int. J. Mol. Sci.* **2014**, *15*, 13437–13460. [CrossRef]
62. Vilar, J.M.; Batista, M.; Morales, M.; Santana, A.; Cuervo, B.; Rubio, M.; Cugat, R.; Sopena, J.; Carrillo, J.M. Assessment of the effect of intraarticular injection of autologous adipose-derived mesenchymal stem cells in osteoarthritic dogs using a double blinded force platform analysis. *BMC Vet. Res.* **2014**, *10*, 143. [CrossRef] [PubMed]
63. Harman, R.; Carlson, K.; Gaynor, J.; Gustafson, S.; Dhupa, S.; Clement, K.; Hoelzler, M.; McCarthy, T.; Schwartz, P.; Adams, C. A prospective, randomized, masked, and placebo-controlled efficacy study of intraarticular allogeneic adipose stem cells for the treatment of osteoarthritis in dogs. *Front. Vet. Sci.* **2016**, *3*, 81. [CrossRef] [PubMed]
64. Yun, S.; Ku, S.K.; Kwon, Y.S. Adipose-derived mesenchymal stem cells and platelet-rich plasma synergistically ameliorate the surgical-induced osteoarthritis in Beagle dogs. *J. Orthop. Surg. Res.* **2016**, *11*, 9. [CrossRef] [PubMed]
65. Vilar, J.M.; Cuervo, B.; Rubio, M.; Sopena, J.; Domínguez, J.M.; Santana, A.; Carrillo, J.M. Effect of intraarticular inoculation of mesenchymal stem cells in dogs with hip osteoarthritis by means of objective force platform gait analysis: Concordance with numeric subjective scoring scales. *BMC Vet. Res.* **2016**, *12*, 223. [CrossRef]
66. Kriston-Pál, É.; Czibula, Á.; Gyuris, Z.; Balka, G.; Seregi, A.; Sükösd, F.; Süth, M.; Kiss-Tóth, E.; Haracska, L.; Uher, F.; et al. Characterization and therapeutic application of canine adipose. *Can. J. Vet. Res.* **2017**, *81*, 73–78.
67. Shah, K.; Drury, T.; Roic, I.; Hansen, P.; Malin, M.; Boyd, R.; Sumer, H.; Ferguson, R. Outcome of allogeneic adult stem cell therapy in dogs suffering from osteoarthritis and other joint defects. *Stem Cells Int.* **2018**, *2018*, 7. [CrossRef]
68. Srzentić Dražilov, S.; Mrkovački, J.; Spasovski, V.; Fazlagić, A.; Pavlović, S.; Nikčević, G. The use of canine mesenchymal stem cells for the autologous treatment of osteoarthritis. *Acta Vet. Hung.* **2018**, *66*, 376–389. [CrossRef]
69. Olsen, A.; Johnson, V.; Webb, T.; Santangelo, K.S.; Dow, S.; Duerr, F.M. Evaluation of Intravenously Delivered Allogeneic Mesenchymal Stem Cells for Treatment of Elbow Osteoarthritis in Dogs: A Pilot Study. *Vet. Comp. Orthop. Traumatol.* **2019**, *32*, 173–181. [CrossRef]
70. Wits, M.I.; Tobin, G.C.; Silveira, M.D.; Baja, K.G.; Braga, L.M.M.; Sesterheim, P.; Camassola, M.; Nardi, N.B. Combining canine mesenchymal stromal cells and hyaluronic acid for cartilage repair. *Genet. Mol. Biol.* **2020**, *43*, 1–8. [CrossRef]
71. Brown, G.S.; Harman, J.R.; Black, L.L. Adipose-derived stem cell therapy for severe muscle tears in working German shepherds: Two case reports. *Stem Cell Discov.* **2012**, *02*, 41–44. [CrossRef]
72. Gibson, M.A.; Brown, S.G.; Brown, N.O. Semitendinosus myopathy and treatment with adipose-derived stem cells in working German shepherd police dogs. *Can. Vet. J.* **2017**, *58*, 241–246. [PubMed]
73. Escalhão, C.C.M.I.; Ramos, I.P.; Hochman-Mendez, C.; Brunswick, T.H.K.; Souza, S.A.L.; Gutfilen, B.; Dos Santos Goldenberg, R.C.; Coelho-Sampaio, T. Safety of allogeneic canine adipose tissue-derived mesenchymal stem cell intraspinal transplantation in dogs with chronic spinal cord injury. *Stem Cells Int.* **2017**, *2017*, 11. [CrossRef] [PubMed]
74. Bach, F.S.; Rebelatto, C.L.K.; Fracaro, L.; Senegaglia, A.C.; Fragoso, F.Y.I.; Daga, D.R.; Brofman, P.R.S.; Pimpão, C.T.; Engracia Filho, J.R.; Montiani-Ferreira, F.; et al. Comparison of the Efficacy of Surgical Decompression Alone and Combined With Canine Adipose Tissue-Derived Stem Cell Transplantation in Dogs With Acute Thoracolumbar Disk Disease and Spinal Cord Injury. *Front. Vet. Sci.* **2019**, *6*, 383. [CrossRef] [PubMed]
75. Mrkovački, J.; Srzentić Dražilov, S.; Spasovski, V.; Fazlagić, A.; Pavlović, S.; Nikčević, G. Case Report: Successful Therapy of Spontaneously Occurring Canine Degenerative Lumbosacral Stenosis Using Autologous Adipose Tissue-Derived Mesenchymal Stem Cells. *Front. Vet. Sci.* **2021**, *8*, 1055. [CrossRef]
76. Chen, C.C.; Yang, S.F.; Wang, I.K.; Hsieh, S.Y.; Yu, J.X.; Wu, T.L.; Huang, W.J.; Su, M.H.; Yang, H.L.; Chang, P.C.; et al. The Long-Term Efficacy Study of Multiple Allogeneic Canine Adipose Tissue-Derived Mesenchymal Stem Cells Transplantations Combined With Surgery in Four Dogs With Lumbosacral Spinal Cord Injury. *Cell Transplant.* **2022**, *31*, 09636897221081487. [CrossRef]
77. Zubin, E.; Conti, V.; Leonardi, F.; Zanichelli, S.; Ramoni, R.; Grolli, S. Regenerative therapy for the management of a large skin wound in a dog. *Clin. Case Reports* **2015**, *3*, 598–603. [CrossRef]
78. Villatoro, A.J.; Hermida-Prieto, M.; Fernández, V.; Fariñas, F.; Alcoholado, C.; Isabel Rodríguez-García, M.; Mariñas-Pardo, L.; Becerra, J. Allogeneic adipose-derived mesenchymal stem cell therapy in dogs with refractory atopic dermatitis: Clinical efficacy and safety. *Vet. Rec.* **2018**, *183*, 654. [CrossRef]

79. Enciso, N.; Avedillo, L.; Fermín, M.L.; Fragío, C.; Tejero, C. Cutaneous wound healing: Canine allogeneic ASC therapy. *Stem Cell Res. Ther.* **2020**, *11*, 261. [CrossRef]
80. Kaur, G.; Ramirez, A.; Xie, C.; Clark, D.; Dong, C.; Maki, C.; Ramos, T.; Izadyar, F.; Najera, S.O.L.; Harb, J.; et al. A double-blinded placebo-controlled evaluation of adipose-derived mesenchymal stem cells in treatment of canine atopic dermatitis. *Vet. Res. Commun.* **2022**, *46*, 251–260, 10. [CrossRef]
81. Villatoro, A.J.; Fernández, V.; Claros, S.; Rico-Llanos, G.A.; Becerra, J.; Andrades, J.A. Use of adipose-derived mesenchymal stem cells in keratoconjunctivitis sicca in a canine model. *Biomed Res. Int.* **2015**, *2015*, 527926. [CrossRef]
82. Bittencourt, M.K.W.; Barros, M.A.; Martins, J.F.P.; Vasconcellos, J.P.C.; Morais, B.P.; Pompeia, C.; Bittencourt, M.D.; Evangelho, K.D.S.; Kerkis, I.; Wenceslau, C.V. Allogeneic Mesenchymal Stem Cell Transplantation in Dogs with Keratoconjunctivitis Sicca. *Cell Med.* **2016**, *8*, 63–77. [CrossRef] [PubMed]
83. Sgrignoli, M.R.; Silva, D.A.; Nascimento, F.F.; Sgrignoli, D.A.M.; Nai, G.A.; da Silva, M.G.; de Barros, M.A.; Bittencourt, M.K.W.; de Morais, B.P.; Dinallo, H.R.; et al. Reduction in the inflammatory markers CD4, IL-1, IL-6 and TNFα in dogs with keratoconjunctivitis sicca treated topically with mesenchymal stem cells. *Stem Cell Res.* **2019**, *39*, 101525. [CrossRef] [PubMed]
84. Falcão, M.S.A.; Brunel, H.D.S.S.; Peixer, M.A.S.; Dallago, B.S.L.; Costa, F.F.; Queiroz, L.M.; Campbell, P.; Malard, P.F. Effect of allogeneic mesenchymal stem cells (MSCs) on corneal wound healing in dogs. *J. Tradit. Complement. Med.* **2020**, *10*, 440–445. [CrossRef] [PubMed]
85. Pérez-Merino, E.M.; Usón-Casaús, J.M.; Zaragoza-Bayle, C.; Duque-Carrasco, J.; Mariñas-Pardo, L.; Hermida-Prieto, M.; Barrera-Chacón, R.; Gualtieri, M. Safety and efficacy of allogeneic adipose tissue-derived mesenchymal stem cells for treatment of dogs with inflammatory bowel disease: Clinical and laboratory outcomes. *Vet. J.* **2015**, *206*, 385–390. [CrossRef]
86. Teshima, T.; Matsumoto, H.; Michishita, M.; Matsuoka, A.; Shiba, M.; Nagashima, T.; Koyama, H. Allogenic adipose tissue-derived mesenchymal stem cells ameliorate acute hepatic injury in dogs. *Stem Cells Int.* **2017**, *2017*, 12. [CrossRef] [PubMed]
87. Gardin, C.; Ferroni, L.; Bellin, G.; Rubini, G.; Barosio, S.; Zavan, B. Therapeutic potential of autologous adipose-derived stem cells for the treatment of liver disease. *Int. J. Mol. Sci.* **2018**, *19*, 4064. [CrossRef]
88. Yan, Y.; Fang, J.; Wen, X.; Teng, X.; Li, B.; Zhou, Z.; Peng, S.; Arisha, A.H.; Liu, W.; Hua, J. Therapeutic applications of adipose-derived mesenchymal stem cells on acute liver injury in canines. *Res. Vet. Sci.* **2019**, *126*, 233–239. [CrossRef]
89. Anonymus. European Medicines Agency: Questions and Answers on Allogeneic Stem Cell-Based Products for Veterinary Use. 2019. Available online: https://www.ema.europa.eu/en/documents/scientific-guideline/questions-answers-allogeneic-stem-cell-based-products-veterinary-use-specific-questions-sterility_en.pdf (accessed on 31 January 2022).
90. Anonymous. European Medicines Agency: Concept Paper on the Development and Data Requirements of Potency Tests for Cell-Based Therapy Products and the Relation to Clinical Efficacy. 2022. Available online: https://www.ema.europa.eu/en/documents/scientific-guideline/concept-paper-development-data-requirements-potency-tests-cell-based-therapy-products-relation_en.pdf (accessed on 3 February 2022).
91. Lynne Boxer. Food and Drug Administration: Guidance for Industry Cell-Based Products for Animal Use. 2015. Available online: https://www.fda.gov/media/88925/download?source=govdelivery&utm_medium=email&utm_source=govdelivery (accessed on 21 December 2021).

animals

MDPI

Communication

Isolation and Characterization of Cat Olfactory Ecto-Mesenchymal Stem Cells

Marie-Laure Mollichella [1,*], Violaine Mechin [1], Dany Royer [2], Patrick Pageat [3] and Pietro Asproni [1]

[1] Tissue Biology and Chemical Communication Department, Institute of Research in Semiochemistry and Applied Ethology, 84400 Apt, France; v.mechin@group-irsea.com (V.M.); p.asproni@group-irsea.com (P.A.)
[2] Independent Researcher, 84440 Robion, France; dany.royer.drv@gmail.com
[3] Research and Education Board, Institute of Research in Semiochemistry and Applied Ethology, 84400 Apt, France; p.pageat@group-irsea.com
* Correspondence: ml.mollichella@group-irsea.com

Simple Summary: Cat's health is impacted by several diseases and lesions for which cell therapy could be an interesting treatment. Mesenchymal stem cells or adult stem cells are found in developed tissue. Olfactory mucosa contains stem cells called olfactory ecto-mesenchymal stem cells which have already been isolated from various animals as dogs and horses. The aim of this study was to evaluate the feasibility of collecting olfactory ecto-mesenchymal stem cells in cats. For that purpose, four cats were biopsied; the cells were collected and characterized. They show stemness features and differentiation capabilities as all the other mammals previously studied. Therefore, olfactory ecto-mesenchymal stem cells could be a promising tool for feline regenerative medicine.

Abstract: The olfactory mucosa contains olfactory ecto-mesenchymal stem cells (OE-MSCs) which show stemness features, multipotency capabilities, and have a therapeutic potential. The OE-MSCs have already been collected and isolated from various mammals. The aim of this study was to evaluate the feasibility of collecting, purifying and amplifying OE-MSCs from the cat nasal cavity. Four cats were included in the study. Biopsies of olfactory mucosa were performed on anesthetized animals. Then, the olfactory OE-MSCs were isolated, and their stemness features as well as their mesodermal differentiation capabilities were characterized. Olfactory mucosa biopsies were successfully performed in all subjects. From these biopsies, cellular populations were rapidly generated, presenting various stemness features, such as a fibroblast-like morphology, nestin and MAP2 expression, and sphere and colony formation. These cells could differentiate into neural and mesodermal lineages. This report shows for the first time that the isolation of OE-MSCs from cat olfactory mucosa is possible. These cells showed stemness features and multilineage differentiation capabilities, indicating they may be a promising tool for autologous grafts and feline regenerative medicine.

Keywords: olfactory stem cells; isolation; stemness; differentiation; characterization; regenerative medicine

Citation: Mollichella, M.-L.; Mechin, V.; Royer, D.; Pageat, P.; Asproni, P. Isolation and Characterization of Cat Olfactory Ecto-Mesenchymal Stem Cells. *Animals* **2022**, *12*, 1284. https://doi.org/10.3390/ani12101284

Academic Editors: Eleonora Iacono, Barbara Merlo and Emir Hadzijusufovic

Received: 23 March 2022
Accepted: 13 May 2022
Published: 17 May 2022

1. Introduction

Mesenchymal stem cells (MSCs) are plastic-adherent cells that show self-renewal and high proliferative capabilities and can differentiate into the mesodermal lineage under standard in vitro differentiating conditions [1,2]. Due to these abilities, MSCs have been described as having therapeutic potential in several diseases, such as cancer [3], traumatic brain injury [4], chondral defects [5] or cardiovascular diseases [6], even if these statements should be confirmed by more extensive investigations.

Stem cell-based regenerative medicine is used in veterinary medicine to repair damaged tissue by a disease or injury. Even if this kind of therapy still needs to be more extensively investigated, in the future stem cells may be an alternative treatment in some cases for which the conventional medicines cannot repair the damages tissues. MSCs have differentiation potential but also immune regulatory properties, influence on vascularization, apoptosis, fibrosis and inflammation [7–9]. Regenerative cell therapy is used in degenerative diseases (heart failure), immune mediated diseases (feline asthma, canine atopic dermatitis, feline chronic gingivostomatitis) [10], inflammatory diseases (wound healing defect). MSCs induce immune enhancing response and have an anti-inflammatory effect [10].

In cats, MSCs have been first isolated in 2002 from bone marrow [11], then from fat, fetal fluid, peripheral blood and amniotic membranes [12,13]. Feline MSCs have been already used in trial against gingivostomatitis, enteropathies, chronic kidney disease, asthma, feline eosinophilic keratitis, neurological ailments, cardiomyopathy. These studies gave variable therapeutics results. Concerning the gingivostomatitis, grafts with autologous or allogenic adipose stem cells led to complete remission in 3/7 and 2/7 cats, respectively, with 2/7 cats that presented a substantial improvement in both cases [14,15]. In acute and chronic asthma, Trzil and colleagues showed that the graft of allogenic adipose stem cells can induce an improvement in different parameters, among which airway eosinophilia [16,17]. Good results were also shown in chronic enteropathy treatment with allogenic stem cells administration, with an improvement of clinical signs in 5 of the 7 treated cats [18]. Finally, concerning chronic and acute kidney disease treatment, the results seem less promising, going from a mild decrease in serum creatinine [19–21] to no improvement in acute or chronic disease [22,23]. This heterogeneity suggests that more studies are needed to optimize the MSCs administration and sources [12,13].

Among the various sources of MSCs, the olfactory mucosa is a promising candidate for both humans and animals [24,25]. Indeed, the olfactory mucosa contains olfactory ecto-mesenchymal stem cells (OE-MSCs) that are easily accessible and collectable due to their localization [26]. OE-MSCs also show stemness and multipotency capabilities, resulting in therapeutic potential, which has already been demonstrated in several diseases, such as hearing loss [27], cerebral ischemia [28] and Parkinson's disease [29].

We previously showed that it is possible to collect and isolate olfactory stem cells from various mammals [26], but the feasibility has never been evaluated in cats. Since cats may suffer from several diseases and lesions impacting their welfare [12], stem cells could also be an interesting tool for cell therapies in this species, and OE-MSCs identified in the olfactory mucosa are promising candidates for autologous grafts [24].

The aim of this study was to investigate the feasibility of isolating OE-MSCs from cats and characterizing these cells.

2. Materials and Methods

2.1. Ethics Statement

This study was conceived and performed in accordance with French (2013-118) and European law (2010/63/EU) on the protection of animals used for scientific purposes. This protocol was approved by the Ministry of Higher Education, Research and Innovation of France and by the IRSEA's Ethics committee C2EA125 (approval number: UE-2018-EU0552).

2.2. Biopsy of Olfactory Mucosa and Isolation and Expansion of OE-MSCs

Four healthy cats from IRSEA's facilities were included in the study (2 males, 2 females; 6 ± 4.6 years). Since the cats belonged to our facilities, the health status was daily monitored by our veterinary team. Sedation was performed with ketamine (Imalgene 1000, Merial SAS, Lyon, France) (10–20 mg/kg, sc), medetomidine (Domitor, vetoquinol, Lure, France) (50 µg/kg, sc) and butorphanol (Dolorex, MSD Santé Animale, Beaucouzé, France) (0.4 mg/kg, sc). Then, anesthesia was induced with propofol (Propovet, Zoetis, Malakoff, France) (1.2 mg/kg) and maintained with isoflurane (Belamont) (2%). The biopsies of olfactory mucosa were performed by nasal cavity exploration with a common rigid biopsy forceps on anesthetized animals. To reach the olfactory mucosa, as for the other domestic species, the forceps were inserted into the nasal cavity until its caudal limit. For each cat, 2 biopsy samples (1 per side) were obtained and placed at 4 °C in culture medium Dulbecco's Modified Eagle's Medium/Ham's F12 (DMEM/F12, 1% GlutaMAX, Pan Biotech, Aidenbach, Germany, cod. P04-41150) supplemented with 10% serum (fetal bovine serum, (FBS, Dutscher, Bernolsheim, France), 2% penicillin and streptomycin 100X (P/S) (Dutscher, Bernolsheim, France, cod. L0022-100), and 2.5 mg/mL amphotericin B (Hyclone, Marlborough, MA, USA, cod. SV30078.01) until culturing. The olfactory mucosa biopsies were washed in DMEM/F12 medium and were mechanically dissociated using 25-gauge needles to obtain pieces of a few square millimeters. Each pieces of biopsy were placed in a 2 cm^2 culture well coated with poly-L-lysine (PLL, Sigma-Aldrich, Saint-Louis, MO, USA cod. P1274) with 200 μL of the culture medium described above for 1 week. When the explants adhered to the plate, the wells were filled with 400 μL of culture medium. Two weeks after plating, the concentration of antibiotic and amphotericin B were halved. The medium was renewed every two to three days. When confluence was reached, the cells were detached, dissociated with trypsin EDTA solution (0.25%, Dutscher, Bernolsheim, France, cod. L0931-100), pooled, centrifuged at 300× *g* for 5 min and replated at lower density.

2.3. Generation of Spheres

Cells were counted on Kova slides (Dutscher, Bernolsheim, France, cod. 050126) and, plated at a density of 30,000 cells/cm^2 in PLL-coated dishes (5 μg/cm^2) and fed with serum-free DMEM/F12 culture medium supplemented with 1% P/S, 1% insulin, transferrin, selenium (ITS-X, Gibco, cod. 51300044), 50 ng/mL epidermal growth factor (EGF, Gibco) and 50 ng/mL fibroblast growth factor 2 (FGF, Gibco, cod. PHG0311L). This culture medium was renewed every two days. After one week of treatment, spheres were observed with an inverted microscope.

2.4. In Vitro Neural Lineage Differentiation Assays

For neuronal differentiation, OE-MSCs after sphere generation were grown under two culture conditions as described previously [30,31]. The spheres were dissociated with trypsin EDTA solution and plated at a density of 15,000 cells per cm^2 in a culture well (2 cm^2) on glass coverslip coated with PLL. The cells were cultured in two different media: DMEM/F12 Glutamax, 1% P/S, 1% FBS, 2% B-27 Supplement (Gibco, cod. 17504044), 1% N-2 Supplement (Gibco, cod. 17502048), 10 ng/mL EGF, 20 ng/mL FGF or DMEM/F12 Glutamax, 1% P/S, 1% FBS, 2% B-27 Supplement, 1 mM Valproic acid (Sigma-Aldrich, Saint-Louis, MO, USA, cod. P4543). The medium was renewed every two days for one week. For confirmation of the differentiation, the cells were fixed in paraformaldehyde solution (4%, Alfa Aesar, Haverhill, MA, USA, cod. J61984), and immunocytochemistry (ICC) of the Glial fibrillary acidic protein (GFAP) and Microtubule Associated Protein 2 (MAP2) proteins was performed as described in Section 2.9 with the antibodies in Table 1.

Table 1. Antibodies used for immunocytochemistry.

Antibody	Target	Host	Supplier	Reference	Dilution	Secondary Antibody
Anti-nestin	Stemness marker	Rabbit	Abcam	ab7659	1/500	Alexa Fluor 488
Anti-GFAP	Neural marker	Chicken	Abcam	ab4674	1/500	Alexa Fluor 488
Anti-MAP2	Neural marker	Chicken	Abcam	ab5392	1/500	Alexa Fluor 488
Anti-tenomodulin	Tenoblast marker	Rabbit	Abcam	ab81328	1/250	Alexa Fluor 488
Anti-scleraxis	Tenoblast marker	Rabbit	Abcam	ab58655	1/250	Alexa Fluor 488

2.5. Expression of Nestin

OE-MSCs (passage 6) were plated on glass coverslips in a 24-well plate at a density of 15,000 cells per cm^2 in growth medium (DMEM, 10% FCS, 1% P/S, 1.25 mg/mL amphotericin B) for approximately 48 h. The cells were then fixed in a paraformaldehyde solution (4%) and ICC was performed as described in Section 2.9 with the antibodies that are reported in Table 1.

2.6. Clonal Efficiency Assay

OE-MSCs (passage 7) were plated in 6-well plates at a density ranging from 10 to 320 cells/well in triplicate. After plating, the dishes were placed at 37 °C, in a humidified, 5% CO2 atmosphere for 7 days. The culture medium (DMEM, 10% FCS, 1% P/S, 1.25 mg/mL amphotericin B) was renewed every two days. The colonies were paraformaldehyde-fixed during 15 min at room temperature (RT). Colonies were stained for 30 min using crystal violet, rinsed with tap water bath and let dry at RT. Then, the colonies were observed with an inverted microscope and manually counted. For each sample, clonal efficiency (% of clonogenicity) was calculated as follows:

$$(\text{mean number of colonies}/\text{total number of seeded cells}) \times 100$$

When too many colonies overlapped, counting was not performed.

2.7. In Vitro Proliferation Assay

The assay was performed on OE-MSCs 2 months (10 passages) and 3 months (20 passages) after the initial plating. The cells were seeded in 96-well plates in triplicate and counted with CellTiter 96 Aqueous One Solution Reagent (Promega, Madison, WI, USA, cod. G3580) according to the manufacturer's protocol at 8 h, 24 h, 48 h, 72 h, and 96 h after seeding. Briefly, 20 µL of CellTiter was added for 100 µL of culture medium. The plate was incubated at 37 °C in a humidified, 5% CO2 atmosphere for 1 h to 4 h. The absorbance at 490 nm was recorded with a plate reader. The population doubling time (PDT) was calculated as follows:

$$\text{Duration} \times \ln(2)/\ln(\text{FinalConcentration}) - \ln(\text{InitalConcentration})$$

2.8. In Vitro Mesodermal Differentiation Assays

For osteogenic differentiation, OE-MSCs (passage 8) were grown in DMEM/F12 Glutamax, 10% FBS, 0.1 µM dexamethasone (Sigma-Aldrich, Saint-Louis, MO, USA), 0.15 mM l-ascorbic acid (Sigma-Aldrich, Saint-Louis, MO, USA cod. A92902), and 1 mM Sodium Phosphate Monobasic (Sigma-Aldrich, Saint-Louis, MO, USA) for 21 days. The culture medium was renewed every two days. For analysis of osteogenic differentiation, cell cultures were fixed in a paraformaldehyde solution (4%) for 15 min and stained with von Kossa (Bio-Optica, Milano Italy, cod. 04-170801) or Alizarin Red stain (ScienCell, Carlsbad, CA, USA cod. 8678) according to the manufacturer's instructions. For chondrogenic differentiation, the cells were grown in pellets in 15 mL polypropylene tube in DMEM/F12 Glutamax, 1% P/S, 0.1 µM dexamethasone, 0.15 mM l-ascorbic acid, 0.35 mM proline (Sigma-Aldrich, Saint-Louis, MO, USA), 1 mM sodium pyruvate (Sigma-Aldrich, Saint-Louis, MO, USA, cod. S8636), 1% ITS, and 10 ng/mL transforming growth factor beta-3 (TGF-β3, Invitrogen, cod. RP-8600) for 21 days and fixed in 10% buffered formalin (pH 7.4), routinely processed and paraffin embedded. Four-micrometer-thick sections were cut and stained with Alcian blue/PAS (Bio-Optica, Milano Italy, cod. 04-163802) according to the manufacturer's instructions.

For tenogenic differentiation, 30,000 OE-MSCs were grown in 24-well plates on a 5 µg/cm^2 collagen-I matrix (Gibco, cod. A1064401) in DMEM/F12 Glutamax without FBS, 50 ng/mL Growth Differentiation Factor 5 (GDF-5, R&D Systems, Minneapolis, MN, USA, cod. 8340-G5-050), 50 ng/mL Growth Differentiation Factor 5 (GDF-7, R&D Systems, Minneapolis, MN, USA, cod. 8386-G7-050) and 20 ng/mL TGF-B3 (Invitrogen, cod. RP-8600) for 7 days. The culture medium was renewed every two to three days. For evaluation of tenogenic differentiation, the cells were paraformaldehyde fixed, and ICC was performed against the tenomodulin and scleraxis proteins.

2.9. Immunocytochemistry

Immunocytochemistry was carried out to assess the expression of nestin, the neural proteins GFAP and MAP2, and the tenoblast proteins tenomoduline and scleraxis, with the appropriate primary antibody (Table 1).

Paraformaldehyde fixed cells were incubated for 1 h at RT with blocking solution (3% bovine serum albumin (BSA, Sigma Aldrich Saint-Louis, MO, USA, cod. A7030), and 0.1% Triton X-100, (Sigma-Aldrich, Saint-Louis, MO, USA, cod. T8787), 5% goat serum (Dutscher, Bernolsheim, France) in phosphate-buffered saline (PBS, Hyclone, Marlborough, MA, USA, cod. SH30264.01) solution. Glass coverslips were then incubated over-night at RT with the appropriate primary antibody diluted in the staining solution PBS 3% BSA, 5% goat serum). The cells were then rinsed 3 times in PBS and incubated for 3 h with the appropriate AlexaFluor 488-conjugated polyclonal secondary antibody. After several washes in PBS, cells were counterstained with 0.5 μg/mL Hoechst blue (33,258, Sigma-Aldrich, Saint-Louis, MO, USA) for 10 min and mounted with anti-fading medium (ProLong Diamond, Invitrogen, cod. P36965). Negative control conditions were carried out by omitting the primary antibody.

2.10. Image Acquisition

Pictures were acquired with an inverted microscope EVOS®® FL Auto Imaging System (Thermofisher, Waltham, MA, USA) and negative controls were used to adjust image acquisition parameters. ICC pictures were acquired with monochrome camera on DAPI (357/447 nm) fluorescence channel for Hoechst staining and GFP (470/525 nm) fluorescence channel for Alexa 488 staining.

Non fluorescent images were acquired with color brightfield image mode.

3. Results

3.1. Biopsy of Olfactory Mucosa and Isolation and Expansion of OE-MSCs

Olfactory mucosa biopsies were successfully obtained from the 4 anesthetized cats. The only undesirable effect observed immediately after the biopsies was nasal bleeding that was rapidly stopped by applying a sterile gauze upon the nostrils. The animals recovered from anesthesia with no other unwanted side effects. One to two weeks after the biopsies, we observed adherent cells with fibroblastic morphology growing from the explants and forming a homogenous monolayer (Figure 1A).

3.2. Stemness and Immature Features

The OE-MSCs displayed nestin protein expression, and under specific culture conditions, these cells could generate spheres, as shown in Figure 1B,C.

3.3. Clonal Efficiency Assay

The OE-MSCs formed colonies at a low cell density (20 to 320 cells/well). The average clonal efficiency for the feline OE-MSCs was 10.64% ± 10.48% (mean ± SD).

3.4. In Vitro Proliferation Assay

The population doubling times of the feline olfactory stem cells were examined at 2 months (P10: 50.07 h ± 43.13 h (mean ± SD)) and 3 months (P20: 80.69 h ± 22.42 h (mean ± SD)) after the biopsies. The population doubling time increased from P10 to P20

3.5. In Vitro Neural and Mesodermal Differentiation Assays

Before neural lineage differentiation, the GFAP and MAP2 proteins were expressed in cells in the basal state (Figure S1A,B in Supplementary Materials). Within in vitro differentiation conditions, ICC demonstrated that the expression of these proteins was increased (Figure 1D,E). Under the appropriate culture conditions, cells expressed biochemical features specific to osteoblasts, chondroblasts and tenoblasts. The differentiated cultures showed Alizarin Red (Figure 1F) and von Kossa (photo not shown) staining after osteogenic differentiation. In chondrogenic differentiation culture conditions, the cells aggregated, and their histological sections were positive for Alcian blue/PAS staining (Figure 1G). The OE-MSCs expressed the tenomodulin protein under tenogenic differentiation conditions (Figure 1H). Consequently, feline OE-MSCs could differentiate into the mesodermal lineage.

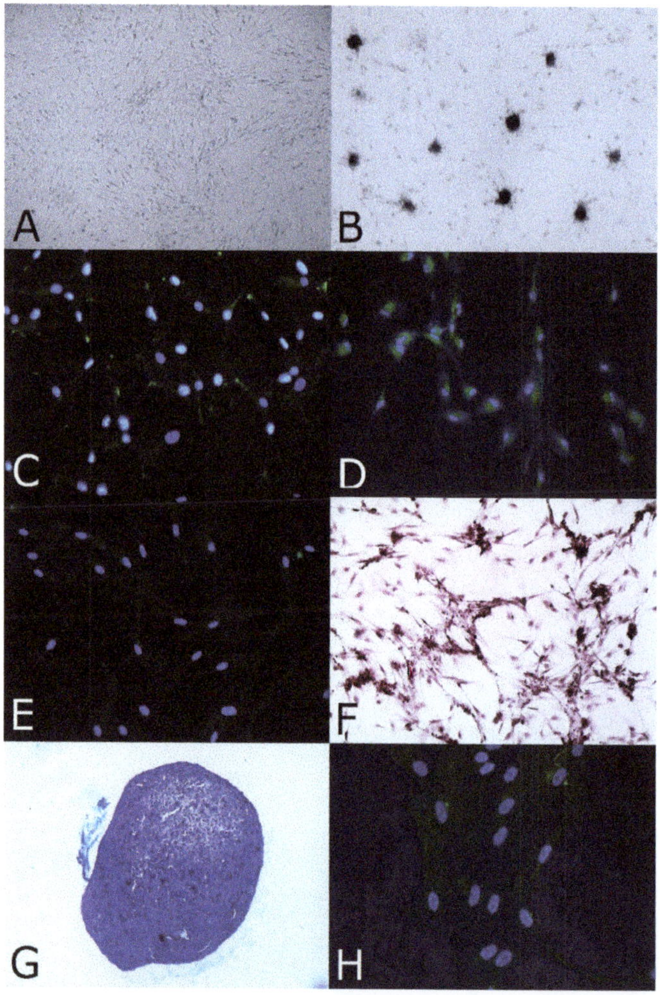

Figure 1. Morphology, stemness features, and assessment of neural and mesodermal differentiation abilities of OE-MSCs in vitro. (**A**) In growth culture medium, the OE-MSCs formed adherent cells with fibroblastic morphology (ob. × 40). (**B**) Grown in specific culture conditions, the OE-MSCs could generate spheres (ob. × 100). (**C**) Cells expressed the nestin protein (in green, ob. × 200). Neural lineage differentiation was assessed with ICC against GFAP (**D**) and MAP2 (**E**) (in green, ob. × 200). (**F**) Osteogenic differentiation was assessed with Alizarin Red, and calcium deposits were positively labeled in brown (ob. × 100). (**G**) OE-MSCs in chondrogenic differentiation medium were positively labeled with Alcian blue/PAS staining (in purple-blue, ob. × 40). (**H**) Tenogenic markers were assessed with ICC against tenomodulin (in green, ob. ×200). For ICC, cells were colabeled with Hoechst (blue).

4. Discussion

Our study showed for the first time that OE-MSCs can be extracted from cat olfactory mucosa. This tissue was easily accessible in the nasal cavity of anesthetized animals, and the sampling presented few technical issues [26]. The Feline OE-MSCs presented fibroblastic-like morphology and two stemness and immaturity features previously described in human OE-MSCs [24], such as nestin protein expression, and the ability to form spheres, even if they were smaller than those in dogs, horses and rabbits [26]. The feline OE-MSCs formed colonies at a low density, which is a characteristic of stem cells. The number of colonies is less high than in the eight mammalian genera already characterized [26]. Clonal efficiency assay or colony-forming unit fibroblast (CFU-f) is used to quantify the number of MSCs progenitors in bone marrow samples. Feline OE-MSCs have better clonal efficiency than MSCs from bone marrow [32] but inferior compared to adipose-derived stem cells [33]. These cells also showed differentiation into neural and mesodermal lineages under appropriate specific culture conditions.

The feline OE-MSCs expanded and amplified rapidly, even if they proliferated slower at P20 than P10 which reveals a decrease of the self-renewal capacity of OE-MSCs. The feline OE-MSCs still have high proliferative capabilities in the early passages even if the PDT is lower than that of dogs, rabbits and horses due to species specific diversity [26]. Moreover, the proliferation rate was the same [33] or better [34–36] than that of feline MSCs from other tissues at P10. The feline MSCs seems to have short proliferation capabilities with time and suffer of early senescence [10]. Indeed, feline adipose tissue derived MSCs show a significant increase of the PDT after 4–5 passage [34,35]. Cat peripheral blood MSCs stop proliferate at passage between 7 and 9 [36]. While these MSCs were evaluated only until P10, OE-MSCs showed to be able to proliferate also at P20, suggesting that they possess a longer duration in tested culture conditions. However, in previous studies, feline MSCs have been transplanted much earlier than passage 20, commonly between P2 and P5 [14,17]. In our study we did not evaluate the karyotype of OE-MSCs at P20, thus further analyses should aim to assess if OE-MSCs possess any kind of alteration that make these cells unsuitable for graft at this point.

Similar to those of rats, rabbits, dogs and horses, the feline OE-MSCs expressed GFAP and MAP2 in the basal state, which is a known stemness feature [26]. The expression of these proteins increased after in vitro differentiation (Figure S1A,B in Supplementary Materials), indicating that these cells could differentiate into neural lineages. This finding may open the way for further studies aiming to evaluate if the feline OE-MSCs could represent a potential treatment for brain or neural lesions [28,37].

Our analyses showed that OE-MSCs could also be induced in osteoblast-like, chondroblast-like and tenoblast-like cells under the appropriate differentiation conditions. Even if these are only in vitro findings, they may suggest that future studies could investigate if OE-MSCs may also have a potential role in the treatment of bones, cartilage and tendon lesions. MSCs demonstrated their efficacy in equine tendinopathy [38,39]. They have also shown benefic effect in bone healing in canine, ovine and caprine clinical model [40]. Canine MSCs have cartilage regenerative effect in dog with osteoarthritis [41].

Compared to bone marrow stem cells, OE-MSCs are easier to collect. Indeed, olfactory mucosa biopsy could be performed during a routine intervention requiring simple anesthesia. Since the olfactory mucosa is easily accessible, sampling is safer and less painful than bone marrow biopsy. On the other hand, the adipose stem cells are the most largely studied in cats, since they are easy to collect and possess a high proliferative ability [12], even if this capability seems to be reduced after 4–5 passages [34,35]. However, the aim of this study was to explore and propose another source of feline MSCs and to expand the knowledge on feline stem cells.

5. Conclusions

This study showed for the first time that the olfactory mucosa is a source of MSCs in cats. These cells can be easily isolated and amplified. Feline OE-MSCs display stemness characteristics and differentiation capabilities. These results pave the way for further studies that should evaluate if OE-MSCs could be a promising tool for feline autologous stem cell therapy and for veterinary regenerative medicine.

Supplementary Materials: The following supporting information can be downloaded at: https://www.mdpi.com/article/10.3390/ani12101284/s1, Figure S1: Negative controls of histochemical and immucytochemical analyses.

Author Contributions: Conceptualization, M.-L.M., P.P. and P.A.; methodology, M.-L.M., D.R. and V.M.; validation, P.A.; formal analysis, M.-L.M. and P.A.; investigation, M.-LM., D.R. and V.M.; resources, P.P.; data curation, M.-L.M. and P.A.; writing—original draft preparation, M.-L.M. and P.A.; writing—review and editing, D.R., V.M. and P.P.; supervision, P.A.; project administration, M.-L.M. and P.A.; funding acquisition, P.P. All authors have read and agreed to the published version of the manuscript.

Funding: This research received no external funding.

Institutional Review Board Statement: This work involved the use of experimental animals and the study therefore had ethical approval from an established committee as stated in the manuscript. This study was conceived and performed in accordance with French (2013-118) and European law (2010/63/EU) on the protection of animals used for scientific purposes. This protocol was approved by the Ministry of Higher Education, Research and Innovation of France and by the IRSEA's Ethics committee C2EA125 (approval number: UE-2018-EU0552).

Informed Consent Statement: Not applicable.

Data Availability Statement: The data presented in this study are available in the article and in the Supplementary Materials (Figure S1).

Acknowledgments: We want to thank the IRSEA's Animal Health Service Sanitary Management of IRSEA, particularly the animal-keepers, for caring for the cats. We want to thank C. Bienboire-Frosini for her suggestions. We are also grateful to IRSEA's Ethical Committee for their comments and the approval of the protocol.

Conflicts of Interest: The authors declare no conflict of interest.

References

1. Baksh, D.; Song, L.; Tuan, R.S. Adult mesenchymal stem cells: Characterization, differentiation, and application in cell and gene therapy. *J. Cell Mol. Med.* **2004**, *8*, 301–316. [CrossRef] [PubMed]
2. Dominici, M.; Le Blanc, K.; Mueller, I.; Slaper-Cortenbach, I.; Krause, D.S.; Deans, R.J.; Keating, A.; Prockop, D.J.; Horwitz, E.M. Minimal criteria for defining multipotent mesenchymal stromal cells. The International Society for Cellular Therapy position statement. *Cytotherapy* **2006**, *8*, 315–317. [CrossRef] [PubMed]
3. Hmadcha, A.; Martin-Montalvo, A.; Gauthier, B.R.; Soria, B.; Capilla-Gonzalez, V. Therapeutic Potential of Mesenchymal Stem Cells for Cancer Therapy. *Front. Bioeng. Biotechnol.* **2020**, *8*, 43. [CrossRef]
4. Willing, A.E.; Das, M.; Howell, M.; Mohapatra, S.S.; Mohapatra, S. Potential of mesenchymal stem cells alone, or in combination, to treat traumatic brain injury. *CNS Neurosci. Epub.* **2020**, *26*, 616–627. [CrossRef] [PubMed]
5. Kyriakidis, T.; Iosifidis, M.; Michalopoulos, E.; Melas, I.; Stavropoulos-Giokas, C.; Verdonk, R. Good mid-term outcomes after adipose-derived culture-expanded mesenchymal stem cells implantation in knee focal cartilage defects. *Knee Surg. Sports Traumatol. Arthrosc.* **2020**, *28*, 502–508. [CrossRef]
6. Goradel, N.H.; Ghiyami-Hour, F.; Negahdari, B.; Malekshahi, Z.V.; Hashemzehi, M.; Masoudifar, A.; Mirzaei, H. Stem Cell Therapy: A New Therapeutic Option for Cardiovascular Diseases. *J. Cell Biochem.* **2018**, *119*, 95–104. [CrossRef]
7. Rajabzadeh, N.; Fathi, E.; Farahzad, R. Stem cell-based regenerative medicine. *Stem Cell Investig.* **2019**, *6*, 19. [CrossRef]
8. Viktorova, V.; Savchenkova, I.P. Multipotent mesenchymal stem cells in clinical veterinary practice. *IOP Conf. Ser. Earth Environ. Sci.* **2020**, *548*, 072072. [CrossRef]
9. Vikartovska, Z.; Humenik, F.; Maloveska, M.; Farbakova, J.; Hornakova, L.; Murgoci, A.N.; Cizkova, D. Adult Stem Cells Based Therapies in Veterinary Medicine. *Arch. Vet. Sci. Med.* **2020**, *3*, 40–50. [CrossRef]
10. Dias, I.E.; Pinto, P.O.; Barros, L.C.; Viegas, C.A.; Isabel Ribeiro Dias, I.R.; Carvalho, P.P. Mesenchymal stem cells therapy in companion animals: Useful for immune-mediated diseases? *BMC Vet. Res.* **2019**, *15*, 358. [CrossRef]
11. Martin, D.R.; Cox, N.R.; Hathcock, T.L.; Niemeyer, G.P.; Baker, H.J. Isolation and characterization of multipotential mesenchymal stem cells from feline bone marrow. *Exp. Hematol.* **2002**, *30*, 879–886. [CrossRef]
12. Quimby, J.M.; Borjesson, D.L. Mesenchymal stem cell therapy in cats: Current knowledge and future potential. *J. Feline Med. Surg.* **2018**, *20*, 208–216. [CrossRef] [PubMed]
13. Gugjoo, M.B.; Pal, A. (Eds.) Cat Mesenchymal Stem Cell Characteristics and Potential Applications. In *Mesenchymal Stem Cell in Veterinary Sciences*; Springer: Singapore, 2020; pp. 197–212. [CrossRef]

14. Arzi, B.; Mills-Ko, E.; Verstraete, F.J.; Kol, A.; Walker, N.J.; Badgley, M.R.; Fazel, N.; Murphy, W.J.; Vapniarsky, N.; Borjesson, D.L. Therapeutic efficacy of fresh, autologous mesenchymal stem cells for severe refractory gingivostomatitis in cats. *Stem Cells Transl. Med.* **2016**, *5*, 75–86. [CrossRef] [PubMed]
15. Arzi, B.; Kol, A.; Murphy, B.; Walker, N.J.; Wood, J.A.; Clark, K.; Verstraete, F.J.; Borjesson, D.L. Feline foamy virus adversely affects feline mesenchymal stem cell culture and expansion: Implications for animal model development. *Stem Cells Dev.* **2015**, *24*, 814–823. [CrossRef] [PubMed]
16. Trzil, J.E.; Masseau, I.; Webb, T.L.; Chang, C.H.; Dodam, J.R.; Cohn, L.A.; Liu, H.; Quimby, J.M.; Dow, S.W.; Rejnero, C.R. Long-term evaluation of mesenchymal stem cell therapy in a feline model of chronic allergic asthma. *Clin. Exp. Allergy* **2014**, *44*, 1546–1557. [CrossRef] [PubMed]
17. Trzil, J.E.; Masseau, I.; Webb, T.L.; Chang, C.H.; Dodam, J.R.; Liu, H.; Quimby, J.M.; Dow, S.W.; Rejnero, C.R. Intravenous adipose-derived mesenchymal stem cell therapy for the treatment of feline asthma: A pilot study. *J. Feline Med. Surg.* **2016**, *18*, 981–990. [CrossRef]
18. Webb, T.L.; Webb, C.B. Stem cell therapy in cats with chronic enteropathy: A proof-of-concept study. *J. Feline Med. Surg.* **2015**, *17*, 901–908. [CrossRef]
19. Quimby, J.M.; Webb, T.L.; Gibbons, D.S.; Dow, S.W. Evaluation of intrarenal mesenchymal stem cell injection for treatment of chronic kidney disease in cats: A pilot study. *J. Feline Med. Surg.* **2011**, *13*, 418–426. [CrossRef]
20. Quimby, J.M.; Webb, T.L.; Habenicht, L.M.; Dow, S.W. Safety and efficacy of intravenous infusion of allogeneic cryopreserved mesenchymal stem cells for treatment of chronic kidney disease in cats: Results of three sequential pilot studies. *Stem Cell Res. Ther.* **2013**, *4*, 48. [CrossRef]
21. Vidane, A.S.; Pinheiro, A.O.; Casals, J.B.; Passarelli, D.; Hage, M.; Bueno, R.S.; Martins, D.S.; Ambrosio, C.E. Transplantation of amniotic membranederived multipotent cells ameliorates and delays the progression of chronic kidney disease in cats. *Reprod. Domest. Anim.* **2017**, *52* (Suppl. 2), 316–326. [CrossRef]
22. Quimby, J.M.; Webb, T.L.; Randall, E.; Marolf, A.; Martinez, A.V.; Dow, S.W. Assessment of intravenous adipose-derived allogeneic mesenchymal stem cells for the treatment of feline chronic kidney disease: A randomized, placebo-controlled clinical trial in eight cats. *J. Feline Med. Surg.* **2016**, *18*, 165–171. [CrossRef]
23. Rosselli, D.D.; Mumaw, J.L.; Dickerson, V.; Brown, C.A.; Brown, S.A.; Schmiedt, C.W. Efficacy of allogeneic mesenchymal stem cell administration in a model of acute ischemic kidney injury in cats. *Res. Vet Sci.* **2016**, *108*, 18–24. [CrossRef] [PubMed]
24. Delorme, B.; Nivet, E.; Gaillard, J.; Häupl, T.; Ringe, J.; Devèze, A.; Magnan, J.; Sohier, J.; Khrestchatisky, M.; Roman, F.S.; et al. The human nose harbors a niche of olfactory ectomesenchymal stem cells displaying neurogenic and osteogenic properties. *Stem Cells Dev.* **2010**, *19*, 853–866. [CrossRef] [PubMed]
25. Girard, S.D.; Devéze, A.; Nivet, E.; Gepner, B.; Roman, F.S.; Féron, F. Isolating nasal olfactory stem cells from rodents or humans. *J. Vis. Exp.* **2011**, *54*, 2762. [CrossRef] [PubMed]
26. Veron, A.D.; Bienboire-Frosini, C.; Feron, F.; Codecasa, E.; Deveze, A.; Royer, D.; Watelet, P.; Asproni, P.; Sadelli, K.; Chabaud, C.; et al. Isolation and characterization of olfactory ecto-mesenchymal stem cells from eight mammalian genera. *BMC Vet. Res.* **2018**, *14*, 17. [CrossRef] [PubMed]
27. Young, E.; Westerberg, B.; Yanai, A.; Evans, K.G. The olfactory mucosa: A potential source of stem cells for hearing regeneration. *Regen. Med.* **2018**, *13*, 581–593. [CrossRef] [PubMed]
28. Veron, A.D.; Bienboire-Frosini, C.; Girard, S.D.; Sadelli, K.; Stamegna, J.C.; Khrestchatisky, M.; Alexis, J.; Pageat, P.; Asproni, P.; Mengoli, M.; et al. Syngeneic Transplantation of Olfactory Ectomesenchymal Stem Cells Restores Learning and Memory Abilities in a Rat Model of Global Cerebral Ischemia. *Stem Cells Int.* **2018**, *2018*, 2683969. [CrossRef]
29. Murrell, W.; Wetzig, A.; Donnellan, M.; Féron, F.; Burne, T.; Meedeniya, A.; Kesby, J.; Bianco, J.; Perry, C.; Silburn, P.; et al. Olfactory mucosa is a potential source for autologous stem cell therapy for Parkinson's disease. *Stem Cells* **2008**, *26*, 2183–2192. [CrossRef]
30. Ohnishi, Y.I.; Iwatsuki, K.; Ishihara, M.; Shikina, T.; Shinzawa, K.; Moriwaki, T.; Ninomiya, K.; Ohkawa, T.; Umegaki, M.; Kishima, H.; et al. Adult olfactory sphere cells are a source of oligodendrocyte and Schwann cell progenitors. *Stem Cell Res.* **2013**, *11*, 1178–1190. [CrossRef]
31. Chu, T.; Zhou, H.; Wang, T.; Lu, L.; Li, F.; Liu, B.; Kong, X.; Feng, S. In Vitro characteristics of valproic acid and all-trans-retinoic acid and their combined use in promoting neuronal differentiation while suppressing astrocytic differentiation in neural stem cells. *Brain Res.* **2015**, *1596*, 31–47. [CrossRef]
32. Maciel, B.B.; Rebelatto, C.L.K.; Brofman, P.R.S.; Brito, H.F.V.; Patricio, L.F.L.; Cruz, M.A.; Locatelli-Dittrich, R. Morphology and morphometry of feline bone marrow-derived mesenchymal stem cells in culture. *Pesq. Vet. Bras.* **2014**, *34*, 1127–1134. [CrossRef]
33. Panasophonkul, S.; Samart, P.; Kongon, K.; Sathanawongs, A. Phenotypic characteristics of feline adipose-derived stem cells affected by cell passage number. *Pol. J. Vet. Sci.* **2017**, *20*, 651–660. [CrossRef] [PubMed]
34. Kim, H.R.; Lee, J.; Byeon, J.S.; Gu, N.Y.; Lee, J.; Cho, I.S.; Cha, S.H. Extensive characterization of feline intra-abdominal adipose-derived mesenchymal stem cells. *J. Vet. Sci.* **2017**, *18*, 299–306. [CrossRef] [PubMed]
35. Lee, B.Y.; Li, Q.; Song, W.J.; Chae, H.K.; Kweon, K.; Ahn, J.O.; Youn, H.Y. Altered properties of feline adipose-derived mesenchymal stem cells during continuous in vitro cultivation. *J. Vet. Med. Sci.* **2018**, *80*, 930–938. [CrossRef] [PubMed]
36. Sato, K.; Yamawaki-Ogata, A.; Kanemoto, I.; Usui, A.; Narita, Y. Isolation and characterisation of peripheral blood-derived feline mesenchymal stem cells. *Vet. J.* **2016**, *216*, 183–188. [CrossRef]

37. Veron, A.D.; Mengoli, M. Olfactory Stem Cell Therapy in Canine Age-related Disorders Treatment: A Controlled Study. In Proceedings of the 11th International Veterinary Behaviour Meeting, Samorin, Slovakia, 14–16 September 2017; Denenberg, S., Ed.; Langford Vets: Bristol, UK; Chapter 43, p. 118.
38. Smith, R.K.W.; Werling, N.J.; Dakin, S.G.; Alam, R.; Goodship, A.E.; Dudhia, J. Beneficial effects of autologous bone marrow-derived mesenchymal stem cells in naturally occurring tendinopathy. *PLoS ONE* **2013**, *8*, e75697. [CrossRef]
39. Godwin, E.E.; Young, N.J.; Dudhia, J.; Beamish, I.C.; Smith, R.K.W. Implantation of bone marrow-derived mesenchymal stem cells demonstrates improved outcome in horses with overstrain injury of the superficial digital flexor tendon. *Equine Vet. J.* **2012**, *44*, 25–32. [CrossRef]
40. Perez, J.R.; Kouroupis, D.; Li, D.J.; Best, T.M.; Kaplan, L.; Correa, D. Tissue Engineering and Cell-Based Therapies for Fractures and Bone Defects. *Front. Bioeng. Biotechnol.* **2018**, *6*, 105. [CrossRef]
41. Sasaki, A.; Mizuno, M.; Mochizuki, M.; Sekiya, I. Mesenchymal stem cells for cartilage regeneration in dogs. *World J. Stem Cells* **2019**, *11*, 254–269. [CrossRef]

animals

Review

An Update on Applications of Cattle Mesenchymal Stromal Cells

Barbara Merlo [1,2,*], Penelope Maria Gugole [1] and Eleonora Iacono [1,2]

[1] Department of Veterinary Medical Sciences, University of Bologna, 40064 Ozzano Emilia, BO, Italy; penelopemaria.gugol2@unibo.it (P.M.G.); eleonora.iacono2@unibo.it (E.I.)
[2] Interdepartmental Center for Industrial Research in Health Sciences and Technologies, University of Bologna, 40126 Bologna, BO, Italy
* Correspondence: barbara.merlo@unibo.it; Tel.: +39-051-2097581

Simple Summary: Among livestock species, cattle are crucially important for the meat and milk production industry. Cows can be affected by different pathologies, such as mastitis, endometritis and lameness, which can negatively affect either food production or reproductive efficiency. The use of mesenchymal stromal cells (MSCs) is a valuable tool both in the treatment of various medical conditions and in the application of reproductive biotechnologies. This review provides an update on state-of-the-art applications of bovine MSCs to clinical treatments and reproductive biotechnologies.

Abstract: Attention on mesenchymal stromal cells (MSCs) research has increased in the last decade mainly due to the promising results about their plasticity, self-renewal, differentiation potential, immune modulatory and anti-inflammatory properties that have made stem cell therapy more clinically attractive. Furthermore, MSCs can be easily isolated and expanded to be used for autologous or allogenic therapy following the administration of either freshly isolated or previously cryopreserved cells. The scientific literature on the use of stromal cells in the treatment of several animal health conditions is currently available. Although MSCs are not as widely used for clinical treatments in cows as for companion and sport animals, they have the potential to be employed to improve productivity in the cattle industry. This review provides an update on state-of-the-art applications of bovine MSCs to clinical treatments and reproductive biotechnologies.

Keywords: bovine; mesenchymal stromal cells; clinical applications; reproductive biotechnologies

Citation: Merlo, B.; Gugole, P.M.; Iacono, E. An Update on Applications of Cattle Mesenchymal Stromal Cells. *Animals* **2022**, *12*, 1956. https://doi.org/10.3390/ani12151956

Academic Editor: Paul Dyce

Received: 25 June 2022
Accepted: 29 July 2022
Published: 2 August 2022

1. Introduction

Research into stem cells has been very active over the past decade. Due to the increasing number of studies, several breakthroughs have been achieved in this field, and stem cell therapy has gained ground as a modality of regenerative medicine. Mesenchymal stromal cells (MSCs) are present in different body tissues and are characterised as able to adhere to plastic, express specific surface antigens and possess multipotent differentiation potential [1]. Furthermore, they are good candidates for the treatment of various diseases due to characteristics such as low immunogenicity, anti-inflammatory potential and their ability to produce various mediators and molecules that help the regenerative function [2].

Bovine MCSs have been isolated and characterised (Figure 1) from different adult and foetal tissues, including bone marrow (BM) [3–38], endometrium (EN) [39–48], adipose tissue (AT) [29–32,34,37,49–66] and foetal liquid and adnexa, such as umbilical cord blood (UCB) [67–70], Wharton's jelly (WJ) [58,71–73], umbilical cord matrix (UC) [74–82], amnion (AM) [83,84], amniotic fluid (AF) [57,83,85–88] and placenta (PL) [37,89,90]. Less common sources of bovine MSCs have been foetal liver [91], dermal tissue [92], foetal lung tissue [93], embryo yolk sack [94], synovial fluid [95], milk [96], pericardium membrane [97], pancreas [98], tongue epithelium [99], skeletal muscle [65,100] and peripheral blood [48,101].

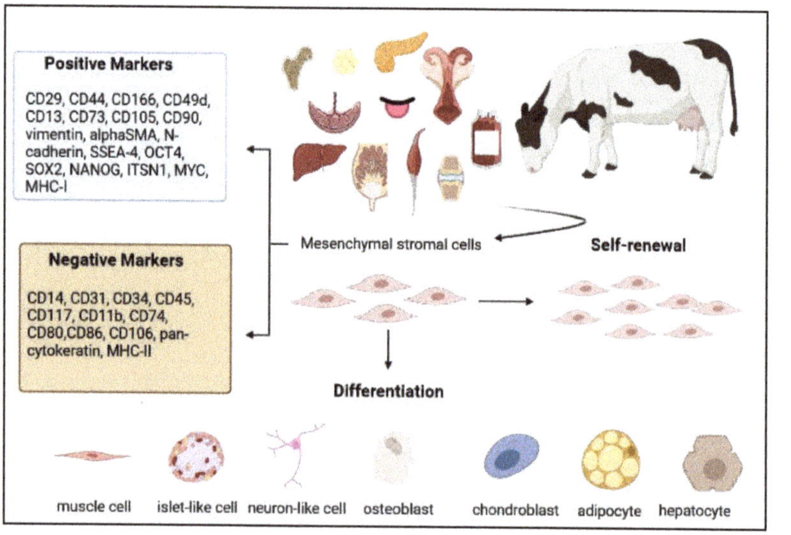

Figure 1. Schematic diagram of the characteristics of bovine mesenchymal stromal cells (created in Biorender.com, accessed on 14 July 2022).

The potential of MSCs for cell-based therapies has originally been based on their typical characteristics, which include the multipotentiality to differentiate in vitro into mesodermal-derived lineages, particularly osteogenic, chondrogenic and adipogenic cells [1]. Furthermore, it has been demonstrated that the paracrine activity of MSCs exerts therapeutical effects involving regeneration, immunomodulation, angiogenesis and antiapoptosis [102–104].

The immunomodulatory activity of MSCs depends on direct cell-to-cell contact and on contact-independent paracrine signalling, with the production of soluble factors regulating proliferation, differentiation, migration and apoptosis of several immune cells [105]. The reduced immunogenicity of MSCs is another aspect that strengthens their potential for cell therapy related in part to the low expression of major histocompatibility complexes I and II (MHC-I and II) and to the absence of expression of T-cell costimulatory molecules (CD40, CD80 and CD86) [106]. Taking together the immune regulatory abilities and reduced immunogenicity, allogeneic MSCs transplanted into recipients are able to escape direct recognition by natural killer cells and prevent activation of T lymphocytes, possibly also reducing the potential activation of the indirect pathway by the presentation of donor-derived MHC-I/II peptides by antigen-presenting cells to B cells and subsequent alloantibodies production [107]. Therefore, low immunogenicity may result in higher efficacy and lower risk of local inflammation following MSCs administration, reducing potential adverse effects [107]. In cattle, it has been demonstrated that foetal AT-MSCs and BM-MSCs respond to inflammatory stimulation with interferon γ (IFNγ) by increasing immune-related gene expression and activity in a dose-dependent manner and upregulating gene expression of IL-6 [30]. However, conditioned medium from IFNγ-stimulated and unstimulated BM-MSCs and AT-MSCs exerts similar suppression of proliferation of alloantigen-activated bovine peripheral blood lymphocytes [30]. Whereas immunomodulatory properties appear to be similar between BM-MSCs and AT-MSCs, higher expression of MHC-I and MHC-II in BM-MSCs suggested that the immunogenic potential of bovine foetal MSCs might be tissue-dependent and that AT-MSCs might be more suitable candidates for allogeneic therapy [30].

Autologous MSCs therapy implies cell isolation and expansion to achieve therapeutic doses. Consequently, there is a lag time between their collection and use, threatening the effectiveness of the treatment. In addition, critical parameters for MSCs isolation include donor variability, tissue of origin, amount of tissue and culture conditions [108]. On the other hand, foetal- and placental-derived MSCs have been found superior to adult MSCs as candidates for allogeneic therapeutic applications due to their lower immunogenicity [109,110]. Cryopreservation represents an efficient method for the preservation and pooling of MSCs to obtain the cell counts required for clinical applications. Samples can be harvested, and then cells can be isolated, expanded and stored for later use, optimising logistics from collection to transplantation. Accordingly, the ability of MSCs to survive long periods of storage and, at the same time, maintain their qualities is critical for the development of allogeneic cell therapies. Upon cryopreservation, it is important to preserve MSCs' functional properties, including immunomodulatory properties and multilineage differentiation ability. Further, a biosafety evaluation of cryopreserved MSCs is essential prior to their clinical applications [111]. Considering cattle, Oyarzo et al. compared PL-MSCs and foetal MSCs originated from AT and BM in order to assess their ability to survive different cryoprotectant solutions exposure [37]. While the apoptotic potential was similar, foetal AT-MSCs and PL-MSCs presented consistently higher percentages of viability than did foetal BM-MSCs [37]. On the other hand, AT-MSCs were more resistant than PL-MSCs, but the latter have the advantage of coming from a readily available tissue usually considered waste, without ethical concerns [37].

Although in veterinary medicine, cell therapies are mainly focused on pets, regenerative medicine applications also involve farm animals, not only for their importance as a food source [112] but also as models [113]. Among livestock species, cows have a high economic impact, and reproductive biotechnologies are routinely applied [114,115]. The dairy and beef industries are essential for food production. Dairy products and ruminant meat provide essential elements for the human diet. According to the Food and Agriculture Organization (FAO), there are almost 1.5 billion cattle in the world. Cows produce 81 per cent of global milk production, and the world demand for beef is projected to increase to 75 million tonnes by 2030 [116]. Animal health is an important issue related not only to animal welfare itself but also to the One Health perspective, in which human, animal, plant and environmental health are interdependent. This review summarises the applications of MSCs in cattle to treat clinical conditions and improve reproductive biotechnologies.

2. Bovine MSCs for Clinical Treatments

So far, MSCs have been used in many experimental instances to treat various diseases in different animal species. Orthopaedic diseases were the primary field of regenerative veterinary medicine, and then the focus rapidly expanded to other areas. Dogs and horses were the species in which stem cell-based therapies were commonly used to treat different diseases of various organ systems, while for cats, they were used for renal, respiratory and inflammatory pathologies [117]. Bovine MSCs can be potentially used in various clinical conditions. Nevertheless, the application of novel MSCs therapies in large ruminants is still limited.

The major obstacles in livestock species are related to a minor interest in treating clinical conditions in these animals compared to pets and the higher maintenance costs in comparison with other animal models [118]. Laboratory animals or small animals are usually preferred as models to start any research for human pathologies due to the reasonable buying and care costs together with easier manageability and housing. However, for a better understanding and a thorough evaluation of cell-based therapies, various animal models are necessary to successfully move from the laboratory bench to human health applications. The development of products for animal use has the advantage that they can be immediately tested in the target species. This aspect not only allows to understand the potential of MSC-based products for clinical application in animals but may also provide models for similar human applications [119].

Although many studies have been published for animal MSCs, it is still not easy to evaluate the efficacy of MSC-based therapies because of the different sources of MSCs and variations in manufacturing processes, inconsistent characterisation and measure of potency, inappropriate controls and a lack of experimental power [119]. MSCs have been isolated from different sources and, depending on the tissue of origin, they may possess different properties, which should be taken into account when choosing the optimal stem cell therapy for a specific pathology in order to achieve successful results. On the other hand, there is no evidence for a favoured tissue as an MSCs source due to the presence of a wide variability between donors [108,120].

2.1. Chronic Wound Healing

In the last years, the application of regenerative medicine to skin lesions has been a focus for both human and veterinary medicine. The physiological healing process of cutaneous wounds is a well-orchestrated complex of molecular and biological activities. Even so, a chronic lesion can develop when the normal process fails. The regenerative potential of stromal cells has also been widely recognised for skin lesion repair [121]. Recent studies support the concept that MSCs can be appropriated for treating chronic wounds [122–124].

Even if the exact functions of stromal cells in wound healing have not yet been completely elucidated, they are involved in the removal of dead cells and necrotic tissue, angiogenesis, reduction in scar tissue formation, contraction of the wound and induction of re-epithelisation [121]. Consequently, wound healing is promoted, and local inflammation is reduced. Table 1 summarises the studies regarding MSC applications for wound healing in cattle.

Table 1. Bovine MSC applications for wound healing.

Source	Application	References
Bone marrow	Autologous treatment of a chronic ulcer in a heifer	[10]
Bone marrow	Autologous treatment of an interdigital chronic ulcerative wound in a cow	[13]
Bone marrow	Autologous treatment of a wound in a hind limb of a bull	[14]

The first report of a case study in which autologous BM-derived MSCs were used to treat a chronic ulcer in a heifer dates back to 2012 [10]. A 2-year-old Jersey heifer had been suffering from a chronic nonhealing ulcerative wound involving full-thickness skin and underlying muscle in the lumbar region for 4 months. Standard therapies were ineffective, so a clinical trial was made with autologous BM-MSCs. Bone marrow was aspirated from the tibia, and MSCs were isolated, expanded and then diluted in saline solution for intradermal and topical implantation in the wound. Various parameters and measures were monitored during the trial. At histopathology, the progression of the healing process was observed since neovascularisation appeared, as well as fibroblasts, sebaceous glands and epithelialisation. The content of collagen was increased after stem cell therapy, and the healed tissue was progressing towards physiological stretchability and tensile strength. The 4-month-old chronic wound healed within 18 days, indicating that MSCs application could be an effective therapeutic approach for nonhealing chronic wounds [10].

Another clinical study of the same research group concerns the successful treatment of an interdigital chronic ulcerative wound in a 6-year-old cross-bred Jersey cow [13]. The animal presented with a 4-month interdigital hoof lesion nonresponding to conventional treatments. Autologous BM-MSCs therapy was also used for this patient. Granulated tissue rapidly grew, and the healing process was completed in 18 days. The parameters analysed to assess the progression of the healing process confirmed the clinical process, and the pain-free walking distance evaluation was gradually increased over the study period [13].

In the last clinical trial [14], a bull was presenting a wound in a hind limb above the hock joint as a consequence of a car accident, which had happened 8 months before.

Different local treatments and antibiotic courses turned out as unsuccessful as chemical and cryocauterisation. Autologous BM-MSCs application was performed. Similar protocols were used for the collection, isolation and expansion of BM-MSCs, but in this case, some cells were intravenously administered in addition to local treatment. Healing was completed within 4 weeks, and the evaluated parameters confirmed the outcome [14].

Despite the lack of controls and large-scale randomised studies and clinical trials, the promising results obtained from the applications of autologous BM-MSCs confirmed the potential of MSC-based therapy for treating chronic nonhealing wounds in bovines.

A weak immunogenicity and a vasculogenic effect are favourable properties for wound healing capacity. Bovine BM-MSCs are the most well-characterised cells, and recently their immunomodulatory properties [30] and proangiogenic potential [31] have been investigated. Comparing bovine foetal MSCs derived from bone marrow and adipose tissue, both upregulated the expression of immunomodulatory genes and showed similar in vitro immunomodulatory ability, while the lower expression of MHC-I and MHC-II suggested that AT-MSCs might be less immunogenic compared with BM-MSCs [30]. Furthermore, BM-MSCs displayed similar migratory ability, higher proliferative capacity and lower proangiogenic potential compared with AT-MSCs [31]. These results might suggest that bovine AT-MSCs could be even more promising than BM-MSCs in enhancing the treatment of chronic wound healing.

2.2. *Mastitis*

In the dairy industry, mastitis is a common problem, which implicates costs to treat the disease, and since antimicrobials are the standard therapy, this increases the possibility of developing antimicrobial resistance. Hence, alternative therapies are required.

The mammary gland contains stromal cells and precursors with high regenerative potential, which apparently are maintained during the productive life of dairy cows. The presence of such cells opens new research perspectives regarding the physiological mechanisms concerned with milk secretion and the possibility of enhancing or prolonging dairy cow production [125]. The presence of a subpopulation of adult stromal cells in the mammary gland was first demonstrated in human and mouse [126,127]. Then, in the cow, three different colony morphologies were isolated, suggesting the existence of different progenitor populations and of an epithelial cell hierarchy in the bovine mammary gland similar to humans [128]. Such stromal/progenitor cells have been largely investigated [125,129].

On the other hand, less research is available for MSCs and bovine mammary glands. As summarised in Table 2, different in vitro studies showed that UC-MSCs could promote milk protein and fat synthesis and the expression of key genes in bovine mammary gland epithelial cells via IGF-1 [75,76,78] and reduce their apoptosis rate [77]. Furthermore, it has been demonstrated that bovine MSCs have antibacterial activity [29]. The conditioned medium from bovine foetal MSCs obtained from bone marrow and adipose tissue showed in vitro antibacterial potential against *S. aureus*, a mastitis-causing pathogen, by reducing about 30% of relative bacterial growth [29]. The mechanisms that regulate the antibacterial activity of bovine MSCs have not been totally elucidated, but the expression of β–defensin 4 A and NK-lysine 1, two antibacterial peptides, was associated with the in vitro effect of such MSCs [29].

Dairy cows were experimentally infected to induce *S. aureus* clinical mastitis in order to evaluate the safety and efficacy of an allogenic MSC-based therapy [62]. Bovine foetal AT-MSCs were intramammary inoculated twice (days 1 and 10) during a 20-day experimental period. No clinical or immunological response was induced in healthy cows, and the bacterial count in milk was reduced in MSC-treated cows compared with controls [62]. A similar decrease in somatic cell count (SCC) in the milk of mastitic animals was observed in cows treated intramammary with a single administration of allogenic AT-MSCs during a 15-day experiment [64]. On days 3 and 7, maximum expression of anti-inflammatory cytokines (IL-6, IL-10), antimicrobial peptides (cathelicidin, lipocalin and cystatin) and angiogenic genes (angiopoietin) was observed [64]. With the aim of preventing subclinical mastitis, UCB-MSCs and extracellular vesicles (EVs) were injected locally and IV on days 0 and 7 in healthy (safety trial) and subclinical mastitis cows [68,69]. Both MSCs and EVs were safe, and all treated cows were cured permanently within 15 days [68]. Treated animals showed a reduced SCC in mastitic milk compared with the control (antibiotic) group, an enhancement in the expression of anti-inflammatory cytokines, antimicrobial peptides and angiogenic genes and a decrease in the expression of proinflammatory cytokines [68,69]. Finally, a conditioned medium from bovine AM-MSCs (2 h coincubation in phosphate-buffered saline (PBS)) was used to treat mastitis in comparison with conventional antibiotics [84]. Milk pH value and titratable acidity were similar between treatments, while the level of ionic calcium concentration decreased 3 days later in MSCs-treated cows compared with antibiotic-treated animals [84]. Moreover, the somatic cell number was similar in both groups, demonstrating that conditioned medium from bovine AM-MSCs has the therapeutic potential to treat bovine mastitis and might replace antibiotics in the future [84].

Table 2. Bovine MSC applications for the mammary gland.

Source	Application	References
Umbilical cord	In vitro effects on mammary gland epithelial cells	[75–78]
Bone marrow, adipose tissue	In vitro effects on *S. aureus*	[29]
Adipose tissue	In vivo effects on *S. aureus*-induced mastitis	[62]
Adipose tissue	In vivo effects on mastitis	[64]
Umbilical cord blood	In vivo effects on subclinical mastitis	[68,69]
Amniotic membrane	In vivo effects of conditioned medium to treat mastitis	[84]

2.3. *Reproductive System*

In the last 50 years, the selection in the dairy industry has led to an improvement in average milk production by a single cow. However, the selection for milk yield has caused some unfavourable effects, such as a decrease in fertility. Despite an improvement in cow fertility in the last two decades, as a consequence of selection for fertility traits in breeding programmes and improvement in animal nutrition and comfort, reproductive performance is not optimal yet [130]. Reproductive disorders are directly correlated with low fertility in dairy cows.

The endometrium is characterised by an elevated and constant regeneration, and mesenchymal progenitor cells have also been identified in the cow endometrium [39]. Progenitor cells were isolated and characterised in cyclic cows [40,42,44] and heifers [41] and were able to respond after challenging with lipopolysaccharide (LPS) [43]. Furthermore, the presence of endometrial MSCs was also confirmed in the postpartum period in both healthy cows and those affected by endometritis [45]. In this period, uterine involution occurs, involving endometrial regeneration [131], and the presence of pathogenic bacteria needs to be controlled in the uterus for fertility restoration. However, pathogenic bacteria are not always rapidly eliminated and often generate uterine disease (metritis and endometritis), leading to reduced fertility [132]. Endometrial MSCs from bovine inflamed uteri showed modified characteristics, especially in clinical than in subclinical endometritis, and the in vitro exposure of endometrial MSCs to PGE2, a mediator of inflammation, modified their transcriptomic profile [45]. Bovine endometrial MSCs have also been immortalised from lines derived in different phases of the oestrous cycle [47]. Immortalised cells maintained mesenchymal and immunomodulatory characteristics, with an increased migratory capacity towards an inflammatory niche but a decreased answer to embryonic cytokine expression at implantation [47]. Interestingly, combined proinflammatory and implantation signals ensured the retention of endometrial MSCs in case of pregnancy, while they showed a mesenchymal to epithelial transition state in the absence of an embryo [47]. Despite research into bovine endometrial MSCs, no report exists about their application in treating cow uterine inflammations. On the other hand, bovine MSCs derived from adipose tissue showed an inhibitory effect on in vitro LPS challenge of endometrial epithelial cells [63]. When used in vivo to treat metritis, allogenic AT-MSCs did not induce any immunological rejection response in treated animals (IV, local, IV + local), and all cows were completely and permanently cured within 30 days after treatment [64]. Polymorphonuclear (PMN) cell count was reduced in cervical vaginal fluid and the expression of IL-6, IL-10, cathelicidin, lipocalin, cystatin and angiopoietin were observed at day 3 in the IV + local group [64]. More recently, UCB-MSCs and their EVs have also been successfully used for metritis treatment by the same research group [70]. Moreover, in this case, a higher decrease in PMN was observed for MSCs and EV-treated cows compared with antibiotic-treated ones, as well as an increase in the expression of anti-inflammatory cytokines [70].

Other pathologies, which can lead to considerable economic loss, are those involving the ovaries. Ovarian dysfunctions in dairy cattle have a high incidence and are responsible for a reduction in reproductive performance. The two major ovarian causes of infertility in dairy cows are inactive ovaries and ovarian cysts [133,134]. Chang et al. transplanted AF-MSCs into cows affected by bilateral ovarian dystrophy in an attempt to restore or improve ovarian function [87]. Each ovary was injected with 50 μL of PBS containing 0.58 million cells, and then cows were monitored for oestrus and inseminated [87]. Half (4/8) of the animals treated with AF-MSCs showed oestrus, and two of them delivered a calf, while no oestrus was observed in control animals, demonstrating that MSCs therapy is a potentially useful treatment to alleviate the impact of ovarian dystrophy in dairy cows [87]. Peng et al. injected PL-MSCs into ovarian cysts with or without fluid drainage and compared them to control animals and GnRH-treated animals [90]. The use of PL-MSCs allowed for recovery and conception [90], indicating a new therapeutic potential of these cells and a possible alternative to hormones in the treatment of cattle ovarian cysts. Finally, the intraovarian injection of MSCs was used to reduce the negative effects of repeated ovum pick-up (OPU) under acute and chronic scenarios in bovines [61]. In fact, this technique is generally considered a safe way to collect oocytes from live donors but inevitably causes trauma to the ovarian tissue, and repeated procedures over years are associated with a progressive decrease in oocyte yield [61]. For the experiment, one ovary was injected with 2.5 million AT-MSCs, and the other one was used as the control [61]. MSCs had beneficial effects on the fertility of acute OPU injured cows, but not in cows with chronic ovarian lesions [61]. In this case, it was speculated that MSCs could no longer restore the compromised follicular population or ovarian physiology in cows with chronic inflammatory processes in the ovaries due to repeated OPU over time [61]. The overall MSC clinical applications for the reproductive system are presented in Table 3.

Table 3. Bovine MSCs from different sources for treatment of reproductive system diseases.

Source	Application	References
Adipose tissue	Metritis	[64]
Umbilical cord blood	Metritis	[70]
Amniotic fluid	Bilateral ovarian dystrophy	[87]
Placenta	Ovarian cysts	[90]
Adipose tissue	Intraovarian injection for repeated OPU lesions	[61]

3. Bovine MSCs for Reproductive Biotechnologies

The first successful nuclear transfer (NT) dates back to 1952, when the nucleus from an early tadpole embryo was transferred into an enucleated frog egg [135]. Then, in 1996, Dolly was the first mammalian cloned using an adult somatic cell as a nucleus donor [136]. Somatic cell nuclear transfer (SCNT) (Figure 2) is an important research tool since it permits a differentiated cell to be reprogrammed to a totipotent state [137]. The donor cell is a key factor in the process, and interest in bovine SCNT led to consider MSCs as appropriate candidates due to their characteristics. Studies using bovine MSCs from different sources for NT were carried out and are summarised in Table 4.

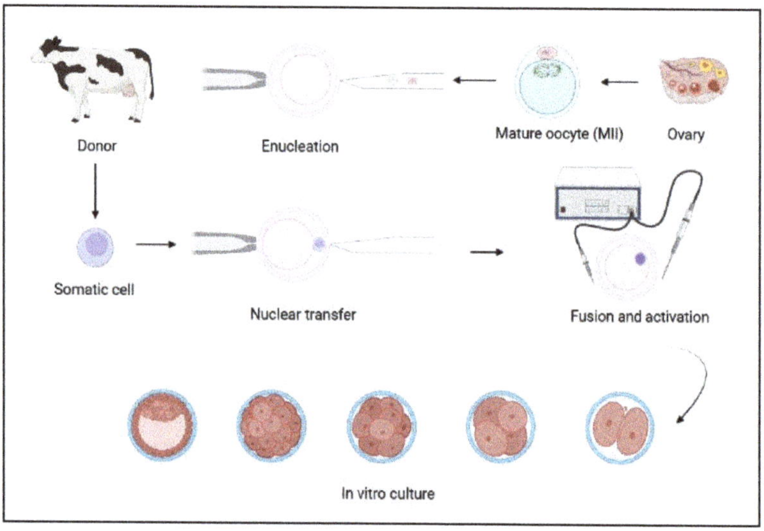

Figure 2. Schematic diagram of the somatic cell nuclear transfer (SCNT) technology (created in Biorender.com, accessed on 14 July 2022).

Table 4. Bovine MSCs from different sources as nucleus donors for nuclear transfer.

Source	References
Bone marrow	[3,5]
Adipose tissue	[52,53,56,57,66,138]
Amniotic fluid	[53,57]
Amniotic membrane	[138]
Wharton's jelly	[72]

Firstly, it was demonstrated that bovine BM-MSCs had developmental totipotency after NT [3] and were better than adult fibroblasts in driving the preimplantation development of cloned embryos efficiently [5]. In another study investigating the epigenetic status of donor cells to improve SCNT [52], it was demonstrated that bovine AT-MSCs at passage 5 had the highest level of multipotency and the lowest level of chromatin compaction. Bovine AF and AT-MSCSs were then successfully used to produce embryos and calves after NT [53], and in vitro development of bovine embryos cloned using less methylated AF and AT-MSCS was improved using trichostatin A [57]. Pregnancies were also obtained after the transfer of blastocysts derived from WJ-MSCs NT [72]. A higher potential for AM and AT-MSCs than adult fibroblasts was observed in terms of blastocysts obtained after oocyte reconstruction [138]. More recently, epigenetic reprogramming events were investigated, and it was observed that the SCNT embryos derived from bovine AT-MSCs endured considerable nuclear reprogramming during early embryo development [56]. Finally, in an attempt to improve NT efficiency, the aggregation of two AT-MSC-derived embryos seemed to positively affect embryo quality, which may improve postimplantation development [66].

Another context of research into cells includes their ability to incorporate exogenous DNA for the production of transgenic animals. Bovine MSCs were transfected with pBC1-anti-CD3 vector, and while those derived from WJ were more sensitive to treatments, AT-MSCs showed a better response to transfection [58].

Bovine MSCs have also been used for in vitro embryo production. The traditional coculture system of bovine embryos with granulosa cells was less efficient than coculture with AT-MSCs [54]. In addition to increasing blastocyst rates, MSCs coculture also improved embryo quality, with an increase in total cell numbers and mRNA expression levels for POU5F1 and G6PDH [54]. It was speculated that the paracrine capacity of MSCs could be responsible for the positive effects observed [54].

Another application of MSCs is to produce germinal cells after differentiation. The in vitro production of germ cell lineages is a new intriguing strategy for obtaining gametes in order to treat infertility, disseminate the genetics of elite animals and preserve endangered species [139]. The in vitro effect of bone morphogenetic protein 4, transforming growth factor β1 and retinoic acid on the potential for germ cell differentiation of bovine foetal BM-MSCs was investigated [27]. The stimulated cells expressed pluripotent markers OCT4, NANOG and male germ cell gene DAZL, demonstrating their potential for early germ cell differentiation [27]. When coculturing bovine foetal BM and AT-MSCS with Sertoli cells, cell morphology modifications were induced, as well as variations in the expression profiles of mesenchymal, pluripotent and germ cell genes, suggesting progression of AT-MSC into early stages of germ cell differentiation and advancement of BM-MSCs into the multipotent state [34].

4. Conclusions

The development of stem cell technologies in species other than bovine can be seen as a useful background for developing and deepening similar advancements in livestock. MSC characteristics make them appealing for their potential in clinical applications, and the lack of ethical concern is the other factor that makes them ideal for laboratory studies. As for humans [140], for successful cell-based therapies, stem cells must be able to differentiate into specific targeting cells or must act via paracrine mechanisms. Their extraction and isolation must be feasible, and transplantation must be effective and safe. Furthermore, ex vivo cell expansion is required since a considerable number of cells is essential to optimise the therapeutic effects. However, the lifespan of MSCs is limited during in vitro culture, and their senescence is a limit from the viewpoint of clinical applications. On the one hand, the limited cell proliferation potency protects them from malignant transformation after transplantation; on the other, senescence can alter various cell functions essential for therapeutic efficacy, such as proliferation, differentiation and migration. Therefore, after in vitro expansion and before therapeutic use, it should be considered whether these cells still possess stemness properties.

The bovine model could be advantageous for the size and physiology when compared with traditional laboratory animals. In cattle, MSCs have been isolated from different tissues, and their pluripotency has been demonstrated, but there is still a lack of clinical applications and studies comparing MSCs from different sources to suggest which one is the best choice for cell therapy or for which specific pathology. The studies presented are promising for the possible applications of MSCs both in veterinary medicine and the livestock industry. However, more studies are required to develop bovine-specific protocols, and further investigation is needed to evaluate clinical responses after cell therapy applications. Attitudes in the livestock industry have shifted towards the preservation of the commercial viability of individual animals with high genetic value, leading, in turn, to an increase in medical expenditure to keep those animals healthy [141]. MSCs treatment has the potential to reduce animal recovery time and reduce economic loss associated with bone and joint injury, reducing the time for repair that can negatively influence milk and meat production and interfere with natural breeding [141]. Nevertheless, orthopaedic applications have not yet been applied clinically in cows. The antimicrobial activity of MSCs and their derivatives has great potential for the treatment of conditions such as mastitis. In addition to the direct impact on milk production in the dairy industry, it would provide an alternative to the use of antimicrobials, reducing the possibility of antimicrobial resistance and the presence of antibiotics in milk. MSCs treatment has the potential to decrease recovery from various diseases affecting production, thus increasing profitability.

Author Contributions: Conceptualisation, B.M.; writing—original draft preparation, B.M.; writing—review and editing, B.M., P.M.G. and E.I. All authors have read and agreed to the published version of the manuscript.

Funding: This review received no external funding.

Institutional Review Board Statement: Not applicable.

Informed Consent Statement: Not applicable.

Data Availability Statement: No new data were created or analysed in this study. Data sharing is not applicable to this article.

Conflicts of Interest: The authors declare no conflict of interest.

Abbreviations

AF	Amniotic fluid
AM	Amnion
AT	Adipose tissue
BM	Bone marrow
EN	Endometrium
EVs	Extracellular vesicles
GnRH	Gonadotropin-releasing hormone
IGF-1	Insulin-like growth factor 1
IL-6	Interleukin 6
IL-10	Interleukin 10
INFγ	Interferon γ
IV	Intravenous
LPS	Lipopolysaccharide
MHC-I	Major histocompatibility complex-I
MHC-II	Major histocompatibility complex-II
MSCs	Mesenchymal stromal cells
NT	Nuclear transfer
OPU	Ovum pick-up
PBS	Phosphate-buffered saline
PGE2	Prostaglandin E2
PL	Placenta
PMN	Polymorphonuclear
SCC	Somatic cell count
SCNT	Somatic cell nuclear transfer
UC	Umbilical cord
UCB	Umbilical cord blood
WJ	Wharton's jelly

References

1. Dominici, M.; Le Blanc, K.; Mueller, I.; Slaper-Cortenbach, I.; Marini, F.C.; Krause, D.S.; Deans, R.J.; Keating, A.; Prockop, D.J.; Horwitz, E.M. Minimal Criteria for Defining Multipotent Mesenchymal Stromal Cells. The International Society for Cellular Therapy Position Statement. *Cytotherapy* **2006**, *8*, 315–317. [CrossRef]
2. Stewart, M.C.; Stewart, A.A. Mesenchymal Stem Cells: Characteristics, Sources, and Mechanisms of Action. *Vet. Clin. Equine Pract.* **2011**, *27*, 243–261. [CrossRef] [PubMed]
3. Kato, Y.; Imabayashi, H.; Mori, T.; Tani, T.; Taniguchi, M.; Higashi, M.; Matsumoto, M.; Umezawa, A.; Tsunoda, Y. Nuclear Transfer of Adult Bone Marrow Mesenchymal Stem Cells: Developmental Totipotency of Tissue-Specific Stem Cells from an Adult Mammal1. *Biol. Reprod.* **2004**, *70*, 415–418. [CrossRef]
4. Bosnakovski, D.; Mizuno, M.; Kim, G.; Takagi, S.; Okumura, M.; Fujinaga, T. Isolation and Multilineage Differentiation of Bovine Bone Marrow Mesenchymal Stem Cells. *Cell Tissue Res.* **2005**, *319*, 243–253. [CrossRef] [PubMed]
5. Colleoni, S.; Donofrio, G.; Lagutina, I.; Duchi, R.; Galli, C.; Lazzari, G. Establishment, Differentiation, Electroporation, Viral Transduction, and Nuclear Transfer of Bovine and Porcine Mesenchymal Stem Cells. *Cloning Stem Cells* **2005**, *7*, 154–166. [CrossRef] [PubMed]

6. Donofrio, G.; Colleoni, S.; Galli, C.; Lazzari, G.; Cavirani, S.; Flammini, C.F. Susceptibility of Bovine Mesenchymal Stem Cells to Bovine Herpesvirus 4. *J. Virol. Methods* **2005**, *127*, 168–170. [CrossRef] [PubMed]
7. Bosnakovski, D.; Mizuno, M.; Kim, G.; Takagi, S.; Okumur, M.; Fujinag, T. Gene Expression Profile of Bovine Bone Marrow Mesenchymal Stem Cell during Spontaneous Chondrogenic Differentiation in Pellet Culture System. *Jpn. J. Vet. Res.* **2006**, *53*, 127–139. [PubMed]
8. Mauck, R.L.; Yuan, X.; Tuan, R.S. Chondrogenic Differentiation and Functional Maturation of Bovine Mesenchymal Stem Cells in Long-Term Agarose Culture. *Osteoarthr. Cartil.* **2006**, *14*, 179–189. [CrossRef]
9. Erickson, I.E.; van Veen, S.C.; Sengupta, S.; Kestle, S.R.; Mauck, R.L. Cartilage Matrix Formation by Bovine Mesenchymal Stem Cells in Three-Dimensional Culture Is Age-Dependent. *Clin. Orthop.* **2011**, *469*, 2744–2753. [CrossRef]
10. Das, J.; Nath, I.; Das, R.K.; Routray, P.; Behera, S.S. Autologous Stem Cell Therapy to Treat Chronic Ulcer in Heifer-A Case Study. Available online: https://scholar.google.com/citations?view_op=view_citation&hl=en&user=V8kdet8AAAAJ&citation_for_view=V8kdet8AAAAJ:UeHWp8X0CEIC (accessed on 15 July 2021).
11. Cortes, Y.; Ojeda, M.; Araya, D.; Dueñas, F.; Fernández, M.S.; Peralta, O.A. Isolation and Multilineage Differentiation of Bone Marrow Mesenchymal Stem Cells from Abattoir-Derived Bovine Fetuses. *BMC Vet. Res.* **2013**, *9*, 133. [CrossRef]
12. Randau, T.M.; Schildberg, F.A.; Alini, M.; Wimmer, M.D.; Haddouti, E.-M.; Gravius, S.; Ito, K.; Stoddart, M.J. The Effect of Dexamethasone and Triiodothyronine on Terminal Differentiation of Primary Bovine Chondrocytes and Chondrogenically Differentiated Mesenchymal Stem Cells. *PLoS ONE* **2013**, *8*, e72973. [CrossRef] [PubMed]
13. Das, J.; Behera, S.S. Successful treatment of inter digital chronic ulcerative wound by mesenchymal stem cell therapy in a cow. *Indian J. Vet. Sci. Biotechnol.* **2013**, *9*, 66–67.
14. Das, J.; Routray, N.P.; Das, R.K. Autologous Bone Marrow Derived Mesenchymal Stem Cells (BM-MSCs) Therapy for Treatment of Chronic Limb Wound in a Bull. *Indian J.* Available online: https://www.indianjournals.com/ijor.aspx?target=ijor:ipo&volume=14&issue=1&article=023 (accessed on 9 December 2021).
15. Dueñas, F.; Becerra, V.; Cortes, Y.; Vidal, S.; Sáenz, L.; Palomino, J.; De los Reyes, M.; Peralta, O.A. Hepatogenic and Neurogenic Differentiation of Bone Marrow Mesenchymal Stem Cells from Abattoir-Derived Bovine Fetuses. *BMC Vet. Res.* **2014**, *10*, 154. [CrossRef]
16. Jeong, J.Y.; Suresh, S.; Park, M.N.; Jang, M.; Park, S.; Gobianand, K.; You, S.; Yeon, S.-H.; Lee, H.-J. Effects of Capsaicin on Adipogenic Differentiation in Bovine Bone Marrow Mesenchymal Stem Cell. *Asian-Australas. J. Anim. Sci.* **2014**, *27*, 1783–1793. [CrossRef] [PubMed]
17. Lu, T.; Huang, Y.; Wang, H.; Ma, Y.; Guan, W. Multi-Lineage Potential Research of Bone Marrow-Derived Stromal Cells (BMSCs) from Cattle. *Appl. Biochem. Biotechnol.* **2014**, *172*, 21–35. [CrossRef] [PubMed]
18. Abou-Easa, K.F.; Ghazy, A.E.; Gewaily, M.S. Characterization of Behavior and Niche of Bovine Marrow Stem Cells In Vitro. *Glob. Vet.* **2015**, *14*, 922–928.
19. Díaz, P.; Cuevas, F.; Peralta, O.A. GFP Labelling and Epigenetic Enzyme Expression of Bone Marrow-Derived Mesenchymal Stem Cells from Bovine Foetuses. *Res. Vet. Sci.* **2015**, *99*, 120–128. [CrossRef] [PubMed]
20. Jeong, J.Y.; Park, M.N.; Cho, E.S.; Jang, H.-J.; Park, S.; Lee, H.-J. Epigallocatechin-3-Gallate-Induced Free-Radical Production upon Adipogenic Differentiation in Bovine Bone-Marrow Mesenchymal Stem Cells. *Cell Tissue Res.* **2015**, *362*, 87–96. [CrossRef]
21. Jeong, J.Y.; Suresh, S.; Jang, M.; Park, M.N.; Gobianand, K.; You, S.; Yeon, S.; Lee, H. Epigallocatechin-3-gallate Suppresses the Lipid Deposition through the Apoptosis during Differentiation in Bovine Bone Marrow Mesenchymal Stem Cells. *Cell Biol. Int.* **2015**, *39*, 52–64. [CrossRef]
22. Fisher, M.B.; Henning, E.A.; Söegaard, N.; Bostrom, M.; Esterhai, J.L.; Mauck, R.L. Engineering Meniscus Structure and Function via Multi-Layered Mesenchymal Stem Cell-Seeded Nanofibrous Scaffolds. *J. Biomech.* **2015**, *48*, 1412–1419. [CrossRef]
23. Goldman, S.M.; Barabino, G.A. Hydrodynamic Loading in Concomitance with Exogenous Cytokine Stimulation Modulates Differentiation of Bovine Mesenchymal Stem Cells towards Osteochondral Lineages. *BMC Biotechnol.* **2016**, *16*, 10. [CrossRef]
24. Ramírez-Espinosa, J.J.; González-Dávalos, L.; Shimada, A.; Piña, E.; Varela-Echavarria, A.; Mora, O. Bovine (Bos Taurus) Bone Marrow Mesenchymal Cell Differentiation to Adipogenic and Myogenic Lineages. *Cells Tissues Organs* **2016**, *201*, 51–64. [CrossRef]
25. McCorry, M.C.; Bonassar, L.J. Fiber Development and Matrix Production in Tissue-Engineered Menisci Using Bovine Mesenchymal Stem Cells and Fibrochondrocytes. *Connect. Tissue Res.* **2017**, *58*, 329–341. [CrossRef]
26. Okamura, L.H.; Cordero, P.; Palomino, J.; Parraguez, V.H.; Torres, C.G.; Peralta, O.A. Myogenic Differentiation Potential of Mesenchymal Stem Cells Derived from Fetal Bovine Bone Marrow. *Anim. Biotechnol.* **2018**, *29*, 1–11. [CrossRef] [PubMed]
27. Cortez, J.; Bahamonde, J.; los Reyes, M.D.; Palomino, J.; Torres, C.G.; Peralta, O.A. In Vitro Differentiation of Bovine Bone Marrow-Derived Mesenchymal Stem Cells into Male Germ Cells by Exposure to Exogenous Bioactive Factors. *Reprod. Domest. Anim.* **2018**, *53*, 700–709. [CrossRef] [PubMed]
28. Zhao, X.-X.; An, X.-L.; Zhu, X.-C.; Jiang, Y.; Zhai, Y.-H.; Zhang, S.; Cai, N.-N.; Tang, B.; Li, Z.-Y.; Zhang, X.-M. Inhibiting Transforming Growth Factor-β Signaling Regulates in Vitro Maintenance and Differentiation of Bovine Bone Marrow Mesenchymal Stem Cells. *J. Exp. Zoolog. B Mol. Dev. Evol.* **2018**, *330*, 406–416. [CrossRef] [PubMed]
29. Cahuascanco, B.; Bahamonde, J.; Huaman, O.; Jervis, M.; Cortez, J.; Palomino, J.; Escobar, A.; Retamal, P.; Torres, C.G.; Peralta, O.A. Bovine Fetal Mesenchymal Stem Cells Exert Antiproliferative Effect against Mastitis Causing Pathogen Staphylococcus Aureus. *Vet. Res.* **2019**, *50*, 25. [CrossRef]

30. Huaman, O.; Bahamonde, J.; Cahuascanco, B.; Jervis, M.; Palomino, J.; Torres, C.G.; Peralta, O.A. Immunomodulatory and Immunogenic Properties of Mesenchymal Stem Cells Derived from Bovine Fetal Bone Marrow and Adipose Tissue. *Res. Vet. Sci.* **2019**, *124*, 212–222. [CrossRef]
31. Jervis, M.; Huaman, O.; Cahuascanco, B.; Bahamonde, J.; Cortez, J.; Arias, J.I.; Torres, C.G.; Peralta, O.A. Comparative Analysis of In Vitro Proliferative, Migratory and pro-Angiogenic Potentials of Bovine Fetal Mesenchymal Stem Cells Derived from Bone Marrow and Adipose Tissue. *Vet. Res. Commun.* **2019**, *43*, 165–178. [CrossRef]
32. Korovina, D.G. The Use of Bovine Multipotent Mesenchymal Stem Cells Isolated from Bone Marrow and Adipose Tissue as Sources to Obtain Muscle Cells in Vitro. *IOP Conf. Ser. Earth Environ. Sci.* **2019**, *315*, 042040. [CrossRef]
33. Pan, S.; Chen, Y.-C.; Zhao, N.; Feng, X.; Yang, D.-D.; Wang, Y.; Jin, Z.-B. A New Subset of Small Stem Cells in Bovine Bone Marrow Stromal Cell Populations. *J. Cell. Biochem.* **2019**, *120*, 13881–13892. [CrossRef] [PubMed]
34. Segunda, M.N.; Bahamonde, J.; Muñoz, I.; Sepulveda, S.; Cortez, J.; De los Reyes, M.; Palomino, J.; Torres, C.G.; Peralta, O.A. Sertoli Cell-Mediated Differentiation of Bovine Fetal Mesenchymal Stem Cells into Germ Cell Lineage Using an in Vitro Co-Culture System. *Theriogenology* **2019**, *130*, 8–18. [CrossRef] [PubMed]
35. Shi, M.; Li, Z.; Miao, Z.; Guo, Y.; Yi, L. Interleukin-15 Inhibits Adipogenic Differentiation of Cattle Bone Marrow-Derived Mesenchymal Stem Cells via Regulating the Crosstalk between Signal Transducer and Activator of Transcription 5A and Akt Signalling. *Biochem. Biophys. Res. Commun.* **2019**, *517*, 346–352. [CrossRef] [PubMed]
36. Chen, J.; Chen, L.; Hua, J.; Song, W. Long-Term Dynamic Compression Enhancement TGF-B3-Induced Chondrogenesis in Bovine Stem Cells: A Gene Expression Analysis. *BMC Genom. Data* **2021**, *22*, 13. [CrossRef] [PubMed]
37. Oyarzo, R.; Valderrama, X.; Valenzuela, F.; Bahamonde, J. Bovine Fetal Mesenchymal Stem Cells Obtained From Omental Adipose Tissue and Placenta Are More Resistant to Cryoprotectant Exposure Than Those From Bone Marrow. *Front. Vet. Sci.* **2021**, *8*, 708972. [CrossRef]
38. Abo-Aziza, F.A.M.; Zaki, A.K.A.; Alajaji, A.I.; Al barrak, S.M. Bone Marrow Mesenchymal Stem Cell Co-Adjuvant Therapy with Albendazole for Managing Toxocara Vitulorum-Rat Model. *Vet. World* **2021**, *14*, 347–363. [CrossRef] [PubMed]
39. Donofrio, G.; Franceschi, V.; Capocefalo, A.; Cavirani, S.; Sheldon, I.M. Bovine Endometrial Stromal Cells Display Osteogenic Properties. *Reprod. Biol. Endocrinol.* **2008**, *6*, 65. [CrossRef]
40. Cabezas, J.; Lara, E.; Pacha, P.; Rojas, D.; Veraguas, D.; Saravia, F.; Rodríguez-Alvarez, L.; Castro, F.O. The Endometrium of Cycling Cows Contains Populations of Putative Mesenchymal Progenitor Cells. *Reprod. Domest. Anim.* **2014**, *49*, 550–559. [CrossRef] [PubMed]
41. Mehrabani, D.; Rahmanifar, F.; Mellinejad, M.; Tamadon, A.; Dianatpour, M.; Zare, S.; Jahromi, I.R.; Ghobadi, F. Isolation, Culture, Characterization, and Adipogenic Differentiation of Heifer Endometrial Mesenchymal Stem Cells. *Comp. Clin. Pathol.* **2015**, *24*, 1159–1164. [CrossRef]
42. de Moraes, C.N.; Maia, L.; Dias, M.C.; Dell'Aqua, C.P.F.; da Mota, L.S.L.S.; Chapwanya, A.; da C. Landim-Alvarenga, F.; Oba, E. Bovine Endometrial Cells: A Source of Mesenchymal Stem/Progenitor Cells. *Cell Biol. Int.* **2016**, *40*, 1332–1339. [CrossRef]
43. de Moraes, C.N.; Maia, L.; de Oliveira, E.; de Paula Freitas Dell'Aqua, C.; Chapwanya, A.; da Cruz Landim-Alvarenga, F.; Oba, E. Shotgun Proteomic Analysis of the Secretome of Bovine Endometrial Mesenchymal Progenitor/Stem Cells Challenged or Not with Bacterial Lipopolysaccharide. *Vet. Immunol. Immunopathol.* **2017**, *187*, 42–47. [CrossRef] [PubMed]
44. Lara, E.; Rivera, N.; Rojas, D.; Rodríguez-Alvarez, L.L.; Castro, F.O. Characterization of Mesenchymal Stem Cells in Bovine Endometrium during Follicular Phase of Oestrous Cycle. *Reprod. Domest. Anim.* **2017**, *52*, 707–714. [CrossRef] [PubMed]
45. Lara, E.; Velásquez, A.; Cabezas, J.; Rivera, N.; Pacha, P.; Rodríguez-Alvarez, L.; Saravia, F.; Castro, F.O. Endometritis and In Vitro PGE_2 Challenge Modify Properties of Cattle Endometrial Mesenchymal Stem Cells and Their Transcriptomic Profile. *Stem Cells Int.* **2017**, *2017*, 4297639. [CrossRef]
46. Lara, E.; Rivera, N.; Cabezas, J.; Navarrete, F.; Saravia, F.; Rodríguez-Alvarez, L.; Castro, F. Endometrial Stem Cells in Farm Animals: Potential Role in Uterine Physiology and Pathology. *Bioengineering* **2018**, *5*, 75. [CrossRef]
47. Calle, A.; López-Martín, S.; Monguió-Tortajada, M.; Borràs, F.E.; Yáñez-Mó, M.; Ramírez, M.Á. Bovine Endometrial MSC: Mesenchymal to Epithelial Transition during Luteolysis and Tropism to Implantation Niche for Immunomodulation. *Stem Cell Res. Ther.* **2019**, *10*, 23. [CrossRef]
48. Calle, A.; Toribio, V.; Yáñez-Mó, M.; Ramírez, M.Á. Embryonic Trophectoderm Secretomics Reveals Chemotactic Migration and Intercellular Communication of Endometrial and Circulating MSCs in Embryonic Implantation. *Int. J. Mol. Sci.* **2021**, *22*, 5638. [CrossRef]
49. Ren, Y.; Wu, H.; Ma, Y.; Cang, M.; Wang, R.; Liu, D. Isolation, cultivation and identification of adipose-derived stem cell in bovines. *Sheng Wu Gong Cheng Xue Bao Chin. J. Biotechnol.* **2010**, *26*, 1645–1651.
50. Lu, T.; Xiong, H.; Wang, K.; Wang, S.; Ma, Y.; Guan, W. Isolation and Characterization of Adipose-Derived Mesenchymal Stem Cells (ADSCs) from Cattle. *Appl. Biochem. Biotechnol.* **2014**, *174*, 719–728. [CrossRef] [PubMed]
51. Sampaio, R.V.; Chiaratti, M.R.; Santos, D.C.N.; Bressan, F.F.; Sangalli, J.R.; Sá, A.L.A.; Silva, T.V.G.; Costa, N.N.; Cordeiro, M.S.; Santos, S.S.D.; et al. Generation of Bovine (*Bos indicus*) and Buffalo (*Bubalus bubalis*) Adipose Tissue Derived Stem Cells: Isolation, Characterization, and Multipotentiality. Available online: https://www.geneticsmr.com/articles/3881 (accessed on 15 July 2021).
52. Abouhamzeh, B.; Salehi, M. DNA Methylation and Histone Acetylation Patterns in Cultured Bovine Adipose Tissue-Derived Stem Cells (BADSCs). *CELL J.* **2015**, *16*, 10.

53. da Silva, C.G.; Martins, C.F.; Cardoso, T.C.; da Cunha, E.R.; Bessler, H.C.; Martins, G.H.L.; Pivato, I.; Báo, S.N. Production of Bovine Embryos and Calves Cloned by Nuclear Transfer Using Mesenchymal Stem Cells from Amniotic Fluid and Adipose Tissue. *Cell Reprogram.* **2016**, *18*, 127–136. [CrossRef]
54. Miranda, M.S.; Nascimento, H.S.; Costa, M.P.R.; Costa, N.N.; Brito, K.N.L.; Lopes, C.T.A.; Santos, S.S.D.; Cordeiro, M.S.; Ohashi, O.M. Increasing of Blastocyst Rate and Gene Expression in Co-Culture of Bovine Embryos with Adult Adipose Tissue-Derived Mesenchymal Stem Cells. *J. Assist. Reprod. Genet.* **2016**, *33*, 1395–1403. [CrossRef] [PubMed]
55. Yue, Y.; Zhang, L.; Zhang, X.; Li, X.; Yu, H. De Novo Lipogenesis and Desaturation of Fatty Acids during Adipogenesis in Bovine Adipose-Derived Mesenchymal Stem Cells. *Vitro Cell. Dev. Biol. Anim.* **2018**, *54*, 23–31. [CrossRef]
56. Salehi, M.; Abouhamzeh, B.; Hosseini, A.; Zare, Z.; Bakhtari, A. Comparison of Epigenetic Modifier Genes in Bovine Adipose Tissue-Derived Stem Cell Based Embryos, as Donors, with In Vitro and Parthenogenesis Embryos. *Cell J.* **2019**, *22*, 149–157. [CrossRef] [PubMed]
57. da Silva, C.G.; Martins, C.F.; Bessler, H.C.; da Fonseca Neto, Á.M.; Cardoso, T.C.; Franco, M.M.; dos S. Mendonça, A.; de O. Leme, L.; Borges, J.R.J.; Malaquias, J.V.; et al. Use of Trichostatin A Alters the Expression of *HDAC3* and *KAT2* and Improves in Vitro Development of Bovine Embryos Cloned Using Less Methylated Mesenchymal Stem Cells. *Reprod. Domest. Anim.* **2019**, *54*, 289–299. [CrossRef] [PubMed]
58. Duarte, F.B.; Brigido, M.D.M.; de O. Melo, E.; Báo, S.N.; Martins, C.F. Strategies for Transfection of Bovine Mesenchymal Stem Cells with PBC1-Anti-CD3 Vector. *Anim. Biotechnol.* **2020**, *30*, 1–11. [CrossRef]
59. HuiNa, L.; QuanBao, F.; WenKang, J.; MingMing, Z.; JingJing, W.; DongZhang, L.; BingYun, W. Isolation, culture and biological characteristics of dairy cow adipose-derived mesenchymal stem cells. *J. South. Agric.* **2020**, *51*, 1462–1469.
60. Jurek, S.; Sandhu, M.A.; Trappe, S.; Bermúdez-Peña, M.C.; Kolisek, M.; Sponder, G.; Aschenbach, J.R. Optimizing Adipogenic Transdifferentiation of Bovine Mesenchymal Stem Cells: A Prominent Role of Ascorbic Acid in *FABP4* Induction. *Adipocyte* **2020**, *9*, 35–50. [CrossRef]
61. Malard, P.F.; Peixer, M.A.S.; Grazia, J.G.; Brunel, H.D.S.S.; Feres, L.F.; Villarroel, C.L.; Siqueira, L.G.B.; Dode, M.A.N.; Pogue, R.; Viana, J.H.M.; et al. Intraovarian Injection of Mesenchymal Stem Cells Improves Oocyte Yield and in Vitro Embryo Production in a Bovine Model of Fertility Loss. *Sci. Rep.* **2020**, *10*, 8018. [CrossRef]
62. Peralta, O.A.; Carrasco, C.; Vieytes, C.; Tamayo, M.J.; Muñoz, I.; Sepulveda, S.; Tadich, T.; Duchens, M.; Melendez, P.; Mella, A.; et al. Safety and Efficacy of a Mesenchymal Stem Cell Intramammary Therapy in Dairy Cows with Experimentally Induced Staphylococcus Aureus Clinical Mastitis. *Sci. Rep.* **2020**, *10*, 2843. [CrossRef] [PubMed]
63. Lu, W.; Xu, Z.-M.; Liu, Q.; Yu, N.-N.; Yu, J.-B.; Li, W.-L.; Mao, Y.-Y.; Du, Z.; Si, L.; Yuan, S.; et al. Inhibitory Effect of Bovine Adipose-Derived Mesenchymal Stem Cells on Lipopolysaccharide Induced Inflammation of Endometrial Epithelial Cells in Dairy Cows. *Front. Vet. Sci.* **2021**, *8*, 726328. [CrossRef]
64. Singh, R.; Bhaskar, V.; Saini, S.; Kumar, A.; Thakur, A.; Kumar, S.; Malakar, D.; Singh, R.; Bhaskar, V.; Saini, S.; et al. 143 Therapeutic Efficacy and Safety of Adipose Tissue-Derived Mesenchymal Stem Cells in Treating Mastitis and Metritis in Dairy Cattle. *Reprod. Fertil. Dev.* **2021**, *33*, 179. [CrossRef]
65. Naraoka, Y.; Mabuchi, Y.; Yoneyama, Y.; Suto, E.G.; Hisamatsu, D.; Ikeda, M.; Ito, R.; Nakamura, T.; Takebe, T.; Akazawa, C. Isolation and Characterization of Tissue Resident CD29-Positive Progenitor Cells in Livestock to Generate a Three-Dimensional Meat Bud. *Cells* **2021**, *10*, 2499. [CrossRef]
66. Savy, V.; Alberio, V.; Vans Landschoot, G.; Moro, L.N.; Olea, F.D.; Rodríguez-Álvarez, L.; Salamone, D.F. Effect of Embryo Aggregation on In Vitro Development of Adipose-Derived Mesenchymal Stem Cell-Derived Bovine Clones. *Cell Reprogram.* **2021**, *23*, 277–289. [CrossRef] [PubMed]
67. Raoufi, M.F.; Tajik, P.; Dehghan, M.M.; Eini, F.; Barin, A. Isolation and Differentiation of Mesenchymal Stem Cells From Bovine Umbilical Cord Blood. *Reprod. Domest. Anim.* **2011**, *46*, 95–99. [CrossRef] [PubMed]
68. Ghai, S.; Verma, V.N.; Ansari, S.; Saini, S.; Thakur, A.; Kumar, A.; Kumar, S.; Malakar, D.; Ghai, S.; Verma, V.N.; et al. 157 Mesenchymal Stem Cells as a Regenerative Therapy for the Prevention of Subclinical Mastitis in Cattle. *Reprod. Fertil. Dev.* **2021**, *34*, 316–317. [CrossRef] [PubMed]
69. Ghai, S.; Saini, S.; Ansari, S.; Verma, V.; Chopra, S.; Sharma, V.; Devi, P.; Malakar, D. Allogenic Umbilical Cord Blood-Mesenchymal Stem Cells Are More Effective than Antibiotics in Alleviating Subclinical Mastitis in Dairy Cows. *Theriogenology* **2022**, *187*, 141–151. [CrossRef]
70. Verma, V.N.; Ghai, S.; Ansari, S.; Saini, S.; Thakur, A.; Kumar, A.; Kumar, S.; Malakar, D. 158 Umbilical Cord Blood-Derived Mesenchymal Stem Cells (UCB-MSC) Used for the Prevention of Metritis in Cattle. *Reprod. Fertil. Dev.* **2021**, *34*, 317. [CrossRef] [PubMed]
71. Cardoso, T.C.; Ferrari, H.F.; Garcia, A.F.; Novais, J.B.; Silva-Frade, C.; Ferrarezi, M.C.; Andrade, A.L.; Gameiro, R. Isolation and Characterization of Wharton's Jelly-Derived Multipotent Mesenchymal Stromal Cells Obtained from Bovine Umbilical Cord and Maintained in a Defined Serum-Free Three-Dimensional System. *BMC Biotechnol.* **2012**, *12*, 18. [CrossRef]
72. da Silva, C.G.; Martins, C.F.; Cardoso, T.C.; da Cunha, E.R.; Bessler, H.C.; McManus, C.M.; Pivato, I.; Báo, S.N. Isolation and Characterization of Mesenchymal Stem Cells Derived from Bovine Wharton's Jelly and Their Potential for Use in Cloning by Nuclear Transfer. *Ciênc. Rural* **2016**, *46*, 1830–1837. [CrossRef]

73. Cardoso, T.C.; Okamura, L.H.; Baptistella, J.C.; Gameiro, R.; Ferreira, H.L.; Marinho, M.; Flores, E.F. Isolation, Characterization and Immunomodulatory-Associated Gene Transcription of Wharton's Jelly-Derived Multipotent Mesenchymal Stromal Cells at Different Trimesters of Cow Pregnancy. *Cell Tissue Res.* **2017**, *367*, 243–256. [CrossRef]
74. Xiong, H.; Bai, C.; Wu, S.; Gao, Y.; Lu, T.; Hu, Q.; Guan, W.; Ma, Y. Biological Characterization of Mesenchymal Stem Cells from Bovine Umbilical Cord. *Anim. Cells Syst.* **2014**, *18*, 59–67. [CrossRef]
75. YanKun, Z.; Wei, S.; LiWen, W.; Xiong, Y. Effect of co-culturing with umbilical cord mesenchymal stem cells on milk fat synthesis and expression of key genes in bovine mammary gland epithelial cells. *Acta Vet. Zootech. Sin.* **2017**, *48*, 846–853.
76. YanKun, Z.; Wei, S.; Xiong, Y. Regulation of umbilical cord mesenchymal stem cells (UC-MSCs) on milk protein synthesis in mammary epithelial cells (BMECs). *Chin. J. Anim. Sci.* **2017**, *53*, 55–59.
77. YanKun, Z.; Wei, S.; ChengLong, L.; KaiLe, W.; Xiong, Y. An in vitro study of umbilical cord mesenchymal stem cells inhibit bovine mammary gland epithelial cells apoptosis by insulin like growth factor-I mediated Janus kinase/signal transducer and activator of transcription signaling pathway. *Chin. J. Anim. Nutr.* **2017**, *29*, 1335–1342.
78. Zhao, Y.; Shao, W.; Luo, C.; Wu, K.; Yu, X. Co-culture with umbilical cord mesenchymal stem cells promotes the synthesis and mechnism of milk protein in bovine mammary epithelial cells. *Xi Bao Yu Fen Zi Mian Yi Xue Za Zhi Chin. J. Cell. Mol. Immunol.* **2017**, *33*, 185–189.
79. LiWen, W.; Xiong, Y.; YanKun, Z.; Yang, L.; Wei, S. Study on the regulation of PI3K/Akt/mTOR signaling pathway mediated by umbilical cord mesenchymal stem cells of cow on the apoptosis of mammary epithelial cell. *Chin. J. Anim. Sci.* **2018**, *54*, 50–53.
80. Debbarma, P.; Mondal, T.; Manna, C.; Kumar, K.; Mukherjee, J.; Das, B.C.; Bag, S.; Das, K. Post-Calving Umbilical Cord Tissue Offcut: A Potential Source for the Isolation of Bovine Mesenchymal Stem Cells. *Vet. World* **2020**, *13*, 2772–2779. [CrossRef] [PubMed]
81. Shimoni, C.; Goldstein, M.; Ribarski-Chorev, I.; Schauten, I.; Nir, D.; Strauss, C.; Schlesinger, S. Heat Shock Alters Mesenchymal Stem Cell Identity and Induces Premature Senescence. *Front. Cell Dev. Biol.* **2020**, *8*, 565970. [CrossRef]
82. Nir, D.; Ribarski-Chorev, I.; Shimoni, C.; Strauss, C.; Frank, J.; Schlesinger, S. Antioxidants Attenuate Heat Shock Induced Premature Senescence of Bovine Mesenchymal Stem Cells. *Int. J. Mol. Sci.* **2022**, *23*, 5750. [CrossRef]
83. Corradetti, B.; Meucci, A.; Bizzaro, D.; Cremonesi, F.; Consiglio, A.L. Mesenchymal Stem Cells from Amnion and Amniotic Fluid in the Bovine. *Reproduction* **2013**, *145*, 391–400. [CrossRef]
84. Ting, W.-J.; Shaw, S.W.; Hii, L.-Y.; Lin, T.-Y.; Chang, S.-C.; Liu, K.-Y.; Shen, P.-C.; Chen, T.-J.; Peng, S.-Y. Therapeutic Effects of Conditioned—DPBS from Amniotic Stem Cells on Lactating Cow Mastitis. *Taiwan. J. Obstet. Gynecol.* **2020**, *59*, 520–526. [CrossRef]
85. Rossi, B.; Merlo, B.; Colleoni, S.; Iacono, E.; Tazzari, P.L.; Ricci, F.; Lazzari, G.; Galli, C. Isolation and in Vitro Characterization of Bovine Amniotic Fluid Derived Stem Cells at Different Trimesters of Pregnancy. *Stem Cell Rev. Rep.* **2014**, *10*, 712–724. [CrossRef] [PubMed]
86. Gao, Y.; Zhu, Z.; Zhao, Y.; Hua, J.; Ma, Y.; Guan, W. Multilineage Potential Research of Bovine Amniotic Fluid Mesenchymal Stem Cells. *Int. J. Mol. Sci.* **2014**, *15*, 3698–3710. [CrossRef] [PubMed]
87. Chang, L.-B.; Peng, S.-Y.; Chou, C.-J.; Chen, Y.-J.; Shiu, J.-S.; Tu, P.-A.; Gao, S.-X.; Chen, Y.-C.; Lin, T.-K.; Wu, S.-C. Therapeutic Potential of Amniotic Fluid Stem Cells to Treat Bilateral Ovarian Dystrophy in Dairy Cows in a Subtropical Region. *Reprod. Domest. Anim.* **2018**, *53*, 433–441. [CrossRef] [PubMed]
88. Nawaz, S.; Özden Akkaya, Ö.; Dikmen, T.; Altunbaş, K.; Yağci, A.; Kibria, A.S.M.G.; Erdoğan, M.; Çelik, H.A. Molecular Characterization of Bovine Amniotic Fluid Derived Stem Cells with an Underlying Focus on Their Comparative Neuronal Potential at Different Passages. *Ann. Anat. Anat. Anz.* **2020**, *228*, 151452. [CrossRef]
89. Peng, S.-Y.; Chou, C.; Kuo, Y.-H.; Shen, P.-C.; Shaw, S.W.S. Potential Differentiation of Islet-like Cells from Pregnant Cow-Derived Placental Stem Cells. *Taiwan. J. Obstet. Gynecol.* **2017**, *56*, 306–311. [CrossRef]
90. Peng, S.-Y.; Wu, T.-H.; Lin, T.-Y.; Hii, L.-Y.; Chan, K.-S.; Fu, T.-Y.; Chang, S.-C.; Shen, P.-C.; Liu, K.-Y.; Shaw, S.W. Application of Cattle Placental Stem Cells for Treating Ovarian Follicular Cyst. *World J. Stem Cells* **2020**, *12*, 1366–1376. [CrossRef]
91. Lu, T.; Hu, P.; Su, X.; Li, C.; Ma, Y.; Guan, W. Isolation and Characterization of Mesenchymal Stem Cells Derived from Fetal Bovine Liver. *Cell Tissue Bank.* **2014**, *15*, 439–450. [CrossRef] [PubMed]
92. Sun, T.; Yu, C.; Gao, Y.; Zhao, C.; Hua, J.; Cai, L.; Guan, W.; Ma, Y. Establishment and Biological Characterization of a Dermal Mesenchymal Stem Cells Line from Bovine. *Biosci. Rep.* **2014**, *34*, e00101. [CrossRef]
93. Hu, P.; Pu, Y.; Li, X.; Zhu, Z.; Zhao, Y.; Guan, W.; Ma, Y. Isolation, In Vitro Culture and Identification of a New Type of Mesenchymal Stem Cell Derived from Fetal Bovine Lung Tissues. *Mol. Med. Rep.* **2015**, *12*, 3331–3338. [CrossRef]
94. Mançanares, C.A.F.; Oliveira, V.C.; Oliveira, L.J.; Carvalho, A.F.; Sampaio, R.V.; Mançanares, A.C.F.; Souza, A.F.; Perecin, F.; Meirelles, F.V.; Miglino, M.A.; et al. Isolation and Characterization of Mesenchymal Stem Cells from the Yolk Sacs of Bovine Embryos. *Theriogenology* **2015**, *84*, 887–898. [CrossRef]
95. Ongaro, A.; Pellati, A.; Setti, S.; Masieri, F.F.; Aquila, G.; Fini, M.; Caruso, A.; De Mattei, M. Electromagnetic Fields Counteract IL-1β Activity during Chondrogenesis of Bovine Mesenchymal Stem Cells. *J. Tissue Eng. Regen. Med.* **2015**, *9*, E229–E238. [CrossRef] [PubMed]
96. Pipino, C.; Mandatori, D.; Buccella, F.; Lanuti, P.; Preziuso, A.; Castellani, F.; Grotta, L.; di Tomo, P.; Marchetti, S.; di Pietro, N.; et al. Identification and Characterization of a Stem Cell-Like Population in Bovine Milk: A Potential New Source for Regenerative Medicine in Veterinary. *Stem Cells Dev.* **2018**, *27*, 1587–1597. [CrossRef] [PubMed]

97. Ferroni, L.; Gardin, C.; Bellin, G.; Vindigni, V.; Mortellaro, C.; Zavan, B. Bovine Pericardium Membrane as New Tool for Mesenchymal Stem Cells Commitment. *J. Tissue Eng. Regen. Med.* **2019**, *13*, 1805–1814. [CrossRef]
98. Gao, F.; Wu, Y.; Wen, H.; Zhu, W.; Ren, H.; Guan, W.; Tian, X. Multilineage Potential Research on Pancreatic Mesenchymal Stem Cells of Bovine. *Tissue Cell* **2019**, *56*, 60–70. [CrossRef]
99. Lee, J.; Byeon, J.S.; Gu, N.-Y.; Lee, S.; Lee, S.-A.; Jeong, D.-U.; Ouh, I.-O.; Cho, I.-S.; Song, J.-Y.; Lee, Y.-H.; et al. Bovine Tongue Epithelium-Derived Cells: A New Source of Bovine Mesenchymal Stem Cells. *Biosci. Rep.* **2020**, *40*, BSR20181829. [CrossRef]
100. Zhang, R.; Deng, Y.; Lv, Q.; Xing, Q.; Pan, Y.; Liang, J.; Jiang, M.; Wei, Y.; Shi, D.; Xie, B.; et al. SQLE Promotes Differentiation and Apoptosis of Bovine Skeletal Muscle-Derived Mesenchymal Stem Cells. *Cell Reprogram.* **2020**, *22*, 22–29. [CrossRef]
101. Calle, A.; Gutiérrez-Reinoso, M.Á.; Re, M.; Blanco, J.; De la Fuente, J.; Monguió-Tortajada, M.; Borràs, F.E.; Yáñez-Mó, M.; Ramírez, M.Á. Bovine Peripheral Blood MSCs Chemotax towards Inflammation and Embryo Implantation Stimuli. *J. Cell. Physiol.* **2021**, *236*, 1054–1067. [CrossRef]
102. Zhou, Y.; Yamamoto, Y.; Xiao, Z.; Ochiya, T. The Immunomodulatory Functions of Mesenchymal Stromal/Stem Cells Mediated via Paracrine Activity. *J. Clin. Med.* **2019**, *8*, 1025. [CrossRef]
103. Maacha, S.; Sidahmed, H.; Jacob, S.; Gentilcore, G.; Calzone, R.; Grivel, J.-C.; Cugno, C. Paracrine Mechanisms of Mesenchymal Stromal Cells in Angiogenesis. *Stem Cells Int.* **2020**, *2020*, 4356359. [CrossRef]
104. Alvites, R.; Branquinho, M.; Sousa, A.C.; Lopes, B.; Sousa, P.; Maurício, A.C. Mesenchymal Stem/Stromal Cells and Their Paracrine Activity—Immunomodulation Mechanisms and How to Influence the Therapeutic Potential. *Pharmaceutics* **2022**, *14*, 381. [CrossRef] [PubMed]
105. Barrachina, L.; Remacha, A.R.; Romero, A.; Vázquez, F.J.; Albareda, J.; Prades, M.; Gosálvez, J.; Roy, R.; Zaragoza, P.; Martín-Burriel, I.; et al. Priming Equine Bone Marrow-Derived Mesenchymal Stem Cells with Proinflammatory Cytokines: Implications in Immunomodulation–Immunogenicity Balance, Cell Viability, and Differentiation Potential. *Stem Cells Dev.* **2017**, *26*, 15–24. [CrossRef] [PubMed]
106. Zhang, J.; Huang, X.; Wang, H.; Liu, X.; Zhang, T.; Wang, Y.; Hu, D. The Challenges and Promises of Allogeneic Mesenchymal Stem Cells for Use as a Cell-Based Therapy. *Stem Cell Res. Ther.* **2015**, *6*, 234. [CrossRef]
107. Consentius, C.; Reinke, P.; Volk, H.-D. Immunogenicity of Allogeneic Mesenchymal Stromal Cells: What Has Been Seen In Vitro and In Vivo? Available online: https://www.futuremedicine.com/doi/epub/10.2217/rme.15.14 (accessed on 19 July 2022).
108. Colleoni, S.; Bottani, E.; Tessaro, I.; Mari, G.; Merlo, B.; Romagnoli, N.; Spadari, A.; Galli, C.; Lazzari, G. Isolation, Growth and Differentiation of Equine Mesenchymal Stem Cells: Effect of Donor, Source, Amount of Tissue and Supplementation with Basic Fibroblast Growth Factor. *Vet. Res. Commun.* **2009**, *33*, 811. [CrossRef]
109. Papait, A.; Vertua, E.; Magatti, M.; Ceccariglia, S.; De Munari, S.; Silini, A.R.; Sheleg, M.; Ofir, R.; Parolini, O. Mesenchymal Stromal Cells from Fetal and Maternal Placenta Possess Key Similarities and Differences: Potential Implications for Their Applications in Regenerative Medicine. *Cells* **2020**, *9*, 127. [CrossRef] [PubMed]
110. Yu, Y.; Valderrama, A.V.; Han, Z.; Uzan, G.; Naserian, S.; Oberlin, E. Human Fetal Liver MSCs Are More Effective than Adult Bone Marrow MSCs for Their Immunosuppressive, Immunomodulatory, and Foxp3+ T Reg Induction Capacity. *Stem Cell Res. Ther.* **2021**, *12*, 138. [CrossRef] [PubMed]
111. Yong, K.W.; Wan Safwani, W.K.Z.; Xu, F.; Wan Abas, W.A.B.; Choi, J.R.; Pingguan-Murphy, B. Cryopreservation of Human Mesenchymal Stem Cells for Clinical Applications: Current Methods and Challenges. *Biopreserv. Biobank.* **2015**, *13*, 231–239. [CrossRef]
112. Du, M.; Huang, Y.; Das, A.K.; Yang, Q.; Duarte, M.S.; Dodson, M.V.; Zhu, M.-J. MEAT SCIENCE AND MUSCLE BIOLOGY SYMPOSIUM: Manipulating Mesenchymal Progenitor Cell Differentiation to Optimize Performance and Carcass Value of Beef Cattle1,2. *J. Anim. Sci.* **2013**, *91*, 1419–1427. [CrossRef]
113. Ribitsch, I.; Baptista, P.M.; Lange-Consiglio, A.; Melotti, L.; Patruno, M.; Jenner, F.; Schnabl-Feichter, E.; Dutton, L.C.; Connolly, D.J.; van Steenbeek, F.G.; et al. Large Animal Models in Regenerative Medicine and Tissue Engineering: To Do or Not to Do. *Front. Bioeng. Biotechnol.* **2020**, *8*, 972. [CrossRef] [PubMed]
114. Strauss, S. Biotech Breeding Goes Bovine. *Nat. Biotechnol.* **2010**, *28*, 540–543. [CrossRef] [PubMed]
115. Binyameen, M.; Saleem, M.; Riaz, A. Recent Trends in Bovine Reproductive Biotechnologies. *CAB Rev. Perspect. Agric. Vet. Sci. Nutr. Nat. Resour.* **2019**, *14*, 1–10. [CrossRef]
116. OECD-FAO Agricultural Outlook 2021–2030. FAO | Food and Agriculture Organization of the United Nations. Available online: https://www.fao.org/publications/oecd-fao-agricultural-outlook/2021-2030/en/ (accessed on 13 July 2022).
117. Voga, M.; Adamic, N.; Vengust, M.; Majdic, G. Stem Cells in Veterinary Medicine—Current State and Treatment Options. *Front. Vet. Sci.* **2020**, *7*, 278. [CrossRef] [PubMed]
118. Gugjoo, M.B.; Amarpal; Fazili, M.R.; Shah, R.A.; Sharma, G.T. Mesenchymal Stem Cell: Basic Research and Potential Applications in Cattle and Buffalo. *J. Cell. Physiol.* **2019**, *234*, 8618–8635. [CrossRef]
119. Devireddy, L.R.; Boxer, L.; Myers, M.J.; Skasko, M.; Screven, R. Questions and Challenges in the Development of Mesenchymal Stromal/Stem Cell-Based Therapies in Veterinary Medicine. *Tissue Eng. Part B Rev.* **2017**, *23*, 462–470. [CrossRef]
120. Bagge, J.; MacLeod, J.N.; Berg, L.C. Cellular Proliferation of Equine Bone Marrow- and Adipose Tissue-Derived Mesenchymal Stem Cells Decline With Increasing Donor Age. *Front. Vet. Sci.* **2020**, *7*, 933. [CrossRef] [PubMed]
121. Lopes, B.; Sousa, P.; Alvites, R.; Branquinho, M.; Sousa, A.; Mendonça, C.; Atayde, L.M.; Maurício, A.C. The Application of Mesenchymal Stem Cells on Wound Repair and Regeneration. *Appl. Sci.* **2021**, *11*, 3000. [CrossRef]

122. Iacono, E.; Merlo, B.; Pirrone, A.; Antonelli, C.; Brunori, L.; Romagnoli, N.; Castagnetti, C. Effects of Mesenchymal Stem Cells Isolated from Amniotic Fluid and Platelet-Rich Plasma Gel on Severe Decubitus Ulcers in a Septic Neonatal Foal. *Res. Vet. Sci.* **2012**, *93*, 1439–1440. [CrossRef]
123. Volk, S.W.; Theoret, C. Translating Stem Cell Therapies: The Role of Companion Animals in Regenerative Medicine. *Wound Repair Regen.* **2013**, *21*, 382–394. [CrossRef]
124. Lanci, A.; Merlo, B.; Mariella, J.; Castagnetti, C.; Iacono, E. Heterologous Wharton's Jelly Derived Mesenchymal Stem Cells Application on a Large Chronic Skin Wound in a 6-Month-Old Filly. *Front. Vet. Sci.* **2019**, *6*, 9. [CrossRef] [PubMed]
125. Martignani, E.; Cravero, D.; Miretti, S.; Accornero, P.; Baratta, M. Bovine Mammary Stem Cells: New Perspective for Dairy Science. *Vet. Q.* **2014**, *34*, 52–58. [CrossRef] [PubMed]
126. Stingl, J.; Eaves, C.J.; Zandieh, I.; Emerman, J.T. Characterization of Bipotent Mammary Epithelial Progenitor Cells in Normal Adult Human Breast Tissue. *Breast Cancer Res. Treat.* **2001**, *67*, 93–109. [CrossRef] [PubMed]
127. Shackleton, M.; Vaillant, F.; Simpson, K.J.; Stingl, J.; Smyth, G.K.; Asselin-Labat, M.-L.; Wu, L.; Lindeman, G.J.; Visvader, J.E. Generation of a Functional Mammary Gland from a Single Stem Cell. *Nature* **2006**, *439*, 84–88. [CrossRef] [PubMed]
128. Martignani, E.; Eirew, P.; Eaves, C.; Baratta, M. Functional Identification of Bovine Mammary Epithelial Stem/Progenitor Cells. *Vet. Res. Commun.* **2009**, *33*, 101–103. [CrossRef] [PubMed]
129. Sharma, N.; Jeong, D.K. Stem Cell Research: A Novel Boulevard towards Improved Bovine Mastitis Management. *Int. J. Biol. Sci.* **2013**, *9*, 818–829. [CrossRef] [PubMed]
130. Brito, L.F.; Bedere, N.; Douhard, F.; Oliveira, H.R.; Arnal, M.; Peñagaricano, F.; Schinckel, A.P.; Baes, C.F.; Miglior, F. Review: Genetic Selection of High-Yielding Dairy Cattle toward Sustainable Farming Systems in a Rapidly Changing World. *Animal* **2021**, *15*, 100292. [CrossRef]
131. Gier, H.T.; Marion, G.B. Uterus of the Cow after Parturition: Involutional Changes. *Am. J. Vet. Res.* **1968**, *29*, 83–96.
132. Sheldon & Owen Postpartum Uterine Infection and Endometritis in Dairy Cattle. Available online: http://www.animal-reproduction.org/journal/animreprod/article/doi/10.21451/1984-3143-AR1006 (accessed on 4 March 2022).
133. Azarbayejani, R.; Mohammadsadegh, M. Glucose, Insulin, and Cortisol Concentrations and Glucose Tolerance Test in Holstein Cows with Inactive Ovaries. *Trop. Anim. Health Prod.* **2020**, *53*, 41. [CrossRef]
134. Rodríguez, F.M.; Gareis, N.C.; Hein, G.J.; Salvetti, N.R.; Amweg, A.N.; Huber, E.; Stassi, A.F.; Ortega, H.H.; Rey, F. Role of Components of the Insulin-like Growth Factor System in the Early Stages of Ovarian Follicular Persistence in Cattle. Elsevier Enhanced Reader. Available online: https://reader.elsevier.com/reader/sd/pii/S0021997517300890?token=CAFD75A62ED09A8BC96BD735E4635151CA835435B23997BF17A2416CA74BA52D7942C35E65FB43FE36FC14EFA9B1DB3F&originRegion=eu-west-1&originCreation=20220505094657 (accessed on 5 May 2022).
135. Briggs, R.; King, T.J. Transplantation of Living Nuclei From Blastula Cells into Enucleated Frogs' Eggs. *Proc. Natl. Acad. Sci. USA* **1952**, *38*, 455–463. [CrossRef]
136. Campbell, K.H.; McWhir, J.; Ritchie, W.A.; Wilmut, I. Sheep Cloned by Nuclear Transfer from a Cultured Cell Line. *Nature* **1996**, *380*, 64–66. [CrossRef]
137. Ogura, A.; Inoue, K.; Wakayama, T. Recent Advancements in Cloning by Somatic Cell Nuclear Transfer. *Philos. Trans. R. Soc. Lond. B Biol. Sci.* **2013**, *368*, 20110329. [CrossRef]
138. Nazari, H.; Shirazi, A.; Shams-Esfandabadi, N.; Afzali, A.; Ahmadi, E. The Effect of Amniotic Membrane Stem Cells as Donor Nucleus on Gene Expression in Reconstructed Bovine Oocytes. *Int. J. Dev. Biol.* **2016**, *60*, 95–102. [CrossRef] [PubMed]
139. Hill, J.R.; Dobrinski, I.; Hill, J.R.; Dobrinski, I. Male Germ Cell Transplantation in Livestock. *Reprod. Fertil. Dev.* **2005**, *18*, 13–18. [CrossRef] [PubMed]
140. Kim, H.J.; Park, J.-S. Usage of Human Mesenchymal Stem Cells in Cell-Based Therapy: Advantages and Disadvantages. *Dev. Reprod.* **2017**, *21*, 1–10. [CrossRef] [PubMed]
141. Hill, A.B.T.; Bressan, F.F.; Murphy, B.D.; Garcia, J.M. Applications of Mesenchymal Stem Cell Technology in Bovine Species. *Stem Cell Res. Ther.* **2019**, *10*, 44. [CrossRef] [PubMed]

animals

MDPI

Article

A Comparative Study of Canine Mesenchymal Stem Cells Isolated from Different Sources

Filip Humenik [1], Marcela Maloveska [1], Nikola Hudakova [1], Patricia Petrouskova [1], Lubica Hornakova [2], Michal Domaniza [2], Dagmar Mudronova [3], Simona Bodnarova [4] and Dasa Cizkova [1,5,*]

1 Centre of Experimental and Clinical Regenerative Medicine, The University of Veterinary Medicine and Pharmacy in Kosice, 041 81 Kosice, Slovakia; filip.humenik@uvlf.sk (F.H.); marcela.maloveska@uvlf.sk (M.M.); nikola.hudakova@student.uvlf.sk (N.H.); patricia.petrouskova@uvlf.sk (P.P.)
2 University Veterinary Hospital, The University of Veterinary Medicine and Pharmacy in Kosice, 041 81 Kosice, Slovakia; lubica.hornakova@uvlf.sk (L.H.); michal.domaniza@student.uvlf.sk (M.D.)
3 Institute of Microbiology and Immunology, The University of Veterinary Medicine and Pharmacy in Kosice, 041 81 Kosice, Slovakia; dagmar.mudronova@gmail.com
4 Department of Pneumology a Phtiseology, Faculty of Medicine, University of Pavol Jozef Safarik, 041 80 Kosice, Slovakia; simona.bodnarova@upjs.sk
5 Institute of Neuroimmunology, Slovak Academy of Sciences, 845 10 Bratislava, Slovakia
* Correspondence: cizkova.dasa@gmail.com

Simple Summary: The present study describes differences in the isolation yield, morphology, presence of surface markers and proliferation capacity but not in the multilineage potential of canine MSCs isolated from bone marrow, adipose tissue and amnion. Among all the MSCs analysed, AT-MSCs showed the highest isolation yield, phenotype homogeneity, proliferation capacity and osteogenic and chondrogenic potential. In addition, for BM-MSCs and AM-MSCs, we uncovered some differences that need to be considered during isolation, expansion and phenotyping prior to their possible application in targeted regenerative veterinary medicine.

Abstract: In this study, we provide comprehensive analyses of mesenchymal stem cells (MSCs) isolated from three types of canine tissues: bone marrow (BM-MSCs), adipose tissue (AT-MSCs) and amniotic tissue (AM-MSCs). We compare their morphology, phenotype, multilineage potential and proliferation activity. The BM-MSCs and AM-MSCs showed fibroblast-like shapes against the spindle shape of the AT-MSCs. All populations showed strong osteogenic and chondrogenic potential. However, we observed phenotypic differences. The BM-MSCs and AT-MSCs revealed high expression of CD29, CD44, CD90 and CD105 positivity compared to the AM-MSCs, which showed reduced expression of all the analysed CD markers. Similarly, the isolation yield and proliferation varied depending on the source. The highest isolation yield and proliferation were detected in the population of AT-MSCs, while the AM-MSCs showed a high yield of cells, but the lowest proliferation activity, in contrast to the BM-MSCs which had the lowest isolation yield. Thus, the present data provide assumptions for obtaining a homogeneous MSC derived from all three canine tissues for possible applications in veterinary regenerative medicine, while the origin of isolated MSCs must always be taken into account.

Keywords: canine mesenchymal stem cells; morphology; phenotype; multilineage potential

Citation: Humenik, F.; Maloveska, M.; Hudakova, N.; Petrouskova, P.; Hornakova, L.; Domaniza, M.; Mudronova, D.; Bodnarova, S.; Cizkova, D. A Comparative Study of Canine Mesenchymal Stem Cells Isolated from Different Sources. *Animals* **2022**, *12*, 1502. https://doi.org/10.3390/ani12121502

Academic Editors: Eleonora Iacono and Barbara Merlo

Received: 31 March 2022
Accepted: 4 June 2022
Published: 9 June 2022

1. Introduction

Mesenchymal stem cells (MSCs) could be described as unspecialised cells with multilineage differentiation potential and self-renewal capacity [1]. In general, they are located in discrete microenvironments, termed perivascular niches, where they play a key role in maintaining tissue homeostasis and healing processes [2–4]. The main sources of mesenchymal stem cells are bone marrow and adipose tissue [5], but these cells can also be obtained from skin, dental pulp, liver, ovarian epithelium, umbilical cord, placenta, amniotic fluid and others [6,7]. Mesenchymal stem cells can differentiate into cells of mesodermal origin, such as chondrocytes, osteoblasts and adipocytes. Several studies also point to the possibility of differentiating into ectodermal or endodermal lineages, such as nerve tissue cells or hepatocytes [1]. However, the transdifferentiation of MSCs from the original mesodermal line into ectodermal cells remains unclear and controversial [8]. Due to the fact of their multilineage differentiating ability, mesenchymal stem cells expand the possibilities of regenerative medicine [9], where their application helps to replace bone tissue and cartilage [5,10]. However, the low survival and transient retention of transplanted MSCs in host tissue [11] indicate that MSCs exert their therapeutic effects via secretion of bioactive factors that provide a favourable microenvironment to facilitate the repair and regeneration of injured tissues [12]. The paracrine effect of MSCs plays an important role, as angiogenesis, neuroprotection, and immunoregulation in the target tissue are affected through the production of important growth and trophic factors or bioactive molecules [13,14]. Due to the lack of understanding of the complexity of secreted bioactive factors, MSC secretome-based therapy in human and veterinary medicine has not yet been fully established [15,16]. One of the obstacles is the variability of the MSC secretome, which is influenced by the donor, tissue source, culture conditions and passage [15,17]. A better understanding of the factors of the biological function that makes up the secretome in relation to its tissue source will allow for the development of more effective and diseases targeted therapy [18].

In principle, according to the minimum criteria required for defining MSCs [19], it is necessary to select the appropriate source of biologically active MSCs capable of releasing the specific factors needed for in situ tissue regeneration [20]. We can currently draw on the amount of information available on MSCs isolated from various tissues and their use for alternative treatments in veterinary medicine [21]. Therefore, in the present study, we compared MSCs isolated from three different canine tissues (i.e., bone marrow, adipose tissue, and neonatal amniotic membrane), which have certain advantages/disadvantages in terms of isolation and yields and their biological activities eligible for the regeneration of specific tissues. Our results are consistent with other works [22–24] and confirm the ability to isolate MSCs from various canine adults as well as perinatal tissues.

2. Materials and Methods

The study was performed after obtaining informed consent from the owners and the approval of the Ethics Committee at the University of Veterinary Medicine and Pharmacy in Kosice on 2 September 2021 (EKVP/2021-01).

2.1. Isolation of MSC from Bone Marrow

The bone marrow was harvested from purebred healthy dogs (n = 3). Donor 1—male, German Shepherd, 35 kg, 3 years old. Donor 2—female, Cane Corso, 70 kg, 4 years old. Donor 3—male, German Shorthaired Pointer, 32 kg. Before bone marrow collection, all donors were examined (clinical examination, biochemical, and haematological parameters were evaluated). Harvesting was performed under general anaesthesia. The place of the collection was prepared according to all principles of sterility and asepsis as during surgery. We selected the proximal part of the humerus as the sampling site. The collections were performed with an 18 G injection needle and a 10 mL syringe. After penetrating the bone marrow cavity, the bone marrow was aspirated with a 10 mL syringe containing 3 mL of flushing medium. Flushing medium composition: Dulbecco's modified Eagle's medium/Nutrient Mixture F12 (DMEM-F12), 10% foetal bovine serum (FBS), 1% antibiotic-antimycotic (ATB+ATM; penicillin–streptomycin–amphotericin B) and 1% glutamine (all Biowest). The aspirate was then centrifuged twice for 10 min at 400× g. Cells were plated in cultivation flask T25, at a seeding density of 10^6 cells/flask, and were cultured in media containing DMEM-F12 + 10% FBS + 2% ATB + ATM at 37 °C and 5% CO_2. After 48 h of incubation, nonadherent cells were removed.

2.2. Isolation of MSCs from Adipose Tissue

Adipose tissue was harvested from purebred healthy dogs (n = 3) and was collected under general anaesthesia, maintaining the principles of sterility and asepsis, from the subcutaneous tissue in the scapular area. Adipose tissue was collected from identical donors as for bone marrow (Section 2.1). We isolated 5–7 g of adipose tissue from each donor. The tissue was then washed with phosphate-buffered saline (PBS; Biowest) containing 2% ATB + ATM, then mechanically dissociated and enzymatically digested with 0.05% collagenase type I and IV (Sigma, Burlington, MA, USA) at 37 °C for 1 h. At the end of the incubation period, the digested tissue was filtered (through a 100 μm cell strainer) to remove tissue fragments, centrifuged at 400× g for 10 min, and the obtained stromal vascular fraction (SVF) pellet was resuspended in culture medium consisting of DMEM-F12 containing 10% FBS and 2% ATB + ATM and plated in a T25 tissue culture flask at a concentration of 10^6 cells/flask and incubated in cultivation medium (DMEM-F12 + 10% FBS + 2% ATB + ATM) at 37 °C and 5% CO_2. After 48 h, non-adherent cells were removed and, subsequently, the medium was changed twice a week.

2.3. Isolation of MSCs from Amniotic Tissue

Amniotic tissues were obtained during caesarean section (new-born puppies, n = 6, 62nd day of pregnancy) under strictly sterile conditions. Donors—German rottweiler, weight of puppies approximately 350 g, sex ratio—3:3. The amnions were then washed with PBS containing 2% ATB + ATM; then, the tissue was mechanically dissociated and enzymatically digested using 0.05% collagenase type I at 37 °C for 30 min. At the end of the incubation period, the digested tissue was filtered (through a 100 μm cell strainer) to remove tissue fragments, and the obtained fraction was centrifuged at 400× g for 10 min. The obtained pellet was resuspended in DMEM-F12 culture medium + 10% FBS + 2% ATB + ATM. Cells were plated in a T25 culture flask at a concentration of 10^6 cells/mL and incubated in culture medium (DMEM-F12) + 10% FBS + 2% ATB + ATM at 37 °C and 5% CO_2 Nonadherent cells were removed, and the medium was subsequently changed twice a week.

2.4. Passaging Cells

When the cultivated cell population reached a confluence of approximately 75–80%, we proceeded to passage. To separate the cells from the surface of the culture flask, we used an enzymatic trypsinization method using 0.25% Trypsin EDTA (Biowest), which acted on the cells depending on the level of confluence for 5–7 min at 37 °C. To inactivate the trypsin, FBS was used in a 1:1 ratio, and the whole suspension was subsequently centrifuged at 400× g for 10 min. The supernatant was removed, and the cell population was plated on T75 culture flasks at a concentration of 1.5×10^6 cells/flask.

2.5. Expression of Surface Markers

Samples were analysed for mesenchymal stromal cell markers (i.e., CD29, CD44, CD90 and CD105) and for hematopoietic stem cells marker (i.e., CD45). Each sample was diluted to a final concentration of 2×10^5 cells/mL and centrifuged at 400× g/5 min. Subsequently, the supernatant was removed and the cell pellet was resuspended in 100 μL of PBS containing 3–5 μL of CD90 ((YKIX337.217, allophycocyanin; APC), CD29 ((MEM-101, phycoerythrin; PE), CD44 (MEM-263, APC), CD105 ((MA1-19594, fluorescein isothiocyanate; FITC) and CD45 (YKIX716.13, PE)—all ThermoFisher, Waltham, MA, USA—and incubated for 60 min at 4 °C in the dark. At the end of the incubation period, the samples were centrifuged again at 400× g/5 min, the supernatant was removed and the sample was washed in 200–500 μL of washing solution (1% FBS in PBS + 0.1% Sodium Azide (SevernBiotech Ltd., Kidderminster, UK). Cytometric analysis was performed on a BD FACSCanto® flow cytometer (Becton Dickinson Biosciences, San Jose, CA, USA) equipped with a blue (488 nm) and red (633 nm) laser and 6 fluorescence detectors. The percentage of cells expressing individual CD traits was determined by a histogram for the respective fluorescence. The data obtained via measurement were analysed in BD FACS Diva™ analysis software. As a negative control, we used the same type of non-labelled MSCs for autofluorescence control. A gating strategy for flow cytometry was performed by forward/sideward scatter and sideward scatter/sideward scatter pulse height for the elimination of debris and doublets.

2.6. Multilineage Potential

To confirm the multilineage potential of the MSCs, we used the StemProMultilineage differential Kit (Gibco) according to the recommended protocol attached. The cells used for multilineage differentiation were from passage 3 (P3). Cells were cultured in 24-well plates with an initial density of 4×10^4 cells/well for osteocytes, 8×10^4 cells/well for adipocytes and 8×10^4 cells per micromass/well for chondrocytes. Each micromass was a single drop of 5 µL/8×10^4 cells, which was placed in the centre of the well and then incubated at 37 °C and 5% CO_2 for 2 h for better adherence to the surface, and then 500 µL chondrogenic medium was added. After the recommended culture time (21 days, day in vitro; DIV 21), the cells were fixed using 4% paraformaldehyde (PFA), and the individual populations were stained with the Alizarin red staining method (Sigma) for evidence of calcium deposits in the osteoblast population; Alcian blue (Sigma) for the detection of proteoglycans in the chondroblast population and Oil red (Sigma) for the staining of fat vacuoles in the adipocyte population.

2.7. Proliferation Activity of MSC

For the description of cell proliferation, we used the MTT cell proliferation assay kit (Invitrogen). The cells from each population in P3 were plated in 96-well plates, seeding density 1×10^4 cell/well and cultivated in standard cultivation medium (DMEM-F12 w/o phenol red + 10% FBS + 2% ATB + ATM) at 37 °C and 5% CO_2 for 24, 48, 168 and 240 h. After the cultivation period, we removed the medium and replaced it with 100 µL of fresh culture medium. In the next step, 10 µL of 12 mM MTT stock solution to each well was added and incubated at 37 °C for 4 h. At the end of incubation time, we added 100 µL of SDS-HCl solution, mixed it and incubated. After 12 h, the content of each well was mixed carefully by the pipette and the absorbance was measured at 572 nm by Perkin Elmer Victor3 Multilabel Plate Reader. Statistical analyses were processed via two-way ANOVA, followed by the Tukey test, with the mean considered from five measurements.

2.8. Freezing Protocol

At the end of study, MSCs were detached from the culture using 0.25% Trypsin EDTA, washed in FBS and suspended in a prechilled freezing medium consisting of 10% dimetylsulfoxid (DMSO; CryoSure, Wak-Chemie Medical GmbH) + 40% FBS and 50% DMEM-F12. Cells were dispended in 1.8 mL cryovials tubes in concentration 1×10^6 cells/mL. Cryovials were slowly cooled in a Mr. Frosty freezing container and placed in −80 °C freezer before transferring to the vapor of liquid nitrogen at ≤ -140 °C.

3. Results

3.1. Isolation of Canine MSCs from Bone Marrow, Adipose Tissue and Amnion

Using the abovementioned protocol, we were able to isolate and cultivate a homogeneous population of canine MSCs from bone marrow, adipose tissue and amnion. The yield of isolated cells varied between 1 and 7×10^6 cells/mL (BM-MSC—1×10^3 cells/mL of bone marrow aspirate, AT-MSC—2.5×10^6 cells/g of adipose tissue and AM-MSC—5.6×10^6 cells/g of amniotic tissue) shown in Table 1. Bone marrow and amniotic mesenchymal stem cells showed a fibroblast-like shape (Figure 1A,C and Figure 2(A1,A2,C1,C2)), which is typical for MSCs, in contrast to adipose tissue mesenchymal stem cells, which were longer (approximately 20 µm) and thinner, revealing a spindle-shaped morphology (Figures 1B and 2(B1,B2)).

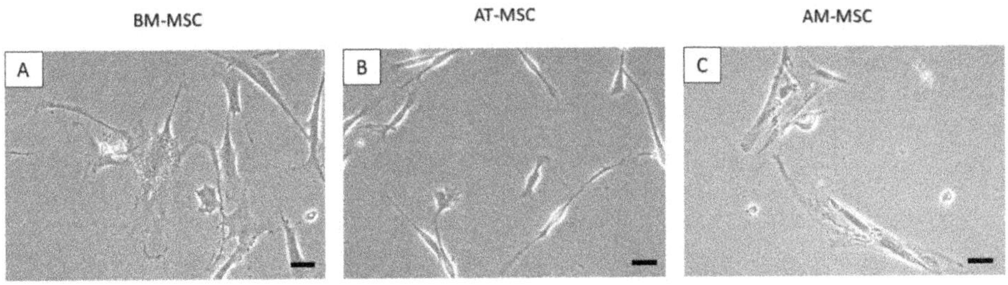

Figure 1. Detailed morphology of canine MSCs at passage 0 (P0) day in vitro (DIV) 3 from different sources. The BM-MSCs (**A**) and AM-MSCs (**C**) revealed a fibroblast-like morphology, while the AT-MSCs (**B**) showed a spindle-like morphology. Scale bars = 20 μm.

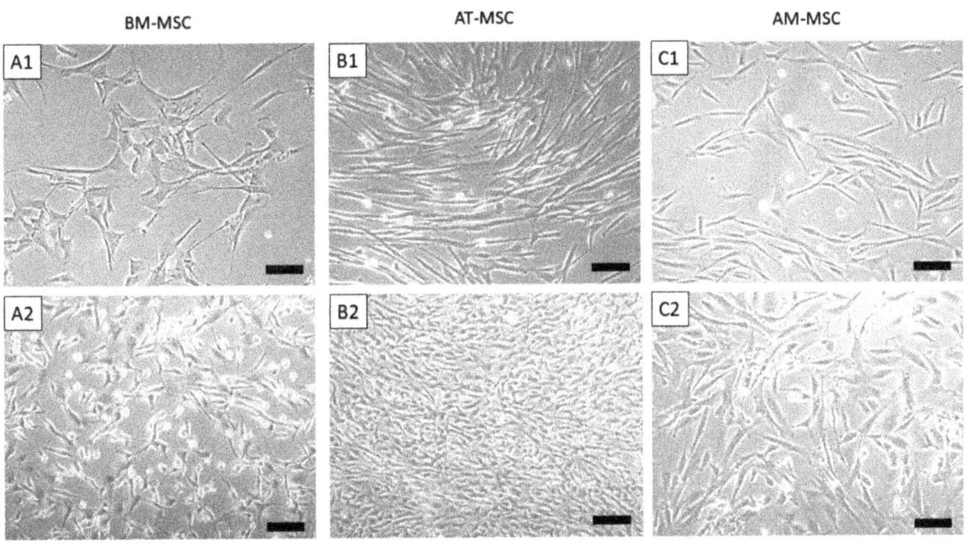

Figure 2. Morphology of canine MSCs from different sources. BM-MSCs at passage 0 (P0) day in vitro (DIV) 5 (**A1**) and DIV15 (**A2**); AT-MSCs at passage P0 DIV5 (**B1**) and DIV15 (**B2**); AM-MSCs at passage P0 DIV5 (**C1**) and DIV15 (**C2**). Scale bars = 50 μm.

Table 1. Comparative analysis of the common MSC properties from canine MSCs isolated from bone marrow, adipose tissue and amnion. +++, ++, + and – indicate high, moderate, low or absence of mentioned properties, respectively.

Results of the Comparative Study of Canine MSCs			
	BM-MSCs	**AT-MSCs**	**AM-MSCs**
Invasiveness of tissue collection	+++	+++	–
Yield	+	++	+++
Homogeneity	+	+++	+
Osteogenic potential	+++	+++	+++
Chondrogenic potential	+++	+++	+++
Adipogenic potential	–	+	–
Proliferation capacity	++	+++	+

3.2. CD Characterization of Canine MSC

Results of CD analyses (Figure 3) show, that passaging is a suitable tool for obtaining high homogeneity and uniformity of the population during cultivating mesenchymal cells, even in the low passage (all results from passage 3). BM-MSC showed high expression of CD29 (98.7% ± 1.5%), CD44 (97.3% ± 1.0%), CD90 (76.3% ± 2.6%) and CD105 (99.8% ± 3.1%), but low expression of CD45 (3.3% ± 0.3%) on the other hand. AT-MSC were positive for CD29 (99.3% ± 0.7%), CD44 (99.1% ± 0.5%), CD90 (85.9% ± 0.6%), CD105 (99.7% ± 1.2%) and negative for CD45 (1.0% ± 0.2%). Results of different phenotype, AM-MSC, showed positivity for CD29 (72.3% ± 0.8%), CD44 (71.5% ± 1.6%), CD90 (5.8% ± 0.9%), CD105 (90.5% ± 2.9%) and CD45 (3.1% ± 0.4%). We also observed differences in the autofluorescence. The BM-MSCs showed minimal auto autofluorescent florescent positivity for PE (1.1% ± 0.1%), FITC (1.2% ± 0.05%) and APC (2.5% ± 0.5%). The AT-MSCs showed a high number of autofluorescent cells for APC (4.0% ± 0.2%) and a minimum of PE-autofluorescent cells (0.3% ± 0.05%) and FITC (0.6% ± 0.01%). We observed the highest number of autofluorescent cells in AM-MSCs: PE (1.8% ± 0.3%), APC (4.5% ± 0.9%) and FITC (2.9% ± 0.5%). The gating strategy—forward/sideward scatter and sideward scatter/sideward scatter pulse height for the elimination of debris and doublets (Figure 4.). The viability of the observed cells varied between 85 and 93%.

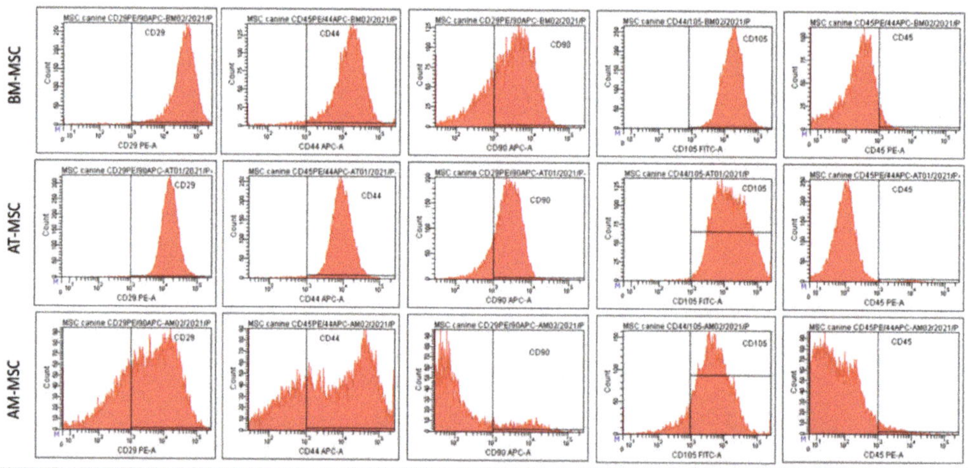

	BM-MSCs	AT-MSCs	AM-MSCs
CD29	98.7%±1.5%	99.3%±0.7%	72.3%±0.8%
CD44	97.3%±1.0%	99.1%±0.5%	71.5%±1.6%
CD90	76.3%±2.6%	85.9%±0.6%	5.8%±0.9%
CD105	99.8%±3.1%	99.7%±1.2%	90.5%±2.9%
CD45	3.3%±0.3%	1.0%±0.2%	3.1%±0.4%

Figure 3. Results of the CD analyses from passage 3 (P3). The BM-MSCs showed positivity for CD29 (98.7% ± 1.5%), CD44 (97.3% ± 1.0%), CD90 (76.3% ± 2.6%), CD105 (99.8% ± 3.1%) and low expression for CD45 (3.3% ± 0.3%). The AT-MSCs showed high expression of CD29 (99.3% ± 0.7%), CD44 (99.1% ± 0.5%), CD90 (85.9% ± 0.6%), CD105 (99.7% ± 1.2%) and low expression of CD45 (1.0% ± 0.2%). The AM-MSCs showed positivity for CD29 (72.3% ± 0.8%), CD44 (71.5% ± 1.6%), CD90 (5.8% ± 0.9%), CD105 (90.5% ± 2.9%) and low expression of CD45 (3.1% ± 0.4%).

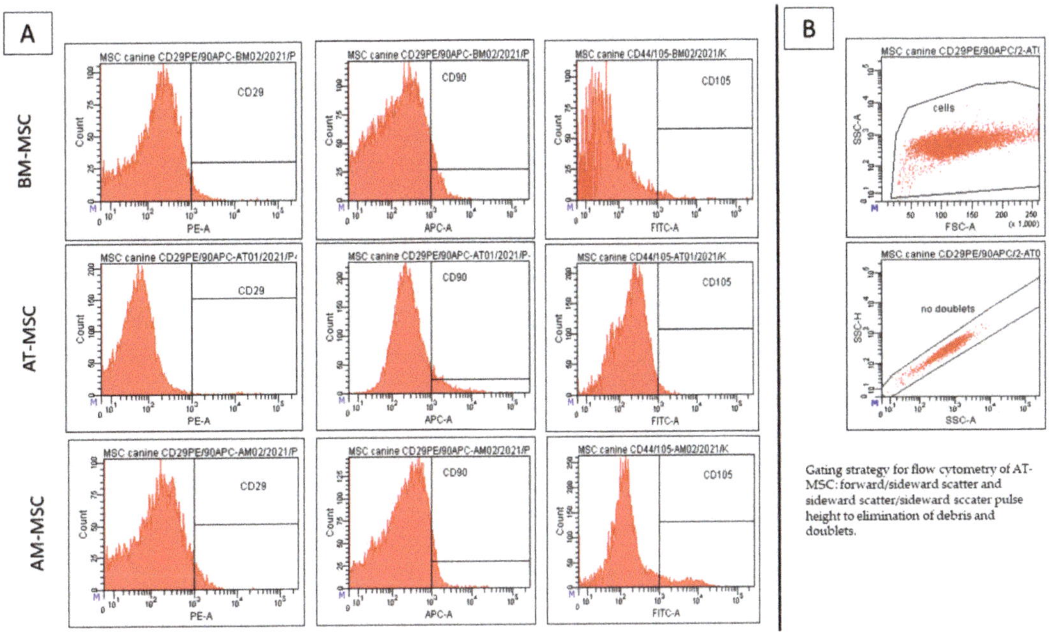

Figure 4. Results of the CD analyses from passage 3 (P3)—negative control (**A**) and Gating strategy (**B**). The BM-MSCs showed minimal autofluorescent positivity for PE (1.1% ± 0.1%), FITC (1.2% ± 0.05%) and APC (2.5% ± 0.5%). The AT-MSCs showed a high number of autofluorescent cells for APC (4.0% ± 0.2%) and a minimum of PE-autofluorescent cells (0.3% ± 0.0 5%) and FITC (0.6% ± 0.01%). We observed the highest number of autofluorescent cells in AM-MSCs: PE (1.8% ± 0.3%), APC (4.5% ± 0.9%) and FITC (2.9% ± 0.5%). Gating strategy—forward/sideward scatter and sideward scatter/sideward scatter pulse height for the elimination of debris and doublets (Figure 4).

3.3. Multilineage Potential

Using a multilineage differentiation kit and the recommended culture protocol and staining methods, we confirmed the ability of MSCs isolated from canine bone marrow, adipose tissue and amnion to differentiate into osteocytes, chondrocytes and only a weak ability to differentiate into adipocytes (Figure 5). All types of isolated canine MSCs (BM-MSCs, AT-MSCs and AM-MSCs) showed high osteogenic and chondrogenic potential; however, the adipogenic differentiation capacity was limited.

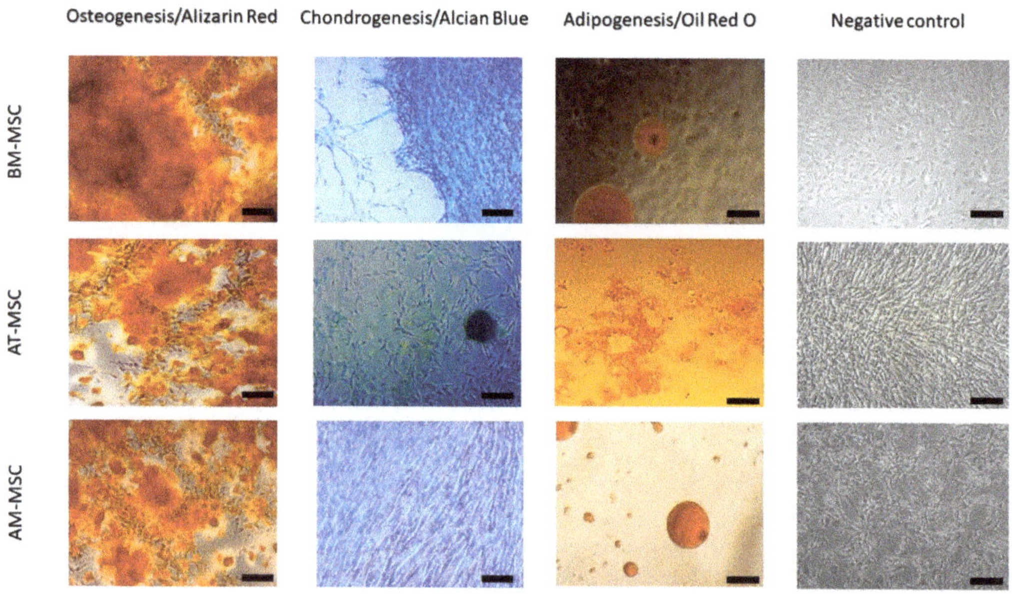

Figure 5. Multilineage potential of canine MSCs. All types of canine MSCs showed high osteogenic (presence of calcium deposits detected by Alizarin red) and chondrogenic potential (presence of glycoproteoglycanes detected by Alcian blue staining); however, there was the absence (BM-MSCs and AM-MSCs) or low (AT-MSCs) level of the adipogenic potential (triglycerides detected by Oil Red O staining). Scale bars = 50 µm.

3.4. *MTT Assay*

The MTT assay showed that MSCs isolated from all sources have good proliferation capacity; however, we could observe differences in cultivation periods and types of MSCs as well. The best proliferation capacity in vitro was observed in AT-MSCs and at 168 h. The lowest capacity was shown in AM-MSCs. All types of MSCs reached the maximum capacity at 168 h, after this time period, the proliferation capacity decreased (Figure 6).

We also observed differences when comparing the growth rate of cell populations from individual sources after isolation. We achieved the fastest confluent population in adipose tissue-derived MSCs (day 11 of culture). On the contrary, the slowest growth was seen in the bone marrow in which we reached a confluent population on the 29th day of cultivation.

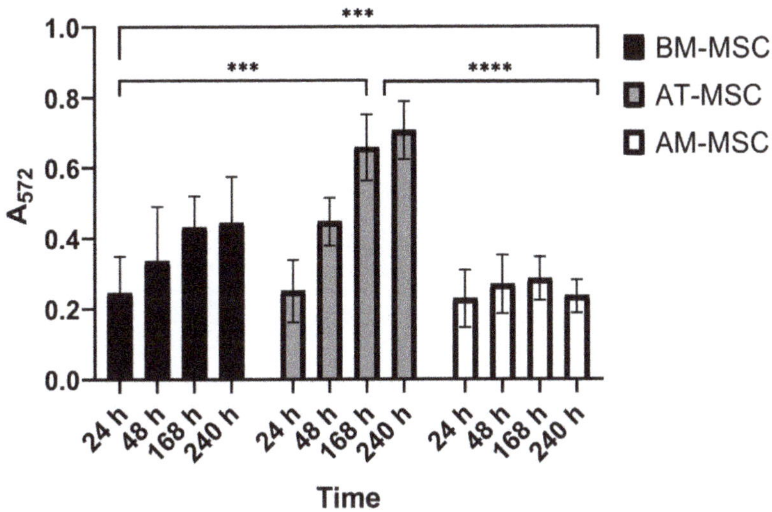

Figure 6. Results of the MTT assay. Graph representing absorbance measured after 24, 48, 168, and 240 h at 572 nm, which correlates with the proliferation capacity of examined types of MSCs. *** $p = 0.0007$ and **** $p < 0.0001$.

4. Discussion

The main goal of this study was to compare the yield, morphology, phenotype, multilineage potential, and proliferation activity of mesenchymal stem cells isolated from different canine tissues—bone marrow, adipose tissue and amniotic tissue. We also met the criteria of the International Stem Cell Research Society (ISSCR) [6,19].

For isolation of MSCs from bone marrow, we used a simple method of centrifugation and size separation using a cell strainer (100 μm). To isolate MSCs from adipose tissue and amnion, we used a combined method of mechanical disruption and enzymatic digestion. As previously described, prolonged digestion can damage the cells [25]. Therefore, we optimised the enzymatic process within the range 25–45 min, depending on the amount and size of the digested fraction at a temperature of 37 °C. For the comparison of isolation yield, the AM-MSCs showed the highest number of isolated cells (up to 5.6×10^6 cells/g). However, the isolation yield of BM-MSCs and AT-MSCs was also not negligible (1×10^3 cells/mL BM-MSC and 2.5×10^6 cells/g AT-MSC). This observation was also confirmed by other studies [26–30]. Concluding from the abovementioned facts, amnion tissue represents a rich source of stem cells. Its collection and retrieval did not cause any added stress and suffering for the animal, and since it had no further use in normal veterinary practice, it would end up as biological waste. Therefore, this tissue source certainly deserves attention in the field of regenerative medicine.

Differences were observed in morphology, as well. While BM-MSCs were characterised by a fibroblast-like shape and a size of 100–120 μm, MSCs isolated from adipose tissue were longer and thinner (100–140 μm), and MSCs isolated from amnion tissue were similar to BM-MSCs (with a fibroblast-like shape) but at first sight morphologically different (oval-shaped) and smaller (80 μm). It was interesting to discover the multilineage differentiation ability of isolated cell populations. After reaching confluence in the P3 and subsequent trypsinization, we tested the cells from each population to confirm the ability to differentiate into osteogenic, chondrogenic, and adipogenic lines. MSCs isolated from canine bone marrow, adipose tissue and amnion showed a very good ability to differentiate into osteogenic and chondrogenic lines but repeatedly very little or lacked the ability to differentiate into adipogenic lines. This variation in the inability of MSCs to differentiate into an adipogenic line has also been described in other studies [31,32]. This may be because cells isolated from the same tissue type but from different collection sites sometimes need up to twice the time to differentiate. For example, the differentiation capacity of bone-marrow-derived MSCs may require a longer time for adipogenic differentiation in comparison to adipose-derived MSCs or amniotic-derived MSCs. Therefore, this has to be optimised for each cell type processed for differentiation [16]. In addition, osteogenesis-mediating and adipogenesis-suppressing bioactive molecules, such as Runx2, Wnt10b and RhoA, or adipogenesis-enhancing bioactive molecules suppressing osteogenesis, including PPARγ, P2X6, LIF, sFRP-1 and BMPs, also play an important role [33]. Thus, osteogenic and chondrogenic differentiation are among the key properties required for hard tissue regeneration in veterinary medicine. This has been confirmed in our study in all MSC types, which nominates them as candidates for possible treatment application [34]. The expression of CD surface markers in MSC populations was tested using flow cytometry. During the analysis, we found a lower percentage of CD29+, CD44+ and CD105+ and almost no positivity for CD90 in AM-MSCs compared with BM-MSCs and AT-MCs, where the expression of all four CD markers was high [35]. The majority of canine papers have demonstrated either a single alternative MSC marker (CD29 or CD44) or both, which is more consistent than any of the classic MSC markers [15,23,24,36]. High expression of CD29 and CD105 is superior to strong chondrogenic potential by regulating TGF-β/Smad2 [37]. The same can be said about the expression of CD44 describing chondrogenic potential via the Smad and ERK signalling pathways [38]. All of these facts correlate to our results of chondrogenic differentiation. Both CD29 and CD44 expression were found to be involved with MSC adhesion, migration and engraftment in vivo [15]. We were able to confirm low or no adipogenic potential in the studied MSCs. This could be caused by high number of CD29 positive cells in the observed population, because CD29 and CD90 reduce adipogenic capacity [39]. The absence of or low positivity for CD90, in the case of canine AM-MSCs, may be associated with their low immunosuppressive properties; similarly, it has been reported in human MSCs [40]. Another study referred to the loss of CD90 (THY-1) as to a process that may increase MSC differentiation potential, which is often associated with a decrease in CD44 expression [41]. Thus, although AM-MSCs are a rich source with good differentiation potential, we have to keep in mind possible differences due to the difference in their neonatal origin. Finally, it should be noted that the actual expression of surface markers also depends on the MSC isolation source [42] as well as on the donor age and cell passage [43]. Despite a previous study [44], higher expression of CD45 in the observed population of BM-MSCs and AM-MSCs (>3%) had no pronounced impact on multilineage potential compared to AT-MSCs. In addition, obvious differences in morphology, phenotype, and proliferative activity in individual cell populations were also observed in a study led by Bearden [23].

The autofluorescence of MSCs at 488 nm correlates with cell granularity and the expression of CD90. Increased autofluorescence was negatively associated with the expression of CD90, which was also confirmed in our study [45]. Furthermore, cells revealing high granularity (i.e., AM-MSCs) showed low expression of CD90 and high autofluorescence.

In this study, differences in the proliferation capacity of MSCs from different sources were noticed using the MTT assay. The AT-MSCs showed the highest capacity, while AM-MSCs showed a high yield of cells but revealed the lowest proliferation activity. All three types of MSCs reached their proliferation peak between 168 and 240 h of cultivation. After this cultivation period, the proliferation capacity decreased. This fact could correlate with the confluency of the given population and cell density. Similar results were published recently [26,28].

All of the abovementioned facts create preconditions for the use of MSCs isolated from all three tissues in cell therapy in veterinary medicine [11]. However, it should be emphasised that each MSC originating from different tissues should be thoroughly characterised and, afterward, carefully considered and adapted to the patient's condition and nature.

5. Conclusions

In this study, we described the effective protocols for the isolation of MSCs from different canine tissues—bone marrow, adipose tissue and amnion membrane. Our results presented differences in yield of isolation, morphology, phenotype, multilineage potential and proliferation activity. All types of isolated canine MSCs expressed CD105 surface markers, but not CD45, which correlated to their high chondrogenic capacity. The AT-MSCs showed the best phenotype homogeneity, proliferation capacity, multilineage potential and even high yield of isolation (Table 1). All the presented data create the foundation for their further investigation and potential use in regenerative therapy.

Author Contributions: D.C. and F.H., conceived of presented idea; F.H., L.H. and M.D., tissue harvesting; F.H., D.C. and M.M., isolation of MSCs and multilineage potential confirmation; F.H., D.M. and N.H., phenotypization of MSCs and figure preparation; D.C., F.H. and P.P., MTT assay; S.B., statistical analyses; D.C., supervision. All authors have read and agreed to the published version of the manuscript.

Funding: This research was funded by Slovak Research and Development Agency (APVV-19-0193) and Scientific Grant Agency of the Ministry of Education Slovak Republic (VEGA 1/0285/22).

Institutional Review Board Statement: The study was performed after obtaining the approval of the Ethics Committee at the University of Veterinary Medicine and Pharmacy in Kosice on 2 September 2021 (EKVP/2021-01).

Informed Consent Statement: The study was performed after obtaining informed consent from the owners.

Data Availability Statement: The data presented in this study are available upon request from the corresponding author.

Conflicts of Interest: The authors declare no conflict of interest.

References

1. Ullah, I.; Subbarao, R.B.; Rho, G.J. Human Mesenchymal Stem Cells—Current Trends and Future Prospective. *Biosci. Rep.* **2015**, *35*, e00191. [CrossRef] [PubMed]
2. Stzepourginski, I.; Nigro, G.; Jacob, J.-M.; Dulauroy, S.; Sansonetti, P.J.; Eberl, G.; Peduto, L. CD34$^+$ Mesenchymal Cells Are a Major Component of the Intestinal Stem Cells Niche at Homeostasis and after Injury. *Proc. Natl. Acad. Sci. USA* **2017**, *114*, E506–E513. [CrossRef] [PubMed]
3. Méndez-Ferrer, S.; Michurina, T.V.; Ferraro, F.; Mazloom, A.R.; Macarthur, B.D.; Lira, S.A.; Scadden, D.T.; Ma'ayan, A.; Enikolopov, G.N.; Frenette, P.S. Mesenchymal and Haematopoietic Stem Cells Form a Unique Bone Marrow Niche. *Nature* **2010**, *466*, 829–834. [CrossRef] [PubMed]
4. Kunisaki, Y.; Bruns, I.; Scheiermann, C.; Ahmed, J.; Pinho, S.; Zhang, D.; Mizoguchi, T.; Wei, Q.; Lucas, D.; Ito, K.; et al. Arteriolar Niches Maintain Haematopoietic Stem Cell Quiescence. *Nature* **2013**, *502*, 637–643. [CrossRef] [PubMed]
5. Neupane, M.; Chang, C.-C.; Kiupel, M.; Yuzbasiyan-Gurkan, V. Isolation and Characterization of Canine Adipose-Derived Mesenchymal Stem Cells. *Tissue Eng. Part A* **2008**, *14*, 1007–1015. [CrossRef]
6. Crisan, M.; Yap, S.; Casteilla, L.; Chen, C.-W.; Corselli, M.; Park, T.S.; Andriolo, G.; Sun, B.; Zheng, B.; Zhang, L.; et al. A Perivascular Origin for Mesenchymal Stem Cells in Multiple Human Organs. *Cell Stem Cell* **2008**, *3*, 301–313. [CrossRef]
7. Kerkis, I.; Kerkis, A.; Dozortsev, D.; Stukart-Parsons, G.C.; Gomes Massironi, S.M.; Pereira, L.V.; Caplan, A.I.; Cerruti, H.F. Isolation and Characterization of a Population of Immature Dental Pulp Stem Cells Expressing OCT-4 and Other Embryonic Stem Cell Markers. *Cells Tissues Organs* **2006**, *184*, 105–116. [CrossRef]
8. Song, L.; Tuan, R.S. Transdifferentiation Potential of Human Mesenchymal Stem Cells Derived from Bone Marrow. *FASEB J. Off. Publ. Fed. Am. Soc. Exp. Biol.* **2004**, *18*, 980–982. [CrossRef]
9. Pezzanite, L.M.; Fortier, L.A.; Antczak, D.F.; Cassano, J.M.; Brosnahan, M.M.; Miller, D.; Schnabel, L.V. Equine Allogeneic Bone Marrow-Derived Mesenchymal Stromal Cells Elicit Antibody Responses in Vivo. *Stem Cell Res. Ther.* **2015**, *6*, 54. [CrossRef]
10. Kangari, P.; Talaei-Khozani, T.; Razeghian-Jahromi, I.; Razmkhah, M. Mesenchymal Stem Cells: Amazing Remedies for Bone and Cartilage Defects. *Stem Cell Res. Ther.* **2020**, *11*, 492. [CrossRef]
11. Miao, C.; Lei, M.; Hu, W.; Han, S.; Wang, Q. A Brief Review: The Therapeutic Potential of Bone Marrow Mesenchymal Stem Cells in Myocardial Infarction. *Stem Cell Res. Ther.* **2017**, *8*, 242. [CrossRef] [PubMed]
12. Caplan, A.I. Mesenchymal Stem Cells: Time to Change the Name! *Stem Cells Transl. Med.* **2017**, *6*, 1445–1451. [CrossRef] [PubMed]

13. Cizkova, D.; Cubinkova, V.; Smolek, T.; Murgoci, A.-N.; Danko, J.; Vdoviakova, K.; Humenik, F.; Cizek, M.; Quanico, J.; Fournier, I.; et al. Localized Intrathecal Delivery of Mesenchymal Stromal Cells Conditioned Medium Improves Functional Recovery in a Rat Model of Spinal Cord Injury. *Int. J. Mol. Sci.* **2018**, *19*, 870. [CrossRef]
14. The Secretion Profile of Mesenchymal Stem Cells and Potential Applications in Treating Human Diseases | Signal Transduction and Targeted Therapy. Available online: https://www.nature.com/articles/s41392-022-00932-0 (accessed on 2 May 2022).
15. Wright, A.; Arthaud-Day, M.L.; Weiss, M.L. Therapeutic Use of Mesenchymal Stromal Cells: The Need for Inclusive Characterization Guidelines to Accommodate All Tissue Sources and Species. *Front. Cell Dev. Biol.* **2021**, *9*, 632717. [CrossRef]
16. Frontiers | A Set of Grand Challenges for Veterinary Regenerative Medicine | Veterinary Science. Available online: https://www.frontiersin.org/articles/10.3389/fvets.2016.00020/full (accessed on 2 May 2022).
17. Rizk, M.; Monaghan, M.; Shorr, R.; Kekre, N.; Bredeson, C.N.; Allan, D.S. Heterogeneity in Studies of Mesenchymal Stromal Cells to Treat or Prevent Graft-versus-Host Disease: A Scoping Review of the Evidence. *Biol. Blood Marrow Transplant. J. Am. Soc. Blood Marrow Transplant.* **2016**, *22*, 1416–1423. [CrossRef] [PubMed]
18. Vizoso, F.J.; Eiro, N.; Cid, S.; Schneider, J.; Perez-Fernandez, R. Mesenchymal Stem Cell Secretome: Toward Cell-Free Therapeutic Strategies in Regenerative Medicine. *Int. J. Mol. Sci.* **2017**, *18*, E1852. [CrossRef]
19. Dominici, M.; Le Blanc, K.; Mueller, I.; Slaper-Cortenbach, I.; Marini, F.; Krause, D.; Deans, R.; Keating, A.; Prockop, D.; Horwitz, E. Minimal Criteria for Defining Multipotent Mesenchymal Stromal Cells. The International Society for Cellular Therapy Position Statement. *Cytotherapy* **2006**, *8*, 315–317. [CrossRef]
20. Munoz-Perez, E.; Gonzalez-Pujana, A.; Igartua, M.; Santos-Vizcaino, E.; Hernandez, R.M. Mesenchymal Stromal Cell Secretome for the Treatment of Immune-Mediated Inflammatory Diseases: Latest Trends in Isolation, Content Optimization and Delivery Avenues. *Pharmaceutics* **2021**, *13*, 1802. [CrossRef]
21. Sasaki, A.; Mizuno, M.; Ozeki, N.; Katano, H.; Otabe, K.; Tsuji, K.; Koga, H.; Mochizuki, M.; Sekiya, I. Canine Mesenchymal Stem Cells from Synovium Have a Higher Chondrogenic Potential than Those from Infrapatellar Fat Pad, Adipose Tissue, and Bone Marrow. *PLoS ONE* **2018**, *13*, e0202922. [CrossRef]
22. Rashid, U.; Yousaf, A.; Yaqoob, M.; Saba, E.; Moaeen-ud-Din, M.; Waseem, S.; Becker, S.K.; Sponder, G.; Aschenbach, J.R.; Sandhu, M.A. Characterization and Differentiation Potential of Mesenchymal Stem Cells Isolated from Multiple Canine Adipose Tissue Sources. *BMC Vet. Res.* **2021**, *17*, 388. [CrossRef]
23. Bearden, R.N.; Huggins, S.S.; Cummings, K.J.; Smith, R.; Gregory, C.A.; Saunders, W.B. In-Vitro Characterization of Canine Multipotent Stromal Cells Isolated from Synovium, Bone Marrow, and Adipose Tissue: A Donor-Matched Comparative Study. *Stem Cell Res. Ther.* **2017**, *8*, 218. [CrossRef] [PubMed]
24. Kisiel, A.H.; McDuffee, L.A.; Masaoud, E.; Bailey, T.R.; Esparza Gonzalez, B.P.; Nino-Fong, R. Isolation, Characterization, and in Vitro Proliferation of Canine Mesenchymal Stem Cells Derived from Bone Marrow, Adipose Tissue, Muscle, and Periosteum. *Am. J. Vet. Res.* **2012**, *73*, 1305–1317. [CrossRef] [PubMed]
25. Mushahary, D.; Spittler, A.; Kasper, C.; Weber, V.; Charwat, V. Isolation, Cultivation, and Characterization of Human Mesenchymal Stem Cells. *Cytometry A* **2018**, *93*, 19–31. [CrossRef] [PubMed]
26. Hass, R.; Kasper, C.; Böhm, S.; Jacobs, R. Different Populations and Sources of Human Mesenchymal Stem Cells (MSC): A Comparison of Adult and Neonatal Tissue-Derived MSC. *Cell Commun. Signal. CCS* **2011**, *9*, 12. [CrossRef] [PubMed]
27. Petrenko, Y.; Vackova, I.; Kekulova, K.; Chudickova, M.; Koci, Z.; Turnovcova, K.; Kupcova Skalnikova, H.; Vodicka, P.; Kubinova, S. A Comparative Analysis of Multipotent Mesenchymal Stromal Cells Derived from Different Sources, with a Focus on Neuroregenerative Potential. *Sci. Rep.* **2020**, *10*, 4290. [CrossRef] [PubMed]
28. Wu, C.; Chen, L.; Huang, Y.; Huang, Y.; Parolini, O.; Zhong, Q.; Tian, X.; Deng, L. Comparison of the Proliferation and Differentiation Potential of Human Urine-, Placenta Decidua Basalis-, and Bone Marrow-Derived Stem Cells. *Stem Cells Int.* **2018**, *2018*, e7131532. [CrossRef] [PubMed]
29. Tsuji, W.; Rubin, J.P.; Marra, K.G. Adipose-Derived Stem Cells: Implications in Tissue Regeneration. *World J. Stem Cells* **2014**, *6*, 312–321. [CrossRef]
30. Brozovich, A.; Sinicrope, B.J.; Bauza, G.; Niclot, F.B.; Lintner, D.; Taraballi, F.; McCulloch, P.C. High Variability of Mesenchymal Stem Cells Obtained via Bone Marrow Aspirate Concentrate Compared with Traditional Bone Marrow Aspiration Technique. *Orthop. J. Sports Med.* **2021**, *9*, 23259671211058460. [CrossRef]
31. Kern, S.; Eichler, H.; Stoeve, J.; Klüter, H.; Bieback, K. Comparative Analysis of Mesenchymal Stem Cells from Bone Marrow, Umbilical Cord Blood, or Adipose Tissue. *Stem Cells Dayt. Ohio* **2006**, *24*, 1294–1301. [CrossRef]
32. Kozlowska, U.; Krawczenko, A.; Futoma, K.; Jurek, T.; Rorat, M.; Patrzalek, D.; Klimczak, A. Similarities and Differences between Mesenchymal Stem/Progenitor Cells Derived from Various Human Tissues. *World J. Stem Cells* **2019**, *11*, 347–374. [CrossRef]
33. Zhang, J.; Nuebel, E.; Daley, G.Q.; Koehler, C.M.; Teitell, M.A. Metabolic Regulation in Pluripotent Stem Cells during Reprogramming and Self-Renewal. *Cell Stem Cell* **2012**, *11*, 589–595. [CrossRef] [PubMed]
34. Jin, Y.-Z.; Lee, J.H. Mesenchymal Stem Cell Therapy for Bone Regeneration. *Clin. Orthop. Surg.* **2018**, *10*, 271–278. [CrossRef] [PubMed]
35. Tsuji, K.; Ojima, M.; Otabe, K.; Horie, M.; Koga, H.; Sekiya, I.; Muneta, T. Effects of Different Cell-Detaching Methods on the Viability and Cell Surface Antigen Expression of Synovial Mesenchymal Stem Cells. *Cell Transplant.* **2017**, *26*, 1089–1102. [CrossRef] [PubMed]

36. Krešić, N.; Prišlin, M.; Vlahović, D.; Kostešić, P.; Ljolje, I.; Brnić, D.; Turk, N.; Musulin, A.; Habrun, B. The Expression Pattern of Surface Markers in Canine Adipose-Derived Mesenchymal Stem Cells. *Int. J. Mol. Sci.* **2021**, *22*, 7476. [CrossRef]
37. Fan, W.; Li, J.; Wang, Y.; Pan, J.; Li, S.; Zhu, L.; Guo, C.; Yan, Z. CD105 Promotes Chondrogenesis of Synovium-Derived Mesenchymal Stem Cells through Smad2 Signaling. *Biochem. Biophys. Res. Commun.* **2016**, *474*, 338–344. [CrossRef]
38. Xu, Y.; Wang, Y.-Q.; Wang, A.-T.; Yu, C.-Y.; Luo, Y.; Liu, R.-M.; Zhao, Y.-J.; Xiao, J.-H. Effect of CD44 on Differentiation of Human Amniotic Mesenchymal Stem Cells into Chondrocytes via Smad and ERK Signaling Pathways. *Mol. Med. Rep.* **2020**, *21*, 2357–2366. [CrossRef]
39. Davies, O.G.; Cooper, P.R.; Shelton, R.M.; Smith, A.J.; Scheven, B.A. Isolation of Adipose and Bone Marrow Mesenchymal Stem Cells Using CD29 and CD90 Modifies Their Capacity for Osteogenic and Adipogenic Differentiation. *J. Tissue Eng.* **2015**, *6*, 2041731415592356. [CrossRef]
40. Campioni, D.; Rizzo, R.; Stignani, M.; Melchiorri, L.; Ferrari, L.; Moretti, S.; Russo, A.; Bagnara, G.P.; Bonsi, L.; Alviano, F.; et al. A Decreased Positivity for CD90 on Human Mesenchymal Stromal Cells (MSCs) Is Associated with a Loss of Immunosuppressive Activity by MSCs. *Cytometry B Clin. Cytom.* **2009**, *76B*, 225–230. [CrossRef]
41. Moraes, D.A.; Sibov, T.T.; Pavon, L.F.; Alvim, P.Q.; Bonadio, R.S.; Da Silva, J.R.; Pic-Taylor, A.; Toledo, O.A.; Marti, L.C.; Azevedo, R.B.; et al. A Reduction in CD90 (THY-1) Expression Results in Increased Differentiation of Mesenchymal Stromal Cells. *Stem Cell Res. Ther.* **2016**, *7*, 97. [CrossRef]
42. Maleki, M.; Ghanbarvand, F.; Reza Behvarz, M.; Ejtemaei, M.; Ghadirkhomi, E. Comparison of Mesenchymal Stem Cell Markers in Multiple Human Adult Stem Cells. *Int. J. Stem Cells* **2014**, *7*, 118–126. [CrossRef]
43. Schachtele, S. Markers and Methods to Verify Mesenchymal Stem Cell Identity, Potency, and Quality. 13. Available online: https://resources.rndsystems.com/images/site/wp-msc-13763.pdf?_ga=2.169946848.1123720049.1654517715-902278695.1654517715 (accessed on 4 June 2022).
44. Ahmed, N.; Vogel, B.; Rohde, E.; Strunk, D.; Grifka, J.; Schulz, M.B.; Grässel, S. CD45-Positive Cells of Haematopoietic Origin Enhance Chondrogenic Marker Gene Expression in Rat Marrow Stromal Cells. *Int. J. Mol. Med.* **2006**, *18*, 233–240. [CrossRef] [PubMed]
45. Bertolo, A.; Baur, M.; Guerrero, J.; Pötzel, T.; Stoyanov, J. Autofluorescence Is a Reliable in Vitro Marker of Cellular Senescence in Human Mesenchymal Stromal Cells. *Sci. Rep.* **2019**, *9*, 2074. [CrossRef] [PubMed]

animals

Article

Shipping Temperature, Time and Media Effects on Equine Wharton's Jelly and Adipose Tissue Derived Mesenchymal Stromal Cells Characteristics

Eleonora Iacono [1,2,*], Aliai Lanci [1], Penelope Gugole [1] and Barbara Merlo [1,2]

1 Department of Veterinary Medical Sciences, University of Bologna, Via Tolara di Sopra 50, Ozzano Emilia, 40064 Bologna, Italy; aliai.lanci2@unibo.it (A.L.); penelopemaria.gugol2@unibo.it (P.G.); barbara.merlo@unibo.it (B.M.)
2 Interdepartmental Centre for Industrial Research in Health Sciences and Technologies, University of Bologna, Via Tolara di Sopra 41/E, Ozzano Emilia, 40064 Bologna, Italy
* Correspondence: eleonora.iacono2@unibo.it

Simple Summary: Today, the use of horse adipose tissue and Wharton's jelly-derived mesenchymal stromal cells in veterinary regenerative medicine represents a promising tool. Cells need to be isolated and expanded in vitro in the laboratory to obtain a sufficient amount for clinical application and its characterization. In many cases, laboratories and clinics where the therapy will be performed are in different and far-flung facilities, and the cells must therefore be shipped by a courier. The authors evaluated the effects of different storage conditions, in terms of temperature, time of storage and storage solutions on cell viability, cell growth, differentiation potential and molecular characteristics. The aim was to state the most appropriate storage conditions for transporting adipose tissue and Wharton's jelly-derived stromal cells, ensuring the maintenance of the stemness features for therapeutic application in horses.

Abstract: To use Mesenchymal Stromal Cells (MSCs) in equine patients, isolation and expansion are performed in a laboratory. Cells are then sent back to the veterinary clinic. The main goal of storage conditions during cell transport is to preserve their biological properties and viability. The aim of this study was to evaluate the effects of storage solutions, temperature and time on the characteristics of equine adipose tissue and Wharton's jelly-derived MSCs. We compared two different storage solutions (plasma and 0.9% NaCl), two different temperatures (4 °C and room temperature) and three time frames (6, 24, 48 h). Cell viability, colony-forming units, trilineage differentiation, the expression of CD45 and CD90 antigens and adhesion potentials were evaluated. Despite the molecular characterization and differentiation potential were not influenced by storage conditions, viability, colony-forming units and adhesion potential are influenced in different way, depending on MSCs sources. Overall, this study found that, despite equine adipose tissue MSCs being usable after 24 h of storage, cells derived from Wharton's jelly need to be used within 6 h. Moreover, while for adipose cells the best conservation solutions seems to be plasma, the cell viability of Wharton's jelly MSCs declined in both saline and plasma solution, confirming their reduced resistance to conservation.

Keywords: mesenchymal stromal cells; equine; adipose tissue; Wharton's jelly; storage

Citation: Iacono, E.; Lanci, A.; Gugole, P.; Merlo, B. Shipping Temperature, Time and Media Effects on Equine Wharton's Jelly and Adipose Tissue Derived Mesenchymal Stromal Cells Characteristics. *Animals* **2022**, *12*, 1967. https://doi.org/10.3390/ani12151967

Academic Editor: Paul Dyce

Received: 1 July 2022
Accepted: 31 July 2022
Published: 3 August 2022

1. Introduction

During the last twenty years, Mesenchymal Stromal Cells (MSCs) have received, both in human and veterinary medicine, considerable attention because of their potential use for promoting tissue regeneration, not only because of their differentiation potential, but likely also because of their trophic, anti-inflammatory and immunomodulatory abilities [1,2].

In the last twenty years, in equine medicine, bone marrow has been the most used source of autologous MSCs. Alternatively, adipose tissue-derived MSCs have been used [3]. However, for recovering these tissues an invasive procedure is required and a large variability in the cell yield related to the donor has been demonstrated, for both eBM (equine bone marrow) and eAT (equine adipose tissue) [4]. As demonstrated by different authors [5–9], fetal adnexa represent an important source of MSCs for equine regenerative medicine. These tissues can be easily procured without invasive procedures, both for mare and foal, and fetal adnexa-derived MSCs preserve some characteristics typical of primitive native layers and have been defined as an intermediate between embryonic and adult MSCs [10]. In equine, among fetal adnexal tissues, the major source of MSCs is eAM (equine amniotic membrane) [5] and Wharton's jelly (eWJ) [11]. Recently, Iacono et al. [12], comparing eAMMSCs and eWJMSCs, found that cells isolated from different matrices have different morphological and molecular features and different differentiation potentials. Particularly, data recovered by the Authors show that eWJ could be considered as a more viable and convenient MSCs source for autologous or allogeneic regenerative therapies.

Due to the results obtained after the in vivo use of MSCs in equines, the interest of horse owners and veterinarians is progressively increasing. For in vivo use, MSCs are cultured, expanded and prepared in a specialized laboratory, then they are transferred to the clinic. However, most of the time, the laboratory and the clinic are located far from each other, and some authors hypothesize that one of the reasons why only 24% of the injected cells are found in the lesion site after 24 h is the reduced cell viability resulting from the transposing of cells over long distances [13,14].

In this context, different authors have attempted to find the best conditions (media, temperature, hours of transport, etc.) for shipping equine MSCs from laboratory to clinic. In 2012, Bronzini et al. [15] analyzed the influence of media, temperature and hours of storage on equine peripheral blood MSCs, finding that neither different media nor temperature were able to maintain cell viability during shipping period; in fact, cell mortality was around 30–40% in every experimental group. The same was demonstrated successively by Mercati et al. [16], by Espina et al. [17] and Garvican et al. [14] on equine adipose tissue MSCs, bone marrow and ePBMSCs (equine peripheral blood MSCs), respectively. Stored eATMSCs and eBMMSCs have retained differentiation and clonogenic potential [16,17], while the ability of ePBMSCs to maintain their molecular features is controversial [15]. Indeed, while the level of expression of CD44 and CD105 were constant in fresh and stored cells, CD90 expression decreased after 9 and 12 h of storage, suggesting, as reported by the authors, an inability of these cells to retain the properties of MSCs after this period of storage [15].

Despite the interest in using cells derived from fetal adnexa, eWJMSCs in particular, and the results obtained after their in vivo application [8], to our knowledge no study has reported on the storage and shipping effects on eWJMSCs viability, in vitro growth and molecular features and their differentiation potential. In the present study the effects of storage temperature, time and solution on eWJMSCs characteristics are determined. In order to define which type of cells are the most resistant in case long-distance transport is needed, in the present study the effects of shipping conditions on adult and Wharton's jelly-derived MSCs have been determined.

2. Materials and Methods

Chemicals were obtained from Sigma-Aldrich (Merck); type I collagenase, DMEM (Dulbecco's modified Eagle's medium) low glucose medium with Glutamine and FBS (fetal bovine serum) are branded GIBCO (ThermoFisher Scientific, Waltham, MA, USA). Plastics were from Falcon™ unless otherwise stated.

2.1. Samples

During the colic surgery of horses spontaneously referred by the owners to the Department of Veterinary Medical Sciences (DIMEVET), University of Bologna, intra-abdominal

AT was collected (n = 3). For the use of removed tissue for research purposes owners gave a written consent. Experimental procedures were approved by the Ethics Committee on Animal Use of the University of Bologna (Prot. 55948-X/10).

Wharton's jelly was isolated from the umbilical cord (UC; n = 3) recovered after foal physiological birth, born from Standardbred mares, and housed at the DIMEVET. For Wharton's jelly sampling, experimental procedures were approved by the Ethics Committee on Animal Use, University of Bologna (Prot. 55948-X/10), and a written consent was given by the owners to allow tissue recovery for research purposes.

2.2. Cell Isolation and Culture

AT and UC samples, stored in DPBS (Dulbecco's polyphosphate buffer solution) plus antibiotics (100 IU/mL penicillin, 100 g/mL streptomycin), were kept at 4 °C until processing. Under a laminar flow hood, the richest portion of WJ was immediately isolated from the cord tissue.

For both tissues, MSCs were isolated as previously described by Iacono et al. [12]. Briefly, AT and WJ were washed with repeated dives in DPBS, weighed, and cut into 0.5 cm pieces by sterile scissors. Samples were transferred into a 50 mL polypropylene tube and digested by a 0.1% collagenase type I solution in DPBS (1 mL solution/1 g tissue). The suspension was kept in a 37 °C water bath for at least 30 min and vigorously mixed every 10 min. For inactivating the collagenase, the suspension was diluted 1:1 with DPBS plus 10% FBS. The solution was filtered through a stainless steel strainer for discarding undigested tissue and centrifuged at 470× *g* at 25 °C for 10 min. The pellet was re-suspended in DMEM Low Glucose + 10% FBS + 100 U/mL penicillin + 100 g/mL streptomycin. Cells were plated into 25 cm^2 culture flasks and incubated in 5% CO_2 at 38.5 °C, in humidified atmosphere (Passage 0). After 48 hrs, the culture medium was completed replaced and non-adherent cells were removed. Culture medium was changed every three days until cell growth reached 80 to 90% confluence. At 80–90% of confluence, cells were dissociated using a 0.25% trypsin EDTA (Ethylenediaminetetraacetic acid) solution and counted and cryopreserved as described by Merlo et al. [18]. Briefly, cells in 0.5 mL of FBS were put in a 1.5 mL cryogenic tube (Sarstetd Inc., Nümbrecht, Germany) at 4 °C. After 10 min, cell suspension was diluted 1:1 with FBS + 16% DMSO (Dimethyl sulfoxide; final concentration 8% DMSO) and maintained for a further 10 min at 4 °C. Then the cryogenic tube was set to −80 °C for 24 h in a "Mr. Frosty" (Nalgene) and finally stored in liquid nitrogen. AT and WJMSCs were thawed at 37 °C in 20 mL DMEM + 10% FBS, then centrifuged at 470 g at 25 °C for 10 min. The pellet was re-suspended in 1 mL of culture medium and cell concentration and viability were evaluated by staining cells with 4% eosin solution and using a Neubawer improved chamber. Cells were plated in a 25 cm^2 flask (5000 cells/cm^2) as "Passage 1" (P1).

2.3. Study Design

When the confluence of 80–90% at P3 was reached, cells from all three AT and WJ samples were detached from the flask and viability and concentrations were determined as described above. For studying the effects of transport conditions, 6×10^6/mL live AT and WJ cells were stored in equine plasma (P) or 0.9% NaCl solution (S), tested for in vivo use, for 6 (T6), 24 (T24) and 48 (T48) hours at refrigeration (4 °C) and room temperature (RT = 20 °C).

At any time, viability was determined by eosin stain and cells were sown to determining colony forming unit ability, adhesion, and tri-lineage in vitro differentiation potential. Furthermore, CDs expression by PCR was also determined. Un-stored cells, namely T0, were considered as control. For all samples (3 AT and 3 WJ), each test was carried out in three replicates.

2.4. CFU (Colony Forming Unit) Assay

For determining the ability of cells to form colonies, 1×10^2 cells at T0, T6, T24 and T48 for both storage conditions were cultured for 8 days in a 30 mm petri dish. Colonies were fixed in 4% paraformaldehyde at RT for 1 h and stained with Giemsa 0.1% stain (15 min). Using an inverted light microscope (Eclipse TE 2000u, Nikon, Tokyo, Japan), the operator counted colonies formed by at least 16–20 nucleate cells.

2.5. Spheroid Formation Assays

To determine whether stored cells preserved their adhesion capability, spheroid formation was performed. Differently from the cell-substratum adhesion, performed on monolayer cultures adherent to rigid substrates, this test gives information about the direct cell–cell adhesion architecture found in normal tissues.

Cells were cultured in a multiwell Corning 96-well Black/Clear Round Bottom Ultra-Low Attachment Spheroid Microplate (5000 cells/25 μL drop). Images were acquired after 24 and 48 hrs of culture by a CCD camera (DS-Fi2, Nikon, Tokyo, Japan) mounted on an inverted light microscope (Eclipse TE 2000u, Nikon, Tokyo, Japan).

2.6. Multi Lineage In Vitro Differentiation

The osteogenic, adipogenic, and chondrogenic in vitro differentiation potential of control and stored AT- and WJ-MSCs were determined. As reported in Table 1, 5000 cells/cm^2 were cultured for two weeks under specific induction media. The same number of cells was cultured in culture medium, ss negative control.

Table 1. Composition of induction media [19,20].

Adipogenic	Chondrogenic	Osteogenic
DMEM	DMEM	DMEM
10% FBS	1% FBS	10% Rabbit Serum
0.5 mM IBMX (removed after 3 days)	6.25 μg/mL insulin	50 μM AA2P
1 μM DXM (removed after 6 days)	50 nM AA2P	0.1 μM DXM
10 μg/mL insulin	0.1 μM DXM	10 mM BGP
0.1 mM indomethacin	10 ng/mL hTGF-β1	

IBMX: isobutylmethylxanthine, DXM: dexamethasone, hTGF: human transforming growth factor, AA2P: ascorbic acid 2-phosphate, BGP: beta-glycerophosphate.

For differentiation evaluation, cells were fixed with 4% paraformaldehyde at RT for 1 h; Oil Red O, Alcian Blue, and Alizarin Red were used for staining adipogenic vacuoles, deposits of glycosaminoglycans, and calcium, respectively. An inverted light microscope (Eclipse TE 2000u, Nikon, Tokyo, Japan) was used to observe stained cells.

2.7. RT-PCR

For molecular characterization, RNA was extracted from snap-frozen cells and using Nucleo Spin® RNA kit (Macherey-Nagel) following the manufacturer's instructions. cDNAs were synthesized by RevertAid RT Kit and used directly in PCR reactions, following the instructions of Maxima Hot Start PCR Master Mix.

The expression of genes coding for MSC marker, CD90, and hematopoietic markers CD45 was determined. To ensure the proper expression of samples, GAPDH was used as a housekeeping gene. Primers are listed in Table 2. PCR products were visualized with ethidium bromide on a 2% (w/v) agarose gel.

Table 2. Primers sequences for PCR analysis.

Primers	References	Sequences (5′→3′)	bp
MSC marker CD90	[21]	FW: TGCGAACTCCGCCTCTCT RW: GCTTATGCCCTCGCACTTG	93
Ematopoietic markers CD45	[21]	FW: TGATTCCCAGAAATGACATGTA RW: ACATTTTGGGCTTGTCCTTAAC	101
Housekeeping GAPDH	[22]	FW: GTCCATGCCATCACTGCCAC RW: CCTGCTTCACCACCTTCTTG	262

2.8. Statistical Analysis

Data were analyzed for normal distribution, using a Shapiro–Wilk test. One-way ANOVA, followed by Student–Newman–Keuls' test when F values indicated significance, was used for analyzing the mean number of colonies and cell viability. Cell viability and CFU are expressed as mean ± SD. Statistical analyses were performed using IBM SPSS Statistics 25 (IBM Corporation). Significance was assessed for $p < 0.05$.

3. Results

3.1. Vitality and CFU Assay

The mean ± SD of eATMSCs and eWJMSCs at T0 were 90 ± 0% and 95 ± 5%. No statistically significant differences ($p > 0.05$) were observed between vitality at T0 and T6 for eATMSCs stored in saline solution at 4 °C (67.3 ± 24.1%) and in plasma at 4 °C (75.7 ± 14.4%) and RT (71.6 ± 11.5%). Only cells stored for 6 hrs in saline solution at RT showed a significantly lower vitality (19.0 ± 22.6%; $p < 0.05$). At T24, a statistically significant difference was found only in the viability of cells stored in plasma at 4 °C (45.8 ± 12.6%) and cells stored in saline solution at RT (17.8 ± 19.9%; $p < 0.05$). The same was also observed at T48 (P4vsS20: 21.4 ± 6.3% vs. 2.2 ± 3.3%; $p < 0.05$). At 48 the same difference was found also between cells stored in plasma at RT and saline solution at RT (25.1 ± 19.9 vs. 2.2 ± 3.3%; $p < 0.05$). Data are shown in Figure 1.

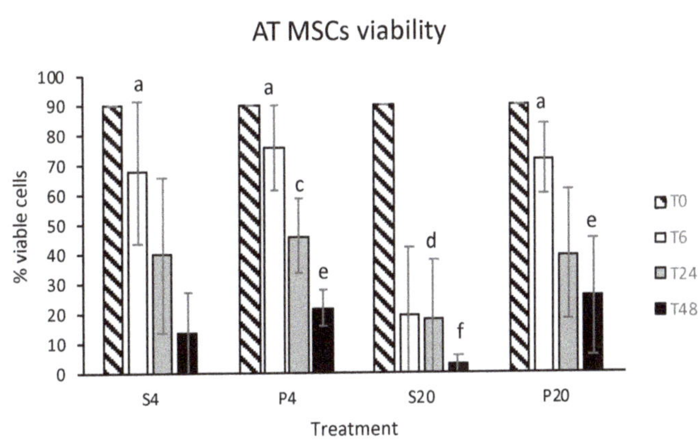

Figure 1. eATMSCs viability after storage for 6 (T6), 24 (T24) and 48 (T48) hrs in saline solution (S) and plasma (P) at 4 °C (4) and room temperature (20): a vs. b; c vs. d; e vs. f: $p < 0.05$.

Data regarding eWJMSCs viability after storage are shown in Figure 2. Different from eATMSCs, all storage groups showed a statistically different viability from T0 ($p < 0.05$), but no differences were observed between groups stored for 6–24–48 hrs in plasma and saline solution at 4 °C and RT ($p > 0.05$).

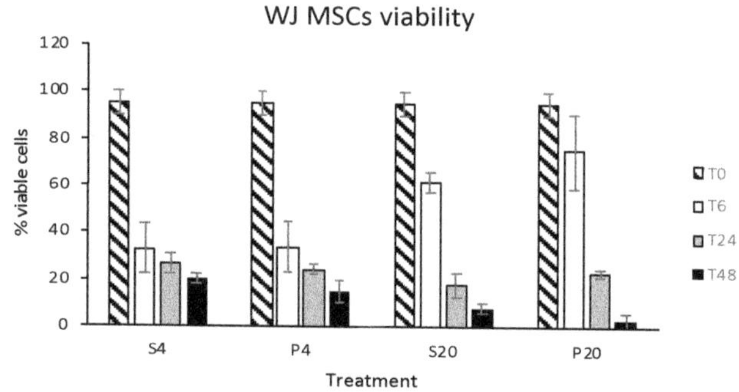

Figure 2. eWJMSCs viability after storage for 6 (T6), 24 (T24) and 48 (T48) hrs in saline solution (S) and plasma (P) at 4 °C (4) and room temperature (20). No statistical differences have been found among groups.

Stored eATMSCs and eWJMSCs also preserved the ability to form CFU when cultured in vitro; in both cell lines this statistically decreased as storage hrs increased, except for eATMSCs stored for 6 hrs in saline solution at 4 °C. Data are shown in Figures 3 and 4.

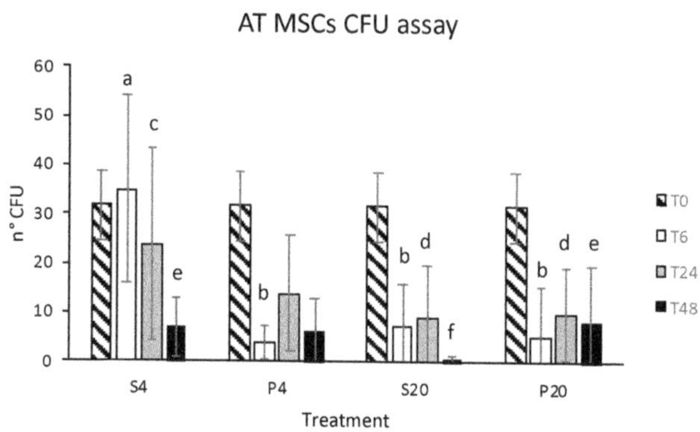

Figure 3. Number of CFU formed by eATMSCs after storage for 6 (T6), 24 (T24) and 48 (T48) hrs in saline solution (S) and plasma (P) at 4 °C (4) and room temperature (20): a vs. b; c vs. d; e vs. f: $p < 0.05$.

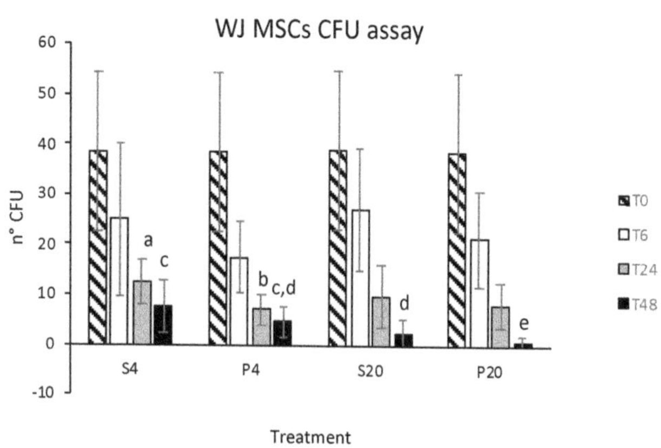

Figure 4. Number of CFU formed by eWJMSCs after storage for 6 (T6), 24 (T24), and 48 (T48) hrs in saline solution (S) and plasma (P) at 4 °C (4) and room temperature (20): a vs. b; c vs. d vs. e: $p < 0.05$.

3.2. Spheroid Formation Assays

eATMSCs and eWJMSCs at T0 were able to form spheroids when cultured for 24 hrs in a multi-well Corning 96-well Black/Clear Round Bottom Ultra-Low Attachment Spheroid Microplate (Figure 5A,B). Stored eATMSCs were able to form spheroids only when they have been preserved for 6 hrs in saline solution, both at 4 °C and RT (Figure 5C). The same cells, stored for 24 and 48 hrs in saline solution RT, after 24 hrs of hanging drop in vitro culture, formed small, separated spheroids (Figure 5E,G). On the other hand, cells from all AT samples, which were stored for the 6–24–48 hrs in plasma at 4 °C and RT, did not form spheroids; after 24 hrs of culture, in a multi-well Corning 96-well Black/Clear Round Bottom Ultra- Low Attachment Spheroid Microplate, they adhered to the plate, returning to the spindle-shape (Figure 5I).

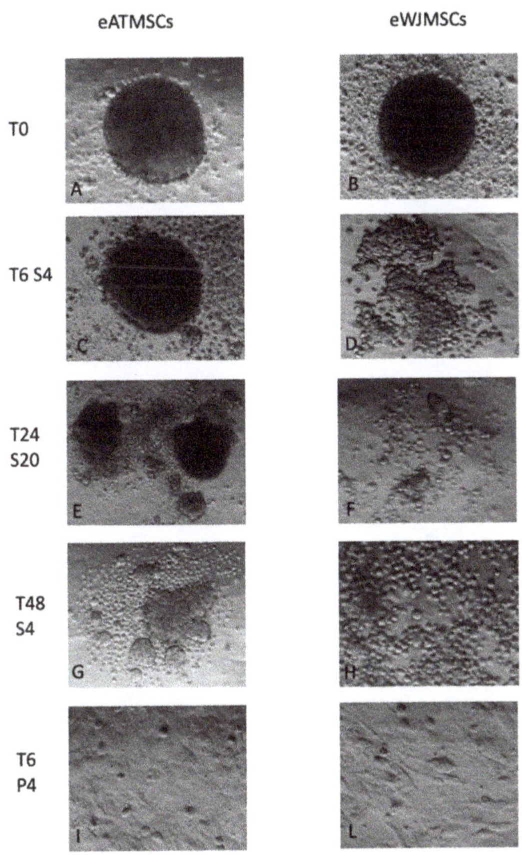

Figure 5. Spheroid formation assay performed with eATMSCs and eWJMSCs. (**A**,**B**) spheroids at T0. As showed in the pictures eATMSCS stored for 6 hrs in saline solution were able to form spheroids (**C**), but after 24 and 48 hrs of storage they formed smaller and fragmented spheroids (**E**,**G**). On the contrary in saline solution stored eWJMSCs were not able to form spheroids (**D**,**F**,**H**). Both cell lines stored in plasma lost their ability to form spheroids and grew adherent to plastic with a fibroblast like form (**I**,**L**). Magnification 10×.

The same was observed also for eWJMSCs stored in plasma for 6–24–48 hrs at 4 °C and RT (Figure 5L). Unlike eATMSCs, eWJMSCs after storage in saline solution for 6–24–48 hrs at 4 °C and RT were not able to form compacted spheroids, as shown in Figure 5D,F,H.

In particular, while after 6 hrs of storage, cells were still able to form small and non-compact spheroids, at T24 and T48, cells were round and in suspension.

3.3. Multi Lineage In Vitro Differentiation and RT-PCR

As requested by ICST [23], in the present study, in order to evaluate the ability of stored cells to preserve MSCs' characteristics, we cultured cells in osteogenic, adipogenic, and chondrogenic induction medium for at least 15 days.

Despite the decline in vitality and in the spheroid formation ability as the number of storage hours increased, both eATMSCs and eWJMSCs maintained the ability to differentiate in vitro toward osteogenic, adipogenic, and chondrogenic lineages at any timepoint. In fact, similar to cells at T0, after culturing in induction medium and staining with Alcian Blue, Oil red O, and Alizarin red, stored eATMSCs and eWJMSCs were able to differentiate and accumulate glycosaminoglycan, calcium, and lipid droplet deposition, as demonstrated in Figures 6 and 7, respectively.

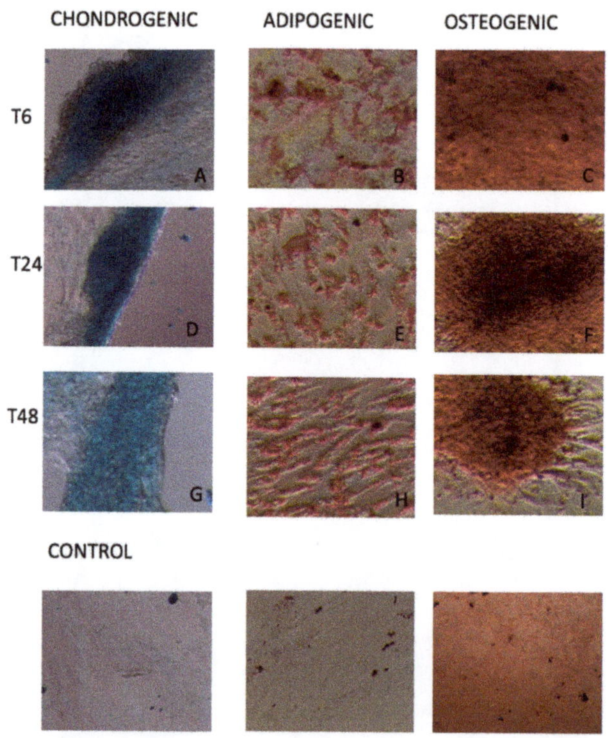

Figure 6. Stored eATMSCs, cultured in induction medium for two weeks and stained for chondrogenic, adipogenic and osteogenic differentiation. T6: (**A**) cells stored in plasma at 4 °C; (**B**) cells stored in plasma at RT; (**C**) cells stored in saline solution at RT. T24: (**D**) cells stored in saline solution at 4 °C; (**E**) cells stored in plasma at 4 °C; (**F**) cells stored in plasma at RT. T48: (**G**) cells stored in saline solution at RT; (**H**) cells stored in saline solution at 4 °C; (**I**) cells stored in plasma at RT. Control groups are reported. Magnification 10×.

At P3 of in vitro culture, fresh cell populations (T0) expressed MSC-associated markers, CD90, but were negative for the hematopoietic marker, CD45. After storage, at any point in time and in both storage solution and temperature, the molecular characterization was preserved by eWJMSCs and eATMSCs.

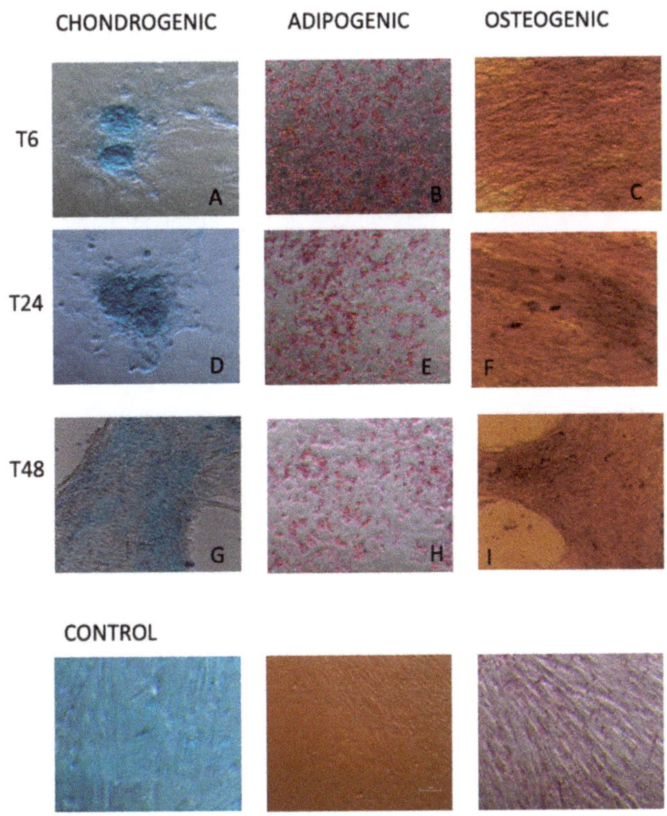

Figure 7. Stored eWJMSCs, cultured in induction medium for two weeks and stained for chondrogenic, adipogenic and osteogenic differentiation. T6: (**A**) cells stored in plasma at 4 °C; (**B**) cells stored in plasma at RT; (**C**) cells stored in saline solution at RT. T24: (**D**) cells stored in saline solution at 4 °C; (**E**) cells stored in plasma at 4 °C; (**F**) cells stored in plasma at RT. T48: (**G**) cells stored in saline solution at RT; (**H**) cells stored in saline solution at 4 °C; (**I**) cells stored in plasma at RT. Control groups are reported. Magnification 10×.

4. Discussion

Equine Mesenchymal Stromal Cells are increasingly used for clinical application. For their isolation and expansion, a laboratory is mandatory. From the laboratory, cells are then sent back to attending clinicians. Preserving MSCs characteristics en route from the laboratory to the clinic is fundamental for the success of the therapy. Due to the importance of this topic for equine regenerative medicine, in the last 10 years, different storage solution, temperatures, and hours have been tested [14–17]. Despite the increasingly recognized importance of MSCs derived from Wharton's jelly and fetal adnexa in the context of equine regenerative therapy, no papers are present in the literature on the best choice for their storage. Moreover, before now, in addition to plasma, authors tested only solution authorized for in vitro use, such as PBS or FBS.

In this context, in the present paper, the effects of saline solution authorized for intravenous administration and equine plasma at 4 °C and RT for 6–24–48 hrs were tested on eATMSCs and on eWJMSCs for the first time.

Unlike the data reported by Garvican et al. [14], in our storage conditions, cell vitality appeared reduced after 6 hrs of storage in both eAT and eWJMSCs. Particularly, cells derived from Wharton's jelly seem to be more sensitive to storage, especially at refrigeration temperatures; indeed, after 6 hrs of storage, eWJMSCs viability is higher for cells maintained in saline solution and plasma membranes at RT. Bronzini et al. [15] observed the same using ePBMSCs, while eATMSCs seem to withstand refrigeration temperature storage better, as already stated by Mercati et al. [16]. However, after 6 hrs of storage cell vitality was reduced in all study groups, in line with data already reported by different authors [14–17].

Observing the data of cell vitality, plasma seems to be the best solution for shipping cells from the laboratory to the clinic. However, the number of CFUs recorded both for eATMSCs and eWJMSCs stored in plasma are lower than that recorded for cells stored in saline solution. Moreover, cells stored in plasma, both at 4 °C and RT, are unable to form spheroids. As reported by Iacono et al. [12], the ability to form spheroids is related to the in vitro differentiation potential in the cartilaginous sense. In the present study, cells stored in plasma lost the ability to form spheroids but retained their potential to differentiate in vitro toward three lineages, as requested by ISCT [23]. These findings could be related to in vitro reaction between plasma and culture medium; in fact, in the culture plates of cells stored in plasma, after 24 hrs of in vitro culture, we observed the formation of a fibrin clot, in which the cells are most likely trapped without being able to form colonies and spheroids. The formation of clots could be avoided by adding heparin to plasma for shipping. However, the impact of heparin and its concentration could be studied in eATMSCs and WJMSCs. Indeed, in human MSCs doses of heparin between 100 and 1000 μg/mL of culture medium inhibit cell growth and changes in gene expression [24].

As a quality control tests, we also used in vitro differentiation and molecular characterization. As previously reported by Mercati et al. [16] for eATMSCs, the present study confirms that for both eATMSCs and eWJMSCs maintained their differentiation potential in vitro.

In the storage conditions used in this study molecular characterizations have also been preserved. Indeed, both eATMSCs and eWJMSCs expressed mesenchymal markers, CD90, and were negative for hematopoietic markers, CD45, as well as control group T0, unlike in ePBMSCs observed by Bronzini et al. [15], indicating that storage conditions can probably modify the characteristics of MSCs in different ways depending on their origin.

As previously reported, data reported in the present study confirm that MSC therapy could be administered as soon as possible after cell preparation. Different from results reported by Mercati et al. [16], in the case of therapy with eWJMSCs, this has to be administered within 6 hrs, while for eATMSCs, despite a decrease in vitality at 24 h of storage, it is possible to use them 24 hrs after their preparation. Despite the molecular characterization and differentiation potential were not influenced by storage conditions, both in eWJMSCs and eATMSCs, viability, CFU, and adhesion potential are influenced in different way, depending on MSCs sources. Overall, this study stated that, despite eATMSCs being able to be used after 24 hrs of storage, eWJMSCs need to be used within 6 hrs. Moreover, while for eATMSCs the best conservation solutions seems to be plasma, cell viability of eWJMSCs declined in both saline and plasma solution, confirming their reduced resistance to conservation.

Data recovered in vitro in the present study need to be compared with results obtained in vivo using cells shipped under tested conditions and with data obtained using frozen cells implanted directly, immediately after thawing.

5. Conclusions

In the present study we demonstrated that different types of MSCs react differently to the storage conditions frequently used for shipping them from the laboratory to the clinic. These conditions influence viability and, depending on the cell type, they can also influence different MSCs characteristics. Particularly, according to the data recovered in the present study, eWJMSCs need to be used quickly to maintain their vitality characteristics.

In conclusion, as different MSCs doses are being shipped every day, further studies are necessary to find the best shipping conditions for each cell type.

Author Contributions: Conceptualization, E.I. and B.M.; methodology, E.I., B.M., A.L. and P.G.; validation, E.I. and B.M.; data curation, E.I., B.M., A.L. and P.G.; writing—original draft preparation, E.I.; writing—review, B.M.; project administration, E.I.; funding acquisition, E.I. All authors have read and agreed to the published version of the manuscript.

Funding: This research was funded by the Fondazione Cassa di Risparmio di Imola, project: "CELLULE STAMINALI MESENCHIMALI PER LA TERAPIA RIGENERATIVA CUTANEA: L'IMPORTANZA DELLA MEDICINA VETERINARIA".

Institutional Review Board Statement: The study was approved by the Ethics Committee on Animal Use of the University of Bologna (Prot. 55948-X/10). Written informed consent was obtained from the owners.

Informed Consent Statement: Written informed consent has been obtained from owner of all patients to publish this paper.

Acknowledgments: The authors wish to thank the staff of Neonatology and Reproduction Unit and Surgery Unit, Equine Clinical Service, DIMEVET, University of Bologna, for equine umbilical cord and adipose tissue samples recovery.

Conflicts of Interest: The authors declare no conflict of interest.

Abbreviations

AF	Amniotic Fluid
eAFMSCs	Equine Amniotic Fluid Mesenchymal Stromal Cells
AM	Amniotic Membrane
AT	Adipose Tissue
eATMSCs	Equine Adipose Tissue Mesenchymal Stromal Cells
BM	Bone Marrow
eBMMSCs	Equine Bone Marrow Mesenchymal Stromal cells
CD	Cluster Designation
DMEM	Dulbecco's Modified Eagle's Medium
DMSO	dimethyl sulfoxide
EV	Extracellular Vesicles
F cells	Fibroblastic cells
FBS	Fetal Bovine Serum
Hrs	Hours
IL	Interleukin
ISCT	International Society For Cytotherapy
MHC	Major Histocompatibility Complex
MSCs	Mesenchymal Stromal Cells
NSAIDs	Non-Steroidal Anti-Inflammatory Drug
P	In vitro culture Passage
DPBS	Dulbecco's Phosphate Buffer Solution
ePBMSCs	Equine Peripheral Blood Mesenchymal Stromal Cells
ePMSCs	Equine Placenta Mesenchymal Stromal Cells
PRP	Platelet Rich Plasma
RT	Room Temperature
RT-PCR	Real Time Polymerase Chain reaction
TGF	Transforming Growth Factor
UC	Umbilical Cord
UCB	Umbilical Cord Blood
eUCBMSCs	Equine Umbilical Cord Blood Mesenchymal Stromal Cells
VEGF	Vascular Endothelial Growth Factor
WJ	Wharton's Jelly
eWJMSCs	Equine Wharton's Jelly Mesenchymal Stromal Cells

References

1. Meirelles, L.d.S.; Fontes, A.M.; Covas, D.T.; Caplan, A.I. Mechanisms Involved in the Therapeutic Properties of Mesenchymal Stem Cells. *Cytokine Growth Factor Rev.* **2009**, *20*, 419–427. [CrossRef]
2. Barrachina, L.; Romero, A.; Zaragoza, P.; Rodellar, C.; Vázquez, F.J. Practical Considerations for Clinical Use of Mesenchymal Stem Cells: From the Laboratory to the Horse. *Vet. J. Lond. Engl. 1997* **2018**, *238*, 49–57. [CrossRef] [PubMed]
3. Iacono, E.; Merlo, B.; Romagnoli, N.; Rossi, B.; Ricci, F.; Spadari, A. Equine Bone Marrow and Adipose Tissue Mesenchymal Stem Cells: Cytofluorimetric Characterization, In Vitro Differentiation, and Clinical Application. *J. Equine Vet. Sci.* **2015**, *35*, 130–140. [CrossRef]
4. Colleoni, S.; Bottani, E.; Tessaro, I.; Mari, G.; Merlo, B.; Romagnoli, N.; Spadari, A.; Galli, C.; Lazzari, G. Isolation, Growth and Differentiation of Equine Mesenchymal Stem Cells: Effect of Donor, Source, Amount of Tissue and Supplementation with Basic Fibroblast Growth Factor. *Vet. Res. Commun.* **2009**, *33*, 811–821. [CrossRef] [PubMed]
5. Lange-Consiglio, A.; Tassan, S.; Corradetti, B.; Meucci, A.; Perego, R.; Bizzaro, D.; Cremonesi, F. Investigating the Efficacy of Amnion-Derived Compared with Bone Marrow-Derived Mesenchymal Stromal Cells in Equine Tendon and Ligament Injuries. *Cytotherapy* **2013**, *15*, 1011–1020. [CrossRef]
6. Iacono, E.; Merlo, B.; Pirrone, A.; Antonelli, C.; Brunori, L.; Romagnoli, N.; Castagnetti, C. Effects of Mesenchymal Stem Cells Isolated from Amniotic Fluid and Platelet-Rich Plasma Gel on Severe Decubitus Ulcers in a Septic Neonatal Foal. *Res. Vet. Sci.* **2012**, *93*, 1439–1440. [CrossRef] [PubMed]
7. Corradetti, B.; Correani, A.; Romaldini, A.; Marini, M.G.; Bizzaro, D.; Perrini, C.; Cremonesi, F.; Lange-Consiglio, A. Amniotic Membrane-Derived Mesenchymal Cells and Their Conditioned Media: Potential Candidates for Uterine Regenerative Therapy in the Horse. *PLoS ONE* **2014**, *9*, e111324. [CrossRef]
8. Lanci, A.; Merlo, B.; Mariella, J.; Castagnetti, C.; Iacono, E. Heterologous Wharton's Jelly Derived Mesenchymal Stem Cells Application on a Large Chronic Skin Wound in a 6-Month-Old Filly. *Front. Vet. Sci.* **2019**, *6*, 9. [CrossRef]
9. Lange-Consiglio, A.; Funghi, F.; Cantile, C.; Idda, A.; Cremonesi, F.; Riccaboni, P. Case Report: Use of Amniotic Microvesicles for Regenerative Medicine Treatment of a Mare With Chronic Endometritis. *Front. Vet. Sci.* **2020**, *7*, 347. [CrossRef]
10. De Coppi, P.; Bartsch, G.; Siddiqui, M.M.; Xu, T.; Santos, C.C.; Perin, L.; Mostoslavsky, G.; Serre, A.C.; Snyder, E.Y.; Yoo, J.J.; et al. Isolation of Amniotic Stem Cell Lines with Potential for Therapy. *Nat. Biotechnol.* **2007**, *25*, 100–106. [CrossRef]
11. Merlo, B.; Teti, G.; Lanci, A.; Burk, J.; Mazzotti, E.; Falconi, M.; Iacono, E. Comparison between Adult and Foetal Adnexa Derived Equine Post-Natal Mesenchymal Stem Cells. *BMC Vet. Res.* **2019**, *15*, 277. [CrossRef]
12. Iacono, E.; Pascucci, L.; Rossi, B.; Bazzucchi, C.; Lanci, A.; Ceccoli, M.; Merlo, B. Ultrastructural Characteristics and Immune Profile of Equine MSCs from Fetal Adnexa. *Reprod. Camb. Engl.* **2017**, *154*, 509–519. [CrossRef]
13. Becerra, P.; Valdés Vázquez, M.A.; Dudhia, J.; Fiske-Jackson, A.R.; Neves, F.; Hartman, N.G.; Smith, R.K.W. Distribution of Injected Technetium(99m)-Labeled Mesenchymal Stem Cells in Horses with Naturally Occurring Tendinopathy. *J. Orthop. Res. Off. Publ. Orthop. Res. Soc.* **2013**, *31*, 1096–1102. [CrossRef]
14. Garvican, E.R.; Cree, S.; Bull, L.; Smith, R.K.; Dudhia, J. Viability of Equine Mesenchymal Stem Cells during Transport and Implantation. *Stem Cell Res. Ther.* **2014**, *5*, 1. [CrossRef]
15. Bronzini, I.; Patruno, M.; Iacopetti, I.; Martinello, T. Influence of Temperature, Time and Different Media on Mesenchymal Stromal Cells Shipped for Clinical Application. *Vet. J. Lond. Engl. 1997* **2012**, *194*, 121–123. [CrossRef]
16. Mercati, F.; Pascucci, L.; Curina, G.; Scocco, P.; Tardella, F.M.; Dall'aglio, C.; Marini, C.; Ceccarelli, P. Evaluation of Storage Conditions on Equine Adipose Tissue-Derived Multipotent Mesenchymal Stromal Cells. *Vet. J. Lond. Engl. 1997* **2014**, *200*, 339–342. [CrossRef]
17. Espina, M.; Jülke, H.; Brehm, W.; Ribitsch, I.; Winter, K.; Delling, U. Evaluation of Transport Conditions for Autologous Bone Marrow-Derived Mesenchymal Stromal Cells for Therapeutic Application in Horses. *PeerJ* **2016**, *4*, e1773. [CrossRef]
18. Merlo, B.; Pirondi, S.; Iacono, E.; Rossi, B.; Ricci, F.; Mari, G. Viability, in vitro differentiation and molecular characterization of equine adipose tissue-derived mesenchymal stem cells cryopreserved in serum and serum-free medium. *Cryo Lett.* **2016**, *37*, 243–252.
19. Mizuno, H.; Hyakusoku, H. Mesengenic Potential and Future Clinical Perspective of Human Processed Lipoaspirate Cells. *J. Nippon Med. Sch. Nippon Ika Daigaku Zasshi* **2003**, *70*, 300–306. [CrossRef]
20. Iacono, E.; Brunori, L.; Pirrone, A.; Pagliaro, P.P.; Ricci, F.; Tazzari, P.L.; Merlo, B. Isolation, Characterization and Differentiation of Mesenchymal Stem Cells from Amniotic Fluid, Umbilical Cord Blood and Wharton's Jelly in the Horse. *Reprod. Camb. Engl.* **2012**, *143*, 455–468. [CrossRef]
21. Mohanty, N.; Gulati, B.R.; Kumar, R.; Gera, S.; Kumar, P.; Somasundaram, R.K.; Kumar, S. Immunophenotypic Characterization and Tenogenic Differentiation of Mesenchymal Stromal Cells Isolated from Equine Umbilical Cord Blood. *In Vitro Cell. Dev. Biol. Anim.* **2014**, *50*, 538–548. [CrossRef] [PubMed]
22. Desmarais, J.A.; Demers, S.-P.; Suzuki, J.; Laflamme, S.; Vincent, P.; Laverty, S.; Smith, L.C. Trophoblast Stem Cell Marker Gene Expression in Inner Cell Mass-Derived Cells from Parthenogenetic Equine Embryos. *Reprod. Camb. Engl.* **2011**, *141*, 321–332. [CrossRef] [PubMed]

23. Dominici, M.; Le Blanc, K.; Mueller, I.; Slaper-Cortenbach, I.; Marini, F.; Krause, D.; Deans, R.; Keating, A.; Prockop, D.; Horwitz, E. Minimal Criteria for Defining Multipotent Mesenchymal Stromal Cells. The International Society for Cellular Therapy Position Statement. *Cytotherapy* **2006**, *8*, 315–317. [CrossRef] [PubMed]
24. Ling, L.; Camilleri, E.T.; Helledie, T.; Samsonraj, R.M.; Titmarsh, D.M.; Chua, R.J.; Dreesen, O.; Dombrowski, C.; Rider, D.A.; Galindo, M.; et al. Effect of Heparin on the Biological Properties and Molecular Signature of Human Mesenchymal Stem Cells. *Gene* **2016**, *576*, 292–303. [CrossRef]

MDPI
St. Alban-Anlage 66
4052 Basel
Switzerland
Tel. +41 61 683 77 34
Fax +41 61 302 89 18
www.mdpi.com

Animals Editorial Office
E-mail: animals@mdpi.com
www.mdpi.com/journal/animals